Thomas Beatty

Applied Mathematics: 100 Must-Know Math Methods

Also of Interest

Mathematics of Deep Learning
An Introduction to Foundational Mathematics of Neural Nets
Leonid Berlyand, Pierre-Emmanuel Jabin, 2026
ISBN 978-3-11-914411-7, e-ISBN (PDF) 978-3-11-221821-1

Variational Methods in Nonlinear Analysis
Optimization and Partial Differential Equations Applications
Dimitrios C. Kravvaritis, Athanasios N. Yannacopoulos, 2026
ISBN 978-3-11-133325-0, e-ISBN (PDF) 978-3-11-133329-8

Biomathematical Modeling
Methods and Software
Olcay Akman, Christopher Hay-Jahans, 2025
ISBN 978-3-11-160952-2, e-ISBN (PDF) 978-3-11-160956-0

Applications of Complex Variables
Asymptotics and Integral Transforms
Foluso Ladeinde, 2024
ISBN 978-3-11-135090-5, e-ISBN (PDF) 978-3-11-135117-9

Differential Equations
Solving Ordinary and Partial Differential Equations with Mathematica®
Marian Mureşan, 2024
ISBN 978-3-11-141109-5, e-ISBN (PDF) 978-3-11-141139-2

Mathematical Cryptology
Based on Elementary Number Theory and Essential Numerical Algorithms
Dimitrios Poulakis, 2026
ISBN 978-3-11-222751-0, e-ISBN (PDF) 978-3-11-222752-7

Thomas Beatty

Applied Mathematics: 100 Must-Know Math Methods

DE GRUYTER

Mathematics Subject Classification 2020
Primary: 00A06

Author
Thomas Beatty
Florida Gulf Coast University

ISBN 978-3-11-150661-6
ISBN 978-3-11-150665-4 (PDF)
ISBN 978-3-11-150671-5 (E-PUB)
DOI https://doi.org/10.1515/9783111506654

Library of Congress Control Number: 2026937907

Bibliographic information published by the Deutsche Nationalbibliothek
The Deutsche Nationalbibliothek lists this publication in the Deutsche Nationalbibliografie; detailed bibliographic data are available on the Internet at https://dnb.dnb.de.

De Gruyter and Walter de Gruyter GmbH are part of De Gruyter Brill.
www.degruyterbrill.com

Questions about General Product Safety Regulation: productsafety@degruyterbrill.com

Cover illustration: Jackie Niam / iStock / Getty Images Plus

Dedicated to Martie, Brett, & MacKenzie and The Memory of Helmut H. Schaefer

Contents

Preface

This book is intended to present a selection of important mathematical methods that would be useful for STEM practitioners of any age needing an introduction to or refreshment of mid-level precalculus topics. Consistent with that objective and wishing to keep the size of the text manageable, I have sidestepped using calculus unless the proper explanation of a subject makes it unavoidable, and then only in very basic terms. These few indiscretions are flagged in the text. I have tried to create a reference that is friendlier and more informative than the usual dry "handbook" style, which can be comprehensive but unhelpful with regard to particular details of application. At the same time, I did not allow the presentation to grow into a textbook, which would have become unwieldy and defeated the purpose of a quick to-the-point reference. It is hoped the reader can accept the resulting text as a reasonable compromise. I would like to thank Dr. Faysal Chowdhury for his many helpful comments and suggestions during the preparation of the manuscript, and especially Ms. Daniela Lamelas, who created all the artwork from my inscrutable chicken scratches. I would also like to thank Ute Skambraks, Adelinde Luyken, and Kristina Grigorjeva at DeGruyter-Brill for their patience and efficiency in bringing this book to life.

Ft. Myers (2026) Thomas Beatty

https://doi.org/10.1515/9783111506654-204

0 Introduction

We believe the best way to learn and recall mathematics is to engage and solve problems. One tried-and-true technique when first encountering a solution method or algorithm is to work out a simple representative problem in detail and try to internalize all the important steps and their logical flow toward the solution. Memorization is a pedagogically unfashionable term these days, but the cold, hard truth is that you have a much better chance to succeed with problem solving if you can readily bring your own previous experience to bear, the broader and deeper, the better. At the very least, you will have some idea how to classify a problem and then a reference such as this can help you tackle it. The model problem that you put in your mental toolkit should be typical of a class of problems, and not so simple that it allows you to sidestep the challenging features that make it a problem to begin with. When you encounter a similar problem in the wild you will have a point of reference that gives you both confidence and a rough idea how to proceed.

Our approach is organized with this philosophy in mind. The template for our discussion is to identify key methods, explain why they work, then apply them to an array of practical problems which are curated with respect to difficulty. The purpose is to promote a deeper understanding of illustrative examples in order to develop skill in adapting them to specific needs. Each chapter has a bibliography, listing references with comments and suggested advanced reading.

0.0 Preliminaries

The focus of this book is on mid-level applied math methods that are widely useful but not dependent on calculus. In order to present our narrative efficiently, we will freely use the common notational conventions and definitions of elementary mathematics. If you feel you have command of the basics, by all means feel free to skip the preliminaries in Chapter 0, which provides a review of these fundamentals. We have not avoided calculus altogether, but we mention it sparingly and only when it contributes to the broader understanding of an application. These sections will be marked by the symbol $\int$. We may introduce additional concepts or terminology as needed for specific problems from time to time, and these will be given ample clarification as they appear. Our first foundational subject is set theory, which supports the rest of mathematics.

0.1 Sets

Formal set theory is a relatively new branch of mathematics. It largely sprang from the imagination of the German mathematician Georg Cantor in the 1870s. Cantor's original conception of set theory was less formal, some would say naïve, than what it eventually

https://doi.org/10.1515/9783111506654-001

became around the turn of the twentieth century. Certain logical difficulties with the theory were encountered, and in an effort to put the subject on a firmer footing, mathematicians introduced axioms reminiscent of the axioms for Euclidean geometry. Today set theory is alive and well, as a branch of mathematics with its own extensive literature and broad research activity. Our use of set-theoretic ideas is strictly at an informal level. Set notation is very compact and precise, so we will use it routinely to formulate mathematical statements.

If we want to consider mathematical objects a, b, and c to be aggregated together for some purpose, we may write $S = \{a, b, c\}$. This means that the set S consists of the **elements** a, b, and c. The elements **belong** to or are **members of** the set S, and S **contains** them. Changing the order in which the elements are listed or duplicating them does not change the specification of the set, so $\{c, a, b\} = \{a, b, c, b, a, b\} = \{a, b, c\}$. The manner in which we are conveying the contents of the set S is called the **roster method**. A properly written roster of the set contents would list each element just once in any order inside "curly brackets." The stylized Greek epsilon "$\in$"(or sometimes just the traditional letter ε) is used to indicate that an element belongs to a set. So $a \in S$ means a belongs to S. Two sets are equal if they have exactly the same elements.

Sets can have finitely or infinitely many elements. We often express an infinite set by writing a few elements and then giving a general rule of formation for the other elements. If S is the set of counting numbers, for example, we would write $S = \{1, 2, 3, \dots, n, \dots\}$. It is understood that n is a general entry and the ellipsis after n means the pattern is to be extended indefinitely. The obviousness of the pattern formation should always be established.

Another way of indicating the contents of a set is to use the following format: $S = \{x \in X : P(x)\}$. This can be translated as "S consists of those elements x which already belong to some other set X and in addition can be determined to have some property abbreviated by P." An example would be $S = \{x \in \mathbb{N} : x \text{ is odd}\}$. So S is being formed by taking all the counting numbers, collectively denoted by $\mathbb{N}$, and then selecting only those members of $\mathbb{N}$ which satisfy the property P which means "is odd." Logicians would call P a predicate, and say $P(x)$ is true if x has property P and false if x does not have property P. The reason we have to start out by specifying $x \in X$ before we check to see if $P(x)$ is true or false has to do with avoiding one of the logical gremlins that plagued early informal set theory.

Some sets arise in mathematics so frequently that they deserve special abbreviations or nicknames. Each set contains all those listed prior:

- $\varnothing$ the **null set**, having no elements;
- $\mathbb{N}$ the set of **natural (whole** or **counting) numbers** $\{1, 2, 3, \dots, n, \dots\}$. If we want to allow zero into this set, we write $\mathbb{N}_0$;
- $\mathbb{Z}$ these are the **integers** $\{\dots, -n, \dots, -2, -1, 0, 1, 2, \dots, n, \dots\}$. Here we let the ellipses go in both directions. We capture the natural numbers, their negatives, and zero;

- $\mathbb{Q}$ these are pairs of integers (m, n) where the order of appearance in the pair matters. We identify these pairs with the fractions $\frac{m}{n}$, where n is never zero. These are the **rational numbers**;
- $\mathbb{R}$ these are the **real numbers**, consisting of all the rationals as well as irrationals like π and $\sqrt{2}$. They can be constructed from $\mathbb{Q}$ by a rigorous bootstrapping argument;
- $\mathbb{C}$ these are the **complex numbers** expressible as ordered pairs (a, b), where a and b are real numbers. For calculating purposes, we write (a, b) as $a + b\sqrt{-1}$.

Suppose we have a set S and another set T where every element of T is also an element of S. We call T a **subset** of S and use the horseshoe symbol "$\subset$" to indicate the relationship. The curved side of the horseshoe faces the subset, so we would write $T \subset S$. Conversely, S is then a **superset** of T. We can flip the horseshoe and write $S \supset T$ to signify the same relationship. If T really does have all the elements of S, we would say it is an **improper subset**, otherwise it is **proper**. If we are concerned about the exact structure of a subset relation between two sets, we can employ a more refined symbolism, namely $T \subseteq S$ means T is a subset of S, possibly improper, while $T \subsetneq S$ means T is a subset of S, definitely proper. The flipped versions of these symbols are also available. The null (**void** or **empty**) **set**, symbolized by $\varnothing$ is a subset of every set and any set is an improper subset of itself. The number of elements in a set is its **cardinality**. If S has n elements, or is an n**-set**, this is often indicated by $|S| = n$. For infinite sets, we write $|S| = \infty$.

0.1.1 Operations on sets

Given two sets S and T, we may wish to construct a set I which has exactly the elements common to both S and T. We call this new set the **intersection** of S and T and write it $I = S \cap T$. Another possibility is to form the set U consisting exactly of the elements in either S or T. This is called the **union** of S and T and is denoted as $U = S \cup T$. The horseshoe is back in two more configurations. So, for example, if $S = \{1, 3, 5, 7, 8, 9, 10\}$ and $T = \{2, 4, 6, 8, 10\}$, we would have $S \cap T = \{8, 10\}$ and $S \cup T = \{1, 2, 3, 4, 5, 6, 7, 8, 9, 10\}$. Of course, $(S \cap T) \subset (S \cup T)$ for all S and T.

If there is some set X which has S as a subset, we can define the notion of complementation. The **relative complement** of S in X is everything that is in X but not in S. This is usually written as $X - S$, where context ensures the minus sign is not confused with numerical subtraction. If X is understood to be the so-called "**universe of discourse**," that is to say, the set of all objects under consideration, we simply write the **universal complement** of S as S^c, although this is not a standard notation throughout the literature.

Another set operation that is fairly common is the **symmetric difference** of two sets S and T, written as $S \triangle T$. The symmetric difference of S and T is everything that is in S but not in T unified with everything that is in T but not in S. Symbolically, $S \triangle T =$

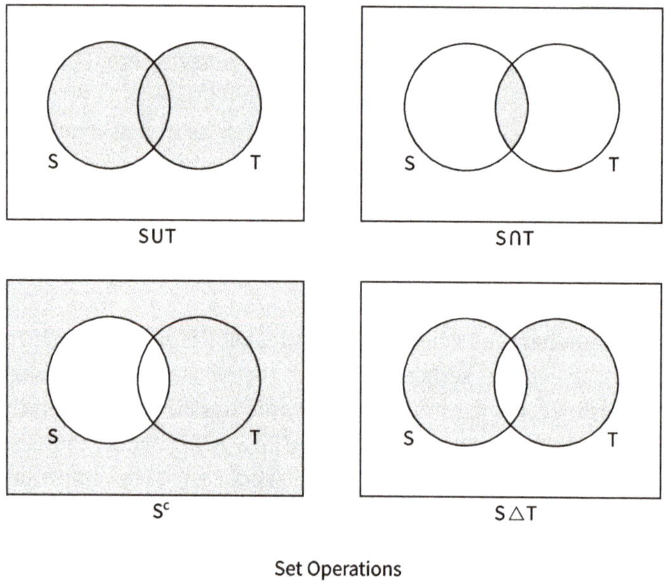

Figure 0.1: Four examples of intersecting circles representing general sets S and T. They depict the four set operations: union, intersection, complementation, and symmetric difference, as described in the text.

$(S \cup T) - (S \cap T)$. You can characterize this as "everything that is in one or the other, but not both." Venn diagrams are the traditional way of depicting these operations. The shaded areas in the Figure 0.1 represent the result of applying the various operations to sets S and T.

The operations of set union and intersection obey several convenient rules. For subsets R, S, and T of X, we have:

Table 0.1: Properties of Set Operations.

1) Associativity of union	$(R \cup S) \cup T = R \cup (S \cup T)$
2) Associativity of intersection	$(R \cap S) \cap T = R \cap (S \cap T)$
3) Commutativity of union	$S \cup T = T \cup S$
4) Commutativity of intersection	$S \cap T = T \cap S$
5) Distributivity of ∪ over ∩	$R \cup (S \cap T) = (R \cup S) \cap (R \cup T)$
6) Distributivity of ∩ over ∪	$R \cap (S \cup T) = (R \cap S) \cup (R \cap T)$

These properties are easily proved from the definitions. Associativity can be generalized so that expressions like $R \cup S \cup T \cup \cdots \cup Z$ and $R \cap S \cap T \cap \cdots \cap Z$ are well defined without the use of parentheses. We may symbolize a union of many sets as $\bigcup_{k=1}^{n} S_k$. This means that for each positive integer k from 1 to n inclusive, we are unifying the sets S_k with index (subscript) k. A similar expression can be formed based on intersection.

Occasionally we may substitute the infinity symbol "∞" for n to signify that a union or intersection is to be taken over a countably infinite collection of sets. Once we have expressions with mixed union and intersection signs we have to be careful to communicate our intended meaning by using parentheses appropriately. For example, suppose $R \cap S = \varnothing$. Then $(R \cap S) \cup T = \varnothing \cup T = T$. However, $R \cap (S \cup T) = (R \cap S) \cup (R \cap T) = \varnothing \cup (R \cap T) = R \cap T$ and $R \cap T$ need not equal T. This is reminiscent of the order of operations in simple arithmetic, where parenthetical expressions are evaluated internally first.

Complementation has its own special rule:

7) $(S^c)^c = S$.

There are two very useful additional rules that combine all three operations, i. e., union, intersection, and complementation. They are named in honor of the British mathematician Augustus DeMorgan:

8) $(S \cup T)^c = S^c \cap T^c$;
9) $(S \cap T)^c = S^c \cup T^c$.

So the mnemonic is "flip the horseshoe and distribute the complementation." These have generalizations:

8*) $(\bigcup_{k=1}^n S_k)^c = \bigcap_{k=1}^n S_k^c$;
9*) $(\bigcap_{k=1}^n S_k)^c = \bigcup_{k=1}^n S_k^c$.

We can form the so-called **Cartesian product** of sets. Given sets A and B, their Cartesian product, written $A \times B$ is the set $\{(a, b) : a \in A, b \in B\}$. In words, the Cartesian product of A and B is the set of all ordered pairs (a, b) where the first entry is from A and the second from B. It is possible to dispense with the first/second language here and define $(a, b) = \{a, \{a, b\}\}$ where the order in the pair is automatically implied. Similarly, $(b, a) = \{b, \{a, b\}\}$. But this is unnecessary for our practical purposes. This construction generalizes as $S_1 \times S_2 \times \cdots \times S_n = \prod_{i=1}^n S_i = \{(s_1, s_2, \ldots, s_n) : s_i \in S_i\}$. Later we will need the **Cartesian plane** $\mathbb{R}^2 = \mathbb{R} \times \mathbb{R}$, **Cartesian 3-space** $\mathbb{R}^3 = \mathbb{R} \times \mathbb{R} \times \mathbb{R}$, and even **Cartesian *n*-space** $\mathbb{R}^n$ defined similarly. The Cartesian plane $\mathbb{R}^2$ is the familiar **xy-plane**, and a point in the plane is uniquely identified by an (x, y) pair, where $x \in \mathbb{R}$ is the **first coordinate** and $y \in \mathbb{R}$ is the **second coordinate**. Points in $\mathbb{R}^3$ and $\mathbb{R}^n$ are specified in an analogous way. We will also use the Cartesian product to define relations and functions below in Section 0.3.

The preceding set concepts and definitions will give us a solid foundation to understand and use the many helpful methods of applied math that benefit all quantitative scientific disciplines.

0.2 Numbers & how they work

In 1886, the German mathematician Leopold Kronecker supposedly said in a lecture "Die ganzen Zahlen hat der liebe Gott gemacht, alles andere ist Menschenwerk," or "God created the integers, all else is the work of man." Man has been busy. Referring to the number sets in Section 0.1, the construction of $\mathbb{Q}$ from $\mathbb{Z}$ is fairly straightforward. We create all possible fractions from the integers, avoiding zero as a denominator. One can see that reduction of fractions to lowest terms eliminates a lot of duplication in expressions for rational numbers. The real numbers $\mathbb{R}$ can be bootstrapped from the rational numbers by a process that requires some subtle mathematical reasoning beyond the scope of our text. But we must somehow use and appreciate the special properties of $\mathbb{R}$; so for our purposes it is enough to say that we can draw a horizontal line on a piece of paper, pretend that the line extends indefinitely in both directions, label one point as "zero," and impose a scale to measure distance from that central zero. This is the real number line and we will accept on faith that any point on the line represents, by its signed distance from zero, some real number. Conversely, any real number can be attached to a point on the line. Finally, the complex numbers $\mathbb{C}$ can be created from the real numbers simply by adding a real number to an arbitrary real multiple of the square root of minus one. This is an easy construction, but the novelty lies in the use of a decidedly nonreal number. Historically, mathematicians were slow to accept negative numbers and even slower to consider imaginary numbers like i in their work. If the solution to an equation could technically be both a real number and an imaginary one, the early attitude was to accept the real solution, that with some demonstrably physical meaning, as the only solution and to ignore the other.

0.2.1 Rules of arithmetic

The five sets of numbers described in Section 0.1 have various operational rules available for combining them. These rules are very similar in their action to the rules for manipulating sets in Section 0.1.1. Before we go further, we need to distinguish $\mathbb{Q}$, $\mathbb{R}$, and $\mathbb{C}$ from $\mathbb{N}$ and $\mathbb{Z}$. The elements of $\mathbb{Z}$, namely the integers, can be added, subtracted, and multiplied, but they cannot be divided. The elements of $\mathbb{N}$ are even less cooperative. They can be added and multiplied but subtraction is possible only in special cases ($a - b \in \mathbb{N}$ only if a is greater than b). On the other hand, the sets $\mathbb{Q}$, $\mathbb{R}$, and $\mathbb{C}$ all have addition, subtraction, multiplication, and division available within them. This means performing one of the arithmetic operation on elements within the set gives a result back in the set. This qualifies them as **fields** and the relevant operational principles for fields are as follows.

Let a generic field be denoted by $\mathbb{F} = \mathbb{Q}, \mathbb{R}$, or $\mathbb{C}$ and suppose a, b, and c are arbitrary elements of $\mathbb{F}$. The operations on $\mathbb{F}$ are denoted by: "+" for addition, "−" for subtraction,

"×", "·", or simply juxtaposition for multiplication, and "÷" or a fraction bar "/" for division. Then the operations obey these rules:

Table 0.2: Rules of arithmetic.

1) Associativity of addition	$a + (b + c) = (a + b) + c$
2) Associativity of multiplication	$a \cdot (b \cdot c) = (a \cdot b) \cdot c$
3) Commutativity of addition	$a + b = b + a$
4) Commutativity of multiplication	$a \cdot b = b \cdot a$
5) Distributivity of multiplication over addition	$a \cdot (b + c) = a \cdot b + a \cdot c$
6) Existence of an additive identity	$a + 0 = a$
7) Existence of a multiplicative identity	$a \cdot 1 = a$
8) Existence of additive inverses	$a + (-a) = 0$
9) Existence of multiplicative inverses	$a \cdot \frac{1}{a} = 1 \ (a \neq 0)$

Subtraction is defined as $a - b = a + (-b)$. Division is defined as $a \div b = \frac{a}{b} = a \cdot \frac{1}{b}$. Also, $-(-a) = a$ and $\frac{1}{\frac{1}{a}} = a$.

Note that associativity of subtraction fails: $a-(b-c) = a-b+c \neq (a-b)-c = a-b-c$. Associativity of division fails as well: $a \div (b \div c) = \frac{ac}{b} \neq (a \div b) \div c = \frac{a}{bc}$. Likewise, there is no commutativity of subtraction or division: $a - b \neq b - a$ and $a \div b \neq b \div a$. Be careful with expressions that blend operations. The convention is that parenthetical expressions should be evaluated first, followed by indicated exponentiations (see Section 0.2.2), multiplications, divisions, additions, and subtractions in that order. Of course, in evaluating any parenthetical expressions, we must follow this order of operations inside as well. For example, $(a \div b + c) = \frac{a}{b} + c$, not $\frac{a}{b+c}$. For clarity, you can never have enough parentheses.

As we mentioned, not all of the operational rules above apply to $\mathbb{N}$ and $\mathbb{Z}$. All but rule #9 (multiplicative inverses) apply to $\mathbb{Z}$. Rules #1 through #5 and #7 apply to $\mathbb{N}$. If we consider $\mathbb{N}_0$ (see Section 0.1) rule #6 is also valid. Technically, rule #8 does not apply to $\mathbb{N}$ even though a form of subtraction is available as long as the result stays within $\mathbb{N}$.

0.2.2 Exponents

When a number $a \in \mathbb{Z}$ is multiplied by itself n times, it is customary to write $a \cdot a \cdot a \cdots a = a^n$. This notation seems so natural and obvious that we might be surprised to find out that it was not always used. Isaac Newton occasionally wrote xxx for x^3, but he eventually adopted the superscript more consistently. This is the multiplicative analogue of writing $a + a + \cdots + a = na$. Considering a^n, we would say that this is the "nth **power** of a" or that the "**exponent** of a is n," where a is the **base** for the exponentiation. There are a few simple rules that exponents obey. In Table 0.3(4), we must agree that $m \neq 0$ and if $a < 0$ then m is odd, since we cannot take even roots of negative numbers.

Table 0.3: Rules for Exponents. For $a, m, n \in \mathbb{Z}$.

1) $a^n \cdot a^m = a^{n+m}$	$2^2 \cdot 2^3 = 2^5 = 32$
2) $a^n \div a^m = a^{n-m}$	$3^7 \div 3^4 = 3^3 = 27$
3) $a^{nm} = a^{nm} = a^{mn} = (a^m)^n$	$(2^3)^2 = 2^6 = 64$
4) $(a^n)^{1/m} = a^{n/m} = (a^{1/m})^n$	$(2^6)^{1/3} = 2^2 = (2^{1/3})^6$

Although the justification is beyond the scope of this text, it is possible to extend the idea of bases and exponents to $\mathbb{R}$ and even $\mathbb{C}$. Moreover, we will allow exponents to be fractions. So an expression like $2^{1/3}$ follows the rules in Table 0.3. What do we mean by $2^{1/3}$? Apply rule #3 with exponent $m = 3$. We get $(2^{1/3})^3 = 2^1 = 2$. We recognize $2^{1/3}$ is the cubic root of 2, usually written as $\sqrt[3]{2}$. In general, the nth root of $a \in \mathbb{R}$ is a number which, when raised to the nth power, is a. A weird looking, but perfectly valid, expression is $2^{\frac{1}{\pi}}$. This would be the number, which when taken to the πth power, is just 2. **Fractional exponents** are generally easier to work with than **radicals** because the formulas in Table 0.3 expedite calculations. As an example, cube 32 and find the fifth root of the result. As a radical, this is $\sqrt[5]{32^3}$. Writing it in fractional exponent notation and using rule #4, we are looking for $32^{3/5}$. Note that $32^{1/5} = 2$, so we immediately have the answer $2^3 = 8$. Of course, modern calculators make short work of exponentiation and root extraction. Even a grocery-store calculator can tell us that $2^{\frac{1}{\pi}} \approx 1.247$. But we should at least understand what the calculator is doing.

There is a chance for confusion in calculating a^{b^c}. Successive exponentiation is not associative like multiplication. This expression means b^c is first calculated, then the result is applied as an exponent to a. For example, $2^{3^2} = 2^9 = 512$, not $(2^3)^2 = 8^2 = 64$. The rule is to start at the top of the tower of exponents and work downwards.

0.2.3 Logarithms

Logarithms (logs, for short) are famous for intimidating beginning students of mathematics, especially those who may have had a prior unsatisfactory introduction to them. If the dog has bitten us, we are going to be reluctant to pat him again. Here is some advance reassurance if you are among those with lingering insecurities about working with logs: *Logarithms are just special exponents and if you know how exponents work, you automatically know how logarithms work.*

Suppose b is a number greater than zero, but not one. If $b^x = A$, we call the exponent x the **logarithm** of the number A to the **base** b. This is written $x = \log_b A$. Originally, $b = 10$ was an obvious choice. If $10^x = A$, the conventional notation omits the base explicitly and we write just $x = \log A$. Logs to the base 10 are called **common logs**. For theoretical reasons having to do with calculus, another base is often used in scientific applications. This is the famous base of **natural logs** called e, honoring the Swiss mathematician Leonhard Euler. Now e to five decimals is 2.71828, and it is a reasonable question why anyone would choose such a strange number for calculating logarithms. Suffice it to say that e tends to come up unusually often in many useful formulas we will discuss. The Latin for "natural logarithm" is **l**ogarithmus **n**aturalis, and if $A = e^x$ we write $x = \ln A$. Any base other than 10 or e requires the user of the notation to be explicit and declare the base as a subscript, such as $\log_b A$.

It should be apparent why 1 would not be a satisfactory base, as every power would just be 1 again. Negative bases pose a different but still disqualifying issue. If we are planning to use fractional exponents, signifying that a root is being extracted, we will have trouble with an expression like $(-2)^{1/2}$. This is an imaginary number, and although the theory of logarithms can be extended to accommodate them, our applications will be limited to the real numbers. You may encounter an expression like $(-8)^{1/3}$, which is superficially similar, but actually a real number -2. Denominators that are even integers in any fractional powers cause this difficulty, so we will avoid negative bases and be confident that any fractional power can be safely used in our calculations.

Logarithms first emerged about 1500 CE in connection with commercial purposes, basically accounting, because they facilitated rapid calculations. You will see in a moment how logs convert multiplication to addition, division to subtraction, and taking powers or roots to a single multiplication or division. In the days before calculators, this was a great time-saver. The basic idea was to have a comprehensive table of logarithms calculated to a desired precision and indexed beforehand. In the case of multiplication, the two factors to be multiplied would be separately converted to logs from the table, the logs added, then the resulting sum converted back to the product of the original numbers. High school algebra texts prior to 1970 often devoted quite a few pages in an appendix to presenting log tables. The watershed event that swept this into the dustbin of history was the ready availability of reasonably powerful hand calculators. So logs have lost their practical computational utility for the most part, but their theoretical importance is still very relevant. Our purpose in studying the mechanics of logarithms is to develop a perspective for understanding their influence on the structure of formulas that come up in applications.

Each exponent law in Table 0.3 spawns a corresponding logarithm law. We will show how this works in detail. Let $b^x = A$ and $b^y = B$. By rule (i) in Table 0.3, we have $A \cdot B = b^x \cdot b^y = b^{(x+y)}$. The logarithm of the product $A \cdot B$ is defined as the power you would have to raise the base b to in order to recover $A \cdot B$. Evidently that power is $(x+y)$. But $x = \log_b A$ and $y = \log_b B$. This means that the "log of a product is the sum of the logs of the factors." So $\log_b AB = \log_b A + \log_b B$. As with all the log laws, it is easier to remember them in words than as formulas. On to division: $\frac{A}{B} = \frac{b^x}{b^y} = b^{x-y}$, so $\log_b \frac{A}{B} = \log_b A - \log_b B$, or "the log of a quotient is the difference of logs" (numerator log minus denominator log). Now suppose z is a power we wish to apply to A. Note that $A^z = (b^x)^z = b^{xz}$. Evidently $\log_b A^z = z \log_b A$. In words, "the log of the power is the power times the log." Finally, what if we need the log of the zth root of A. Recall this just the $\frac{1}{z}$ power of A. We write $\sqrt[z]{A} = A^{1/z} = (b^x)^{1/z} = b^{x/z}$. It follows that $\log_b \sqrt[z]{A} = \frac{x}{z} = \frac{1}{z} \log_b A$, or "the log of the root is the log divided by the index." These rules are summarized below in Table 0.4.

Table 0.4: Rules for Logarithms.

1) $\log_b AB = \log_b A + \log_b B$
2) $\log_b \frac{A}{B} = \log_b A - \log_b B$
3) $\log_b A^z = z \log_b A$
4) $\log_b \sqrt[z]{A} = \frac{1}{z} \log_b A$

0.3 Relations

The idea of a relation between two sets is fundamental to establishing a clear definition of a function. The corresponding notion of a relation between a set and itself allows us to define what we mean by equality and also the order of elements in a set. These concepts are foundational for just about everything we will do. A (binary) **relation R between sets A and B** is some subset of the Cartesian product $A \times B$. Relation R could be all, some, or none of $A \times B$, although the void relation is usually not considered. In the event $A = B$, we call $R \subset A \times A$ a **relation on** A. If the pair (a, b) belongs to a relation R between A and B, we say "a is R-related to b" and occasionally denote this by $a \sim b$.

0.3.1 Equivalence relations

The following relation on a set A is arguably the most important type of relation in all of mathematics. This relation models our intuitive notion of equality. Given $R \subset A \times A$, we say R is an **equivalence relation** whenever the following three conditions hold:
1) **reflexivity** $(a, a) \in R$
2) **symmetry** $(a, b) \in R$ implies $(b, a) \in R$
3) **transitivity** $(a, b) \in R$ and $(b, c) \in R$ imply $(a, c) \in R$

Ordinary equality of numbers, congruence of triangles, similarity of triangles, and equality of more complicated objects which we will subsequently define and discuss, such as polynomials, vectors, and matrices are all examples of equivalence relations. Less mathematically, being siblings with the same parents is an equivalence relation. Being cousins is not, since you are not your own cousin, and the reflexivity condition fails.

Given any nonvoid set, we may invent a condition that some elements share and others do not. In general, the condition can be quite arbitrary. For example, given $\mathbb{N}$ there are some numbers that are even and some odd. We can declare that all even numbers are equivalent and all odd numbers are equivalent. So if m is even, then $m \sim n$ if n is also even but $m \nsim n$ if n is odd. It is easy to see that "$\sim$" defines an equivalence relation on $\mathbb{N}$. The two subsets of $\mathbb{N}$ defined by this relation are called **equivalence classes**. In general, a family of equivalence classes on a set induce a **partition** of the set. A partition of a set S is a collection of subsets of S, called **blocks** labeled S_i with the following properties: (i) $S = \bigcup_i S_i$, (ii) $S_i \neq S_j$ if $i \neq j$, and (iii) $S_i \neq \varnothing$ for all i. In words, the blocks are exhaustive, disjoint, and nonvoid. Conversely, any partition of a set defines an equivalence relation on it. Two elements are considered equivalent if they belong to the same block.

0.3.2 Order relations

Another important type of relation on a set is the one that models relative position in an implied queue. This is called an **order** relation. There are several types, the most basic of which is called a (weak) **partial order**. Given $R \subset A \times A$, we say R is a partial order relation if the following conditions hold:

1) **reflexivity** $(a, a) \in R$
2) **antisymmetry** $(a, b) \in R$ and $(b, a) \in R$ imply $a = b$ (so we need an equivalence relation in place)
3) **transitivity** $(a, b) \in R$ and $(b, c) \in R$ imply $(a, c) \in R$

A good example of a partial order is the containment relation defined on pairs of subsets from some universal set. Certainly, a set is a subset of itself, so reflexivity holds. If set $U \subseteq V$ and also $V \subseteq U$, it must be that $U = V$. Similarly, if $U \subseteq V$ and $V \subseteq W$, then $U \subseteq W$, showing transitivity. Note that, given any two arbitrary sets, there is no guarantee that one is a subset of the other.

We can add the property of **comparability** to the specification of a partial order: (iv) either $(a, b) \in R$ or $(b, a) \in R$ for all a and b in A. Partial orders satisfying the comparability condition are called **total (linear)** orders. The relation "is less than or equal to" among real numbers, written $a \leq b$, is a typical example of a linear order. We may wish to distinguish the order symbolized by "≤", which we call **weak inequality**, from the one symbolized by "<", which we call **strong inequality**. The possibility that elements may be equal is being explicitly removed. Our previous conditions for a (weak) partial order have to be amended. Reflexivity is out since $a < a$ is impossible. Antisymmetry is replaced by (ii)' either $(a, b) \in R$ or $(b, a) \in R$ but not both. This condition is called **asymmetry**. Finally, transitivity goes through without modification. So a (strong) partial order has only asymmetry and transitivity. Note that comparability is baked into the asymmetry condition. Note that the symbols "⊇", "≥", and ">" really do not add anything new to the discussion, since they are defined by their opposites. Notation $V \supseteq U$ means $U \subseteq V$, $b \geq a$ means $a \leq b$, and $b > a$ means $a < b$.

A set with a partial order defined on it may exhibit some special elements. Given elements a and $b \in A$ as above, let us use the stylized symbol $a \preceq b$ to mean $(a, b) \in R$, a partial order on A. If there is an element $M \in A$ with the property that for all $a \in A$ it is true that $a \preceq M$, then we call M the **maximum (greatest) element** of A. Note that not all elements of A have to be comparable with one another, just with M. If there is an element $M' \in A$ such that if $M' \preceq X$ then $M' = X$ for any $X \in A$, we say M' is a **maximal element** of A. Note that A can have many maximal elements, but only one maximum element, which, of course, is also maximal. A partially ordered set may have a maximum and no other maximal elements, no maximum but some maximal elements, or no maximum and no maximal elements. Think of $\mathbb{N}$ with the order "$\leq$" for this last case. At the other end of the spectrum, there may be an element $m \in A$ with the property that for all $a \in A$ it is true that $m \preceq a$. This m is the **minimum (least) element** of A. Similarly, if there is an element $m' \in A$ such that if $x \preceq m'$ then $x = m'$ for any $x \in A$, we say that m' is a **minimal element** of A. In complete analogy with the case for maximum and maximal elements, a partially ordered set may have a unique minimum element or perhaps none at all. It may also have one or more minimal elements or none at all. In some contexts, it makes sense to call a least element the first and the greatest element the last.

Another type of partial order, actually a total order, turns out to be important in validating a powerful method of mathematical proof called **induction**. We will explore this method in some detail subsequently, but at the moment it is enough to characterize the underlying special order. A **well-ordering** of the set A is a total order on A with the additional property that any nonempty set of elements selected from A has a least element. This automatically forces the order to be total, since any two element set has a least element and the two are therefore comparable. Some sets are easily seen to be well ordered. Using the usual order "$\leq$" in the following examples, it is easy to see that the natural numbers $\mathbb{N}$ are well ordered. The integers $\mathbb{Z}$ would seem to be as well, but they are not, e. g., the subset $\{n \in \mathbb{Z} : z \leq 0\}$ has no least element. Neither are the rational numbers $\mathbb{Q}$. Consider the set $\{r \in \mathbb{Q} : \pi \leq r\}$. The number π is famously irrational, so this set includes all the rational numbers greater than a particular irrational number. The trouble is that there is always a rational number slightly greater than π no matter how small a tolerance you demand. There is no least one. Mathematicians who regularly work with these ordered sets have given them nicknames: **poset** for a partially ordered set, **toset** for a totally ordered set, and **woset** for a well-ordered set.

We should record for future reference the rules for using arithmetic operations on the real numbers endowed with their natural order. Given reals a and b, we officially define $a < b$ to mean $a \leq b$ and $a \neq b$. So strong inequality is weak inequality without the equality. Note that $|a|$ means a if $a > 0$ and $-a$ if $a < 0$ and $|0| = 0$. We call $|a|$ the **absolute value** of a. The following apply:

Table 0.5: Order Rules for Arithmetic.

1) if $a > 0$ and $b > 0$, then $a \cdot b > 0$ and $a \div b > 0$
2) if $a < 0$ and $b < 0$, then $a \cdot b > 0$ and $a \div b > 0$
3) if either $a < 0$ and $b > 0$, or $a > 0$ and $b < 0$, then $a \cdot b < 0$ and $a \div b < 0$
4) if $a > b > 0$, then $0 < \frac{1}{a} < \frac{1}{b}$
5) $\|a\| \geq a$
6) $\|a + b\| \leq \|a\| + \|b\|$, this is called the triangle inequality
7) $\|a - b\| \geq \|\|a\| - \|b\|\|$, this is the difference version

"Minus times minus is plus. The reasons for this we must not discuss" – W. H. Auden (English poet) via Joseph Gallian

Why don't we talk about order in connection with the complex numbers? Suppose the usual order on $\mathbb{R}$ was extended to $\mathbb{C}$. Then the imaginary unit $i = \sqrt{-1}$ needs to be assigned an order in relation to 0. Certainly $i \neq 0$, so either $i > 0$ or $i < 0$. Suppose we take $i > 0$. Then by rule (i) of the above table, $i \cdot i = -1 > 0$. This cannot be, so we conclude $i < 0$. Now by rule (ii) we have $i \cdot i = -1 > 0$. Oops! Evidently, there is no way to integrate i into the natural order of the reals, and this means there is no way to impose an order on the complex numbers that when restricted to the reals would yield the usual order. There is an old theorem which states that any set can have a well order constructed for it. So we could, in principle, impose an order on $\mathbb{C}$. The problem is that the induced order on $\mathbb{R} \subset \mathbb{C}$ would not be the usual order we recognize.

0.4 Functions

Returning to the notion of a relation between sets, we can formulate a general definition of function. Elementary texts usually introduce the idea of function as "a rule that associates one number uniquely with another," perhaps give a formula, and leave it at that. That is certainly true, but we can be more precise. Given sets A and B, let us call the relation $f \subset A \times B$ a **function (mapping)** if whenever $(a, b) \in f$ and $(a, c) \in f$ then $b = c$. More concisely, we would call f a function from the set A to the set B, and write $f : A \to B$ to suggest a dynamic process that uniquely associates an element from A with a corresponding element in B. Here A is called the **domain** of the function and B is called the **codomain**. Since the same rule of association might be used with different domains and codomains, we will agree that two functions are to be considered the same only if they are the same subset of the same Cartesian product. If it is necessary to emphasize all the data of a function, we can write it as an ordered triple $\langle f, A, B\rangle$. Usually, the domain and codomain are clear from context and we will just refer to the function as f.

If $(x,y) \in f$, we also write $f(x) = y$ and often refer to x as the **argument** of the function. The rule embodied by f is being applied to $x \in A$ with the result $y \in B$. The intent of writing $f(x)$ is to make it clear that the action of f applies to one element of the domain at a time. It is common to "stretch" the notation and write $f(U) = \{f(x) \in B : x \in U\}$. So we are using f to signify both a point function (consistent with the relational definition) and by extension also a set function. This should not get us into any trouble if we pay attention to the context. If D is the domain of f and C is its codomain, we may, in principle, calculate $f(D) \subset C$. We call $f(D)$ the **range** of f and note that it need not be all of C. The range of f is the set of points in the codomain that have a predecessor, relative to the action of f, in the domain. It is always assumed that the function f may be applied to any element of its domain D, so if we have some subset $V \subset D$, the set $f(V) \subset C$ is well defined. In this case we call $f(V)$ the **forward image** or **direct image** of V under f.

It may (and often does) happen that a function $g \subset B \times A$ exists that reverses the pointwise action of f. We denote the **composition** of g and f by $g \circ f$ and say that g

is **postcomposed** with f or that f is **precomposed** with g. The terminology is justified by the convention that composite functions operate from right to left. So $(g \circ f)(x)$ is understood to mean g applied to $f(x)$, or $g(f(x))$. The extension to multiple compositions is clear. So if g reverses the action of f, we would have $(g \circ f)(x) = g(y) = x$. This situation means that $g \circ f$ is the identity or "do nothing" function. Accordingly, we change our notation for g to f^{-1} and note that f and f^{-1} act as mutual **inverses**. We would also say that f is **invertible**.

Sometimes a function does not have an inverse. Here's an example: $f : \mathbb{R} \to \mathbb{R}$ given by $f(x) = 0$ for all $x \in \mathbb{R}$. This is the **constant function** that maps every real number to zero. The range of f is a single point. If f had an inverse, it could only map zero back to a single real number in order to comply with the definition of function. What about all the other numbers in the domain of f, now the codomain of our proposed f^{-1}? Nothing gets mapped to them. So f^{-1} is not the inverse of f.

Let us not discard the symbol f^{-1} prematurely if we know f has no pointwise inverse. Given $f : A \to B$ and $W \subset B$, we might be interested in the set of elements in A whose forward image is W. We write $f^{-1}(W) = \{x \in A : f(x) \in W\} \subset A$ as the **inverse image** or **preimage** of W in A. Even though f has no inverse as a point function, it has an inverse as a set function. Now we really have to pay attention to the context. Some authors go to the trouble of modifying the notation for a point function f to $\hat{f}$ or some other distinction when it is used for the corresponding set function. We are going to use f to indicate the forward image of either a point or a set and f^{-1} to indicate the inverse image of a set or (if f has a pointwise inverse) the inverse image of a point.

The interplay between set operations and set function operations is not completely obvious. Let X and Y be subsets of A and $f : A \to B$ not necessarily invertible, except as a set function. Then we have the following rules for the **commutation** (order reversal) of functions with set operations:

Table 0.6: Commutation Rules.

1) $f(A \cup B) = f(A) \cup f(B)$
2) $f^{-1}(A \cup B) = f^{-1}(A) \cup f^{-1}(B)$
3) $f(A \cap B) \subseteq f(A) \cap f(B)$
4) $f^{-1}(A \cap B) = f^{-1}(A) \cap f^{-1}(B)$
5) $f(A^c) \neq (f(A))^c$
6) $f^{-1}(A^c) = (f^{-1}(A))^c$

The general rule is that preimages of sets commute with all basic set operations, and forward images of sets commute with the union, but not with intersection or complementation. To see why forward images misbehave with regard to the intersection, consider the sets $A = \{x \in \mathbb{R} : x \geq 0\}$ and $B = \{x \in \mathbb{R} : x \leq 0\}$ where $f(x) = x^2$. Note $f(A) = A$ and also $f(B) = A$, so $f(A) \cap f(B) = A$. On the other hand, $A \cap B = \{0\}$, so $f(A \cap B) = \{0\} \neq A$.

Likewise for the complementation, $A^c = \{x \in \mathbb{R} : x < 0\}$, so $f(A^c) = \{x \in \mathbb{R} : x > 0\}$. But $(f(A))^c = \{x \in \mathbb{R} : x \geq 0\}^c = \{x \in \mathbb{R} : x < 0\} \neq f(A^c)$. For any set A in the domain of f, if $(f^{-1} \circ f)(A) = A$, we call A f**-saturated**. The set $A = \{x \in \mathbb{R} : x \geq 0\}$ is not saturated for the squaring function $f(x) = x^2$ because $(f^{-1} \circ f)(A) = \mathbb{R}$. We can alleviate this by imposing a **restriction** on the domain of f, namely redefining it to be just the nonnegative reals.

Functions can display two important properties. If the function f is invertible, we say it is synonymously **injective** or **one-to-one**. If the range of f is identically its codomain, then f is said to be a **surjective** or **onto** function. The most convenient functions to work with are both injective and surjective. This blend of properties makes the function **bijective**. Note that any function can be made surjective by redefining its codomain to be its range.

Our work in subsequent articles will focus on the so-called **elementary functions**. These include polynomial, exponential, logarithmic, trigonometric, and hyperbolic functions. This is a standard classification. Some of them may not seem so elementary to deal with, but we will explain their properties in detail and try to defang them as necessary.

References & suggested further reading

There are many historical comments woven into the overall narrative to provide context for the development of various mathematical ideas. Most of these are derived from occasional historical remarks appearing in the technical references listed in each chapter-specific bibliography. A comprehensive and current reference that subsumes all of this is the book *History of Mathematics* (3rd ed.) by Boyer and Merzbach (2011), Wiley.

1 Counting methods

The more sophisticated word for counting is **enumeration**. What we want to enumerate are configurations or ways of arranging things. For example, how many different top-three finishes can there be in a ten horse race? Or how many different subcommittees of eight senators can be formed if there are 100 senators available. These kinds of problems all have the feature that we have a pool of objects, horses, or senators (Caligula's horse was both), and we have to distribute them among places (three available winning positions or eight available committee seats) subject to additional conditions. These and more complicated situations are usually modeled by idealized thought experiments where we assign objects from one set to available containers or cells in another set. Traditionally, the objects and containers were imagined to be balls and urns, respectively. Let us modernize this and simply count the ways we can assign **objects** to **cells**.

1.1 The Twelve-Fold Way

The theory of enumeration was not systematized until fairly recently. For at least a millennium, there were various formulas available for counting configurations like permutations, combinations, partitions and the like, but the first modern attempt to pull all of this together was the book *Combinatory Analysis* by the English mathematician P. A. MacMahon, published in 1916. More recently, the Italian–American mathematician Gian-Carlo Rota popularized a systematic approach called the Twelve-Fold Way. Physicists have their Eight-Fold Way, so mathematicians, being competitive, developed their Twelve-Fold Way. This is a logical framework for organizing the major formulas of enumerative combinatorics. The key idea is to view a specific placement of objects into cells as a function from the finite set of objects to the finite set of cells, so each placement or configuration corresponds to a function. If we can figure out how to count the functions, we can count the configurations.

The system allows three types of functions – unrestricted, injective, and surjective – and four options based on whether the objects or cells are separately to be considered distinguishable or not. Three functions times four distinguishability options gives 12 cases for counting allowed configurations. There have been generalizations of the Twelve-Fold Way based on increasing the granularity of the cases considered, but we will stick with the original and outline methods to solve an assortment of the most common counting problems.

In this chapter we are counting functions $f : K \to N$ where K is the set of objects, of which there are $|K| = k$, and N is the set of cells, of which there are $|N| = n$. We have used the double bar notation $|\cdot|$ before to indicate absolute value. In this set theory context, the notation means the **cardinality** or number of elements in the set. Two general counting principles are at the root of most enumeration formulas. To describe them, we assume the sets A and B are **disjoint**.

https://doi.org/10.1515/9783111506654-002

- **Multiplicative Principle.** If there are m ways to select an object from set A and there are n ways to select an object from set B, then there are $m \cdot n$ ways to select a pair of objects, one from A and the other from B.
- **Additive Principle.** If there are m ways to select an object from set A and there are n ways to select an object from set B, then there are $m+n$ ways of selecting an object from either A or B.

So "and" goes with the product and "or" goes with the sum.

1.1.1 Distinguishable objects into distinguishable cells unrestricted case n^k

For each object in K, there are n choices for the cell into which it is placed. Then by the product principle above, we have n ways of making the first choice and n ways of making the second choice, giving $n \cdot n = n^2$ choices for the first two objects. Continuing in this way, we will make n choices k times to place all the objects. The result of this process is to create an ordered selection of cells. Such a sequence of k selections from a set of n cells is called a **list**, or in this case specifically a k-**list**. The same cell can be selected repeatedly, but its position or positions in the list are well defined. We conclude that the overall number of ways to create a k-list from an n-set is n^k. A list is an ordered multiset, which is a set that allows multiple copies of the same element.

Example 1.1. How many subsets are there for a set with k elements (do not forget the null set)?

Example 1.2. A professional licensing examination has 40 questions and each question allows five possible answer selections a through e. How many different answer sheets are possible?

Example 1.3. How many ways can five different flavored doughnuts be distributed to three meeting participants, Al, Bob, and Carol?

Injective case (one object per cell max) $\frac{n!}{(n-k)!}$

First of all, we note that this cannot be done at all unless the number of cells is at least as great as the number of objects, $n \geq k$. Provided this is the case, for the first object in K there are n choices of cell assignment. But for the second object in K, there are now only $n-1$ choices because one cell has been filled. Then for the third, only $n-2$ choices, and so forth. Eventually, the kth object has only $n-(k-1)$ choices left. By the product principle, the total number of possible configurations is $n(n-1)\cdots(n-k+1)$. We could leave it at that, but a better way to write this is $\frac{n!}{(n-k)!}$, where we have used the **factorial** notation, $n! = 1{\cdot}2{\cdot}3\cdots(n-1){\cdot}n$. This situation has a very intuitive interpretation. If you think of the objects in K as first prize, second prize, and so forth, we are "awarding" k prizes to some (possibly all) of the cells in N. In effect, we are taking k of the n cells and assigning a rank to them. This is called a **permutation** of the n cells taken k at a time. Other notations for this in common use are ${}_nP_k$ and $P(n,k)$. A notation that is gaining popularity is $n^{\underline{k}}$. This is read as "n falling factorial k" and is very descriptive of $n(n-1)\cdots(n-k+1)$. A permutation of n objects taken n at a time is a shuffle of those objects, and by our formula there should be $\frac{n!}{(n-n)!} = n!$ possible shuffles or rearrangements of N. Do not worry, $0! = 1$ by convention.

Example 1.4. How many different top-three finishes (win/place/show) could there be in an eight horse race?

Example 1.5. Dance competition pairs are to be formed from a pool of three men and six women. How many ways can this be done?

Another viewpoint is that man #1 can be paired with any of the six women, leaving man #2 any of five candidates, and finally man #3 any of the remaining four women. Again this gives $6\cdot5\cdot4 = 120$ ways. Note that we cannot turn this argument around without sacrificing injectivity. However, we could select three of the six women and match them with the three men. For each distinct subset of three women, there would be six such matchups. Then we would need to determine how many subsets of three women could be formed from the original six. We will learn shortly that this can be done in 20 ways. By the product principle, there would be $6\cdot20 = 120$ ways to form the pairs as before.

Example 1.6. The numbers 1 through 50 are painted on ping-pong balls. A lottery machine selects six balls in order. If your ticket has those numbers in the correct order, you win. What is the number of possible outcomes for the lottery selection, or equivalently, what is the total number of possible distinct tickets?

Example 1.7. What fraction of five-character passwords drawn from A–Z and 0–9 have repeat characters?

Surjective case (all cells used) $n!\cdot S(k,n)$

As with the case in Section 1.1.2, there is a constraint imposed, this time by surjectivity. For a surjective function $f: K \to N$ to exist, we must have $k \ge n$. It is unfortunate that this case is somewhat more difficult to grasp intuitively. Suppose $N = \{y_1, y_2, \ldots, y_n\}$. The preimage $f^{-1}(y_i)$ of an arbitrary element in N cannot be the void set by surjectivity. At least one element of K has to be mapped to each y_i. Let us call that preimage $X_i \subset K$ and note that $\bigcup_{i=1}^{n} X_i = K$. Certainly, $X_i \cap X_j = \varnothing$ if $i \ne j$. The set of preimages therefore satisfies the conditions for a partition of K, specifically a partition of K into n blocks. We could write a typical partition of K into blocks like this: $\{x_1, x_2|x_3, x_4, x_5|x_6, \ldots|\ldots|\ldots, x_k\}$

where the n successive blocks are separated by the delimiter |. Now we ask how many distinct partitions of a k-set into n blocks are there. The answer is $S(k,n)$, a so-called **Stirling number of the second kind**. There is no simple formula for Stirling numbers, so for the moment the following chart will have to do. After we have some additional machinery, we can develop the formula that allows us to calculate the entries in the chart.

Table 1.1: Stirling Numbers of the 2nd Kind. $S(k,n)$ is the number of partitions of a k-set into n blocks.

$k \downarrow n \rightarrow$	0	1	2	3	4	5	6	7	8	9	10
0	1	*	*	*	*	*	*	*	*	*	*
1	0	1	*	*	*	*	*	*	*	*	*
2	0	1	1	*	*	*	*	*	*	*	*
3	0	1	3	1	*	*	*	*	*	*	*
4	0	1	7	6	1	*	*	*	*	*	*
5	0	1	15	25	10	1	*	*	*	*	*
6	0	1	31	90	65	15	1	*	*	*	*
7	0	1	63	301	350	140	21	1	*	*	*
8	0	1	127	966	1701	1050	266	28	1	*	*
9	0	1	255	3025	7770	6951	2646	462	36	1	*
10	0	1	511	9330	34105	42525	22827	5880	750	45	1

Our strategy for finding the number of surjective functions $f : K \rightarrow N$ is to count up the number of ways we can map the blocks of a particular partition of K to the elements of N and then (by the product principle) multiply by the total number of partitions available. Since there are $n!$ permutations of the elements of N, there must be $n!$ mappings from the n blocks of one particular partition of K to N. Now multiply this by the number of partitions $S(k,n)$ and we get the number of possible surjective functions. So at last we have found the number of surjective maps from K to N to be $n!S(k,n)$.

Example 1.8. Al, Bob, and Charlie carry a big suitcase, a little suitcase, a briefcase, a computer, and a backpack into the airport. Everyone carries at least something. How many ways can this be done.

Example 1.9. Eight copy machine repairmen (Al, Bob,...– the usual crew) are assigned to five clients (ABC Corp., etc.) who have same-day service agreements. Assuming the repair jobs take all day, how many ways could the work assignments be covered?

1.1.2 Indistinguishable objects into distinguishable cells injective case (one object per cell max) $\frac{n!}{k!(n-k)!}$

What exactly do we mean by indistinguishable objects? In the present case it means that the elements of the domain set K are to be considered interchangeable. If they are people, they will not have names. If they are objects, they will be generic. There will not be any labels to discriminate among elements. More technically, suppose f and g are two functions from K to N. If there is some permutation $\pi : K \to K$ such that $f(x) = g \circ \pi(x)$ for all $x \in K$, then we say f and g are **equivalent with K indistinguishable**. More succinctly, one function is equivalent to another if some prior shuffling of the given domain for one function gives the same result as the other function. We can do the same trick with the range set. Again suppose f and g are two functions from K to N. If there is some permutation $\sigma : N \to N$ such that $f(x) = \sigma(g(x))$ for all $x \in K$, we declare f and g to be **equivalent with N indistinguishable**. Nothing prevents us from considering both the domain and range elements indistinguishable. In that case two functions f and g would be **equivalent with both K and N indistinguishable** if the permutations π and σ exist so that $f(x) = \sigma(g(\pi(x)))$. Here we shuffle the domain before and the range after application of the function g.

We can get a formula for the present case easily by using a result of Section 1.1.1. Suppose we calculate $P(n,k) = \frac{n!}{(n-k)!}$. This is the number of injective mappings of K into N where we have been careful to preserve the individual identities of the elements of K. But we want to relax that condition here – each element of K is now considered interchangeable with any other. So $P(n,k)$ overcounts the number of distinct mappings. To correct for the overcount, we need to divide $P(n,k)$ by the number of ways the elements of K can be rearranged among themselves, namely $P(k,k) = k!$ ways. Finally, we see that the number of distinct functions from a set of k indistinguishable objects to a set of n distinguishable cells is $\frac{1}{k!}P(n,k) = \frac{n!}{k!(n-k)!}$. This number is customarily denoted as $\binom{n}{k}$ (read "n choose k") and is called a **binomial coefficient**. It has a bigger role in mathematics than this derivation suggests, and we will meet it again. Since $P(n,k)$ counts the number of arrangements of k elements selected from a pool of n elements, if we suppress the distinctiveness of the k elements and treat them as simply markers, we see that $\binom{n}{k}$ just counts the number of possible k-subsets of an n-set. Other notations for $\binom{n}{k}$ are ${}_nC_k$ and $C(n,k)$. The latter emphasize that we are taking combinations of n objects, k at a time. So $P(n,k)$ counts **ordered selections** or **permutations**, and $C(n,k)$ counts **unordered selections** or **combinations**. Summarizing, $C(n,k) = \binom{n}{k} = \frac{n!}{k!(n-k)!}$. This is not in the usual order of presenting the Twelve-Fold Way formulas, but we need the injective case now to facilitate understanding of the unrestricted case.

Example 1.10. How many ways can four bananas be selected from 15?

Example 1.11. If three diodes in a lot of ten are known to be defective and we randomly select five, what is the ratio of "all good" selections to the number of total selections?

Example 1.12. What is the coefficient of x^4 in the expression $(x+1)^9$?

Example 1.13. Generalizing the previous example, write $(1+x)^n$ as a sum.

Unrestricted case $\left(\!\binom{n}{k}\!\right) = \binom{n+k-1}{k}$

Ordinarily, if we encounter a set such as $\{a, a, b, c, c, c, d, d\}$, we would cull the repeated elements and reduce it to $\{a, b, c, d\}$. However, for some purposes we might want to put the "no repeat elements" rule in abeyance and work with so-called **multisets**, where repeats are allowed. Technically, a multiset is a function β from some (proper) set $S = \{a, b, c, \dots\}$ to $\mathbb{N}_0$ where $\beta(a) = n_a$ means a is repeated n_a times in the multiset, and so forth. For this section we will need to know how many multisets of a certain size can be manufactured from a given set. We will regard the set K of indistinguishable elements as markers, and a function from K to N will distribute these markers among the elements of N. The number of markers mapped to a particular element of N will tell us how many times that element is to be repeated in a multiset and the multiset itself will have $|K| = k$ entries. The number of multisets of size k that can be formed from a set of size n is denoted by $\left(\!\binom{n}{k}\!\right)$. You can read this as "$n$ multichoose k" or "n choose k with replacement." Note that unlike the binomial coefficient $\binom{n}{k}$ where $k \le n$, the multiset number allows $k > n$ because of allowed reselection from N. We need a formula. But first, a simple example.

Suppose we form the multiset $\{a, a, b, b, b, d\}$ from the set of available elements $\{a, b, c, d\}$. Note c has not been selected. We could visualize this multiset by using dots and vertical bars: $\cdot\cdot \mid \cdot\cdot\cdot \mid\mid \cdot$. Consider the four zones from the left to the right created by the three bars: one space to the left of the first bar, one space to the right of the last bar, and one space between any two adjacent bars. These zones are filled with six dots according to the specification of the multiset. So the initial two dots mean a has been selected twice, then three dots mean b is selected three times, no dots mean no c's, and the final dot means d once. Every multiset of size six on four elements can be represented this way, so if we could count the number of possible configurations of this type, we would immediately be able to evaluate $\left(\!\binom{4}{6}\!\right)$.

There are nine symbols in our representation. Imagine a string of nine "wild card" symbols, $* * * * * * * * *$. To represent any multiset of six elements assembled from the four available, we would need three wild cards declared to be bar delimiters creating four zones and the remaining six wild cards to be dots populating the zones. How many ways can three of the wild cards be converted to bars? The other six will then automatically be considered dots. The answer is $\binom{9}{3} = 84$, the number of 3-subsets of a 9-set. So $\left(\!\binom{4}{6}\!\right) = 84$.

Generalizing this to the number of k multisets of an n-set, we see that $n-1$ delimiters would create n zones for k dots. There would be $(n-1)+k$ wild cards of which $n-1$ would be converted to delimiters. We know there are $\binom{n-1+k}{n-1}$ ways to do this, so $\left(\!\binom{n}{k}\!\right) = \binom{n+k-1}{n-1} = \frac{(n+k-1)!}{k!(n-1)!} = \binom{n+k-1}{k}$.

Example 1.14. How many ways can ten identical cookies be distributed among four children, Al, Bob, Charlie, and Dave?

Example 1.15. How many solutions in nonnegative integers are there to the equation $x + y + z = 10$?

Example 1.16. How many solutions in positive integers are there to the equation $x + y + z = 10$? Compare with the previous example.

Example 1.17. How many ways could $100,000 be invested in stocks, bonds, real estate, or gold in $5,000 increments.

Example 1.18. How many distributions are possible in the preceding example if at least a quarter of the total must be invested in gold?

Surjective case (all cells used) $\left(\!\binom{n}{k-n}\!\right)$

Again we will regard the set K of indistinguishable objects as markers and a function from K to N will distribute these markers among the elements of N, but this time ensuring that each cell receives at least one object. Note that K must be at least as big as N for surjectivity to be possible, so we assume $k \geq n$. Since the objects in K are indistinguishable, we may start by assigning one of them to each cell in N. There is essentially only one way to do this because any collection of n objects from K is identical to any other. Any surjective function will require this to happen. Now if $k > n$, we have $(k - n)$ remaining elements of K that need to be assigned to cells in N. There is no restriction whatsoever on how this can be done, so an arbitrary distribution of the leftover elements corresponds exactly to a multiset of size $(k - n)$ drawn from the n target cells. By the multiplicative principle, the number of overall mappings of K onto N is $1 \cdot \left(\!\binom{n}{k-n}\!\right) = \left(\!\binom{n}{k-n}\!\right)$.

Example 1.19. How many ways can eight identical candy bars be distributed to four kids with each kid getting at least one?

Example 1.20. How many ways could five equivalent prizes be distributed to three contestants so that at least one contestant doesn't get a prize?

Example 1.21. How many solutions in positive integers are there to the equation $x + y + z + w = 15$?

1.1.3 Distinguishable objects into indistinguishable cells unrestricted case $B(k)$

Now we consider the objects in K to be distinguishable from one another, but the cells in N to be identical. We will need the Stirling numbers again. The basic idea is to partition the set K into various blocks and map those blocks to the cells in N. The number of blocks can be anywhere from 1 to n. Because the cells in N can be shuffled, there is only one distinct mapping for each partition of K. So the overall number of mappings from K to N is $S(k,1) + S(k,2) + \cdots + S(k,n) = \sum_{j=1}^{n} S(k,j)$. Of course, if $k < n$, $S(k,n) = 0$. The sum $\sum_{j=1}^{n} S(k,j)$ is the grand total number of partitions of a k-set. This number is denoted by $B(k)$ and is called a **Bell number** after Eric Temple Bell, the mathematician and novelist. Despite the difficulty of calculating the individual Stirling numbers, there is a surprisingly simple algorithm for computing Bell numbers called the Bell triangle. We construct a left-justified triangle inside a square array with typical element $A_{m,n}$ sitting in the mth row and nth column. Initially, all $A_{m,n} = 0$. The algorithm is seeded by setting $A_{1,1} = 1$ and $A_{2,1} = 1$. Now follow the rule for row $m \geq 2$ that for $n = 2$ to m set $A_{m,n} = A_{m-1,n-1} + A_{m,n-1}$. Then set $A_{m+1,1} = A_{m,m}$ and repeat the rule for row $m + 1$. Then $A_{m,m}$ will be $B(m)$. Here is the Bell triangle for $B(4)$: $\begin{pmatrix} 1 & 0 & 0 & 0 \\ 1 & 2 & 0 & 0 \\ 2 & 3 & 5 & 0 \\ 5 & 7 & 10 & 15 \end{pmatrix}$. Successive Bell numbers starting with $B(0)$ are given in column 1 (or the main diagonal starting with $B(1)$), so $B(4) = 15$.

Example 1.22. How many ways can nine different gumdrops be distributed among five identical bowls?

Example 1.23. How many ways can a Toyota, a Ford, a Chevy, and a VW go through any of three equivalent lanes of a toll station? Only the lane and not the sequence matters.

Injective case (one object per cell max) 1 or 0

This is an easy case. If K is bigger than N, there is no way to set up an injective map. But if N is at least as big as K, there is exactly one injective map. The cells can be shuffled so that any assignment of the objects in K to the cells in N is essentially the same.

Example 1.24. One hundred distinguished and distinguishable patrons have purchased tickets to a play to be performed in a venue with 90 perfectly equivalent seats. How many seating configurations are possible without someone sitting in someone else's lap?

Example 1.25. What if there are 110 seats in the preceding example?

Surjective case (all cells used) $S(k,n)$

We revisit the Stirling numbers once again. The task is to partition K into exactly n blocks so that each equivalent cell in N can be matched to one of the blocks. The statistic $S(k,n)$ gives us the number of partitions satisfying this condition, hence the number of surjective functions from K to N is just $S(k,n)$. Note that if $k < n$ there are not enough objects to populate a partition with n blocks, so in those cases $S(k,n) = 0$.

Example 1.26. How many ways can seven different candy bars be given to four kids with each kid getting at least one?

Example 1.27. How many ways can ten different gift cards be put in five envelopes with one envelope having at least three cards and all of them at least something?

1.1.4 Indistinguishable objects into indistinguishable cells unrestricted case $\sum_n p_n(k)$

Since both the domain and range of these functions are indistinguishable sets, we are free to shuffle the domain elements before applying the map and also free to shuffle the range elements after application of the map. In general, these cases result in fewer distinct mappings than those where one or both of the domain and range are distinguishable sets. The model for a function or configuration in this category is some placement of identical objects into identical cells. Since K has k elements we can place k_1 objects in the first cell, k_2 objects in the second cell, and so forth, until we run out of objects at the nth cell, so $k_n \neq 0$ but for $i > n$ we have $k_i = 0$. We can certainly rearrange and renumber the cells so that their contents appear in weakly descending order: $k_1 \geq k_2 \geq k_3 \geq \cdots \geq k_n > 0$. Since $\sum_{i=1}^{n} k_i = k$, what we have done is to partition the integer k into exactly n parts. If $k \geq n$, this is always feasible. A **partition of the integer** $k \geq 1$ into n parts is a set of positive integers $k_1, k_2, \ldots, k_n$ that sum to k. Order does not matter, but for the sake of definiteness many authors do list the parts in either increasing or decreasing order. We choose to list them in (weakly) decreasing order. So $(6,4,3,3,2,1,1) \vdash 20$ is a partition of 20. The symbol "$\vdash$" means "partitions." Duplicate terms are allowed while zeroes are not, in order to preserve the uniqueness of the representation. Here are the partitions of 5: (5), $(4,1)$, $(3,2)$, $(3,1,1)$, $(2,2,1)$, $(2,1,1,1)$, and $(1,1,1,1,1)$. So $p_2(5) = p_3(5) = 2$, and so forth. Our claim is that the number of unrestricted functions from K to N, where the range consists of n available cells in N, is precisely $\sum_{n=1}^{k} p_n(k)$.

It is important to realize that the partition of an integer is a *completely different thing* than the partition of a set. But just like the Stirling number $S(k,n)$ counts how many partitions of a k-set into n blocks exist, the **integer partition number** $p_n(k)$ counts how many partitions of k into exactly n parts exist. Regrettably, just like the Stirling numbers, the partition numbers are not easily obtained, so we will appeal to Table 1.2. *Note carefully* that the table entries $p_m(k)$ are for the number of partitions of k into *at most* m parts, not the number of partitions of k into exactly m parts. The statistic we want for the Twelve-fold Way is the number of partitions of k into *exactly* n parts. This is

Table 1.2: Integer Partitions By Maximum Parts. $p_m(k)$ counts number of partitions of k with at most m parts.

$k \downarrow m \rightarrow$	0	1	2	3	4	5	6	7	8	9	10
0	1	1	1	1	1	1	1	1	1	1	1
1	0	1	1	1	1	1	1	1	1	1	1
2	0	1	2	2	2	2	2	2	2	2	2
3	0	1	2	3	3	3	3	3	3	3	3
4	0	1	2	3	5	5	5	5	5	5	5
5	0	1	3	5	6	7	7	7	7	7	7
6	0	1	4	7	9	10	11	11	11	11	11
7	0	1	4	8	11	13	14	15	15	15	15
8	0	1	5	10	15	18	20	21	22	22	22
9	0	1	5	12	18	23	26	28	29	30	30
10	0	1	6	14	23	30	35	38	40	41	42

found easily as follows: set n equal to m and compute $p_m(k)-p_{m-1}(k)$ from the table. This gives $p_n(k)$, the number of partitions of k into exactly n parts. For example, if we want $p_4(5)$, the number of partitions of 5 into exactly 4 parts, we go into the table at the row $k = 5$ and compute $p_4(5)-p_3(5) = 6-5 = 1$. This lone partition is $(2, 1, 1, 1)$. It is reasonable to ask why the table is set up this way. The general information about various partitions is more readily available from the “at most” table than the “exactly” table, even though they are equivalent. Computing an exact number of partitions involves subtraction of adjacent entries in the “at most” table; computing an “at most” number would entail adding up an entire row of entries in the “exact” table. Some texts switch our $p_m(k)$ and $p_n(k)$, so be sure of the convention.

Example 1.28. Five nameless people buy eight identical loaves of bread at a bakery. How many ways can this happen?

Example 1.29. How many ways can the previous example occur if no one buys more than two loaves?

Injective case (one object per cell max) 1 or 0

The argument in Section 1.1.3 applies verbatim to this case. So if $k > n$, no injections. If $k \leq n$, one distinct injection.

Example 1.30. Five nameless people buy five identical loaves of bread at a bakery. How many ways can this happen injectively?

Example 1.31. Five nameless people buy five out of eight identical loaves of bread at a bakery. How many ways can this happen injectively?

Surjective case (all cells used) $p_n(k)$

This case reprises the unrestricted one with the additional constraint that the partition of k must be into exactly n parts to ensure surjectivity. So the number of surjections is $p_n(k)$.

Example 1.32. Six identical gemstones are to be used to make four rings. Each ring could hold up to five stones, but the designer of the jewelry obviously wants at least one stone on each ring. How many sets of four rings could he make satisfying this criterion?

Sections 1.1.1 through 1.1.4 comprise the Twelve-Fold Way. We summarize the entries here for easy reference.

Table 1.3: Twelve-fold way.

# functions $K \to N$	arbitrary	injective	surjective
K distinct N distinct	n^k	$(n)_k = P(n,k)$ $= n(n-1)\cdots(n-k+1)$	$n!S(k,n)$
K indistinct N distinct	$\left(\!\binom{n}{k}\!\right) = \binom{n+k-1}{k}$	$\binom{n}{k} = \frac{n!}{k!(n-k)!} = C(n,k)$	$\left(\!\binom{n}{k-n}\!\right)$
K distinct N indistinct	$\sum_{n=1}^{k} S(k,n) = B(k)$	1 if $k \le n$ 0 if $k > n$	$S(k,n)$
K indistinct N indistinct	$\sum_{n=1}^{k} p_n(k)$	1 if $k \le n$ 0 if $k > n$	$p_n(k)$

There are some additional useful enumerative formulas that fall outside the scope of the Twelve-Fold Way.

1.2 Circular permutations – distinct objects permuted with circular symmetry $(k-1)!$

Let K be a set of people and N be a set of chairs arranged around a circular table. We will give the people names and mark the chairs with numbers. Each person is sitting in a chair and there are no chairs left over, so $k = n$. How many distinct ways could the people rearrange themselves around the table? "Distinct" here means two seating configurations are regarded as the same if everyone's neighbors to the right and left are the same. What we want is the number of **circular permutations** $P_{\text{circ}}(k)$ of k objects necessarily taken k at a time. We know from Section 1.1.1 that the number of linear permutations is $P(k,k) = k!$, but that does not quite apply to the circular situation. If we designate one person as the anchor of a configuration, we can permute everyone else relative to them. There would be $P(k-1, k-1) = (k-1)!$ such permutations. None of these would generate a new relative permutation if the anchor decided to sit in a different chair. So taking circular symmetry into account, $P_{\text{circ}}(k) = (k-1)!$

Example 1.33. King Arthur had 12 knights at the Round Table. If Sir Galahad always sat to Arthur's right, how many different ways could the king and 12 knights seat themselves?

Example 1.34. What adjustment to the circular permutation formula would be necessary if $n > k$.

1.3 Partitioned permutations $\frac{(n+k-1)!}{(n-1)!}$

Imagine we have the four cars from the last example in Section 1.1.3: a Toyota (T), a Ford (F), a Chevy (C), and a VW (V). This time we are interested in how many ways they can go through the three distinct lanes 1,2, and 3 of a toll station where the order in which they go through each lane also matters. Intuitively, we expect many more possible patterns than we found in Section 1.1.3. There are three lane choices for the Toyota. Suppose it goes through lane 1. We can sketch the result writing the car after the lane it uses: 1-T-2-3. What about the Ford? There are four possible diagrams: 1-F-T-2-3, 1-T-F-2-3, 1-T-2-F-3, and 1-T-2-3-F. The first two of these diagrams indicate that the Toyota and Ford went through lane 1 in a different sequence. You can see that there are now five choices for the Chevy. A typical outcome might be 1-T-2-C-3-F. Clearly, there would then be six choices for locating the VW. As the our diagrams get longer they give more choices for the next step. We started with a choice of three lanes for the first car, so we had three possible diagrams. Then each of those spawned four possibilities for the next car, then five, and finally six. How many of our little diagrams do we have in all? Evidently, $3 \cdot 4 \cdot 5 \cdot 6 = 360$. Note that each one of the diagrams is unique. We could have considered the cars in any order, but the set of 360 resulting diagrams would have been the same.

We would like to have a general formula for enumerating situations of this type. If we have k distinct objects and we map the first object to n cells, then the second object to $n+1$ cells, and so forth, until the last object is mapped to $n+k-1$ cells, where the cells are distinct at every stage, then there are $n \cdot (n+1) \cdots (n+k-1) = \frac{(n+k-1)!}{(n-1)!}$ such overall mappings. The number of cells increases by one each time an object is mapped to a cell because the object effectively partitions that cell into two cells.

Example 1.35. There are five flag poles in front of a UN mission. The custom is to fly the same 12 national flags each day. A single pole could hold all 12 flags if necessary, even though that would be considered undiplomatic. How many different ways could the 12 flags be displayed on the flagpoles?

Example 1.36. Captain Morgan is stranded on an island. He managed to save five of the different signal flags from his ship before it sank. He finds three palm trees in a row and decides to send a message using all the flags tied to the trees. How many different messages can he send? Position of a flag on a tree counts. Message is read left to right.

1.4 Inclusion/exclusion principle

Suppose we have a set of objects X where $B \subset X$ is the subset of blue objects and $H \subset X$ is the subset of heavy objects. We might want to know how many objects in X are neither blue nor heavy. A naïve estimate is $|X| - |B| - |H|$. This would be incorrect if there were blue objects that were also heavy. A heavy blue object would be deducted twice from $|X|$, once for being blue and once for being heavy. So we need to add a correction factor. This would be the number of heavy blue objects counted once. So our revised and correct formula would be $N_=(\emptyset) = |X| - |B| - |H| + |B \cap H|$, where $N_=(\emptyset)$ stands for the number of objects that possess none of the properties in the set {*blue, heavy*}. What happens if $S \subset X$ is the subset of shiny objects? Now the calculation of $N_=(\emptyset)$ becomes more involved because we can have elements that are shiny and blue but not heavy, shiny and heavy but not blue, and so forth, for every combination of the three properties. Eventually, we discover that the correct formula is $N_=(\emptyset) = |X| - (|B| + |H| + |S|) + (|B \cap H| + |B \cap S| + |H \cap S|) - |B \cap H \cap S|$. It is unwieldy enough for three properties, and it only gets worse as the number of properties grows, so we need a more transparent approach. It is worth the effort to have a general result.

There are two tasks ahead of us to achieve this goal. The first is to settle on a notation that is compact to work with and sufficiently expressive to allow us to easily follow a proof of the final formula. The second is to justify our formula. We start by letting X be the set of objects under consideration and P be the set of n properties that they collectively possess. Note that an individual object may possess any number from none to all of the properties in P. Suppose M is a particular subset of P. We will write the number of objects that have just the properties in M as $N_=(M)$ and the number of objects that have at least all the properties in M as $N_\geq(M)$. This is a useful notation since it allows us to avoid hordes of subscripted indices. Also $N_=(\emptyset)$ is the number of objects with no properties at all. We can now state the principle in terms of this notation.

Inclusion/exclusion formula $N_=(\emptyset) = \sum_{M \subset P} (-1)^{|M|} N_\geq(M)$

Let us parse this rather odd-looking formula before showing why it is true. Recall that $N_=(\emptyset)$ is the number of objects in X that have none of the properties in P. The sum is indexed by all subsets $M \subset P$. This means that for every subset M of properties, we have to find how many elements there are in M and compute the "parity" factor $(-1)^{|M|}$. You can get a glimpse of this working in the introductory case with three properties above.

Then we multiply the parity factor by the number of objects which have *at least* all the properties in M. Finally, we add all these terms together.

Here is why the formula works. Let us unpack and organize the sum by the increasing size of the subsets M. So we start with $|M| = 0$. We recognize this M as the null set. The contribution to the sum is $(-1)^0 N_{\geq}(\varnothing) = 1 \cdot |X|$ because every object in X has at least zero properties. Now let $|M| = 1$. We find all the subsets of P that have just one property. There should be $|P|$ of them. Each one of these singleton subsets $M \subset P$ contributes $(-1)^1 N_{\geq}(M)$ to the sum. Here $N_{\geq}(M)$ means the number of objects that have at least the property in the singleton subset M. The example above shows this happening clearly for the three singleton subsets B, H, and S. At the next level, we are back to $(-1)^2 N_{\geq}(M)$ where now $N_{\geq}(M)$ means the number of objects having at least two of the properties in the subset $M \subset P$ where $|M| = 2$. Every pair of properties in P gives rise to one of these subsets and each subset contributes $(1)N_{\geq}(M)$ to the overall sum. We continue in this way increasing $|M|$ until we reach $|M| = n = |P|$. Along the way we expect to shed objects that have fewer than $|P|$ properties, but we can be assured that stopping at $|M| = |P|$ will account for all the objects in X. The last term we will add to the sum is $(-1)^{|P|} N_{\geq}(P) = (-1)^n N_{=}(P)$. There is only one subset at this level and it is all of P.

Let us summarize what we have at this point and rewrite the inclusion/exclusion formula in terms of the progressively larger sets of properties with the appropriate parity factors ±1. We have

$$\begin{aligned} N_{=}(\varnothing) = |X| - \sum_{|M|=1} N_{\geq}(M) + \sum_{|M|=2} N_{\geq}(M) - \sum_{|M|=3} N_{\geq}(M) + \cdots + (-1)^k \sum_{|M|=k} N_{\geq}(M) + \cdots \\ + (-1)^n \sum_{|M|=n} N_{=}(M). \end{aligned}$$

Now pick an object in X that has k of the possible n properties and calculate how many times it is counted in the expression on the right-hand side. If $k = 0$, it has no properties and is counted once in X and never again in any subsequent sums. If $k = 1$, it is counted once in X and then immediately deducted once in the sum $\sum_{|M|=1} N_{\geq}(M)$ when M has exactly the property it displays. Then it is never counted subsequently. Now here is where the logic of the proof starts to show. If $k = 2$, it is counted once in X, deducted twice in $\sum_{|M|=1} N_{\geq}(M)$, and finally added back once in $\sum_{|M|=2} N_{\geq}(M)$ when M has exactly the pair of properties it displays. Then it is never counted subsequently. So the net result is that the expression on the right-hand side does not count it at all. What happens to an object with k properties? Marching through the terms on the right, it is counted $\binom{k}{0} - \binom{k}{1} + \binom{k}{2} - \binom{k}{3} + \cdots + (-1)^k \binom{k}{k}$ times and never subsequently. Surprisingly, perhaps, this alternating sum of binomial coefficients is zero. We will show this in a moment, but the proof of the inclusion/exclusion formula is now clear. If an object has no properties it is counted once by the right-hand side, and if it has any positive number of properties it is not counted at all. So the net result is $N_{=}(\varnothing)$.

Now why is $\sum_{i=0}^{k}(-1)^i\binom{k}{i} = 0$? From Example 1.13 in Section 1.1.2, the binomial theorem states that $(1+x)^k = \sum_{i=0}^{k}\binom{k}{i}x^i$, for a positive integer k. Substitute $-x$ for x and get $(1-x)^k = \sum_{i=0}^{k}\binom{k}{i}(-x)^i$, which equates to 0 after setting $x = 1$.

Example 1.37 (The drunken hat check girl problem). This is a famous old problem having to do with derangements. A **derangement** is a permutation or shuffle of n objects where no object is returned to its place in the original order. So a derangement of $1, 2, 3, 4$ is $4, 1, 2, 3$. French nobles used to play a card game where they had two identical decks of shuffled cards and then each fellow simultaneously turned over a card from the top of his deck, which they compared. If there was no match in rank, they turned over another pair of cards. If at any point the cards matched in rank, the noble who bet on a match occurring won. If they went all the way through their decks without a match, the other noble won. Clearly, if one deck was a derangement of the other, no matches would ever come up. Empirically, betting on one outcome seemed to be a winner. We can let the hat check girl assist us in determining the smart play.

Here is the problem: Ten gangsters go to a meeting at their private club. Being gangsters, each one wore a fedora which they left with the hat check girl at the entrance. The meeting went on interminably and the girl was bored so she kept asking the bartender for martinis. Eventually, the meeting concluded and the gangsters filed out of the club past the hat check girl who returned their hats. She had forgotten to mark the hats and she could not remember which hat went with which gangster. Not wanting to upset any of them, they were gangsters after all, she confidently returned the hats at random. What is the chance that nobody got the correct hat back?

Example 1.38. Where do the Stirling numbers of the second kind come from? This is a classic application of the inclusion/exclusion principle.

Let M_i be the property that a function excludes at least i elements of N. How many such functions are there? A typical function will map K to a subset of N with no more than $(n-i)$ elements, since it avoids at least i elements. These functions are otherwise unrestricted and we know from Section 1.1.1 that there are $(n-i)^k$ of them. Now how many distinct "avoidance subsets" of i elements are there? Since $|N| = n$, from Section 1.1.4 we know there must be $\binom{n}{i}$ such subsets. Evidently, the $N_{\geq}(M_i)$ needed for our inclusion/exclusion formula is $\binom{n}{i}(n-i)^k$ by the product principle. We can now set up that formula:

$$N_{=}(\varnothing) = n!S(k,n) = \sum_{i=0}^{n}(-1)^i\binom{n}{i}(n-i)^k.$$

It follows that

$$S(k,n) = \frac{1}{n!}\sum_{i=0}^{n-1}(-1)^i\binom{n}{i}(n-i)^k.$$

Note that $i = n$ contributes nothing, so the sum may be stopped at $(n-1)$. All practical cases have both positive n and k. Our formula generates the entries in Table 1.1. For example,

$$S(4,3) = \frac{1}{3!}\left[\left(\binom{3}{0}(3-0)^4\right) - \left(\binom{3}{1}(3-1)^4\right) + \left(\binom{3}{2}(3-2)^4\right)\right] = \frac{1}{6}(81 - 3\cdot 16 + 3\cdot 1) = 6,$$

which agrees with the table.

1.5 Permutations of indistinguishable objects within distinguishable types $\binom{n}{k_1,k_2,\ldots,k_m}$

Occasionally we encounter the situation where we need to know how many ways distinguishable types of objects can be permuted but within the type the objects are indistinguishable. A typical case might ask how many ways can you arrange ten books in a row on a shelf where three are history books, five literature, and two science. We are interested in the arrangement of distinct subjects but not the particular books within the subjects. The solution is to artificially distinguish all the books, calculate the number of possible permutations, then deflate that number by dividing out the possible permutations just within each subject. We would argue that there are $\frac{10!}{2!5!3!}$ distinct arrangements by subject. There are 10! overall permutations of individual books. But within each such permutation, the three history books are interchangeable so that reduces the number of distinct configurations to $\frac{1}{3!}\cdot 10!$. But so are the five history books, so that cuts it down to $\frac{1}{5!}\cdot(\frac{1}{3!}\cdot 10!)$. And finally, the two interchangeable science books reduce it further to $\frac{1}{2!}\cdot(\frac{1}{5!}\cdot\frac{1}{3!}\cdot 10!) = \frac{10!}{2!5!3!} = 2520$.

For the general formula, we have k_i indistinguishable objects of distinguishable type i, and $\sum_i^m k_i = n$ objects total. Then the number of linear arrangements where only the object type matters is $\frac{n!}{k_1!k_2!\cdots k_m!}$. This expression is symbolized as $\binom{n}{k_1,k_2,\ldots,k_m}$ and is called a **multinomial coefficient**. The name is explained by the first example below.

Example 1.39. Find the coefficient of x^2y^3 in the expression $(1+x+y)^{12}$.

Example 1.40. A curator has room for eight paintings on a wall in a line. She has two Rembrandts (this is a high class exhibition), three Vermeers, and three Rubens'. How many different ways can she display the paintings if only the identity of the artists matters?

Example 1.41. In the previous example, what if the two Rembrandts had to appear side-by-side with no restrictions on the others?

1.6 Counting with symmetries

Problems that involve an enumeration of distinct configurations or patterns may require us to compensate for possible underlying symmetries in what we are counting. For example, suppose we have a little cube of wood and also four different colors with which to paint each face solid. How many distinct ways can this be done? Two patterns are to be considered the same if there is some motion of the cube that would allow one pattern to coincide perfectly with the other. One pattern might be a single face painted white and all the others blue. There are clearly six ways to do this with a stationary cube; however, we would certainly not have six unique patterns as the cube could be picked up and rotated so that each of the other five patterns coincides exactly with the original paint scheme. Finding the number of *distinct* patterns when the object supporting the patterns has axes of symmetry is the goal of this section. A cube seems innocent enough in this regard at first glance, but the number of possible motions which map the cube back to itself may surprise you. A good way to see this is to imagine the cube in a close-fitting cubical box with no lid. If we take the cube out of the box, we may return it by putting any of the six faces on the floor of the box. As we do this, we may rotate the cube through any of the four multiples of 90° about the axis joining the centers of the top and bottom faces. This gives us $6 \cdot 4 = 24$ possible **symmetry operations**. There is no other way to put the cube back into the box, so we are confident that this is the totality of symmetry motions.

Figures displaying geometric regularity like the cube or even the icosahedron with 20 faces have various axes that define possible symmetry operations on them. Turning the figure through just the right angle about one of these axes returns it so that it looks exactly the same as before, even though some constituent parts of the figure have been interchanged. It is much easier to visualize and discuss symmetry operations that happen in a plane, so let us start by focusing on regular polygons. The family of symmetry operations for a given figure form a mathematical system called the **symmetry group** of the figure. The theory of groups is an enormous and important topic in mathematics. We plan to nibble at the edges of this theory just enough to be able to solve some counting with symmetry problems.

1.6.1 Groups

Formally, a **group** is a set of elements G together with a binary operation $*$ on those elements that satisfies the following axioms:

Table 1.4: Axioms for a group $\langle G, *\rangle$.

1) $(a * b) * c = a * (b * c)$ for all $a, b, c \in G$	$*$ is associative
2) there exists $e \in G$ such that $e * a = a * e = a$ for all $a \in G$	an identity exists
3) for each $a \in G$ there exists a^{-1} such that $a * a^{-1} = a^{-1} * a = e$	all inverses exist

In addition to the above, although not guaranteed, we may also have 4) $a * b = b * a$ for all $a, b \in G$. In that case we say the group is **commutative** or **abelian**. There is never an issue with the binary operation being closed since it is built in to the definition. We will express symmetry operations as permutations of the defining features, usually vertices, of various geometrical objects. The permutations we consider will be physically realizable in the sense that they will correspond to possible symmetry motions. An arbitrary permutation might require us to tear and reassemble an object and therefore not be a simple rotation. Our group operation $*$ will be composition of permutations executed from right to left to conform to regular functional composition. First, a basic example:

Example 1.42. Show that the rotations of an equilateral triangle in its plane through 0°, 120°, and 240° constitute a group.

There is nothing wrong with expressing the symmetry operations in Example 1.42 as rotations through various angles, but there is a much better method available to characterize general symmetry operations by their effect on the defining parts of a figure. So label the vertices of the triangle as 1, 2, and 3 in the clockwise order. How might we symbolize a 120° rotation using the labels on the vertices? We could write $1 \to 2 \to 3 \to 1$ to indicate vertex 1 replaces 2, 2 replaces 3, and 3 replaces 1. This is a clockwise rotation of 120°. How about 240°? That would be $1 \to 3 \to 2 \to 1$. An improvement on our ad hoc notation would be to remove the arrows and express the 120° rotation as (123) and the 240° rotation as (132) with the understanding that the last vertex in the parentheses replaces the first via a "wrap around." This allows us to communicate the symmetry operation by writing each vertex label only once. What about the 0° rotation? We symbolize that as (1)(2)(3). Each vertex is mapped right back to itself. To economize further, we may write this identity operation as (1), so it implies all of (1)(2)(3).

Parenthetical expressions which imply permutations like (abc) or $(xyzw)$ are called **cycles**. They are a shorthand convention for the action of a permutation on the letters that appear within. The **length** of a cycle is the number of letters/elements that appear in the parentheses, so (123) has length 3. A cycle which interchanges two elements is called a **transposition**. Since cycles represent permutations, we can chain them together into composition products, remembering to apply them right to left. Ordinarily, composition of functions is **order dependent** and cycles which have elements in common share this property. Cycles that have no elements in common are called **disjoint cycles**, and products made up of them are conveniently **order independent** or commute. We can always convert a product of nondisjoint cycles to an equivalent product of disjoint cycles with an obvious algorithm.

Example 1.43. Express the product (123)(345)(146) as a product of disjoint cycles.

Now that we have cycle notation, we can easily characterize the symmetry group of the triangle we began to look at in Example 1.42.

Example 1.44. Show that the three in-plane rotations in Example 1.42 together with the "flips" about the three altitudes of the triangle form a symmetry group for the triangle.

Here is a table of the possible 36 composition products among the symmetry operations. Since every possible permutation of $\{1, 2, 3\}$ is represented and the group axioms in Table 1.4 are satisfied, Table 1.5 is actually the group table, or **Cayley table**, for the **full symmetric group** on three symbols, known as S_3. The entries represent the result of postcomposing a permutation α in the leftmost column with a permutation β in the top row. Note that, in general, we cannot rely on commutativity, and in this case it is clear that $\alpha \circ \beta = \beta \circ \alpha$ is not true across the board.

Table 1.5: Group table for S_3/D_6.

		β					
	$\alpha \circ \beta$	(1)	(123)	(132)	(1)(23)	(2)(13)	(3)(12)
	(1)	(1)	(123)	(132)	(1)(23)	(2)(13)	(3)(12)
	(123)	(123)	(132)	(1)	(3)(12)	(1)(23)	(2)(13)
α	(132)	(132)	(1)	(123)	(2)(13)	(3)(12)	(1)(23)
	(1)(23)	(1)(23)	(2)(13)	(3)(12)	(1)	(123)	(132)
	(2)(13)	(2)(13)	(3)(12)	(1)(23)	(132)	(1)	(123)
	(3)(12)	(3)(12)	(1)(23)	(2)(13)	(123)	(132)	(1)

A property of group tables is that each entry can appear only once in each row and column in the body of the table. This is the **Latin square** property and it serves as a quick check on the validity of the table. Note that the three-element group consisting of rotational-only symmetries in Example 1.42 appears as a subgroup of the full symmetry group in the nine cells in the upper left. Table 1.3 could have been constructed simply by considering all the permutations of $\{1, 2, 3\}$ without any reference to the triangle at all. It is a coincidence that this "flying by instruments" approach mirrors the result that follows from analyzing all the physical symmetries of the triangle. Groups associated with the symmetries of regular polygons are called **dihedral groups**, denoted by D_{2n}, where n is the number of sides of the polygon. Half of the elements in a dihedral group are rotations (through n multiples of the form $\frac{360k^\circ}{n}$, $0 \le k \le n-1$), and the other half are flips about n available axes. Groups formed from all of the permutations of n symbols are called full symmetric groups, denoted by S_n, and have $n!$ elements. The only time $2n = n!$ for $n \ge 3$ is when $n = 3$. So $D_6 \cong S_3$ and then never again. The symbol "$\cong$" means the two groups are essentially the same, or **isomorphic**, even though they were formed by different constructions. Of course, the dihedral groups will be essentially the same as certain **subgroups** of symmetric groups. A subgroup is a set of elements in a group that satisfy the group axioms on their own. In fact, Arthur Cayley showed that any group whatsoever can be represented as a subgroup of some group of permutations.

1.6.2 Pólya's method

The basic version of Pólya theory can be applied to counting problems where some sort of physical symmetry reduces the number of nominal cases to a population of distinct cases. We described a typical case above involving painting the faces of a cube. It is a complicated theory, so we are going to accept its guidance on faith. The Hungarian mathematician George Pólya developed this method in the late 1930s as a powerful extension of some group-theoretic results from the late nineteenth century.

Suppose we are tasked with enumerating the number of distinct ways an object can be colored and the object has some self-symmetries that we can code as motions in three dimensions. The notion of "colorings" is traditional and could represent any system for distinguishing parts of a symmetric object. So we are given a palette consisting of n colors, any of which can be applied to any of the distinguishing parts. The first order of business is to determine the nature of the group of possible symmetries. Specifically, we need the group table with all of the symmetry operations expressed in disjoint cycle notation. We are going to be very interested in the distribution of cycle lengths in each product. Suppose a particular symmetry operation is coded as a product of α_1 cycles of length 1, α_2 cycles of length 2, and so forth, up to α_k cycles of length k, which is the longest cycle length appearing in the product. Now – here comes the black box – we plan to create a multinomial term that is going to be ultimately summed with similar terms from all the other cycle products. We are going to use k variables each written z_i, for $1 \le i \le k$. The term itself is $z_1^{\alpha_1} z_2^{\alpha_2} \cdots z_k^{\alpha_k}$. Let us stop and look at the multinomial term that would correspond to (1)(234)(56)(78). It would be $z_1 z_2^2 z_3$.

Now, let G be the entire symmetry group and $\pi \in G$ be a typical symmetry operation. Define $f(\pi) = z_1^{\alpha_1} z_2^{\alpha_2} \cdots z_k^{\alpha_k} \cdots z_t^{\alpha_t}$ where t is the length of the longest cycle in any of the products in G. Note that if the longest cycle in π is k, then the exponents $\alpha_{k+1}, \alpha_{k+2}, \ldots, \alpha_t$ are all zero. Finally, we define the **cycle index** $Z = \frac{1}{|G|} \sum_{\pi \in G} f(\pi)$. The notation $|G|$ means the number of elements in G. This will be a little clearer when we have a concrete example. After that we will see how knowing the cycle index solves our counting problems. It should be clear that the cycle index depends only on the symmetry group under investigation.

Example 1.45. Find the cycle index for the dihedral group of the square.

Our final step in applying Pólya's method is to construct a **pattern inventory** from the cycle index. This is where the details of the palette are accounted for. Suppose we have three colors, black, red, and gold. Symbolize them as b, r, and g. We let $z_1 = b + r + g$, $z_2 = b^2 + r^2 + g^2$, $z_3 = b^3 + r^3 + g^3$, and $z_4 = b^4 + r^4 + g^4$. Now expand $Z(z_1, z_2, z_3, z_4)$ with these substitutions. This can involve a lot of algebra, so a CAS (computer algebra system) is invaluable. The result will be the pattern inventory $P(b, r, g)$, which is a multinomial in b, r, and g. It will have terms like $Cb^u r^v g^w$ where $u + v + w = 4$ and C gives the number of distinct colorings of the vertices of the square such that u are black, v are red, and w are gold. The total number of distinct patterns or configurations is summarized in the form of the multinomial pattern inventory. Polynomials or multinomials, where the coefficients are the important thing and enumerate various configurations that are coded as exponents applied to the variables, are called **generating functions**. So let us explore a practical example.

Example 1.46. A jeweler wants to make a pendant fashioned from a small square of silver with a precious stone at each corner. He has diamonds, rubies, and emeralds available. The obverse and reverse of the metal are indistinguishable. How many distinct pendants can he make?

Example 1.47. Returning to the original problem posed in Section 1.6: How many ways can we paint the faces of a cube with four colors available?

Example 1.48. Continuing Example 1.47, how many of the 240 overall patterns have no blue faces?

Example 1.49. Continuing Example 1.47, how many patterns have two blue faces, two white faces, and one face each green and red?

Example 1.50. Continuing Example 1.47, how many patterns contain exactly k blue faces for $0 \leq k \leq 6$?

Example 1.51. Continuing Example 1.47, how many patterns contain at least one red and one white face?

1.7 Problem solving tips

Once we recognize a problem is fundamentally a counting problem, the best of all possible outcomes is that its data fit a ready-made formula from the Twelve-Fold Way or one of the supplementary formulas. This may not occur as often as we would like. Here are some insights that may help.

1) When you are setting up a Twelve-Fold Way argument, be clear on what the objects are and what the cells are that define the functions you are counting. This includes being sure of the distinguishability of the objects and cells. These facts can be obscured in the statement of the problem.

2) If you do not see a way to do an enumeration directly, see if what you need to count can be associated bijectively with something you already know how to count. If you are trying to count configurations with a certain property, it might be easier to count total configurations and deduct those without the property.

Example 1.52. Find the sum of the squares of the binomial coefficients from $\binom{n}{0}$ to $\binom{n}{n}$.

3) Several enumeration formulas may be needed for a single problem.

Example 1.53. How many ways can a dozen different donuts be distributed to seven meeting participants so that at least one person receives no donut?

4) With permutation problems, if two or more objects need to be contiguous, consider them as a single object in the computation.

Example 1.54. A baseball manager has two players that he wants to keep together in the batting lineup of nine players, wherever they appear. How many lineups are possible?

5) Sometimes it is useful to impose an arbitrary distinction on otherwise indistinguishable objects or cells, use the appropriate formula to enumerate distributions, then relax the distinction by dividing out possible permutations.

Example 1.55. See Section 1.5 (Example 1.39).

References & suggested further reading

1) *Enumerative Combinatorics, Vol. 1* by Richard P. Stanley (1986), Wadsworth & Brooks – *sophisticated presentation of the Twelve-Fold Way and much beyond*
2) *Introduction to Combinatorial Mathematics* by C. L. Liu (1968), McGraw-Hill – *comprehensive treatment of enumerative methods*
3) *Combinatorics, A Guided Tour* by David R. Mazur (2010), Mathematical Assoc. of America – *modern, basic, and very readable discussion of enumeration*
4) *Combinatorial Theory* by Marshall Hall, Jr. (1986), Wiley-Interscience – *higher level presentation by a famous theorist*

Solutions to Chapter 1 examples

Example 1.1

Let K be the set with k elements (objects). Create two cells, so $|N| = 2$. The first is the "YES" cell, the second the "NO" cell. If all the objects of K are assigned to these two cells, we may view the contents of the YES cell as the elements selected as a subset of K. Every assignment corresponds to a subset, so there are 2^k assignments, hence 2^k subsets. Note that the null set is that where all the elements of K are sent to the NO cell.

Example 1.2

The objects are the questions and the cells are the answer selections. So $k = 40$ and $n = 5$. We are assigning each of the 40 questions to any one of the five possible responses, so we have $5^{40} \approx 9.1 \times 10^{27}$ possible different answer sheets. We have counted the number of 40-lists drawn from a 5-set.

Example 1.3

The K doughnuts are distinct objects and the N people are distinct cells. A choice of $n = 3$ assignments is made $k = 5$ times without restriction, so the number of doughnut distributions is $3^5 = 243$. This includes the possibility that one person gets all the doughnuts. Sometimes life is not fair.

Example 1.4

The prizes are K and the horses are N, so $P(8,3) = \frac{8!}{(8-3)!} = 8 \cdot 7 \cdot 6 = 336$. Another way to look at problems like this is to see that any of the eight horses could be the first, leaving seven horses available the second, and finally six for the third place. By the product principle, the total number of ways this could occur would be $8 \cdot 7 \cdot 6 = 336$.

Example 1.5

The men are K and the women are N. An injective function from K to N uniquely creates three pairs. There are $P(6,3) = \frac{6!}{(6-3)!} = 6 \cdot 5 \cdot 4 = 120$ ways to form three pairs.

Example 1.6

The positions in the selection scheme are #1 through #6. This is K. These positions are mapped to N, the set of 50 numbered balls injectively. Each such mapping is an outcome for the lottery. There are $P(50,6) = \frac{50!}{(50-6)!} = \frac{50!}{44!} = 15{,}890{,}700$ lottery outcomes, so picking a winner is unlikely.

Example 1.7

There are 36 symbols available to make words of length 5, so the total number of five-character passwords is $36^5 = 60{,}466{,}176$. The number of five-character passwords where no symbol is repeated is $P(36,5) = 45{,}239{,}040$. So the number of passwords with repeat characters is 15,227,136 and this accounts for about 25.2 % of all passwords.

Example 1.8

The set of objects K being carried is partitioned into three blocks and the blocks assigned to the people N. So we need $S(5,3)$ and we will multiply that by $3!$. From the table, $S(5,3) = 25$, so the number of ways is $3! \cdot 25 = 150$.

Example 1.9

The set of repairmen K is partitioned into five blocks to handle the clients N. From the table, $S(8,5) = 1050$. So the number of ways is $5! \cdot 1050 = 126{,}000$.

Example 1.10

We are choosing four from 15 total, so the binomial coefficient almost writes itself: $\binom{15}{4} = \frac{15!}{4!(15-4)!} = \frac{15!}{4! \cdot 11!} = \frac{15 \cdot 14 \cdot 13 \cdot 12}{2 \cdot 3 \cdot 4} = 1{,}365$.

Example 1.11

All the good selections would have to come from picking five out of the seven good diodes. There are $\binom{7}{5} = 21$ such selections. On the other hand, there are $\binom{10}{5} = 252$ possible selections of five overall. When we discuss classical probability, we will form the ratio $\frac{21}{252} = 0.083$ and call this the probability of selecting five good diodes out of ten at random where three are known defective.

Example 1.12

Short of actually multiplying everything out, we can reason as follows. Writing $(x+1)^9 = (x+1)(x+1)\cdots(x+1)$ with nine identical factors, ask how x^4 would be produced. Evidently, we would need to select four factors and multiply the x's in those factors and then multiply the 1's in the unselected factors. Clearly, each selection like this contributes one x^4 to the overall expansion. So how many ways can we select four factors from the nine available? The answer is $\binom{9}{4} = 126$. So the desired coefficient is 126.

Example 1.13

With the argument in the preceding example, we have

$$(1+x)^n = \sum_{k=0}^{n} \binom{n}{k} x^k = \binom{n}{0} x^0 + \binom{n}{1} x^1 + \cdots + \binom{n}{k} x^k + \cdots + \binom{n}{n} x^n.$$

This is called the **Binomial theorem**.

Example 1.14

We are distributing ten indistinguishable cookies into four cells, so the relevant statistic is $\left(\!\binom{4}{10}\!\right) = \binom{13}{10} = 286$.

Example 1.15

There is certainly no unique solution to this equation. For example, $x = 3, y = 3, z = 4$ and $x = 5, y = 2, z = 3$ are two solutions. We reason as follows: there are ten identical 1's to be distributed among the cells x, y, and z. This corresponds to $\left(\!\binom{3}{10}\!\right) = \binom{12}{10} = 66$.

Example 1.16

Give each variable a 1. This ensures they are all positive. Now we have seven 1's to distribute among three distinct cells. There are $\left(\!\binom{3}{7}\!\right) = \binom{9}{7} = 36$ possible solutions. Each solution corresponds to an ordered sum of 10. An ordered sum of n consisting of k positive integers is called a **k-composition of n.**

Example 1.17

There are 20 equivalent dollar amounts to be distributed among four investment opportunities. So the number of possible distributions is $\left(\!\binom{4}{20}\!\right) = \binom{23}{20} = 1771$.

Example 1.18

Set aside \$25,000 in the gold account. Now calculate $\left(\!\binom{4}{15}\!\right) = \binom{18}{15} = 816$.

Example 1.19

Here $n = 4$ and $k = 8$, so $\left(\!\binom{4}{8-4}\!\right) = \left(\!\binom{4}{4}\!\right) = \binom{7}{4} = 35$.

Example 1.20

Here $k = 5$ and $n = 3$. The total number of possible distributions is $3^5 = 243$ (see Section 1.1.1). We want to deduct those where everyone gets at least one prize. That number would be $\left(\!\binom{3}{5-3}\!\right) = \left(\!\binom{3}{2}\!\right) = \binom{4}{2} = 35$, so $243 - 35 = 208$ prize distributions have either one or two people getting nothing.

Example 1.21

We distribute 15 1's to the four variables surjectively to make certain that x, y, z, and w are not zero. So the number of surjections is $\left(\!\binom{4}{11}\!\right) = \binom{14}{11} = 364$.

Example 1.22

The set of gumdrops is K and the set of bowls is N. From the table (Appendix 1): $\sum_{j=1}^{5} S(9,j) = 1 + 255 + 3025 + 7770 + 6951 = 18{,}002$.

Example 1.23

The cars are distinguishable, and the lanes are not. So $k = 4$ and $n = 3$. We have again from the table $\sum_{j=1}^{3} S(4,j) = 1 + 7 + 6 = 14$.

Example 1.24

Here $k > n$, so none.

Example 1.25

This time $k \leq n$, so one.

Example 1.26

Here $k = 7$ and $n = 4$, so from the Stirling table we have $S(7, 4) = 350$.

Example 1.27

Give one envelope three cards at the outset. Then there are two cases: (1) the envelope with three cards gets no more when the remaining seven are distributed, or (2) the envelope with three cards is eligible to receive more cards. These situations are mutually exclusive, so by the additive principle we will find the number of ways we can do either and then add them together. For the first case, we have seven cards to be distributed surjectively to four envelopes. There are $S(7, 4)$ ways to do this. For the second case, we have seven cards to be distributed over all five envelopes. Since this is being done surjectively, the three-card envelope will get at least one more. This count is $S(7, 5)$. So the answer to the original question is $S(7, 4) + S(7, 5) = 350 + 140 = 590$.

Example 1.28

Here K is the set of loaves, N is the set of people, so $p_1(8) + p_2(8) + p_3(8) + p_4(8) + p_5(8) = 1 + 4 + 5 + 5 + 3 = 18$ ways.

Example 1.29

The only partitions of 8 into five or fewer parts that satisfy this constraint are $(2, 2, 2, 2)$ and $(2, 2, 2, 1, 1)$, so two ways.

Example 1.30

Here K is the set of loaves, N is the set of people, so $k = n$. There is one distribution that is valid. Each person gets one loaf.

Example 1.31

Here $k \leq n$, so one way. Each person gets one loaf and three are left over (not included in the range of the function).

Example 1.32

If the designer puts at least one stone on each ring (surjectively) then no ring will have more than three stones. There are $p_4(6) = 2$ distributions with this property. We can visualize the two patterns as partitions of 6, namely $(3, 1, 1, 1)$ and $(2, 2, 1, 1)$.

Example 1.33

Treat Arthur and Galahad as a unit. This leaves 12 units free to sit anywhere at the table. So $P_{\text{circ}}(12) = 11! = 39{,}916{,}800$.

Example 1.34

None. As long as the left/right neighbors stay the same, extra chairs could be placed between occupied seats without altering the relative configuration.

Example 1.35

Here $n = 5$ and $k = 12$, so $\frac{(n+k-1)!}{(n-1)!} = \frac{16!}{4!} = 871{,}782{,}912{,}000$.

Example 1.36

Here $n = 3$ and $0 < k \leq 5$. So $\frac{(3+1-1)!}{2!} + \frac{(3+2-1)!}{2!} + \frac{(3+3-1)!}{2!} + \frac{(3+4-1)!}{2!} + \frac{(3+5-1)!}{2!} = 2955$.

Example 1.37

There are 10! possible permutations of ten objects. Some of these are derangements and others have one or more fixed points. A **fixed point** in this context is an object that is actually not moved by a permutation but retains its position in the original order. This would be a gangster getting his correct hat back. The fraction of all these permutations that are derangements would be the probability that no gangster got the right hat back. So how many derangements are there? This is an inclusion/exclusion problem. The universe X consists of the 10! permutations. The properties we are interested in are the fixed points of the permutations. Here $N_{=}(\varnothing)$ represents the number of permutations with no fixed points, i. e., derangements. Hence the probability we are looking for is $P(\text{no hats}) = \frac{N_{=}(\varnothing)}{10!}$. How many permutations have at least one fixed point? There are ten candidates to be a definite fixed point, and once a point is fixed there are 9! possible permutations of the other objects. By the multiplicative principle this gives $10 \cdot 9!$ permutations with at least one property (fixed point). Yes, that does happen to be 10!. Wait for the magic. How about two fixed points? There are $\binom{10}{2}$ candidates for distinct pairs of fixed points. There are 8! permutations of the remaining points for each fixed pair, which yields $\binom{10}{2} \cdot 8!$ permutations with at least two properties. Continuing in this vein, we get $\binom{10}{k} \cdot (10-k)!$ permutations with at least k fixed points. Now we can assemble the pieces into the inclusion/exclusion formula. Evidently, $N_{=}(\varnothing) = 10! - 10! + \binom{10}{2}8! - \binom{10}{3}7! + \binom{10}{4}6! - \binom{10}{5}5! + \binom{10}{6}4! - \binom{10}{7}3! + \binom{10}{8}2! - \binom{10}{9}1! + \binom{10}{10}0!$. We can vastly simplify this expression by expanding the binomial coefficients, dividing by 10!, and harvesting cancellations: $\frac{N_{=}(\varnothing)}{10!} = \frac{1}{2!} - \frac{1}{3!} + \frac{1}{4!} - \frac{1}{5!} + \frac{1}{6!} - \frac{1}{7!} + \frac{1}{8!} - \frac{1}{9!} + \frac{1}{10!}$. It is a fact that we will accept for the moment without proof that $\frac{1}{e} = \sum_{n=0}^{\infty} \frac{(-1)^n}{n!}$. This series representation of the reciprocal of the base of natural logs converges very rapidly and agrees with our inclusion/exclusion formula to one part in 10!, so $\frac{1}{e} \approx 0.3679$ is a good estimate of the derangement fraction of the set of permutations of ten objects. Evidently, there is about one chance in three that nobody got their correct hat back. This is the probability of a derangement. Our advice to the French noble would then be to bet on a match occurring.

Example 1.38

Now that we have the inclusion/exclusion formula, we can develop a formula for the Stirling numbers $S(k,n)$ that we needed for the Twelve-Fold Way. Recall that $n!S(k,n)$ is the number of surjective functions from a k-set to an n-set. To apply inclusion/exclusion, the set of objects X will consist of all the functions $f : K \to N$. The set of properties M will be the various degrees to which a function in X can exclude elements of N. Surjective functions will exclude no elements of N. Nonsurjective functions will exclude one or more elements of N. Perhaps you can already see where this is going, since the $N_=(\varnothing)$ that will emerge from inclusion/exclusion will be precisely the number of functions *with* surjectivity.

Example 1.39

Multiplying 12 identical terms $(1 + x + y)$ together will involve making 12 selections of either a 1 or an x, or a y from each term as we move left to right. We can summarize our selections in the form of a sequence, for example, $(1,1,x,y,y,1,x,x,y,y,y,1)$. In particular, we are interested in the sequences that can be formed by selecting two x's, three y's, and seven 1's. Each such sequence will contribute a 1 to the coefficient of x^2y^3. So temporarily and artificially distinguish the x's as x_1 and x_2, then the y's as y_1, y_2, and y_3, and the 1's similarly. We now have 12 "distinct" characters, so there are 12! possible permutations. Turn off the artificial distinction among the x's. This reduces the number of distinct permutations to $\frac{12!}{2!}$. Now do the y's. This cuts the number to $\frac{12!}{2!3!}$. Finally, do the 1's. The result is $\frac{12!}{2!3!7!} = 7920$ sequences and this will be the coefficient of x^2y^3. Certainly, an easier method than multiplying everything out. We will revisit binomial and multinomial expansion formulas when we discuss polynomials.

Example 1.40

$\binom{8}{2,3,3} = \frac{8!}{2!3!3!} = 560$ ways.

Example 1.41

Treat the Rembrandts as a unit, then $\binom{7}{1,3,3} = \frac{7!}{3!3!} = 140$ ways, now multiply by two for the two orders in which they can appear.

Example 1.42

Associativity is clear since we are effectively adding the degrees of rotation for each motion. The identity rotation is the "do-nothing" rotation through 0°. The inverse of a rotation of x degrees is just the rotation of $360 - x$ degrees. Of course, a 360° rotation has the same effect as a 0° rotation. The order in which rotations are applied does not matter, so this group is also abelian.

Example 1.43

We write out all the elements affected by the product and successively apply the indicated permutations, then track the overall mapping from top to bottom:

	1	2	3	4	5	6
apply (146)	4	2	3	6	5	1
apply (345)	5	2	4	6	3	1
apply (123)	5	3	4	6	1	2

The algorithm is to start with the lowest entry and track its permutation path back to itself, then repeat, if necessary, with the next lowest entry that has not already appeared in the path. Rinse and repeat.

So we have that (15)(2346) is the disjoint cycle product equivalent to (123)(345)(146).

Example 1.44

You can convince yourself that this is true by manipulating a paper triangle (have fun), but we will be more rigorous and develop the entire "multiplication table" for the six possible permutations using cycle notation. Labeling the vertices of the triangle as above, we already know that the identity permutation is (1), the 120° rotation is (123), and the 240° rotation is (132). A flip about the altitude drawn to vertex 1 leaves it alone and interchanges vertices 2 and 3. We symbolize this as (1)(23). Likewise, for the other two altitudes we have (2)(13) and (3)(12). The six possible symmetry operations for the triangle have been converted into cycle notation.

Example 1.45

This is the group D_8 consisting of four rotations of the square in its plane and four flips about the two midpoint axes and the two diagonal axes, respectively. We can characterize these eight symmetry moves in terms of the vertices labeled 1 through 4 clockwise starting top left. The identity (1)(2)(3)(4) is the zero degree rotation. The 90° clockwise rotation is (1234). The 180° rotation interchanges the diagonals, so it is (13)(24). The 270° rotation is the inverse of (1234), so (4321). Now for the flips: about the vertical midpoint axis, we have (13)(24); about the horizontal midpoint axis, (14)(23); about the diagonal connecting 1 and 3, (1)(3)(24); finally, about the diagonal connecting 2 and 4, (2)(4)(13). The longest cycle has length 4. We can organize the eight symmetry operations and their contributions to the cycle index as follows:

cycle product	index term
(1)(2)(3)(4)	z_1^4
(1234)	z_4
(13)(24)	z_2^2
(4321)	z_4
(13)(24)	z_2^2
(14)(23)	z_2^2
(1)(3)(24)	$z_1^2 z_2$
(2)(4)(13)	$z_1^2 z_2$

Adding the terms in the second column, we have the cycle index

$$Z(z_1, z_2, z_3, z_4) = \frac{1}{8}(z_1^4 + 3z_2^2 + 2z_1^2 z_2 + 2z_4).$$

Example 1.46

Since the front and back of the piece are interchangeable, only the distinct configurations of gems at the corners matter. We conclude that all the symmetry operations applicable to the square are in play, so the relevant symmetry group is D_8. By amazing coincidence we have the cycle structure of D_8 completely mapped out. In fact, we already have the cycle index

$$Z(z_1, z_2, z_3, z_4) = \frac{1}{8}(z_1^4 + 3z_2^2 + 2z_1^2 z_2 + 2z_4).$$

We generate the pattern inventory with the substitutions $z_1 = d + r + e$, $z_2 = d^2 + r^2 + e^2$, $z_3 = d^3 + r^3 + e^3$, and $z_4 = d^4 + r^4 + e^4$. After a bit of algebra we determine that

$$\begin{aligned} P(d, r, e) = {} & d^4 + d^3 e + d^3 r + 2d^2 e^2 + 2d^2 er + 2d^2 r^2 + de^3 + 2de^2 r + 2der^2 \\ & + dr^3 + e^4 + e^3 r + 2e^2 r^2 + er^3 + r^4. \end{aligned}$$

If we let $d = r = e = 1$ then $P(1, 1, 1)$ just counts the number of distinct patterns that appear. Evidently, there are $P(1, 1, 1) = 21$ distinct pendants that can be made. If we want to know how many designs would have exactly two emeralds, we look for multinomial terms with a factor of e^2. There appear to be 6. Or if we want to know how many contain at least three diamonds, we look for terms with a factor of d^3 or d^4. We locate three such terms – the all diamond design plus the two designs with three diamonds and either a ruby or emerald as the odd stone. The pattern inventory is a rich source of detailed information.

Example 1.47

The cycle index of the rotation group of the cube is

$$Z(z_1, z_2, z_3, z_4, z_5, z_6) = \frac{1}{24}(z_1^6 + 8z_3^2 + 6z_2^3 + 3z_1^2 z_2^2 + 6z_1^2 z_4).$$

You could work this out with a little effort or just look it up (we did). As a quick check on the cycle index, we calculate $Z(1,1,1,1,1,1) = 1$. Anything other than 1 would mean that the cycle index does not agree with the earlier determination that there are 24 elements overall in the rotation group. Let the colors be red, white, blue, and green. Then $z_i = r^i + w^i + b^i + g^i$. If we wanted a complete picture of the number and type of each possible pattern, we could substitute these various values into the cycle index to obtain the pattern inventory. Even with a computer assisted calculation, we would have a monster multinomial. If we only want the total number of patterns and not the details for each pattern, we could, in principle, let $r = w = b = g = 1$ in $P(r,w,b,g)$. This would give us the sum of all the coefficients in the monster, which is what we are seeking. But this is the same as letting $z_i = 4$ for $1 \le i \le 6$ in $Z(z_1,z_2,z_3,z_4,z_5,z_6)$. So we can avoid wrestling with the monster and compute $Z(4,4,4,4,4,4) = 240$ possible distinct paint jobs for the cube.

Example 1.48

Let $r = w = g = 1$ and $b = 0$ in $P(r,w,b,g)$. This is equivalent to $z_i = 3$ for $1 \le i \le 6$ in $Z(z_1,z_2,z_3,z_4,z_5,z_6)$. So $Z(3,3,3,3,3,3) = 57$.

Example 1.49

We are tempted to adapt the argument in Example 1.51, but this will not work. We have to find the coefficient in the monster of the term b^2w^2rg. This turns out to be 8, so there are 8 paint jobs with two blue faces, two white faces, and one face each green and red.

Example 1.50

Let $r = w = g = 1$ and $b = b$ in $P(r,w,b,g)$. This is equivalent to $z_i = 3 + b^i$ for $1 \le i \le 6$ in $Z(z_1,z_2,z_3,z_4,z_5,z_6)$. Expanding the pattern inventory, we get a smaller monster this time. Indeed,

$$\begin{aligned} P(1,1,b,1) = \frac{1}{24}&((3+b)^6 + 8(3+b^3)^2 + 6(3+b^2)^3 \\ &+ 3(3+b)^2(3+b^2)^2 + 6(3+b)^2(3+b^4)). \end{aligned}$$

This boils down to $P(1,1,b,1) = b^6 + 3b^5 + 12b^4 + 29b^3 + 66b^2 + 72b + 57$. This is the generating function for the number of blue faces on the cube. The coefficient of b^0 is 57, which agrees with the result of Example 1.51. There are apparently 72 paint jobs with exactly one blue face, 66 with two, 29 with three, 12 with four, 3 with five, and (obviously) 1 with all six faces blue. All sorts of questions can be answered about the presence or absence of blue faces and which other colors they may be combined with by examining the coefficients of this multinomial.

Example 1.51

By the symmetry of the pattern inventory function in each color variable, there must be 57 designs with no red face and 57 with no white face. The question at the moment is how many elements of these two sets of designs overlap. Parsing the monster multinomial, we find the following terms which only have blue and green variables: $b^6, b^5g, 2b^4g^2, 2b^3g^3, 2b^2g^4, bg^5$, and g^6. The sum of their coefficients is 10. If we add the "not-red" designs to the "not white" designs, we will have double counted the designs that have neither red nor white. So the correct tally of designs missing either a red or white face is $57 + 57 - 10 = 104$. We conclude that the complementary number $240 - 104 = 136$ gives the number of designs with at least one red face and one white face.

Example 1.52

Restating for clarity, we want

$$\sum_{k=0}^{n} \binom{n}{k}^2 = \binom{n}{0}^2 + \binom{n}{1}^2 + \cdots + \binom{n}{k}^2 + \cdots + \binom{n}{n}^2.$$

Recall $\binom{n}{k} = \frac{n!}{k!(n-k)!} = \binom{n}{n-k}$. What if we imagined $2n$ objects and asked how many subsets of size n could be created? We know that answer immediately $\binom{2n}{n}$. Is there some other way we can build up this collection of subsets? Certainly. We can divide the pile of $2n$ objects into two piles of n. Now select a subset of size k from one and a subset of size $n-k$ from the other. Taken together, we have a subset of size n. How many ways can we select a subset of size k from one pile of n objects? The answer: $\binom{n}{k}$ ways. How many ways can we form a subset of size $(n-k)$ from the other pile? The answer: $\binom{n}{n-k}$ ways. By the multiplicative principle, doing both selections can be done in $\binom{n}{k}\binom{n}{n-k}$ ways. But wait – aren't these two factors the same? By the formula above, they are, so forming a subset of size $k + (n-k) = n$ from the pile of $2n$ objects in this way can be done $\binom{n}{k}^2$ ways. As k goes from 0 to n, we will have constructed every n-subset of a $2n$-set. Now we are justified in stating $\sum_{k=0}^{n} \binom{n}{k}^2 = \binom{2n}{n}$. Dealing with the squares directly seemed fruitless, but interpreting them as an enumeration gave us the answer.

Example 1.53

We can assume the people have names, so Section 1.1.1 applies. There are 7^{12} possible mappings of donuts to people. This includes the surjective maps which we need to deduct. There are $S(12,7)$ distributions where everybody receives at least one donut. So there would be $7^{12} - S(12,7) = 13{,}840{,}659{,}805$ ways to deny at least one person their donut.

Example 1.54

Considering the two players as a unit, there would be 8! possible lineups if the unitized players batted with a fixed order between them. If they can change the order between them multiply by two.

2 Probability methods

> The probable is that which for the most part happens.
>
> – from Aristotle's *Rhetoric*.
> Hard to disagree with that.

A significant motivation for developing the theory of probability was, not surprisingly, the desire to have an advantage in playing games of chance. The theory matured from these mercenary beginnings over the course of several centuries with many famous mathematicians making contributions. Probability theory serves as the basis for statistics, which in turn supports the immense modern scientific enterprise. Who said gambling leads to a life of ruin? There have been and continue to be philosophical differences within the discipline based on the appropriateness of applying the theory to various situations, interpreting the result of such applications, and even the knowability of what we are calling probability. Our description of probability theory is intended to be at a modest practical level. We are going to sidestep the epistemological issues and take the gambler's viewpoint, which more properly might be called classical probability theory. This will involve consideration of idealized thought experiments where we define what we declare to be favorable outcomes of those experiments. The proportion of favorable outcomes relative to total possible outcomes is what we will call the probability of a favorable outcome. Our estimates of probability will be made *in principle* based on the details and assumptions of the thought experiments and not derived by running the experiments repeatedly in real life. By the way, this conflicts with the philosophical stance of the Frequentists, who argue that probabilities can only be truly estimated empirically. In the other camp are the Bayesians, who are more tolerant of subjective estimates. They would allow an a priori estimate of the probability of a meteorite colliding with the Earth as long as it was rationally defensible. That is an experiment we may not want to run at all. At any rate, it is the mathematical consistency of probability theory that we are interested in and fortunately this can be approached from an axiomatic viewpoint, similar to Euclid's axioms and postulates underpinning geometry. In 1933 the Russian mathematician Andrei Kolmogorov published *Foundations of the Theory of Probability*. This seminal book connected probability to the established mathematical theory of measure, where probabilities are interpreted as areas or volumes. This is well beyond the scope of our discussion, but we will at least proceed with the assurance that our axiomatic approach to classical probability is mathematically watertight.

2.1 Basic definitions

An **experiment** is any process which results in some measurable **outcome**. We can flip a coin (the experiment) and the result could be heads or tails (two outcomes). Or we could roll two six-sided dice where the outcomes are the numbers 2 through 12 as mea-

https://doi.org/10.1515/9783111506654-003

sured by the sum of the number of dots on the upward faces of both dice. Or we could select a playing card from a standard deck where there are 52 different outcomes. These are all **discrete** outcomes. We could also measure the voltage in a particular circuit at a given time, or the temperature at a particular time and place in a room. These would be **continuous** outcomes as they would be quantified by real numbers in some interval of possibilities. The set of all possible outcomes for a particular experiment is called the **sample space** for that experiment and each outcome is then an element of that set or a **sample point**. There is usually some discretion available in deciding what the sample space is going to be for a given experiment. For example, if we are selecting cards from a deck, we could set the sample space to {red, black} or {hearts, diamonds, clubs, spades} or {52 individual cards}. Depending on our plan for interpreting the results of this experiment, we may adjust the granularity of the sample space. The general recommendation is to make the granularity finer rather than coarser whenever practical. Sample spaces may also be distinguished by the number of sample points in them. Discrete spaces are usually **finite**, although there are cases where an experiment can yield outcomes that are **countably infinite**. A sample space with continuous outcomes is always **uncountably infinite**. A complication with continuous sample spaces that we can safely ignore in all practical work was discovered in 1905 by the Italian mathematician Giuseppe Vitali. Paradoxically, in any interval of real numbers, there are strange subsets which have no viable definition of measure (length). So our definition of probability in terms of measure will fail for these pathological sets. Not to worry about discrete spaces, though. They are well behaved.

If we denote a typical sample space as S, we can define an **event** as some subset $E \subset S$. So if an experiment were "pick two colors out of {red, green, blue, yellow, purple}," the event of picking red and blue is the subset {red, blue}. The number of 2-subsets of S is $\binom{5}{2} = 10$, so there are 10 possible pick-two events for this experiment. Mathematicians typically write 2^X to mean the collection of all subsets of X. So in general $2^S = \mathcal{E}$, the **event space**. The singleton sets in the event space correspond precisely to the points of the sample space. The sets within the event space obey, of course, the usual rules of set operations. If E_1 and E_2 are events in S, then $E_1 \cap E_2$ represents both E_1 and E_2 occurring together, and if $E_1 \cap E_2 \neq \varnothing$, we say they are **compatible** events. Similarly, $E_1 \cup E_2$ represents either E_1 or E_2 occurring and E_1^c is the event **complementary** to E_1, namely E_1 not occurring. We will discuss discrete spaces first and then continuous spaces, which require just a little bit of calculus. All of the general concepts discussed below apply to both types of spaces, unless otherwise noted.

2.2 Probability spaces

Let $P : \mathcal{E} \to \mathbb{R}$ be the **probability function** which assigns a real number to events (subsets of the sample space S). We call $\langle S, \mathcal{E}, P\rangle$ a **probability space** if it satisfies the following axioms:

Table 2.1: Kolmogorov axioms for a probability space.

1) $P(E) \geq 0$ for any event $E \in \mathcal{E} = 2^S$
2) $P(S) = 1$
3) If $E_i \cap E_j = \varnothing$ whenever $i \neq j$, then $P(\bigcup_{i=1}^{\infty} E_i) = \sum_{i=1}^{\infty} P(E_i)$

For discrete spaces, where S consists of n equally likely (**equiprobable**) outcomes and the event $E \subseteq S$ contains k of those outcomes, Axiom #1 can be replaced by $P(E) = \frac{k}{n}$. This is the core concept of classical probability theory. The probability of E is the ratio of total outcomes in E to total outcomes in S. Also, $P(E)$ can be quoted as a fraction or often as a percentage. The assumption of equiprobability may be problematic for situations where the outcomes are determined by indirect or complex processes. The frequentist solution for this would be to repeat the experiment many times, tabulate the rate of occurrence of various outcomes favorable to the event E, and estimate the limiting value $P(E)$ directly as the number of repetitions increases. Axiom #2 just says the event S is certain – the experiment must have some outcome. Axiom #3 can be paraphrased as "the probability function is **countably additive**" where it is understood the events must be mutually disjoint. Furthermore, P is also sometimes referred to as a **probability measure** Note that finite unions of m events are automatically included in Axiom #3, since for $i > m$ we have $E_i = \varnothing$. Also, since $P(S) + P(\varnothing) = 1$ implies $P(\varnothing) = 0$ by Axiom #2, for $i > m$ we have $P(E_i) = 0$ as well.

We have several immediate consequences from the axioms:

1) Since $E \cup E^c = S, P(E)+P(E^c) = 1$, hence $P(E^c) = 1-P(E)$, or in words, complementary events have complementary probabilities.
2) If $E_2 \subset E_1$ we can write $E_1 = E_2 \cup (E_1 - E_2)$. This is a disjoint union, so it follows that $P(E_1) = P(E_2) + P(E_1 - E_2)$ and since $P(E_1 - E_2) \geq 0$ it must be true that $P(E_1) \geq P(E_2)$.
3) From the preceding, it is clear that for any event E, $\varnothing \subseteq E \subseteq S$ implies $0 \leq P(E) \leq P(S) = 1$. So the range of the probability function is contained in $[0, 1]$.

Axiom #3 gives the addition law for probabilities where the events are pairwise disjoint. This is often called the **special addition law** for probabilities. If we want an addition law for probabilities where the events are not necessarily disjoint, we have to appeal to the principle of inclusion/exclusion (Section 1.4). In that context, consider the arbitrary events $\{E_i \subset S : 1 \leq i \leq m\}$ which may not be disjoint and let $|E_i|$ be the number of sample points of S that are contained in each E_i. The set of properties $\mathcal{P}$ that the sample points may individually have is being in, respectively, none, one, and so forth, up to m of the event sets E_i. Then the number of sample points contained in none of the event sets is $N_=(\varnothing) = |S| - \sum_{|M|=1} N_\geq(M) + \sum_{|M|=2} N_\geq(M) - \cdots + (-1)^n \sum_{|M|=m} N_=(M)$. Consider the term $\sum_{|M|=1} N_\geq(M)$. Here $M \subset \mathcal{P}$ is a set of a single property and $N_\geq(M)$ is the number of sample points that have at least that single property. We are summing over all such sets, so in terms of events, $\sum_{|M|=1} N_\geq(M) = \sum_{i=1}^{m} |E_i|$. Now for $\sum_{|M|=2} N_\geq(M)$ each M is a pair of properties, so $N_{\geq}(M)$ is the number of sample points that have at least that particular pair of properties. In terms of our event sets this translates to $\sum_{i,j=1,j>i}^{m} |E_i \cap E_j|$. The condition $i > j$ ensures that we do not intersect an event set with itself and that we do not double count the pairs of properties. Continuing similarly (and laboriously) in this way, we finally arrive at an expression for the number $N_{=}(\varnothing)$ of sample points in S that do not belong to any of the event sets in $\{E_i \subset S : 1 \leq i \leq m\}$. But that is the same as $|S| - |\bigcup_{i=1}^{m} E_i|$. Equating the two expressions, we have $|S| - |\bigcup_{i=1}^{m} E_i| = |S| - \sum_{i=1}^{m} |E_i| + \sum_{i,j=1,j>i}^{m} |E_i \cap E_j| - \cdots$. Finally, cancelling $|S|$ from both sides and flipping signs, we get $|\bigcup_{i=1}^{m} E_i| = \sum_{i=1}^{m} |E_i| - \sum_{i,j=1,j>i}^{m} |E_i \cap E_j| + \cdots$. This derivation explicitly assumes the probability space is discrete, but the result can be extended to continuous spaces. To restate this in more recognizable terms, here are the extended formulas for two and three event sets:

1) $|E_1 \cup E_2| = |E_1| + |E_2| - |E_1 \cap E_2|$,
2) $|E_1 \cup E_2 \cup E_3| = |E_1| + |E_2| + |E_3| - |E_1 \cap E_2| - |E_1 \cap E_3| - |E_2 \cap E_3| + |E_1 \cap E_2 \cap E_3|$.

It is now an easy matter to divide across the board by $|S|$ to obtain the corresponding

Table 2.2: General addition formulas for probabilities.

1) $P(E_1 \cup E_2) = P(E_1) + P(E_2) - P(E_1 \cap E_2)$
2) $P(E_1 \cup E_2 \cup E_3) = P(E_1) + P(E_2) + P(E_3) - P(E_1 \cap E_2) - P(E_1 \cap E_3) - P(E_2 \cap E_3) + P(E_1 \cap E_2 \cap E_3)$

These formulas are valid for all probability spaces.

2.2.1 Probabilities from experiments

We now have the machinery to calculate probabilities associated with some typical discrete situations. We will introduce another key concept in Section 2.6 that will enable calculation of probabilities in continuous spaces.

Example 2.1. What is the probability that if we flip a fair penny three times at least two heads will appear?

Example 2.2. What is the probability of drawing four cards from a standard deck and having them be the four aces?

Example 2.3. What is the probability of rolling an eight given two fair standard cubical dice? A fair die is defined analogously to a fair coin.

Example 2.4. If we flip a fair coin where heads is worth 1 and tails is worth 2, then we roll a fair die and count the dots, what is the probability that the product of the coin result with the die result is at least 4?

Example 2.5. A man phones a friend and brags about his two new dogs. Then he says the only problem was that the male bit him. What is the probability that the other dog is male?

Example 2.6. A game show contestant has a chance to win a new car by picking one of three doors leading to various prizes. Behind one door is the car. Behind the other two doors are goats. The contestant selects a door, but before it is opened, the show host opens one of the other two doors that has a goat behind it. Then he gives the contestant the option to switch his original selection. Should he do it or not?

2.2.2 Using the addition formulas

We use **chance** and **likelihood** as informal synonyms for probability. Bettors often speak of the **odds** for something happening or not happening. Provided an event E is not certain ($P(E) = 1$), the odds that E will occur is the fraction $\frac{P(E)}{1-P(E)}$. So if the odds for E to occur are two to one, we could set $\frac{P(E)}{1-P(E)} = \frac{2}{1}$ and solve for $P(E) = \frac{2}{3}$. In general, if the odds for E to occur are x to y then by the same argument we find $P(E) = \frac{x}{x+y}$.

There is a seeming paradox in continuous probability spaces where an event with $P(E) = 0$ is nevertheless possible. For example, if we measure the voltage across the terminals of a standard car battery we are going to get a number between 0 and 12. The probability of getting a reading of exactly x volts has to be zero. Otherwise, if the probability is $\varepsilon > 0$ and there are infinitely many exact readings between 0 and 12, we would be forced to conclude by our addition formula that the entire sample space had infinite probability. Different exact readings are disjoint events. So the rule is “impossible events have zero probability, but events with zero probability need not be impossible.” We can be sure that some voltage, after all, is recorded, even if its prior probability is zero.

Example 2.7. A survey taken in a small town reported that a resident had a 35 % probability of being a smoker, 18 % probability of being a drinker, and a 7 % probability of being both. What is the chance that a resident was either a smoker or drinker?

Example 2.8. A business owner knows from experience that the probability of running out of raw materials for a given production run is 10 %. The probability of production machinery failure is 15 % and the chance of staffing shortages is 12 %. The chance that all three happen at once is 3 %. The raw material situation is independent of either machinery failure or staff problems, but there is a 7 % chance that staff shortages and machinery failure occur together because of maintenance issues. What is the probability that the production run will go smoothly?

Example 2.9. Candidate A has odds of seven to five for winning the election. Candidate B has odds of four to three to win. Which one (at least in the eyes of the betting parlor) has the best chance of winning?

2.3 Conditional probability & independence

Suppose A and B are events in the sample space S. Then $P(A)$ and $P(B)$ could be calculated the usual way by comparing the number of sample points in each event with the total in S. Likewise, we could find $P(A \cap B)$ by counting the sample points common to both A and B. It is reasonable to ask what the probability of A would be if it is certain that B will occur as well. It is tempting to read into this some sort of implied time sequence, such as A occurs after B. This is somewhat misleading. All that is being claimed is that if the condition that B is guaranteed to occur, the sample points that contribute to the probability of A must be already present in B, namely those in $A \cap B$. Depending on the specific relationship of the event sets A and B, the original sample points of A may be reduced in number. That number could be zero if $A \cap B = \varnothing$, in which case B occurring forces A to have zero probability. We define the **conditional probability of A given B**, written $P(A|B)$, as the ratio $\frac{P(A \cap B)}{P(B)}$. This can be restated as $P(A|B) \cdot P(B) = P(A \cap B)$. Interchanging A and B, we have the theorem of **joint probability**: $P(A \cap B) = P(A|B) \cdot P(B) = P(B|A) \cdot P(A)$.

Whenever the probability of A as a single event satisfies $P(A) = P(A|B)$, we say A is **independent** of B. The previous formula reduces to $P(A) \cdot P(B) = P(A \cap B)$. On the other hand, since the conditional probability of B given A satisfies $P(B|A) \cdot P(A) = P(A \cap B)$, we compare this with $P(A) \cdot P(B) = P(A \cap B)$ and discover that if A is independent of B, then B is independent of A. So independence is a mutual or symmetric property. If two events are not independent, they are **dependent**. Bear in mind that this is a technical definition and we may conflate an informal notion of independence with the technical one. We highlight this result as the:

Table 2.3: Multiplication formulas for probabilities.

(Special)	Events A and B are independent if and only if $P(A) \cdot P(B) = P(A \cap B)$
(General)	If A and B are dependent events then $P(A|B) \cdot P(B) = P(B|A) \cdot P(A) = P(A \cap B)$

We should note that mutually exclusive events, such as an event and its complement, are dependent. If $P(A)$ and $P(B)$ are both nonzero, but $A \cap B = \varnothing$, then $P(A) \cdot P(B) \neq P(A \cap B) = 0$. On the other hand, if events A and B are independent, so are the events A^c and B^c. This is an immediate consequence of the fact that $P(A^c) \cdot P(B^c) = (1 - P(A)) \cdot (1 - P(B)) = 1 - P(A) - P(B) + P(A)P(B)$. Using De Morgan's laws (Section 0.1.1), we recognize that $P(A^c \cap B^c) = P((A \cup B)^c) = 1 - P(A \cup B)$ and this in turn is equivalent to $1 - [P(A) + P(B) - P(A)P(B)] = 1 - P(A) - P(B) + P(A)P(B)$. Hence $P(A^c) \cdot P(B^c) = P(A^c \cap B^c)$.

Example 2.10. The probability that a starting baseball pitcher throws at least 120 pitches in a game is 75 %. But if he manages to do that the odds that he becomes the winning pitcher are 4 to 1. What is the likelihood that he throws at least 120 pitches and wins the game?

Example 2.11. The event A is the outcome that an ace is randomly drawn from a standard deck. The event K is the outcome that a king is randomly drawn from the deck. Show that drawing the ace and also drawing the king without replacing whichever card was drawn first are dependent events.

Example 2.12. A fair penny is flipped three times. Are the events "exactly two heads occur" and "the third flip is a head" independent?

Example 2.13. Back to the dogs of Section 2.2.1 Example 2.5. There are two dogs. One of them is male and was born on a Sunday. What is the probability that the other dog is male? You may assume that male and female dogs are equally likely to be born.

2.3.1 Generalized independence

In order to extend the idea of independence to more than two events, we need to impose a condition that is more stringent than just requiring multiple events to be independent in pairs. For three events E_1, E_2, and E_3, we begin by requiring $P(E_i \cap E_j) = P(E_i) \cdot P(E_j)$ for $1 \le i < j \le 3$. Now we add the requirement that $P(E_1 \cap E_2 \cap E_3) = \prod_{i=1}^{3} P(E_i)$. Events satisfying these conditions are independent. It is possible to devise situations where the pairwise conditions are met but not the triple condition, hence the necessity of checking that condition as well to verify independence.

In the general case for n events E_i, we give the following definition: the events $\{E_i : 1 \le i \le n\}$ are **independent** if the following conditions hold:

1) $P(E_i \cap E_j) = P(E_i) \cdot P(E_j)$ for $1 \le i < j \le n$,
2) $P(E_i \cap E_j \cap E_k) = P(E_i) \cdot P(E_j) \cdot P(E_k)$ for $1 \le i < j < k \le n$,

$\vdots \qquad \vdots \qquad \vdots \qquad \vdots$

n-1) $P(\bigcap_{i=1}^{n} E_i) = \prod_{i=1}^{n} P(E_i)$.

Conversely, if we know that n events E_i are independent, we can write $P(\bigcap_{i=1}^{n} E_i) = \prod_{i=1}^{n} P(E_i)$. So the probability of the intersection of multiple independent events is just the product of their individual probabilities. If we are not assured of the independence, then we have to appeal to the general multiplication rule and its extension.

Example 2.14. The probability of rolling a standard die and getting a particular number from 1 to 6 is $\frac{1}{6}$. Successive rolls are independent events, so the probability of getting a particular sequence of numbers in three rolls is $(\frac{1}{6})^3 = \frac{1}{216}$. This would be the same for three dice labeled #1, #2, and #3 rolled simultaneously and the result read in the same order as the dice labels. Either way, the number of 3-lists from a 6-set (the numbers 1 to 6) is 216. We do not need to imagine enormous sample spaces to calculate the probability of events created by big single experiments if those experiments can be constructed by repetition of many small experiments. For example, the probability of getting the number 1234567890 by rolling ten sequenced dice all at once can be obtained by calculating the probability of rolling one die ten times and getting a 1, then a 2 and so forth. We know right away that the probability of getting a specific 10-list from a 6-set is $(\frac{1}{6})^{10} \approx 1.654 \times 10^{-8}$.

Example 2.15. A communication system consists of a source with four intermediate repeating stations *A*, *B*, *C*, and *D* connected in series and terminating in a receiver. The event that a station is functioning properly at a given time is independent from any other station. Stations *A* and *C* are working 96 % of the time, station *B* 98 %, and station *D* 95 %. What is the overall reliability of the communication channel? What is the probability that it is working at any moment?

Example 2.16. What is the probability of selecting four aces one by one from a standard deck without replacing each ace after it is drawn?

Example 2.17 (Poker). The probabilities of getting certain hands in five card draw poker (or any similar game) are a ready source of examples that can be explained with the method of the preceding example and some help from the enumeration formulas in Chapter 1. Poker hands are rated from weaker to stronger depending on the relative likelihood of them occurring in a five card deal. Stronger hands are less likely. The weakest possible hands contain cards with no match in rank (2, 3, 4, 5, 6, 7, 8, 9, 10, J, Q, K, A), no sequence of ranks (e. g., 5, 6, 7, 8, 9), or no match of five cards with the same suit (♠, ♣, ♢, ♡). Determine the likelihoods of the various five card poker hands.

Example 2.18. What is the probability that among 30 people in a room at least two will share a birthday (day/month).

2.4 Total probability

An event A may be conditional on a number of other events $B_1, B_2, \ldots, B_n$. If the events B_i are mutually exclusive (pairwise disjoint) and their union constitutes the entire sample space S, it is possible to express the probability of A as a sum, specifically $P(A) = \sum_{i=1}^{n} P(A|B_i)P(B_i)$. We may assume $P(B_i) \neq 0$ or it could be excluded. This formula follows from the expression $A = A \cap S = A \cap \bigcup_{i=1}^{n} B_i = \bigcup_{i=1}^{n}(A \cap B_i)$. Since $P(A \cap B_i) = P(A|B_i)P(B_i)$ and each subevent $A \cap B_i$ is disjoint from any other, we can conclude $P(\bigcup_{i=1}^{n}(A \cap B_i)) = \sum_{i=1}^{n} P(A|B_i)P(B_i)$. Hence $P(A) = \sum_{i=1}^{n} P(A|B_i)P(B_i)$. This is called the theorem of **total probability.**

Example 2.19. Police are looking for a bank robber who fled to the airport and may have escaped. Based on the relative number of daily flights from the airport, the police estimate that there is a 50 % chance he flew Delta, 30 % chance United, and 20 % chance Southwest. Airport security is not as efficient at some gates as others, so there is a 40 % chance he was not recognized and not stopped via Delta, 25 % at United, and 10 % at Southwest. What is the probability that he escaped by air?

Example 2.20. A concert pianist is very good, of course, but in a particularly difficult composition she sometimes hits a key to the left or right of the intended key each with a probability of 1 %. A particular G note appears in the composition 2.4 % of all notes, the F to the left 0.5 %, and the A to the right 1.1 %. What is the probability that she will play the particular G note, either intentionally or not, during the performance?

2.4.1 Bayes' rule

Suppose $\{B_i : 1 \le i \le n\}$ is the set of events as in Section 2.2 and A is an event that occurs with some conditional probability relative to each B_i. We can view the B_i events as possible companion events or coevents of A and there is no real harm in assuming they precede A in time. If we have $P(B_i)$ and $P(A|B_i)$ for $1 \le i \le n$, it is possible to construct a formula for $P(B_i|A)$. This answers the question "Now that A has happened, what is the chance that we got here through B_i?". The method is called Bayes' rule after Rev. Thomas Bayes, an English philosopher and mathematician who discovered the formula in the mid-eighteenth century. Some probabilists and statisticians are philosophically uneasy with the formula, particularly the seeming inversion of causation. But the formula is mathematically impeccable, so we will use it for practical calculations.

We have $P(A \cap B_i) = P(A|B_i)P(B_i) = P(B_i|A)P(A)$ from the theorem of joint probability in Section 2.3. This can be written $P(B_i|A) = \frac{P(A|B_i)P(B_i)}{P(A)}$. We know $P(A) = \sum_{i=1}^{n} P(A|B_i)P(B_i)$ from the theorem of total probability, so substituting this in the denominator for $P(A)$, we finally have

$$P(B_i|A) = \frac{P(A|B_i)P(B_i)}{\sum_{i=1}^{n} P(A|B_i)P(B_i)}.$$

Bayes' rule.

$$P(B_i|A) = \frac{P(A|B_i)P(B_i)}{\sum_{i=1}^{n} P(A|B_i)P(B_i)}$$

Example 2.21. A baseball pitcher throws three types of pitches: a fastball 50 % of the time, a curveball 30 % of the time, and a slider 20 % of the time. A particular hitter gets a hit 27 % of the time off a fastball, only 15 % off a curveball, and is completely fooled by a slider, getting a hit only 10 % of the time. This hitter faces the pitcher and gets a hit on the first pitch. What is the most likely kind of pitch that was thrown?

Example 2.22. Returning to the first example in Section 2.4, given that the robber escaped, what is the most probable airline that he used?

Example 2.23. There are three marksmen. Edmundo hits the bullseye 50 % of the time, Federico 30 % of the time, and Gumercindo 10 % of the time. They shoot at a target and since Gumercindo is so near-sighted, the others give him two shots while they take only one each. A single bullet is dug out of the bullseye. What is the probability that it belonged to Gumercindo?

Example 2.24. There are three boxes, each containing two drawers. In each drawer, there is either a diamond or an emerald. One box has two diamonds, another has two emeralds, and the third has one stone of each type. A box is chosen at random and one of its drawers is opened to reveal a diamond. What is the probability that the other drawer in the same box holds a diamond?

2.5 Random variables

Sample spaces can be very arbitrary. We can flip coins, roll dice, draw playing cards, record performance metrics in all manner of human activity, measure all sorts of physical phenomena – the list is almost endless. No matter what experiment is performed, the rates of occurrence of various outcomes depend ultimately on the natural randomness or unpredictability of some physical process. These outcomes may or may not be readily interpreted as numbers, but if they are not we need to "numericalize" them to take full advantage of our theory. For example, the experiment of repeatedly measuring voltage in a given circuit is certainly already in this form. But the experiment of flipping a coin just yields strings of heads and tails. In order to make sense out of this probabilistically, we need to impose a numerical interpretation. By counting the number of heads in each string, we implicitly define a function from the physical sample space of possible heads/tails strings to the nonnegative integers. This is an example of a random (or stochastic) variable, sometimes abbreviated RV.

By definition, a **random variable** X is a function from a discrete or continuous probability space to the real numbers, $X : \langle S, \mathcal{E}, P\rangle \to \mathbb{R}$. Using capital X for the function may be unsettling, but it is traditional. If X takes on the particular value x, we write that as $X = x$ and the probability of this happening as $P(X = x)$. So X does the job of numericalization by converting arbitrary sample space points into real numbers. This allows us to lift our probability function P from the sample space to the real numbers, the **state space or observation space**, by setting $P(X = x) = P(X^{-1}(x))$. In words this means that the probability that X attains the real value x is equal to the probability of the event in the sample space that is the preimage of x. Revisiting Section 2.2.1 (Example 2.1), we can define the discrete random variable X as the function that counts the number of heads in a string of three successive coin flips. Then $X^{-1}(2) = E = \{\{HHH\}, \{HHT\}, \{HTH\}, \{THH\}\}$. We had calculated that the probability of the event E was $\frac{1}{2}$, so we conclude $P(X = 2) = P(X^{-1}(2)) = P(E) = \frac{1}{2}$. If the experiment is to measure the output voltage of ten identical car batteries, the random variable X is the measured voltage and we know $0 \leq X \leq 12$. Note that if $0 \leq x \leq 12$ then, very much unlike a discrete space, $P(X = x) = 0$. The probability that a specific voltage is measured absolutely precisely is zero. The best we can do to get a nonzero result is to find the probability that the measured voltage is within a certain range. We will develop a way to handle these situations in the next section.

2.6 Probability distributions ($\int$)

Random variables allow us to routinely employ the real number system as a proxy for diverse sample spaces. This locates the theory in a standardized setting where all the tools of real analysis are available. In particular, we have a concise and intuitively appealing way to visualize probabilities by graphing them. A single random variable can be plotted in the plane with x on the horizontal axis and $P(X = x)$ on the vertical axis.

The function $f(x) = P(X = x)$ that this describes is called the **probability distribution** of the random variable X.

Probability distributions obey several rules. These apply to both discrete and continuous probability spaces. Suppose $f(x)$ is the distribution function, then:

Table 2.4: Rules for probability distributions.

1) $f(x) \geq 0$	(distributions are never negative)
2) $\lim_{x \to -\infty} f(x) = \lim_{x \to \infty} f(x) = 0$	(distributions tend to zero for extreme values)
3) for discrete probabilities, $\sum_{x \in \mathbb{R}} f(x) = 1$	(total probability is 1)
4) for continuous probabilities, $\int_{-\infty}^{\infty} f(x)dx = 1$	(total probability is 1)

The **cumulative distribution function** for the random variable X is defined as $F(x) = P(X \leq x)$. The relationship between the distribution and cumulative distribution is straightforward. In the discrete case, $F(x) = \sum_{z \leq x} f(z)$ and in the continuous case $F(x) = \int_{-\infty}^{x} f(z)dz$. Clearly, $F(-\infty) = 0$, $F(\infty) = 1$, and if $x_1 < x_2$ then $F(x_1) \leq F(x_2)$. Assuming $f(x)$ is continuous, the fundamental theorem of calculus implies that $F'(x) = f(x)$. To avoid potential confusion in talking about probability distributions it is helpful to distinguish among the types. We call the ordinary discrete distribution a **probability mass function (pmf)** where we imagine the discrete amounts of probability are attached to specific values of the random variable. The ordinary continuous distribution is called a **probability density function (pdf)**. We will not bother here to distinguish a **cumulative distribution function (cdf)** between discrete and continuous cases, although sometimes the word cumulative is reserved for discrete distributions and the word integral is applied to continuous distributions. In any case it is traditional to use lower case letters for pmf's or pdf's and the corresponding upper case letters for the respective cdf's.

With the concept of distribution function available, we can write down expressions that give the probability that a random variable takes on values in a given interval of real numbers $[a, b]$:

Getting probabilities from distributions

1) (Discrete case)	$P(a \leq X \leq b) = \sum_{a \leq x \leq b} f(x)$
2) (Continuous case)	$P(a \leq X \leq b) = \int_a^b f(x)dx$

Because discrete distributions are **atomic** (finite probability at single points), care must be taken to include or exclude end points of the summation interval to match the desired probability. No such restriction applies to integrals of continuous distributions.

Getting cumulative distributions from distributions

1) (Discrete case)	$F(x) = \sum_{z \le x} f(z)$
2) (Continuous case)	$F(x) = \int_{-\infty}^{x} f(z)dz$

We can also get probabilities directly from cumulative distributions.

Getting probabilities from cumulative distributions

1) (Discrete case)	$P(a < X \le b) = F(b) - F(a)$ – adjust by case
2) (Continuous case)	$P(a \le X \le b) = F(b) - F(a)$ – no restrictions

For example, in the discrete case for $P(a < X < b)$, we would subtract $P(b)$ from $P(a < X \le b)$.

2.7 Common univariate distributions ($\int$)

Let us develop a small library of common ordinary distributions in one variable and work some examples based on them.

Uniform discrete distribution

The probability of the n values of the **uniform discrete random variable** X, specifically $\{x_1, \ldots, x_n\}$, are all equal. So the pmf $f(x_i) = \frac{1}{n}$ if $x = x_i$ for $i \in [1, n]$. The cdf is $F(x_k) = \frac{k}{n}$ for $k \in [1, n]$.

Example 2.25. What is the probability mass function for the random variable X defined by the number of dots on the top face of a single standard die rolled once?

Example 2.26. What is the probability that a single fair die roll will yield an even number?

Uniform continuous distribution

The **uniform continuous random variable** X has a constant pdf, specifically the reciprocal of the length of the interval $[a, b]$ on which it is defined. So the pdf $f(x) = \frac{1}{b-a}$ if $x \in [a, b]$. The cdf

$$F(x) = \int_a^x \frac{dx}{b-a} \quad \text{for } x \in [a, b].$$

Moreover, $P(\alpha \le x \le \beta) = \frac{\beta-\alpha}{b-a}$.

Example 2.27. The exact time Joe gets to work each day is a random variable between 8 : 45 to 9 : 15 a. m. Experience has shown that he is equally likely to arrive at any time during that interval. Determine the probability density function for this continuous random variable stated in minutes with 8 : 45 considered zero. What is the probability that he arrives before 9 : 05?

Example 2.28. A spinner on a board game has 60 equally spaced marks around its circumference. Assuming no mechanical bias is present, what is the probability that a single spin will come up between 15 and 40 or between 50 and 60?

Triangular (Simpson) distribution

This continuous probability distribution is completely determined by three points $\{a, b, c\}$ on the real line, where $a < c < b$. The graph of the probability density function $f(x)$ connects the points $(a, 0)$ with $(c, f(c))$ and $(c, f(c))$ with $(b, 0)$. Note $f(c)$ is determined by the requirement that the area inside the triangle is 1. So $\frac{1}{2}(b-a)f(c) = 1$ implies $f(c) = \frac{2}{b-a}$. Oddly, $f(c)$ does not depend on c directly. If X is a **triangular random variable** with a **triangular continuous distribution**, then $P(\alpha \le x \le \beta) = \int_{\alpha}^{\beta} f(x)dx$.

Example 2.29. The probability distribution function that describes daily foot traffic into a supermarket is triangular. The number N of customers is a discrete random variable that is large enough that we can model it as if it were continuous. Experience has shown that the pdf for N is linear rising from zero at $n = 100$ to 0.002 at $n = 600$ and falling back to zero at $n = 1100$. What is the probability that on a given day fewer than 400 customers will be shopping?

Binomial distribution

This is a discrete distribution that applies to experiments like flipping coins and expecting a certain number of heads H or tails T. In general, we imagine that there is a repeatable process or **trial** that yields a **binary outcome**, often described as either **success** or **failure**, although these are arbitrary technical labels assigned by the experimenter and may not align semantically with the viewpoint of an observer. The experiment consists of repeating the trial n times. The individual trials are independent. The outcome of the entire experiment is a sequence or string of binary values. For example, ten successive coin flips might yield $HHTTTHTHTT$. Let X be the random variable defined by the number of H's in a string of ten flips. We are interested in the overall probability of getting exactly x occurrences of H, given that the probability of getting an H from a single trial is p. Note that the coin may not be fair so p need not be $\frac{1}{2}$. In this example we have four H's and six T's. So by the multiplication rule, the probability of getting four H's is p^4 and the probability of getting six T's is $(1-p)^6$. Then the probability of getting a string of four H's and six T's in some order is $p^4(1-p)^6$. But how many different strings can occur? There are $\binom{10}{4}$ distinct ways to assign four positions in the string out of ten to H. We conclude by the addition rule for disjoint events (different strings) that the probability of getting four heads in ten flips, with probability $P(H) = p$ for an individual flip, is $P(X = x) = \binom{10}{4}p^4(1-p)^6$. Note that we could have gotten the total number of strings by counting those with six positions out of ten for T. This approach would yield $\binom{10}{6}$ distinct strings. Fortunately, $\binom{10}{6} = \binom{10}{4}$, so both alternatives give the same result.

Generalizing the preceding argument, we define X as the number of successes in n binary trials. Then X is a **binomial random variable** with the **binomial distribution** given by $b(x; n, p) = \binom{n}{x}p^x(1-p)^{n-x}$, where $0 \leq x \leq n$. This represents the probability of getting exactly x successes in n trials where the probability of getting a success in a single trial is p. The **cumulative binomial distribution** is $B(x; n, p) = \sum_{k=0}^{x} b(k; n, p)$. This sum can be gruesome for large x and n because it involves many terms with large factorials appearing in the coefficients $\binom{n}{x}$. Lacking a table or computer resource, the sum can be approximated with an integral. We will explain this method in the section covering the normal distribution. The binomial distribution itself can also be computationally difficult for large x and n. We discuss two workarounds (approximating formulas) subsequently.

Example 2.30. What is the probability of getting exactly six heads in 12 flips of a fair coin?

Example 2.31. If the coin in Example 2.30 were secretly altered so that it came up heads 55 % of the time, would we be suspicious getting six heads in twelve flips?

Example 2.32. If $p = \frac{1}{2}$, and n is even, is it true that $B(\frac{n}{2}; n, \frac{1}{2}) = \frac{1}{2}$?

Example 2.33. A ski lodge in the Alps is open during November. There is already an adequate snow pack on November 1st, and on clear days the patrons spend the day on the slopes. On snowy days they tend to congregate in the lodge bar instead of schussing. The lodge proprietor makes an average of 5000 francs on snowy days and only 2000 francs on clear days. He has expenses of 3000 francs per day. It snows one day out of three during the month. What is the probability that he will lose money operating the lodge during November?

Example 2.34. Two soccer teams have played 50 games against each other overall with Team A winning 30 and Team B winning 20. Assuming this record reflects their relative strengths on any given day, is it more likely for Team A to win at least two games out of three against Team B than it is for Team B to win at least three games out of five against Team A?

Pascal / Pólya distribution

This is another discrete distribution that is based on the same setup as the binomial distribution, namely repetitive trials with a binary outcome. With the Pascal distribution, we are interested in the probability that not only k successes occur in n trials, but that the kth success occurs exactly on the nth trial. Suppose we flip a coin ten times and the fourth head comes up on the tenth flip. In order for this to happen, it must be that in the nine preceding flips only three heads occurred. With a probability of p for heads and $(1-p)$ for tails, the binomial distribution gives the probability of this happening as $\binom{9}{3}p^3(1-p)^6$. Now the probability of one more flip yielding heads is just p. So by the multiplicative principle, the probability of getting the fourth head on the tenth try is $p\cdot\binom{9}{3}p^3(1-p)^6=\binom{9}{3}p^4(1-p)^6$.

Generalizing this argument, we define X to be the number of trials required to get the kth success to occur on exactly the last trial. Then X is a **Pascal random variable** with the **Pascal distribution** given by $b^*(x;k,p)=\binom{x-1}{k-1}p^k(1-p)^{x-k}$, where p is the probability for a success. We denote the distribution function by b^* to distinguish it from the regular binomial distribution and remind us that it refers to the "waiting time" for a particular number of successes to appear in a binomial process. In fact, this distribution is also called the **binomial waiting time distribution**.

The Pascal distribution has several aliases depending on the property to be emphasized. It is also called the **negative binomial distribution** for the following reason. We have introduced the binomial theorem $(1+x)^k=\sum_{i=0}^{k}\binom{k}{i}x^i$ for positive integer k. There is a more general version applicable to negative integer or fractional k, provided x is restricted so that the sum converges ($|x|<1$ will be sufficient and that is very convenient for our probabilities). The enhanced version is $(1+x)^k=1+\sum_{i=1}^{\infty}\frac{x^i}{i!}[k(k-1)(k-2)\cdots(k-i+1)]$. This expression terminates if k is a positive integer and then the two versions coincide. If p is the probability of success for some trial, then $q=(1-p)$ is the probability of failure. Now consider $p^{-k}=(1-q)^{-k}$. Expanding this using the generalized binomial theorem, we get $(1-q)^{-k}=1+\sum_{i=1}^{\infty}\frac{(-q^i)}{i!}[(-k)(-k-1)(-k-2)\cdots(-k-i+1)]$. Applying the i powers of -1 to the i factors in the square brackets, all the negative terms disappear and the sum reduces to $1+\sum_{i=1}^{\infty}\frac{q^i}{i!}[(k)(k+1)(k+2)\cdots(k+i-1)]$. The coefficient of q^i in this expression is $\frac{(k)(k+1)(k+2)\cdots(k+i-1)}{i!}$. Multiplying this by $\frac{(k-1)!}{(k-1)!}$ and simplifying, we have $\frac{(k+i-1)!}{i!(k-1)!}$. Now letting i be the number of failures in x trials (note carefully that this x is the Pascal random variable, not the dummy variable in the statement of the binomial theorem), $k+i=x$, so $k+i-1=x-1$. Finally, we get $\frac{(k+i-1)!}{i!(k-1)!}=\frac{(x-1)!}{(x-k)!(k-1)!}=\binom{x-1}{k-1}$, which we recognize as the coefficient in $b^*(x;k,p)$. We could have written $\frac{(k+i-1)!}{i!(k-1)!}$ as $\binom{x-1}{x-k}$. This has an intuitive interpretation – it is the number of ways that $i=x-k$ failures could occur in the sequence of $x-1$ trials before the kth success is achieved on the xth trial.

Example 2.35. What is the probability that we would have to wait for the tenth flip of a fair coin to get six heads?

Example 2.36. What is the probability of rolling "snake eyes" twice with the second occurrence happening on the seventh roll of two standard fair dice. Snake eyes are just "1" on each die.

Geometric distribution

When $k = 1$, the Pascal distribution is called the geometric distribution. If X is the number of trials to get the first success, we say that X is a **geometric random variable** with **geometric distribution** given by $g(x; p) = p(1-p)^{x-1}$.

Example 2.37. In baseball the ratio of hits to "at bats" is essentially a players batting average. If we regard the batting average as a probability that the player will get a hit in a given at bat, and the player has a 0.300 batting average, what are the probabilities that he gets his first hit in a game respectively on the first, second, or third at bat?

Example 2.38. How likely is it that you can flip a fair coin ten times and only get the first head on the last flip?

Hypergeometric distribution

This is a discrete distribution that arises in situations where there is selection without replacement of objects that are regarded as either successes or failures. The binary trials in these cases are certainly not independent. Suppose we have a bag that contains six oranges and ten lemons. We are going to pick – sight unseen – five fruits from the bag without replacing any. What is the probability that we get exactly three oranges? Picking an orange is a "success" and picking a lemon is a "failure." Denote the set of successes by S, failures by F, and total objects by T. So $|S| = 6$, $|F| = 10$, and $|T| = 16$. There are $\binom{6}{3}$ ways of picking the three oranges, $\binom{10}{2}$ ways of picking two lemons, and $\binom{16}{5}$ ways of picking five fruits. By the multiplicative principle, there are $\binom{6}{3} \cdot \binom{10}{2}$ ways of getting three oranges in five picks, so the probability of this exact configuration is $P(3 \text{ oranges}) = \frac{\binom{6}{3} \cdot \binom{10}{2}}{\binom{16}{5}} = \frac{20 \cdot 45}{4368} = 20.6\,\%$.

Generalizing this argument, suppose X is defined to be the number of successes achieved in n selections without replacement from a pool of $|T|$ objects, of which $|S|$ are deemed successes. Then we say X is a **hypergeometric random variable** with **hypergeometric distribution** $h(x; n, |S|, |T|) = \frac{\binom{|S|}{x}\binom{|T|-|S|}{n-x}}{\binom{|T|}{n}}$. We have to restrict x to be no greater than n and $(n - x)$ no greater than $(|T| - |S|)$ so that we do not run out of available selections.

Example 2.39. In a box of candy, there are 15 pieces of which four are caramels. What is the probability that if you take eight pieces you will get all the caramels?

Example 2.40. Among 30 patients in a clinic five have a certain blood disorder. If 10 % of the patients are randomly tested, what is the probability that no one tests positive for the disorder?

Poisson distribution

The Poisson distribution is discrete but it is derived through an infinite limit process where we start with the binomial distribution $b(x; n, p)$ and let n approach infinity while the product $np = \lambda$ is held constant. Clearly, the probability p then approaches zero. To see where this takes us, we need the series expansion of the natural logarithm: $\ln(1-z) = -\sum_{k=1}^{\infty} \frac{z^k}{k}$, where $0 < z < 1$. Now since $p = \frac{\lambda}{n}$, we can rewrite the binomial distribution as $b(x; n, p) = \binom{n}{x}(\frac{\lambda}{n})^x(1 - \frac{\lambda}{n})^{n-x}$. This can be expanded into the form

$$\frac{n!}{x!(n-x)!}\left(\frac{1}{n}\right)^x \lambda^x \left(1 - \frac{\lambda}{n}\right)^n \left(1 - \frac{\lambda}{n}\right)^{-x}.$$

Let us examine the behavior of these factors as n goes to infinity. First, $\frac{n!}{x!(n-x)!} = \frac{n(n-1)\cdots(n-x+1)}{x!}$. There are x factors in the numerator and conveniently enough, there are x factors in $(\frac{1}{n})^x$, so divide each term in the numerator by n and express the result as

$$\frac{1}{x!}\left[(1)\left(1 - \frac{1}{n}\right)\left(1 - \frac{2}{n}\right)\cdots\left(1 - \frac{x-1}{n}\right)\right].$$

Since the limit of each of the finitely many factors in the square bracket as n becomes arbitrarily large is 1, the square bracket itself approaches 1, and we may conclude that the limit of $\frac{n!}{x!(n-x)!}(\frac{1}{n})^x\lambda^x$ is $\frac{\lambda^x}{x!}$. We are half way there. Now look at $(1 - \frac{\lambda}{n})^n$. Taking the natural log and using the series expansion of the log above, we have $\ln(1 - \frac{\lambda}{n})^n = n[-\frac{\lambda}{n} - \frac{\lambda^2}{2n^2} - \cdots - \frac{\lambda^k}{kn^k}\cdots] = [-\lambda - \frac{\lambda^2}{2n} - \cdots - \frac{\lambda^k}{kn^{k-1}}\cdots]$. Note that the only surviving term as n tends to infinity is $-\lambda$. By taking antilogs, we have shown that $\lim_{n\to\infty}(1 - \frac{\lambda}{n})^n = e^{-\lambda}$. Now, consider $\lim_{n\to\infty}(1 - \frac{\lambda}{n})^{-x}$. Since x is fixed and $\lim_{n\to\infty}(1 - \frac{\lambda}{n}) = 1$, we know $\lim_{n\to\infty}(1-\frac{\lambda}{n})^{-x} = 1$. This completes the analysis of the limit behavior of the five factors above. We conclude that the limit form of the binomial distribution subject to $np = \lambda$ is $p(x; \lambda) = \frac{\lambda^x e^{-\lambda}}{x!}$, which we call the **Poisson distribution** and a random variable with this distribution is a **Poisson random variable**. So what is it good for?

Since the Poisson distribution had its provenance in the binomial distribution, we might expect $p(x; \lambda)$ to be an estimator of $b(x; n, p)$ when n is large and p is small. Recall the calculations needed for $b(x; n, p)$ are a computational headache for large x and n. The Poisson expression is much simpler. The conventional wisdom is that for $n \geq 20$

and $p \leq \frac{1}{20}$ the Poisson distribution is a pretty good approximation to the binomial, and if $n \geq 100$ and $p \leq \frac{10}{n}$ the approximation is excellent. For cases where n is large but p is not small, a different asymptotic estimate is sometimes used. This is the **Laplace approximation** to the binomial distribution, which we state here without derivation. Let $q = 1 - p$, then $b(x; n, p) \approx \frac{1}{\sqrt{2\pi npq}} \exp[-\frac{(x-np)^2}{2npq}]$. This approximation is most accurate when p is $\frac{1}{2}$. We write $\exp[A]$ for e^A purely for readability issues when A is a complicated expression.

In addition to its utility as an approximation to the binomial distribution, the Poisson distribution has direct applicability in its own right. The parameter $\lambda = pn$ obscures the fact that the probability given by $p(x; \lambda) = \frac{\lambda^x e^{-\lambda}}{x!}$ depends on n, which is a number of trials. This may seem like an odd state of affairs until we realize n was allowed to approach infinity in our derivation. We can apply the Poisson distribution in situations where events are rare ($p \to 0$) but the number of opportunities for the event to occur are large ($n \to \infty$). Suppose an event E occurs on average λ times with equal likelihood at any time in the interval $(0, T)$. We can imagine partitioning $(0, T)$ into equal subintervals of length $\frac{T}{n}$, where n is chosen large enough to exclude the possibility that E will occur more than once in a given subinterval. This can always be done because n may be taken arbitrarily large. Then the occurrence of E in any subinterval is independent of the occurrence in any other subinterval. Since E happens on average λ times in $(0, T)$, it must happen in λ of the subintervals, so the probability that it happens in a specific subinterval is $P(E) = \frac{\lambda}{n}$. Now we see $\lambda = nP(E)$ and since λ is unaffected by taking the limit as n goes to infinity, the probability of x occurrences of E in the time interval $(0, T)$ is exactly $\frac{\lambda^x e^{-\lambda}}{x!}$. There is nothing magic about time here – we can rephrase this whole argument in terms of space. Suppose some event E' can occur λ times over an area A. We can partition A into equal subregions with area $\frac{A}{n}$ and proceed in complete analogy with the above to conclude that the probability of x occurrences of E' in the area A is given by $\frac{\lambda^x e^{-\lambda}}{x!}$. In applications we are often given the average temporal or spatial frequency ν of an event. To recover λ, simply multiply the frequency by the appropriate time or space interval.

Example 2.41. The frequency of Category 5 hurricanes in Miami is four per century. What is the probability of having two occur in the next 20 years?

Example 2.42. An airplane parts manufacturer packages high-strength bolts 100 to a box. The quality control department estimates that only one bolt in 500 does not meet specifications. What is the probability that there is at least one off-spec bolt in a box?

Example 2.43. Ten thousand bits are written to a disk drive. The error rate is 0.0002. What is the probability of encountering exactly one error?

Example 2.44. Stainless steel panels for appliance manufacture are subject to nicks and small scratches. The supplier guarantees a maximum of one defect per 20 square feet. What is the probability of getting no more than two defects in a shipment of 1000 square feet?

Example 2.45. The probability of someone using a counterfeit ticket to attend a Superbowl is 0.002 %. What is the probability that ten people are using such tickets at a game with 80,000 attendance? Express this as a binomial probability and as a Poisson probability, then do the simpler calculation.

Example 2.46. What is the probability of getting exactly 50 heads in one hundred flips of a fair penny?

Exponential distribution

A Poisson process is one where the events follow a Poisson distribution. The exponential distribution models the time separation between Poisson events. Accordingly, it is a continuous distribution, since two Poisson events can occur with any positive real number time separating them. Let ν be the average event frequency as in Section 2.7 (Poisson distribution), T be the random time between two successive events, and $f(t)$ be the pdf for T. The incremental time Δt can be chosen small enough that only one event can occur during it (we are going to take the limit as $\Delta t \to 0$ after all). Suppose t is the time since the last event occurred. The probability that at the beginning of the time interval Δt no event had occurred, but an event did occur during Δt, is $f(t)\Delta t$. Equivalently, the Poisson formula (restated in terms of frequency) $\frac{(\nu t)^x e^{-\nu t}}{x!}$ at the beginning of Δt is $\frac{(\nu t)^0 e^{-\nu t}}{0!} = e^{-\nu t}$, and at the end of Δt is $\frac{(\nu(\Delta t))^1 e^{-\nu(\Delta t)}}{1!} = (\nu(\Delta t))^1 e^{-\nu(\Delta t)}$. By the multiplicative principle, the probability that exactly one event occurred during Δt is $e^{-\nu t} \cdot \nu e^{-\nu(\Delta t)}\Delta t$. Equating $f(t)\Delta t = e^{-\nu t} \cdot \nu e^{-\nu(\Delta t)}\Delta t$, dividing by Δt, then letting Δt approach zero we have $f(t) = \nu e^{-\nu t}$, where $t \geq 0$. This is the **exponential distribution** and T is an **exponential random variable**.

Not only does the exponential distribution model the time between Poisson events, if we apply it to a Poisson process that has just begun it models the waiting time until the first event. It can also be used to model the time to failure of a component such as a light in a security system or a bearing in a drive mechanism.

Example 2.47. The long term average frequency of phone calls to the complaint desk at the department store is 30 per day. What is the probability that there is an hour that goes by with no complaint calls? Assume that all calls are at random (equiprobable) times from 8 a. m. to 6 p. m. – so they are truly a Poisson process. What is the probability that the person at the complaint desk can take an uninterrupted lunch hour starting at noon?

Weibull distribution

The Weibull distribution is a generalized and somewhat adjustable version of the exponential distribution that uses additional parameters to more faithfully model phenomena where the event frequency is changing in time. For example, mechanical systems that have rotating parts typically tend to have higher rates of failure as their bearings wear with age. Conversely, some products may be subject to a "shake out" period where, perhaps for reasons of manufacturing complexity, the chances of failure earlier rather than later in their life cycle are greater. The Weibull distribution is $w(t) = \frac{\beta}{\delta}(\frac{t}{\delta})^{\beta-1}\exp(-\frac{t}{\delta})^{\beta}$ for $t > 0$. Note that if $\beta = 1$ then $w(t) = \frac{1}{\delta}\exp(-\frac{t}{\delta})$, which is exactly the exponential distribution above with $\nu = \frac{1}{\delta}$. For a given modeling situation, there may be some observational data that suggest certain values of β and δ are more realistic than others.

Normal (Gaussian) distribution

This is arguably the most important of all distributions, since it lies at the heart of modern statistics. The general form of a normal distribution is

$$n(x;\mu,\sigma) = \frac{1}{\sigma\sqrt{2\pi}}\exp\left(-\frac{1}{2}\left(\frac{x-\mu}{\sigma}\right)^2\right),$$

where μ and σ are parameters. We will postpone a detailed discussion until Section 3.3. Normal distributions are different from all the other distributions discussed above in a significant and unfortunate way. They are difficult to work with mathematically. Although they are continuous functions, and therefore integrable, the usual techniques of closed-form integration do not work, so we cannot use them to find areas, hence probabilities (see Section 2.6). We are saved to some extent by the fact that any normal distribution can be reduced by substitution to a unique so-called **standard normal distribution**. Then the cumulative area problem for this "stationary target" can be addressed numerically once and for all. To solve problems involving normal distributions, we will eventually have to appeal to printed or electronic tables of areas.

Lognormal distribution

This is an offshoot of the normal distribution where if X is a normally distributed random variable, then $Y = e^X$ is lognormally distributed. In other words, Y is lognormal if its log X is normal. Specifically, the distribution function is $f(y;\mu,\sigma) = \frac{1}{y\sigma\sqrt{2\pi}}\exp[-\frac{1}{2}(\frac{\ln y-\mu}{\sigma})^2]$. The extra "$y$" in the denominator of the lognormal distribution is a consequence of the transformation law In general, if $y = f(x)$ is an injective

(1 : 1) function from the random variable X to the random variable Y, then there exists an inverse function $x = g(y)$. If $f(x)$ is an increasing function (as it is in the case where $y = e^x$), whenever $X \le x$ it follows that $Y \le y$. That implies $P(X \le x) = P(Y \le y)$. In turn this means the respective cumulative distributions are equal, $F_X(x) = F_y(y)$. Taking differentials, we finally see that $dF_X(x) = p_X(x)dx = dF_y(y) = p_y(y)dy$, where $p_X(x)$ and $p_y(y)$ are the respective probability density functions. Substituting into the differential equality and setting $x = g(y)$, we have $p_y(y) = p_X(g(y))\frac{dg}{dy}$. This is the **general transformation law** for univariate distributions. If $f(x)$ were decreasing, we would need to use $|\frac{dg}{dy}|$ to make sure $p_y(y) \ge 0$. In words this means that the pdf in terms of y can be constructed by taking the pdf in terms of x, substituting $g(y)$ for every occurrence of x, and multiplying the entire expression by $\frac{dg}{dy}$. Now we can build the lognormal distribution for Y. Start with the normal distribution for X, which is given, and substitute $\ln y$ for x. This gives $\frac{1}{\sigma\sqrt{2\pi}}\exp[-\frac{1}{2}(\frac{\ln y-\mu}{\sigma})^2]$. Next multiply that by $\frac{dg}{dy} = \frac{d(\ln y)}{dy} = \frac{1}{y}$. We now have $f(y;\mu,\sigma) = \frac{1}{y\sigma\sqrt{2\pi}}\exp[-\frac{1}{2}(\frac{\ln y-\mu}{\sigma})^2]$.

There are numerous other more specialized univariate distributions that have application to problems in physics, engineering, and economics. The references at the end of this chapter discuss many of these in detail. We have covered the basic ideas that are common to working with all types of distributions. There are several distributions that are most useful in connection with the analysis of statistics, and we will defer their introduction until our discussion of basic statistics.

2.8 Multivariate distributions (∫)

Distributions that involve two or more random variables are called **joint distributions**. They are handled much like the preceding **univariate distributions**, but there are some new ideas. We will discuss the basics in terms of distributions involving just two random variables, or **bivariate distributions**. The generalization to three or more random variables is straightforward.

Suppose X and Y are both discrete random variables. The **joint probability mass function** is the expression $f(x,y) = P(X = x, Y = y)$, where the point (x,y) corresponds to values attainable by X and Y. Kolmogorov's axioms in this context are 1) $f(x,y) \ge 0$ and 2) $\sum_x \sum_y f(x,y) = 1$. The **cumulative joint discrete distribution function** corresponding to $f(x,y)$ is $P(X \le x, Y \le y) = F(x,y) = \sum_{u\le x}\sum_{v\le y} f(u,v)$.

If instead X and Y are both continuous random variables, the axioms become 1) $f(x,y) \ge 0$ and 2') $\iint_{\mathbb{R}^2} f(x,y) = 1$, where $\mathbb{R}^2$ is the xy-plane. In this case $f(x,y)$ is called the **joint probability density function**. Then the corresponding **cumulative joint continuous distribution function** is $P(X \le x, Y \le y) = F(x,y) = \int_{-\infty}^{x}\int_{-\infty}^{y} f(x,y)dydx$. If E is some region of the plane, then $P((x,y) \in E) = \iint_E f(x,y)dydx$.

Since we have two dimensions to work with, we can sum or integrate $f(x,y)$ over all values of x to obtain a new function $g(y)$ called the marginal distribution of Y. Alternatively, we can sum or integrate over y to produce $h(x)$, the marginal distribution of X. In the discrete case it is a **marginal probability mass function**, and in the continuous case a **marginal probability density function**.

Table 2.5: Marginal distribution formulas.

1) Discrete case	$g(y) = \sum_x f(x,y)$
2) Continuous case	$g(y) = \int_{-\infty}^{\infty} f(x,y)dx$
3) Discrete case	$h(x) = \sum_y f(x,y)$
4) Continuous case	$h(x) = \int_{-\infty}^{\infty} f(x,y)dy$

To show how the notion of marginal distribution could be extended to three random variables with continuous joint density function $f(x,y,z)$, we would write $h(x) = \iint_{\mathbb{R}^2} f(x,y,z)dydz$ where $\mathbb{R}^2$ is the yz-plane. And so forth, summing or integrating the multivariate mass or density function over all variables except that defining the marginal distribution.

Another feature possible with multivariate distributions is the idea of a conditional distribution. We saw in Section 2.3 that the probability of event A, given the occurrence of event B was the probability of both A and B happening divided by the probability of just B, provided that was not zero. In symbols, $P(A|B) = \frac{P(A\cap B)}{P(B)}$. Let event A be $X = x$, and event B be $Y = y$. Then $P(X = x|Y = y) = \frac{P(X=x,Y=y)}{P(Y=y)}$. In terms of joint and marginal distributions, this is equivalent to $\widehat{f}(x|y) = \frac{f(x,y)}{g(y)}$. The function $\widehat{f}(x|y)$ is the **conditional distribution** (discrete or continuous as appropriate) of the random variable X given that Y is fixed at y.

Recall from Section 2.3 our definition of independence for two events: A and B are independent if and only if $P(A) \cdot P(B) = P(A \cap B)$. From the preceding conditional distribution formula, $f(x,y) = \widehat{f}(x|y)g(y)$. If $\widehat{f}(x|y)g(y) = P(X = x|Y = y)$ and $P(X = x|Y = y) = P(X = x)$, that is to say, the distribution for X does not depend on a particular y, we can write $f(x,y) = h(x)g(y)$. So the joint distribution is simply the product of the individual marginal distributions whenever X and Y are independent random variables. This gives us an **independence test** for two random variables, and it is easily extended to the condition that $X_1, X_2, \ldots, X_n$ are independent random variables whenever their joint distribution is a product of their marginal distributions: $f(x_1, x_2, \ldots, x_n) = g_1(x_1)g_2(x_2)\cdots g(x_n)$. So if you know two random variables are independent it is immediately possible to get their joint distribution simply by multiplying their individual distributions together. This is a starting point for constructing the distributions corresponding to simple functions of two random variables, such as $X + Y$, $X - Y$, and so forth. This project is beyond the scope of our text, but the suggested reading at the end of this chapter offers detailed guidance.

The following examples illustrate some of these multivariate concepts.

Multinomial distribution

This is a generalization of the binomial distribution. If independent trials are no longer binary, but consist of k possible choices, we use the multinomial distribution to describe the outcome of n trials. Define X_i to be random variable counting the number of times the ith choice is made in n trials. Then $\sum_{i=1}^{k} x_i = n$. Suppose the probability of the ith choice is given by p_i. So $\sum_{i=1}^{k} p_i = 1$. A typical overall outcome after n trials can be written as the string $\{x_1, x_2, \dots, x_k\}$. The probability of this string is $p_1^{x_1} p_2^{x_2} \cdots p_k^{x_k}$. This is not yet the expression for the multinomial distribution because we need to account for all the permutations of the string entries. The multinomial coefficient comes to the rescue (see Section 1.5). The number of ways we can pick x_1 objects of type X_1, x_2 objects of type X_2, and so forth is $\binom{n}{x_1, x_2, \dots, x_k}$. Now we can finally write the **multinomial distribution** as $m(x_1, x_2, \dots, x_k; n, p_1, p_2, \dots, p_k) = \binom{n}{x_1, x_2, \dots, x_k} p_1^{x_1} p_2^{x_2} \cdots p_k^{x_k}$. This is an example of a discrete joint distribution, where the function depends on two or more random variables.

Example 2.48. Suppose there is an election for sheriff and three candidates are running. Polling indicates the voting public prefers candidate #1 with 50 % approval, candidate #2 with 30 % approval, and candidate #3 with 20 % approval. If the poll numbers reflect actual voting probabilities, what is the probability that the result of the election will be 1000 votes for #1, 400 votes for #2, and 600 votes for #3?

Example 2.49. Ten vacationers are traveling to a distant city. The modes of travel are car, bus, and train. The likelihood that a person takes their car is 50 %, a bus 20 %, and the train 30 %. What is the probability that four of the vacationers will take their cars, one the bus, and the rest the train?

Multivariate hypergeometric distribution

This distribution bears the same relation to the multinomial distribution as the hypergeometric bears to the binomial, specifically it models the probability of sampling without replacement among several choices. It is also a discrete joint distribution among k random variables X_i. Mimicking our derivation of the hypergeometric distribution, suppose there is a set T of elements and S_1 is the subset of type 1, S_2 is the subset of type 2, and so forth, up to S_k with type k. Certainly, $|T| = \sum_{i=1}^{k} |S_i|$. Define X_i to be the number of elements selected from S_i when n elements are randomly (equiprobably) selected from T without replacement. There are $\binom{|S_i|}{x_i}$ ways of selecting x_i elements from S_i for each i. Clearly, $\sum_{i=1}^{k} x_i = n$. So the probability of selecting a specific configuration $\{x_1, x_2, \dots, x_k\}$ is $\widehat{m}(x_1, x_2, \dots, x_k; n, |S_1|, |S_2|, \dots, |S_k|) = \frac{\binom{|S_1|}{x_1}\binom{|S_2|}{x_2}\cdots\binom{|S_k|}{x_k}}{\binom{|T|}{n}}$. The k variables X_i are called **multivariate hypergeometric random variables** and their joint distribution is the **multivariate hypergeometric distribution**.

Example 2.50. A 12-person jury is to be selected from a jury pool consisting of five white men, seven white women, three black men, and nine black women. What is the probability that the final jury selection consists of three white men, four white women, two black men, and three black women? No juror challenges allowed, so the selection is random.

Bivariate continuous uniform distribution

Suppose X and Y are random variables with joint density $f(x,y) = k$, a constant. Let $a \le X \le b$ and $c \le Y \le d$. Then since $\int_a^b \int_c^d f(x,y)dydx = 1$, it must be that $k = \frac{1}{(b-a)(d-c)}$. We can find the probability that a point (x,y) lies within the area A by evaluating $\iint_A \frac{dydx}{(b-a)(d-c)}$.

Example 2.51. John and Mary are both scheduled to meet the boss at 1 p. m. John is a bit informal with his punctuality and has an equal chance of getting to the meeting at any time between 12:50 and 1:10 p. m. On the other hand, Mary is more organized and she has an equal chance of getting to the meeting at any time between 12:45 and 1:05 p. m. What is the probability that John will get to the meeting no later than Mary?

Example 2.52. A stick is broken into three pieces. What is the probability that the pieces can be reassembled into a triangle if: 1) the two break points on the stick are first chosen at random, 2) one break point is chosen at random, then after breaking once the second break point is chosen on the longer piece, or 3) one break point is chosen at random and, after breaking, one of the two resulting pieces is selected at random to be broken at a random location again?

Example 2.53. A toothpick of length L is dropped randomly on a wood floor where the seams between the wood strips also happen to be separated by distance L. What is the probability that the toothpick will cross a floor seam?

Marginal and conditional distributions

Example 2.54. A private school gives an entrance examination where X and Y are random variables representing the fraction of correct answers on the quantitative and verbal parts of the exam, respectively. Experience has shown that the joint density function for these scores is $f(x,y) = \frac{1}{4}(3x + 5y)$ where both x and y are modeled as continuous variables between zero and one. 1) Show that $f(x,y)$ is a valid joint distribution, 2) find the marginal distribution for X, 3) find the marginal distribution for Y, 4) find the conditional distribution for X given $y = 0.5$, and 5) decide if X and Y are independent.

Example 2.55. Let X_k be the random variable that counts the number of times a standard die shows six on the kth roll. So $x_k = 0$ or 1. Find the probability mass function for n rolls of the die.

Example 2.56. Suppose the bivariate discrete distribution $f(x,y) = \frac{xy^2}{C}$, where C is a constant, $x = 1, 2, 3$, and $y = 1, 2$. 1) Find C, 2) find the marginal distribution for both X and Y, 3) find the conditional distribution of y given $x = 3$, and 4) decide if X and Y are independent.

Suggested reading

1) *Elements of Applied Probability Theory* by Petr Beckmann (1968), Harcourt, Brace & World – *intended for electrical engineers but very compact and readable;*
2) *John E. Freund's Mathematical Statistics with Applications* (7th ed.) by I. Miller and M. Miller (2004), Pearson – *excellent undergraduate text;*
3) *Applied Statistics and Probability for Engineers* (4th ed.) by D. Montgomery and G. Runger (2007), Wiley;
4) *An Introduction to Probability Theory and Its Applications (Vol. 1)* by William Feller (1950), Wiley – *classic comprehensive treatise;*
5) *The Theory of Probability* by B. V. Gnedenko (translated by G. Yankovsky) (1978), MIR Publishers, Moscow.

Solutions to Chapter 2 examples

Example 2.1

A fair coin (which we will spend a lot of time flipping) has two sides: the obverse is "heads" symbolized by H, and the reverse is "tails" symbolized by T. The coin is fair in the sense that the flipping process results in either H or T with equal probability. We first have to decide what the sample space is going to be. A space with simply H for heads and T for tails is not going to help, but it points the way. What we want are sequences of three H's and T's. There are eight of these $\{HHH\}, \{HHT\}, \{HTH\}, \{THH\}, \{HTT\}, \{THT\}, \{TTH\}$, and $\{TTT\}$. Every possible outcome of three successive flips is represented in this list. Which of these sample points is favorable to the event E where at least two heads appear? It looks like $E = \{\{HHH\}, \{HHT\}, \{HTH\}, \{THH\}\}$. So four of the eight equiprobable sample points are favorable to E. By our definition, $P(E) = \frac{4}{8} = \frac{1}{2}$. The event where at least two heads do not appear is complementary to this, so we know immediately that $P(E^c) = 1 - \frac{1}{2} = \frac{1}{2}$ as well. By symmetry, this is the event where at least two tails appear. Had we asked for the probability that exactly two heads appear, that event is $E' = \{\{HHT\}, \{HTH\}, \{THH\}\}$. Then $P(E') = \frac{3}{8}$.

Example 2.2

We are not interested in the order in which the aces are drawn, just that they are aces. The sample space is going to be all of the four card sets that could be drawn from the deck. How many are there? The number of distinct subsets of size four in a set of size 52 is $\binom{52}{4} = 270{,}725$. There is only one subset of four aces, so $P(4\text{ aces}) = \frac{1}{270725} = 0.00036\,\%$.

Example 2.3

Each die has six faces inscribed with from one to six dots. When a die comes to rest, the number of dots on the topmost face is deemed the value of the roll. For multiple dice in a single roll, the values for all the dice are added. The number of mappings from a 2-set to a 6-set is $6^2 = 36$, so this is the number of possible rolls for two dice. The sample space will be ordered pairs (a, b) where a is the value of die #1 and b the value of die #2. The event “8” is the subset $E = \{(2,6),(3,5),(4,4),(5,3),(6,2)\}$. So $P(E) = \frac{5}{36}$.

Example 2.4

The sample space is again ordered pairs:

$$S = \{(1,1),(1,2),(1,3),(1,4),(1,5),(1,6),(2,1),(2,2),(2,3),(2,4),(2,5),(2,6)\}.$$

We have the favorable event

$$E = \{(1,4),(1,5),(1,6),(2,2),(2,3),(2,4),(2,5),(2,6)\}.$$

Each sample point is equally likely, so $P(E) = \frac{|E|}{|S|} = \frac{8}{12} = \frac{2}{3}$.

Example 2.5

Let F stand for female and M for male. In the absence of any other information, we have possible pairs (dog 1, dog 2): (M, M), (M, F), (F, M), and (F, F). The experiment is to pick a pair. The sample space is not all four pairs, since (F, F) is ruled out by the declaration that a male dog is present. So $S = \{(M, M), (M, F), (F, M)\}$. The event that the other dog is also male is $E = \{(M, M)\}$. So $P(E) = \frac{|E|}{|S|} = \frac{1}{3}$.

Example 2.6

Switch! By switching the contestant is sure to get the car unless it was behind the originally selected door. The probability that the car was behind the originally selected door is clearly $\frac{1}{3}$. This is the probability of getting the car without switching doors. By switching the contestant now enjoys the complementary probability $\frac{2}{3}$. So he has doubled his original chance of getting the car. The mechanics of the contest make switching effectively the selection of two doors at once. This is the famous Monte Hall problem, named after the host of the game show that inspired this problem. It appeared about 1975 and captured the popular imagination. Professional mathematicians and statisticians weighed in publicly on the theory of whether to switch or not, some to their eventual chagrin.

Example 2.7

We want the general addition formula $P(E_1 \cup E_2) = P(E_1) + P(E_2) - P(E_1 \cap E_2)$. Let E_1 be the event of a random resident being a smoker, and E_2 the event being a drinker. Then the probability of a resident being either a smoker or a drinker is $P(E_1 \cup E_2)$. Plugging in the data, $P(E_1 \cup E_2) = 0.35 + 0.18 - 0.07 = 0.36$. Just adding together, $P(E_1) + P(E_2)$ overstates the case, since some residents contribute to both behaviors and would be counted twice.

Example 2.8

Let R be the event "raw material shortage," M be the event "machinery failure," and S be the event "staff shortage." The general addition formula for three events in this context is $P(R \cup M \cup S) = P(R) + P(M) + P(S) - P(R \cap M) - P(R \cap S) - P(M \cap S) + P(R \cap M \cap S)$. Plugging in data, $P(R \cup M \cup S) = 0.10 + 0.15 + 0.12 - 0.07 + 0.03 = 0.33$. Interpretation: this is the chance that at least something goes wrong, so it looks like the business owner has a 67 % chance of the production run going smoothly.

Example 2.9

From the odds formula above, A has probability 0.583 of winning and B has probability 0.571, so it looks like A by a nose.

Example 2.10

Let A be the event that he is the winning pitcher and B the event that he throws at least 120 pitches. Note $A \cap B$ is the event whose probability we need to find. We know $P(B) =$ 75 % and we know that given B, the odds that event A occurs are 4 to 1. So first convert the odds to probability: $P(A|B) = \frac{4}{4+1} = 80\,\%$. From the formula $P(A|B) = \frac{P(A \cap B)}{P(B)}$, we can plug in our data $0.80 = \frac{P(A \cap B)}{0.75}$ and solve for $P(A \cap B) = 0.60$. So there is a 60 % chance that he throws at least 120 pitches and wins. Of course, he could win another way, say by only throwing 90 pitches, but that is outside the scope of the stated question.

Example 2.11

There are four aces in the deck, so getting an ace on a single draw from a full deck would have probability $\frac{4}{52} = \frac{1}{13}$. Likewise for a king. If these events were independent, the multiplication formula would yield $P(A \cap K) = P(A) \cdot P(K) = (\frac{1}{13})^2 = \frac{1}{169}$. But whichever card is taken on the second draw has a probability of $\frac{4}{51}$ of being correct, since the other card was removed from the deck of 52. So the true $P(A \cap K) = P(A) \cdot P(K|A) = P(K) \cdot P(A|K) = \frac{1}{13} \cdot \frac{4}{51} = 0.00603$ but $\frac{1}{169} = 0.00592$, so we conclude that successive draws without replacement are dependent events. Successive draws with replacement are independent.

Example 2.12

Of the $2^3 = 8$ possible sequences of three flips, those where exactly two heads occur are $\{(HHT), (HTH), (THH)\}$. The sequences where the third flip is a head are $\{(HHH), (HTH), (THH), (TTH)\}$. The sequences where both conditions occur are $\{(HTH), (THH)\}$. Let A be the event "exactly two heads" and B be the event "third flip head." Then $P(A) =$

$\frac{3}{8}$, $P(B) = \frac{1}{2}$, and $P(A \cap B) = \frac{1}{4}$. Now $P(A) \cdot P(B) = \frac{3}{8} \cdot \frac{1}{2} = \frac{3}{16} \neq P(A \cap B)$. We conclude the two events are not independent.

Example 2.13

There are two genders and seven days of the week available for assignment to each dog. A sample point in this space S is a pair of pairs: ((dog 1, day 1), (dog 2, day 2)). So there are $2 \cdot 7 = 14$ possible entries for (dog 1, day 1) and the same for (dog 2, day 2). Overall this gives a sample space of $14^2 = 196$ elements. Event A is "dog 1 and dog 2 are both male." Event B is "either dog 1 or dog 2 is a male born on Sunday." We are asked to find the conditional probability $P(A|B)$. Let us first account for the sample points which do not feature a male born on Sunday: there are $7 \cdot 7 = 49$ female/female points, $6 \cdot 6 = 36$ non-Sunday male/non-Sunday male points, $7 \cdot 6 = 42$ female/non-Sunday male points, and $6 \cdot 7 = 42$ non-Sunday male/female points. This gives 169 outcomes that will not figure in the final conditional probability because no male dog born on Sunday appears. In other words, these points are in S but not in B. Now let us count the sample points with a Sunday male. There are $1 \cdot 6 = 6$ Sunday male/non-Sunday male points, and $6 \cdot 1 = 6$ non-Sunday male/Sunday male points. There are $1 \cdot 7 = 7$ Sunday male/female points, and $7 \cdot 1 = 7$ female/Sunday male points. Finally, there is 1 Sunday male/Sunday male point. This second group of 27 outcomes constitutes event B and exhausts all possibilities for a male born on Sunday to appear. Just to check, $169 + 27 = 196\checkmark$. Note that there are 13 of these outcomes where both dogs are also male. So these are the points in $A \cap B$. Therefore, $P(A|B) = \frac{P(A \cap B)}{P(B)} = \frac{13}{27} < \frac{1}{2}$. This problem and variations on it have become popular "brainteasers" due to the counterintuitive influence of a seemingly irrelevant piece of data – one dog's birthday.

Example 2.15

Here $P(A) = P(C) = 0.96$, $P(B) = 0.98$, and $P(D) = 0.95$. The probability that the entire system is working is $P(A \cap B \cap C \cap D)$. By independence, this is $P(A) \cdot P(B) \cdot P(C) \cdot P(D) = (0.96)(0.98)(0.96)(0.95) = 0.858$.

Example 2.16

The probability for getting the first ace is $\frac{4}{52} = \frac{1}{13}$. With that ace removed from the deck, there are three other aces left in a deck of 51 cards. So the probability of getting another ace on the second draw is then $\frac{3}{51} = \frac{1}{17}$. Then the probability for an ace on the third draw $\frac{2}{50} = \frac{1}{25}$. Finally, the probability that the last draw yields an ace is $\frac{1}{49}$. Multiplying these four probabilities together we get $\frac{1}{270{,}725}$. Let $P(A_i)$ be the probability of getting an ace on the ith draw. Note that draws without replacement are certainly dependent, so we are using an extension of the general multiplication rule here. Our calculation is really a sequence of nested conditionals: $P(A_1 \cap A_2 \cap A_3 \cap A_4) = P(A_4|A_1 \cap A_2 \cap A_3) \cdot P(A_3|A_1 \cap A_2) \cdot P(A_2|A_1) \cdot P(A_1)$. To see what is happening, note that $P(A_1 \cap A_2) = P(A_2|A_1) \cdot P(A_1)$, then the next level is $P(A_3 \cap A_1 \cap A_2) = P(A_3|A_1 \cap A_2) \cdot P(A_1 \cap A_2)$, and so forth. We are compensating as we go along for the effect of cards not being replaced as they are drawn. To check our calculation, we could calculate how many subsets of four cards can be assembled from a 52 card deck. That number is $\binom{52}{4} = 270{,}725$, as we found earlier, so the likelihood of one subset being just the four aces is $\frac{1}{270{,}725}$.

Example 2.17

One pair – two cards match in rank. Each rank has four cards, so there are $\binom{4}{2} = 6$ ways to form a pair. Since there are 13 ranks, the deck contains $6 \cdot 13 = 78$ distinct pairs. The other three cards in the hand cannot be the same rank as the pair (or the hand would become three-of-a-kind or four-of-a-kind), so there are 48 cards left to fill those positions. But our selections cannot be arbitrary because we must avoid picking cards which match in some different rank. So there are 48 ways to pick the third card, 44 ways to pick the fourth, and finally 40 ways to pick the last card in the hand. Note we are allowed to shuffle our hand, so we want combinations, not permutations, so there are $\frac{48 \cdot 44 \cdot 40}{3!} = 14{,}080$ sets of three cards eligible to fill out the hand. That gives a total of $78 \cdot 14{,}080 = 1{,}098{,}240$ ways to assemble a hand with exactly one pair. If we divide this by the total number of possible five card hands, we will have the probability of a one-pair hand. But the number of 5-subsets of a 52-set is $\binom{52}{5} = 2{,}598{,}960$. So the probability of a one-pair hand is $\frac{1{,}098{,}240}{2{,}598{,}960} = 0.4226$.

Two pairs – two cards match in rank and two cards match in different rank. As above, there are 6 ways of forming a pair. Also, there are $\binom{13}{2} = 78$ ways of choosing the ranks of the two pairs. We are left with 44 cards available to fill out the fifth position without matching either paired rank. Overall we have $6 \cdot 6 \cdot 78 \cdot 44 = 123{,}552$ hands containing two pairs, hence the probability of that hand is $\frac{123{,}552}{2{,}598{,}960} = 0.04754$.

Three-of-a-kind – three cards match in rank. For each rank, there are $\binom{4}{3} = 4$ ways to form three-of-a-kind. There are 13 ranks, so the deck contains 52 triples. The remaining two cards must avoid the rank of the triple and also not share the same rank. This leaves 48 choices for the fourth card and 44 for the fifth. Again we adjust the $48 \cdot 44$ permutations of the last two cards by dividing by 2! to get distinct combinations. Thus we have $\frac{52 \cdot 48 \cdot 44}{2} = 54{,}912$ three-of-a-kind hands in the deck. The corresponding probability is $\frac{54{,}912}{2{,}598{,}960} = 0.02113$.

Straight – five cards with ranks in sequence not all same suit. Aces are considered low (1) or high (A), so the straight starting with the lowest rank would be 1–2–3–4–5. The straight 10–J–Q–K–A has the highest starting rank, so apparently straights can start anywhere from 1 through 10 inclusive. Hence there are ten starting ranks for straights and four cards for each rank available. This gives $10 \cdot 4^5 = 10240$ possible straights in the deck. We have counted straights that may have all five cards the same suit. These are called straight flushes and need to be deducted from our total to yield ordinary straights. There are $10 \cdot 4 = 40$ of these, so the adjusted count for ordinary straights is 10,200 giving probability $\frac{10{,}200}{2{,}598{,}960} = 0.003925$. If aces are not considered low, only 90 % of the 10240 straights are available.

Flush – five cards same suit, no rank sequence. There are 13 cards in each suit, and 4 suits, so there are $4 \cdot \binom{13}{5} = 5148$ ways of assembling five cards of the same suit. We have counted the straight flushes again, so deducting them as above gives $5148 - 40 = 5108$ ordinary flushes with probability $\frac{5{,}108}{2{,}598{,}960} = 0.001965$.

Full house – three cards match in rank and two remaining match in rank. We know there are 52 triples in the deck. Any rank matching pairs would necessarily avoid the rank of the triple. So there are 12 ranks available for 6 pairs. Evidently, there are $52 \cdot 72 = 3744$ full houses in the deck. This translates to a probability of $\frac{3{,}744}{2{,}598{,}960} = 0.001441$.

Four-of-a-kind – four cards match in rank. There are 13 ranks, hence 13 four-of-a-kind subsets in the deck. The fifth card can be any of 48, so the number of hands is 624. The corresponding probability is a miniscule $\frac{624}{2{,}598{,}960} = 0.0002401$.

Straight and royal flushes – as we calculated above, there are 40 straight flushes of which 4 are 10–J–Q–K–A of one suit. These are the royal flushes. It follows that the probability of a nonroyal straight flush is $\frac{36}{2{,}598{,}960} = 0.00001385$ and for a royal flush $\frac{4}{2{,}598{,}960} = 0.000001539$.

High card – the probability of having a high card only hand is the complement of the sum of all the above probabilities. That sum is 0.4988, so the probability of having "nothing," that is, a hand with no configuration of any higher value than the highest ranked card in it is 0.5012.

Example 2.18

This is a tough problem to tackle directly, so let us go the easier route. Recast the question as "What is the probability that no two people in the room share a birthday?". If we can find that, we will take the complementary probability. Consider any two people in the room. The probability that one person does not have the same birthday as the other is $\frac{364}{365}$ (no leap years). Consider a third person. Given the condition that the first two do not share a birthday, the probability that the third person does not share a birthday with either of the first two is $\frac{363}{365}$. Continue in this way until we reach the 30th person. The probability that they do not share a birthday with the preceding 29, who are assumed not to share any birthdays among themselves, would be $\frac{336}{365}$. Again we have a chain of conditional probabilities and the probability of no birthday matches among the 30 is $\prod_{i=1}^{29} \frac{365-i}{365} = 0.2937$. So the complementary probability is about 70 % that two people will have the same birthday. Repeating this calculation for 23 people, the probability is almost exactly 50 % that a birthday is shared. Speakers sometimes ask attendees in smaller meetings for their individual birthdays as an "icebreaker." With luck, they may get a match, which amazes people because of its counterintuitive nature.

Example 2.19

Let A be the event that he escaped, D the event that he flew Delta, U United, and S Southwest. We have $P(D) = 0.50$, $P(U) = 0.30$, $P(S) = 0.20$, also $P(A|D) = 0.40$, $P(A|U) = 0.25$, and $P(A|S) = 0.10$. Assembling our formula, $P(A) = (0.50)(0.40) + (0.30)(0.25) + (0.20)(0.10) = 0.20 + 0.075 + 0.02 = 0.295$. So there is a 29.5 % chance he escaped by air.

Example 2.20

Let $P(G)$ be the probability she actually plays the G. Let $P(G_c)$ be the probability that the G occurs in the composition, $P(F_c)$ an F in the composition, and $P(A_c)$ an A in the composition The conditional probability that she plays a G intending a G is $P(G|G)$, a G intending an F is $P(G|F)$, and likewise a G intending an A $P(G|A)$. We have $P(G) = P(G|G) \cdot P(G_c) + P(G|F) \cdot P(F_c) + P(G|A) \cdot P(A_c)$. Plugging in our data, $P(G) = (0.98)(0.024) + (0.01)(0.005) + (0.01)(0.011) = 0.02368 = 2.368\,\%$. The discrepancy is almost not enough to notice.

Example 2.21

Let H be the event that the batter got a hit. Let F be the event that the pitcher threw a fastball, C the curveball, and S the slider. We have $P(F) = 0.50$, $P(C) = 0.30$, $P(S) = 0.10$, $P(H|F) = 0.27$, $P(H|C) = 0.15$, and $P(H|S) = 0.10$. We will get three fractions with the same denominator, so let us compute that first: $P(H|F)P(F) + P(H|C)P(C) + P(H|S)P(S) = (0.27)(0.50) + (0.15)(0.30) + (0.10)(0.20) = 0.20$. Now we have $P(F|H) = \frac{0.135}{0.2} = 67.5\,\%$, $P(C|H) = \frac{0.045}{0.2} = 22.5\,\%$, and $P(S|H) = \frac{0.020}{0.2} = 10\,\%$. So if we had to guess which pitch was thrown, we would say fastball.

Example 2.22

We have already calculated the total probability in that example with a result of 0.295. Now $P(A|D)P(D) = 0.20$, so there is a $\frac{0.20}{0.295} = 67.8\,\%$ chance the robber took Delta. Likewise, $P(A|U)P(U) = 0.075$, so there is $\frac{0.075}{0.295} = 25.4\,\%$ chance the robber took United. Finally, $P(A|S)P(S) = 0.02$, so there is a $\frac{0.20}{0.295} = 6.8\,\%$ chance the robber took Southwest. It looks like a combination of most flights with the worst security made Delta the airline of choice.

Example 2.23

We get to a single bullet four ways: (1) Edmundo hits, all others miss – event E, (2) Federico hits, all others miss – event F, (3) Gumercindo hits first shot, all others miss including his second shot – call this event G_1, (4) Gumercindo hits second shot, all others miss including his first shot – call this event G_2. The priors are all unity (everyone gets a shot, including Gumercindo twice). Let S stand for the event "single bullet in bullseye." The conditionals are $P(S|E) = 0.5 \cdot 0.7 \cdot (0.9)^2 = 0.284$, $P(S|F) = 0.5 \cdot 0.3 \cdot (0.9)^2 = 0.122$, and $P(S|G_1) = P(S|G_2) = 0.5 \cdot 0.7 \cdot 0.9 \cdot 0.1 = 0.032$. The proportion of the total conditional probability leading to the event S attributable to Gumercindo is $\frac{2 \cdot 0.032}{0.284+0.122+2 \cdot 0.032} = \frac{0.064}{0.470} = 13.6\,\%$.

Example 2.24

Superficially, it would seem that the box with the two emeralds can be ignored. The box that has been chosen and known to contain at least one diamond could be either the diamond/diamond box or the diamond/emerald box. It is tempting to say that the probability that another diamond is in the chosen box is simply $\frac{1}{2}$. That would be too easy. This is a Bayes argument, and we need to construct the complete picture. There is a $\frac{1}{3}$ probability that we choose any particular box. The conditional probability that we get a diamond from the diamond/diamond box is, of course, one. The conditional probability that we get a diamond from the emerald/emerald box is zero. Finally, the conditional probability that we get a diamond from the diamond/emerald box is $\frac{1}{2}$. The total probability is then $\frac{1}{3}(1 + 0 + \frac{1}{2}) = \frac{1}{2}$. The only way to get a second diamond from a box is to take the selection path through the diamond/diamond box. Then by Bayes' rule, the probability of finding a second diamond in the originally chosen box is $\frac{\frac{1}{3}}{\frac{1}{2}} = \frac{2}{3}$. This is a famous problem from the 1880s first posed by the French probabilist Joseph Bertrand.

Example 2.25

Each face is assumed equiprobable (fairness) to be on top, so each integer x from 1 to 6 is as well. Therefore the probability that $X = x$ is $\frac{1}{6}$ for $1 \le x \le 6$. The pmf for X is $f(x) = \frac{1}{6}$ if $x = 1, 2, \ldots, 6$.

Example 2.26

$P(\text{even}) = P(2) + P(4) + P(6) = 3 \cdot \frac{1}{6} = \frac{1}{2}$.

Example 2.27

The probability density function is $f(x) = \frac{1}{30}$, where x is in minutes and $0 \le x \le 30$. All of the distribution requirements are met and $P(0 \le X \le 20) = \int_0^{20} \frac{dx}{30} = \frac{2}{3}$.

Example 2.28

We do not need to worry about correcting for the endpoints being included or excluded because the probability of a single point with a nonatomic distribution is zero. The pdf is $f(x) = \frac{1}{60}$ for $x \in [0, 60)$. Note that the zero mark is also the 60 mark. $P(15 < X < 40) + P(50 < X < 60) = \int_{15}^{40} \frac{dx}{60} + \int_{50}^{60} \frac{dx}{60} = \frac{25}{60} + \frac{10}{60} = 58.3\,\%$.

Example 2.29

Here $P(0 \leq n < 400) = P(100 \leq n < 400)$ because in this model there are never any less than 100 customers per day. Now formally $P(100 \leq n < 400) = \int_{100}^{400} f(x)dx$, but we can use the properties of triangles to simplify the calculation. The pdf $f(x)$ rises from zero at $x = 100$ to 0.002 at $x = 600$. By similar triangles, $f(400) = \frac{3}{5} \cdot f(600) = 0.0012$. This is the altitude of a triangle with base 300. The previous integral is just the area of this triangle, or $\frac{1}{2}(300)(0.0012) = 0.18$. So the probability of having fewer than 400 customers on a given day is 18 %.

Example 2.30

We are tasked with finding

$$b\left(6; 12, \frac{1}{2}\right) = \binom{12}{6}\left(\frac{1}{2}\right)^6\left(1 - \frac{1}{2}\right)^6 = \binom{12}{6}\left(\frac{1}{2}\right)^{12}.$$

This works out to $(924)(\frac{1}{4096}) = 0.2256$. So the probability is a little less than a quarter.

Example 2.31

We calculate

$$b(6; 12, 0.55) = \binom{12}{6}(0.55)^6(0.45)^6 = (924)(0.02768)(0.00830) = 0.2123.$$

Pretty close – we probably would not challenge the fairness of the coin. It may seem counterintuitive to expect fewer strings with six heads if the coin is biased towards heads. But imagine a coin that somehow came up heads on every flip. The probability of getting just six out of 12 heads would be zero!

Example 2.32

Yes. The pmf is symmetric since

$$\begin{aligned} b\left(k; n, \frac{1}{2}\right) &= \binom{n}{k}\left(\frac{1}{2}\right)^k\left(1 - \frac{1}{2}\right)^{n-k} = \binom{n}{k}\left(\frac{1}{2}\right)^n = \binom{n}{n-k}\left(\frac{1}{2}\right)^n \\ &= b\left(n-k; n, \frac{1}{2}\right), \end{aligned}$$

for $k = 0, 1, \ldots, \frac{n}{2}$. So $B(\frac{n}{2}; n, \frac{1}{2}) = 1 - B(\frac{n}{2}; n, \frac{1}{2}) = \frac{1}{2}$.

Example 2.33

We have $n = 30$. If it snows k days his monthly income $I = 5000k + 2000(30-k) - 30 \cdot 3000 = 3000k - 30000$. If $k = 10$, he breaks even, and if $k \leq 9$ he loses money. So we want the statistic $P(\text{loss}) = B(9; 30, \frac{1}{3})$. We find $B(9; 30, \frac{1}{3}) = \sum_{k=0}^{9} \binom{30}{k} (\frac{1}{3})^k (\frac{2}{3})^{30-k} \approx 0.4319$. So the lodge has about a 43 % chance of losing money during November.

Example 2.34

Let X be the random variable that counts the number of games won by A against B. For A to win at least two out of three against B, we have $P(X \geq 2) = b(2; 3, 0.6) + b(3; 3, 0.6)$. This reduces to $\binom{3}{2}(0.6)^2(0.4) + \binom{3}{3}(0.6)^3 = 0.648$. On the other hand, let Y be the random variable that counts the number of games won by B against A. For B to win at least three out of five against A. we have $P(Y \geq 3) = \binom{5}{3}(0.4)^3(0.6)^2 + \binom{5}{4}(0.4)^4(0.6) + \binom{5}{5}(0.4)^5 = 0.317$. Evidently, the chance that A wins two out of three against B is roughly double the chance that B wins three out of five against A.

Example 2.35

We evaluate

$$b^*\left(10; 6, \frac{1}{2}\right) = \binom{10-1}{6-1}\left(\frac{1}{2}\right)^6 \left(1 - \frac{1}{2}\right)^{10-6} = (126) \cdot \left(\frac{1}{2}\right)^{10} = 12.3\,\%.$$

Example 2.36

There are 36 possible outcomes in rolling two dice together once. Only one of these outcomes is snake eyes, so the probability of a single occurrence is 2.78 %. The statistic we want is

$$b^*(7; 2, 0.0278) = \binom{6}{1}(0.0278)^2(0.9722)^5 = 0.04\,\%.$$

Example 2.37

Since $p = 0.30$, $g(1; 0.30) = 0.30$. Then

$$g(2; 0.30) = (0.30)(0.70) = 0.21 \text{ and } g(3, 0.30) = (0.30)(0.70)^2 = 0.147.$$

This is a good model for illustrating the geometric distribution, but not a very good one for actual game performance, as there are many confounding factors.

Example 2.38

We want $g(10; \frac{1}{2}) = 2^{-10} = 0.000977$. A rare occurrence.

Example 2.39

$|T| = 15$, $|S| = 4$, $n = 8$, and $x = 4$. So

$$h(4; 8, 4, 15) = \frac{\binom{4}{4}\binom{11}{4}}{\binom{15}{8}} = \frac{(1)(330)}{(6435)} = 5.13\,\%.$$

Example 2.40

Three patients are tested. A "success" is not having the disorder, so $|T| = 30$, $|S| = 25$, $n = 3$, and $x = 3$. Then

$$h(3; 3, 25, 30) = \frac{\binom{25}{3}\binom{5}{0}}{\binom{30}{3}} = \frac{(2300)(1)}{(4060)} = 56.7\,\%.$$

Example 2.41

We express the frequency as 0.04 per year, then $\lambda = (0.04)(20) = 0.8$. Plugging into the formula, $p(2; 0.8) = \frac{(0.8)^2 e^{-0.8}}{2!} = 0.144$.

Example 2.42

Let X be the number of off-spec bolts. We want $P(X \geq 1) = 1 - P(X = 0)$. The frequency of bad bolts is 0.002 so $\lambda = 0.002 \cdot 100 = 0.2$. Then $P(X = 0) = p(0; 0.2) = \frac{(0.2)^0 e^{-0.2}}{0!} = e^{-0.2} = 0.819$ and $P(X \geq 1) = 0.181$.

Example 2.43

Here $\lambda = (0.0002)(10000) = 2$. So $p(1; 2) = \frac{(2)^1 e^{-2}}{1!} = 0.271$.

Example 2.44

The defect rate is 0.05 per square foot, so $\lambda = (0.05)(1000) = 5$. We want $P(X < 3) = p(0; 5) + p(1; 5) + p(2; 5)$. Hence $P(X < 3) = e^{-5}(1 + 5 + \frac{25}{2}) = (0.00673)(18.5) = 0.125$. Does not look promising.

Example 2.45

Using the binomial distribution, we want

$$b(10; 80000, 0.00002) = \binom{80000}{10}(0.00002)^{10}(0.99998)^{79990}.$$

Using Poisson, $\lambda = (.00002)(80000) = 1.6$, then $p(10; 1.6) = \frac{(1.6)^{10} e^{-1.6}}{10!} = 0.00061\,\%$.

Example 2.46

As a binomial problem, we are facing the evaluation of

$$b(50; 100, 0.5) = \binom{100}{50}(0.5)^{100}.$$

This is not as bad as in Example 2.45, but it is still a nasty computation. The Poisson approximation does not apply, so let us use the Laplace approximation with

$$\begin{aligned} b(50; 100, 0.5) &\approx \frac{1}{\sqrt{2\pi(100)(0.5)^2}} \exp\left[-\frac{(50-50)^2}{2(100)(0.5)^2}\right] = \frac{1}{\sqrt{2\pi(100)(0.5)^2}} \\ &= \frac{1}{\sqrt{157.08}} = 7.98\,\%. \end{aligned}$$

Note that $p = 0.5$ maximizes the accuracy of the Laplace formula in estimating the binomial.

Example 2.47

Thirty calls per ten hours is three per hour, so $\nu = 3$. We want $P(T > 1) = 1 - \int_0^1 3e^{-3t}\,dt = 1 - 0.950 = 0.05$. This answers both questions.

Example 2.48

If we believe the polls, this seems like a very unlikely outcome. Evidently $n = 2000$, so

$$\begin{aligned} m(1000, 400.600; 2000, 0.5, 0.3, 0.2) &= \binom{2000}{1000, 400, 600}(0.5)^{1000}(0.3)^{400}(0.2)^{600} \\ &= 2.78 \times 10^{-39}. \end{aligned}$$

Confirmed.

Example 2.49

We want

$$m(4, 1, 5; 10, 0.5, 0.2, 0.3) = \binom{10}{4, 1, 5}(0.5)^4(0.2)(0.3)^5 = 3.83.$$

Example 2.50

We want

$$\widehat{m}(3, 4, 2, 3; 12, 5, 7, 3, 9) = \frac{\binom{5}{3}\binom{7}{4}\binom{3}{2}\binom{9}{3}}{\binom{24}{12}} = 3.26.$$

Example 2.51

Each person has a 20-minute window to arrive. Let John's arrival time be the random variable X, and Mary's Y. Use a scale in minutes and set the meeting time (1 p. m.) at the origin for each person's arrival interval. So John can arrive any time between $x = -10$ and $+10$. Mary can arrive any time between $y = -15$ and $+5$. Every joint arrival time occurs in the 20×20 square defined by these intervals. Since the square has area $20 \cdot 20 = 400$ units, we can compute $k = \frac{1}{400}$. Also, the line $y = x$ corresponds to both arrival times being the same. Points (x,y) where $x \leq y$ (John earlier than Mary) are to the left of this line. We do not even have to do an integral in this case. The zone where John arrives no later than Mary is a triangle of area $\frac{15 \cdot 15}{2} = 112.5$ units. So the probability that John gets to the meeting no later than Mary is $P(X \leq Y) = \frac{112.5}{400} = 28.1\,\%$.

Example 2.52

1. Mark two places on the stick, then break. The result is independent of scale due to geometric similarity, so we can assume a stick of length 1. Let X and Y be random variables representing the locations of the two respective breaking points from one end of the stick. Note that X and Y are both uniformly continuously distributed with their joint density $f(x,y) = 1$. For definiteness, we assume $0 < x < y < 1$. It is clear that the three pieces have respective lengths $x, y - x$, and $1 - y$. The condition for assembly into a triangle is that any two sides are at least as long as the third (a degenerate triangle is allowed). This yields three inequations: $x + (y - x) \geq 1 - y$, $x + (1 - y) \geq y - x$, and $(y - x) + (1 - y) \geq x$. Simplifying these, we have $y \geq \frac{1}{2}$, $y - x \leq \frac{1}{2}$, and $x \leq \frac{1}{2}$. Admissible outcomes are points in the unit square above the line $y = x$, since we have assumed $y \geq x$. Outcomes corresponding to triangles, i. e., those satisfying the three constraints, are points below $x = \frac{1}{2}$, above $y = \frac{1}{2}$, and below the line $y = x + \frac{1}{2}$. These triangle-enabling points are easily seen to comprise $\frac{1}{4}$ of the area of all admissible points, and since the joint density is uniform, the probabilities are proportional to area. No integral is required, and we conclude the probability of constructing a triangle from the random pieces is $P(T) = \frac{1}{4}$.
2. Again assume a stick of length 1 and let X be the random variable representing the first break point. Note X is uniformly distributed with range $0 \leq x \leq 1$. So go ahead and break the stick at a random point between 0 and 1. We can assume that the leftmost piece is the longer piece. Otherwise, walk around to the other side of the stick and start over. We see that the longer piece has length $x > 0$ (it was a proper break, after all). Now break that piece at a random point ξ between 0 and x. Call $y = \frac{\xi}{x}$. Note that Y measures the random fraction of the longer piece at which the second break is made, and hence is also uniformly distributed between 0 and 1. We now have three pieces with lengths xy, $x - xy$, and $1 - x$. As in Solution #1, the condition that they may be reassembled into a triangle is that each be less than $\frac{1}{2}$. Now the region in the plane with points that admit a triangle is bounded by $y \leq \frac{1}{2x}$,

$y \geq 1 - \frac{1}{2x}$, and $x \geq \frac{1}{2}$. Since we have agreed to always pick the longer piece for the second break, the sample space is $[\frac{1}{2}, 1] \times [0, 1]$. Hence

$$P(T) = \left(\frac{1}{1/2}\right) \int_{\frac{1}{2}}^{1} \left[\left(\frac{1}{2x}\right) - \left(1 - \frac{1}{2x}\right)\right] dx = 2 \int_{\frac{1}{2}}^{1} \left(\frac{1}{x} - 1\right) dx = 2[\ln x - x]_{\frac{1}{2}}^{1}$$

$$= 2(0.693 - 0.500) = 38.6\,\%.$$

3. Break the stick. Now pick either of the pieces with equal likelihood and break it again. Let S be the event "pick smaller piece" and L be the event "pick larger piece." Then using the preceding result, $P(T) = P(T|S)P(S) + P(T|L)P(L) = (0.00)(0.50) + (0.386)(0.50) = 19.3\,\%$.

Example 2.53

We can assume that the position of the center of the toothpick is a random variable X distributed uniformly in the interval $[0, \frac{L}{2}]$, where a seam has coordinate zero. No matter where the toothpick lands, its center is going to be no further than $\frac{L}{2}$ from some seam. Also, we assume that the angle the toothpick makes with the closest seam is a random variable Θ distributed uniformly in the interval $[0, \pi]$. The toothpick will cross a seam if the projection of half its length perpendicular to the seam is greater than or equal to the distance from its center to the seam. Symbolically, this means $\frac{L}{2} \sin\theta \geq x$ is the condition ensuring a cross. If the center lands close to a seam, the range of angles where the toothpick crosses is relatively large. If the center lands far from the seam, the range of angles allowing a cross will be small. The sample space for this experiment is a rectangle $[0, \pi] \times [0, \frac{L}{2}]$. So the total area of the sample space in the θx plane is $\frac{\pi L}{2}$ and the implied joint uniform distribution is $f(\theta, x) = \frac{2}{\pi L}$. We need to integrate this distribution over the portion of the sample space where $\frac{L}{2} \sin\theta \geq x$. This would be the area under the curve $x = \frac{L}{2} \sin\theta$, so $P(\text{cross}) = \int_0^{\pi} (\frac{L}{2})(\frac{2}{\pi L}) \sin\theta d\theta = \frac{2}{\pi} \approx 63.7\,\%$. The conditions of this example can be generalized to allow both arbitrary toothpick length and seam spacing.

This example has an interesting history. It is known as Buffon's needle problem and was first proposed in 1777 by the Comte de Buffon. Simply inverting the final expression, we get $\pi \approx \frac{2}{P(\text{cross})}$, and this invites an experimental estimate of π by repetitive needle dropping. In fact, the Italian mathematician Mario Lazzarini published the results of an alleged approximation in 1901. He obtained $\pi \approx 3.141592$ after 3408 needle tosses. Recently, a case has been made that this was too good to be true and perhaps it was all a hoax – but in retrospect a successful one.

Example 2.54

1)

$$\int_0^1\int_0^1 \frac{1}{4}(3x+5y)dydx = \frac{1}{4}\int_0^1\left[3xy+\frac{5}{2}y^2\right]_0^1$$
$$= \frac{1}{4}\int_0^1\left(3x+\frac{5}{2}\right)dx = \frac{1}{4}\left[\frac{3}{2}x^2+\frac{5}{2}x\right]_0^1 = 1,$$

so the total probability is correct and the density is clearly nonnegative;

2)

$$h(x) = \int_0^1 \frac{1}{4}(3x+5y)dy = \frac{1}{4}\left(3x+\frac{5}{2}\right);$$

3)

$$g(y) = \int_0^1 \frac{1}{4}(3x+5y)dx = \frac{1}{4}\left(\frac{5}{2}+5y\right);$$

4)

$$\hat{h}(x|y) = \frac{f(x,y)}{g(y)} = \frac{1}{4}(3x+5y)\Big/\frac{1}{4}\left(\frac{5}{2}+5y\right), \quad \text{so } \hat{h}\left(x|\frac{1}{2}\right) = \frac{3x+\frac{5}{2}}{5} = \frac{3}{5}x+\frac{1}{2};$$

5)

$$h(x)g(y) = \left(\frac{1}{4}\right)\left(3x+\frac{5}{2}\right)\left(\frac{1}{4}\right)\left(\frac{5}{2}+5y\right) = \frac{5}{4}(12xy+6x+10y+5) \neq \frac{1}{4}(3x+5y),$$

so *X and Y are not independent.*

Example 2.55

The probability mass function for X_k is $f_k(x_k) = \begin{cases} \frac{1}{6} \text{ if } x_k = 1, \\ \frac{5}{6} \text{ if } x_k = 0. \end{cases}$ For n independent rolls, the overall mass function is $\prod_{k=1}^{n} f_k(x_k)$. So the probability of having a string of n sixes would be $\frac{1}{6^n}$ and the probability of getting no sixes in n rolls would be $(\frac{5}{6})^n$.

Example 2.56

1)

$$\sum_x\sum_y f(x,y) = \frac{1}{C}[1+4+2+8+3+12] = 1$$

implies $C = 30$.

2)

$$h(x) = \sum_y f(x,y) = \frac{1}{30}[x+4x] = \frac{x}{6},$$
$$g(y) = \sum_x f(x,y) = \frac{1}{30}[y^2+2y^2+3y^2] = \frac{y^2}{5}.$$

3)
$$\widehat{g}(y|x) = \frac{f(x,y)}{h(x)} = \frac{1}{30}\frac{xy^2}{\frac{x}{6}} = \frac{y^2}{5}.$$

4) We have two clues: $\widehat{g}(y|x) = g(y)$ and $f(x,y) = \frac{xy^2}{30} = \frac{x}{6} \cdot \frac{y^2}{5} = h(x)g(y)$. So either the lack of dependence on x of the conditional distribution $\widehat{g}(y|x)$ or the product of marginals yielding the joint distribution is enough to confirm the independence of X and Y.

3 Statistical methods

There are lies, damned lies, and statistics

– Mark Twain, whose real name, ironically, was Samuel Langhorne Clemens.

It could be argued that statistics is what happens when probability theory encounters data sets. Statistics is popularly defined as the science of data collection, analysis, interpretation, and presentation. Our interest in this chapter is going to be on the basic mathematics of analysis and interpretation. We are going to leave the comprehensive subjects of collection, including design of experiments, and presentation to the specialized texts, some of which are referenced at the end of the chapter. The more mathematical activities of statistics center around treating data elements as random variables, which are then subject to all the methods of the preceding chapter, as well as some new ones that will allow us to characterize data sets as a whole. This is the area of descriptive statistics. We will also deal with the relationship of a data sample (small subset) to the population (larger set) from which it was drawn and develop methods to assess the "representativeness" of the sample. This will lead us into the area of inferential or inductive statistics, where we quantify the reasonableness of assertions made about the behavior of data. The centerpiece of this chapter will be the normal distribution, which asserts itself in all manner of applications. There are several other probability distributions deferred from Chapter 2 to this chapter that are relevant to the parameters of normally distributed populations, and we will illustrate their application.

A case can be made that mathematical statistics, as distinct from classical probability theory, began with Abraham deMoivre's discovery of the equation for the normal distribution, or bell curve because of the shape of its graph. It became important by application to the theory of measurement errors developed by C. F. Gauss and astronomical calculations by Pierre-Simon Laplace. The first statistician who fully appreciated the fact that statistical methods applicable to one research problem could be generalized and profitably applied to others was Adolph Quetelet, a Belgian social science researcher in the early nineteenth century. The locus of the ongoing development of mathematical statistics turned towards England later in that century with Sir Francis Galton and his younger collaborator Karl Pearson. Galton developed percentiles and worked out the theory of regression in the 1880s. Pearson formalized correlation theory and initiated much of the modern theory of sampling. He was also instrumental in establishing the first academic statistics department at University College London in 1911. Both Galton and Pearson were biostatisticians and their work touched on eugenics, which in their era did not carry the negative connotation it does today. Perhaps Galton's involvement with research in heredity was unavoidable – he was Charles Darwin's cousin.

One of the probability distributions we will encounter is the so-called t-distribution, which is useful in interpreting samples which are smaller than appropriate for the normal distribution and for which the overall population standard deviation is unknown.

https://doi.org/10.1515/9783111506654-004

These terms will be explained shortly. Another Englishman of this era, W. S. Gosset, developed the mathematics behind the *t*-distribution while he was an employee of the Guiness brewery. Gosset's work improved factory and farm level decision-making in manufacturing beer. Guiness considered his method of small sample analysis to be proprietary, so they forbade him to publish his mathematical work under his own name. He responded by publishing as a "Student," so the distribution is often referred to as Student's *t*-distribution. Yet another English statistician, Sir Ronald Fisher, is widely regarded as the "Father of Modern Statistics." Fisher made many contributions including ANOVA (ANalysis Of Variance), experimental design, hypothesis testing, and the F(isher)-test. The Polish mathematician Jerzy Neyman is credited with inventing the theory of confidence intervals and being a cofounder of statistical hypothesis testing. Although statistics is a relatively new branch of mathematics, its broad application to science and engineering has secured its status as a major discipline, especially in the modern era of "big data."

3.1 Basic definitions

The basic raw material of statistics is a numerical or sometimes qualitative data element we call an **observation**. Numerical data could be the weight of a lab rat, and qualitative data might be whether a flipped coin landed "heads or tails." Someone typically writes down a number or records a classification that characterizes some **outcome** of an experiment. All of these observations taken together constitute a **population**. The number of observations in the population is its **size**. The size can be any natural number, assumed large enough to support a meaningful theory. A very large size can, for most practical purposes, be considered infinite. Recall from Section 2.5 that a **random variable** X is a function from a discrete or continuous probability space to the real numbers, $X : \langle S, \mathcal{E}, P \rangle \to \mathbb{R}$. We remarked in Chapter 2 that the point of a random variable was to convert a probabilistic outcome in a physical space (flipping coins or rolling dice or measuring fish) to a bona fide numerical outcome with the same probability as the physical event. Here we are starting with numericalized data that already belong to an **observation space**, which is a subset of $\mathbb{R}$. A **sample** is a nonvoid subset of a population. In principle, it could be the entire population, but the whole point of sampling is to try to characterize the population based on a much smaller piece of it as a practical matter, due to the cost or infeasibility of examining the entire population. Populations and samples will both be referred to generically as **data sets**.

We will have **parameters** that are numbers (not the observations) that in some way summarize or characterize the nature of a population. A **random sample** from a population is a selection of n observations which have the same probability of being chosen as any other n observations. Analogous to parameters, a sample has **statistics** that characterize it. The important difference is that a population is a static collection of observations, so the parameters are fixed, but the statistics for a sample are dependent on the data observed in the exact sample chosen, so statistics themselves are random variables. By convention, if the random variable X is a statistic, the formula that defines it will be written with capital letters, but if we are computing the statistic for a specific sample we will use lower case x. So populations have parameters, but samples have statistics. The connection is that statistics can be used to estimate parameters, and vice versa. One of our goals is to quantify the fidelity of this relationship.

3.1.1 Mathematical expectation

The expected value operator is fundamental in probability and statistics. We can regard the elements of a data set as values of a random variable X with some known discrete probability distribution. The **expected value** of X is defined to be $E[X] = \sum_x x \cdot f(x)$, where x is a particular value of the random variable and $P(X = x) = f(x)$ and the sum is taken over all x. In words, it is the weighted sum of all the numerical data points, where the weights are the probabilities of those data points occurring. There is a calculus-based analog of the expected value operator that can handle continuous random variables, and in the few cases where this comes up we will just accept a known result, since we are trying to minimize our explicit use of calculus. The concept of mathematical expectation will play such a fundamental role in what we do that we immediately establish the basic properties of $E[X]$.

Table 3.1: Basic properties of the expected value operator. X, Y are random variables, $f(x) = P(X = x)$, $g(X)$ is function of X, $g_i(X)$ for $1 \leq i \leq n$.

1) $E[aX + b] = aE[X] + b$, for constants a, b
2) $E[g(X)] = \sum_x g(x) \cdot f(x)$ for $g(X)$ as function of X
3) $E[(X - E[X])^2] = E[X^2] - (E[X])^2$
4) $E[\sum_{i=1}^n c_i g_i(X)] = \sum_{i=1}^n c_i E[g_i(X)]$ for constants c_i
5) $E[X \pm Y] = E[X] \pm E[Y]$
6) $E[XY] = E[X]E[Y]$ if X, Y are statistically independent

Statistical independence means the joint probability of X and Y is the product of their marginal probabilities (see Section 2.3).

Example 3.1. Show that $E[aX + b] = aE[X] + b$, for constants a, b.

Example 3.2. Show that $E[g(X)] = \sum_x g(x) \cdot f(x)$ for $g(X)$ as a function of X.

Example 3.3. Show that $E[(X - E[X])^2] = E[X^2] - (E[X])^2$.

Example 3.4. Show that $E[\sum_x c_i g_i(X)] = \sum_{i=1}^{n} c_i E[g_i(X)]$ for constants c_k.

Example 3.5. Show that $E[X \pm Y] = E[X] \pm E[Y]$.

Example 3.6. Show that if X, Y are statistically independent, then $E[XY] = E[X]E[Y]$.

3.1.2 Measures of central tendency

Given the elements of a numerical data set, we can define a function that identifies a reasonable notion of the "center" of the data. Any function that does this is a **measure of central tendency.** This measure **locates** the data set on $\mathbb{R}$. There are three such measures in common use: 1) the mean or average, 2) the median, and 3) the mode. No assumption is made that the data points, symbolized by x_k, are all distinct.

1) The most important measure of central tendency is the **mean** μ_X of a population of N data points $\{x_1, x_2, \ldots, x_N\}$. In terms of the expected value operator, $\mu_X = E[X] = \sum_{k=1}^{N} x_k \cdot f(x_k)$, where each data point is equiprobable and hence $P(X = x_k) = \frac{1}{N}$. The **population mean** can then be written more simply as $\mu_X = \sum_{k=1}^{N} x_k \cdot \frac{1}{N} = \frac{\sum_{k=1}^{N} x_k}{N}$. For a sample of size n drawn from this population, $\overline{x} = \frac{\sum_{k=1}^{n} x_k}{n}$ is a particular value of the **sample mean** $\overline{X}$, which is a random variable since it depends on the sample chosen. As noted earlier, the population mean μ_X is a parameter, but the sample mean $\overline{X}$ is a statistic.
2) The population **median** $\tilde{\mu}$ is the middle value after the data set $\{x_1, x_2, \ldots, x_N\}$ has been ordered to be either increasing or decreasing. If N is odd then one of the data points is exactly in the middle. If N is even, we take the arithmetic mean of the two middle values (so it may not be realized as a specific x_k). A sample median $\tilde{x}$ is computed the same way.
3) The population **mode** $\hat{\mu}$ is the value of the population data set that has the highest frequency of appearance in the set. It may not exist, or if it does, it may not be unique. If there are two modes the data are called **bimodal** and if there are more **multimodal**. The sample mode $\hat{x}$ is computed similarly.

Each of these measures of central tendency has strengths and weaknesses. The mean is fairly easy to compute but it is subject to being distorted by one or more **outliers** – extreme data points that dominate the calculation and misrepresent the reasonable center of the data. This often happens with real estate listings in a neighborhood, where one or two houses are priced very much higher than the rest and it renders the average meaningless for gauging the affordability of the neighborhood. The median is very simple to compute, but it is not as stable from sample to sample, so if we use it to estimate the median of a population, we may easily mislead ourselves. However, it is a better estimator of neighborhood affordability than the mean. The mode requires no calculation, but it is relatively useless for small samples or even large samples that are multimodal. For larger unimodal samples, it begins to better reflect central tendency. We will drop the subscript from μ_X, for example, when it is clear from context which data set we are talking about.

Example 3.7. Calculate the mean, median, and mode of the following set of battery voltage readings: $\{5.0, 5.4, 5.2, 5.5, 5.8, 5.5, 4.9, 5.3\}$.

Example 3.8. What is the average roll of one fair die?

Example 3.9. A student needs an average of 80 for the semester to earn a "B." She has taken four equally weighted exams with the following scores: 76,71,79,84. What does she need to score on the last exam to get a "B"?

In the context of gambling, the **expected payoff** from a game of chance is the total value of the "pot" (potential winnings) times the probability of winning it. So if you play a game where you flip a fair coin and if it comes up heads you receive five dollars, the expected payoff is $(0.5)(5.00) = 2.50$. No one is going to let you play this game for free, so what is a fair price to play once? If the gambling house charges you x per game, over the long run, say n games, you will pay xn to play, but you will win $2.50n$. The house profit is $xn - 2.50n = n(x - 2.50)$. So if $x > 2.50$, you lose in the long run, and if $x < 2.50$, the house loses. The latter outcome is unlikely. If $x = 2.50$, you and the house both break even. A game with this feature is called **fair**. Realistically, games like this (slot machines, roulette, 21, and so forth) are set up to give the house a slight profit – enough to maintain the casino infrastructure and pay the investors, but not so much that gamblers can sense it is pointless to play despite an occasional adrenaline rush after an infrequent win. If the pot varies from game to game along with the probability of winning, we may still calculate expected payoff X as $E[X] = \sum_{k=1}^{N} x_k \cdot f(x_k)$, where there are N games and the pot of the kth game is x_k with probability of winning that game $f(x_k)$.

Example 3.10. Here is an innocent looking game. You get to flip a fair penny. If it comes up heads for the first time on the kth flip you get 2^k dollars. How much is a fair price to play?

3.1.3 Measures of dispersion

Dispersion or **variability** or **spread** of data is properly measured relative to its central tendency. Consider the following two data sets: $S_1 = \{10, 12, 15, 17, 16, 14, 17, 18, 12, 9\}$ with $\overline{x} = \frac{140}{10} = 14$, and (ii) $S_2 = \{3, 19, 25, 7, 14, 22, 11, 13, 8, 18\}$ with $\overline{x} = \frac{140}{10} = 14$. They both have the same average, but the data in S_2 seem more extreme and to vary more than the data in S_1. We would like to quantify this fact with a measure of dispersion. There are several statistics that capture this effect. The **range** of S_1 is $\max(x \in S_1) - \min(x \in S_1) = 18 - 9 = 9$. In contrast, the range of $S_2 = 25 - 3 = 22$. It is clear that S_2 is more variable than S_1, so we get the correct message from the data, but it is a weak message that actually ignores any central tendency. Only two data points out of ten define the range for each data set. All the rest of the data is invisible. We could change S_1 to $\{10, 12, 15, 1, 32, 14, 17, 18, 12, 9\}$ by altering two data points in a way that did not change the mean but its range would now be 31. So the range formula is a blunt instrument and useful only as a rough indicator of dispersion. If the data points represent some critical measurement in a process over which we are trying to exert control, then limits on the range may be enough to guide acceptance or rejection.

A much better measure of dispersion calculates the absolute distances of data points from the mean, adds them up, and then averages those absolute distances. This statistic is the **mean deviation**. It would be misleading to add up the signed deviations as there could be considerable offsetting cancellations. The formula for mean deviation of the random variable X is $\overline{d}_X = E(|X - \mu|) = \sum_{k=1}^{N} |x_k - \mu| \cdot f(x_k)$, where as usual $f(x_k) = P(X = x_k)$. In the equiprobable case, $\overline{d}_X = \frac{\sum_{k=1}^{N} |x_k - \mu|}{N}$. Regrettably, working algebraically with absolute values is difficult enough that mean deviation is not a popular statistic for measuring dispersion. What *is* popular is a measure that expresses the square of an absolute distance as simply the square of a signed distance. Using $(x_k - \overline{x})^2$ has the double virtue of avoiding offsetting cancellations and avoiding having to work with absolute values. In a real sense, the absolute value of a deviation is already accounted for in the square of that deviation, and squares are tolerated more graciously by algebra.

The **variance** of the population $\{x_1, x_2, \ldots, x_N\}$ is defined as $\sigma_X^2 = E[(X - \mu)^2] = \sum_{k=1}^{N} (x_k - \mu)^2 \cdot f(x_k)$, where the population mean is μ. The **standard deviation** σ of the population is the square root of the variance. In the equiprobable case, $\sigma_X^2 = \frac{\sum_{k=1}^{N} (x_k - \mu)^2}{N}$. Note that if the data set consists of numbers that have some dimensionality (kg, m, ft, etc.), the standard deviation conveniently has the same dimension.

There is a slight issue in computing the variance of a sample, which is the statistic denoted by S^2. Unlike σ^2, which is computed for a fixed population data set, we would expect s^2 to be different for different samples, so it is a random variable. It turns out that if we compute $s^2 = \frac{\sum_{k=1}^{n} (x_k - \mu)^2}{n}$, we have an unbiased estimator of σ^2. This means that if we averaged s^2 over all possible samples drawn from the population, that average would be σ^2. This is important if we do not really know σ^2 exactly and we are using s^2 to estimate it. Now it may be that we do not have an exact value for μ either. So if we are going to have any hope of estimating σ^2, we will have to rely on the population variance formula with $\overline{x}$ substituted for μ. This builds in an underestimate or **bias** for σ^2. So to compensate, we use $s^2 = \frac{\sum_{k=1}^{n} (x_k - \overline{x})^2}{n-1}$. This is not, as the engineers say, a "fudge factor." The compensation is mathematically justified and renders s^2 with this adjusted formula an unbiased estimator for σ^2. We will try to remove the mystery of this adjustment in Example 3.24 below. As a typographical convenience, we may write σ_X^2 as $V(X)$, especially when the subscripts become microscopic. Here are some convenient rules for working with variance.

Table 3.2: Basic properties of the variance operator. X, Y are random variables, a, b are constants.

1) $V[aX + b] = a^2V[X]$
2) $V[X] = E[X^2] - (E[X])^2$
3) $V[X + Y] = V[X] + V[Y]$

Example 3.11. What is the variance and standard deviation of the possible rolls of a fair (equiprobable) die?

Example 3.12. What is the variance and standard deviation of the possible rolls of two fair dice?

Example 3.13. A game show has a wheel-of-fortune with the numbers 1 through 29 equally likely to come up. What is the variance and standard deviation of the possible observations?

Example 3.14. Show that an alternative formula for s^2 is

$$\frac{n\sum_{k=1}^{n} x_k^2 - (\sum_{k=1}^{n} x_k)^2}{n(n-1)}.$$

Example 3.15. Show that $\sigma^2 = E[(X - \mu)^2] = E[X^2] - \mu^2$. This makes calculations a little simpler.

Example 3.16. Show that if $g(X) = X + a$, where a is a constant, then the variances of $g(X)$ and X are the same. So adding the same constant to every data point does not change the population variance.

Example 3.17. Show that if $g(X) = bX$, where b is a constant, then $\sigma_{bX}^2 = b^2\sigma_X^2$.

Example 3.18. Given two statistically independent random variables X and Y, show that $\sigma_{X+Y}^2 = \sigma_X^2 + \sigma_Y^2$.

Example 3.19. Find the variance of the possible rolls of two fair dice by using Examples 3.11 and 3.18.

Example 3.20. The nineteenth century Russian mathematician Pafnuty Chebyshev discovered a theorem that gives a rough, but occasionally useful, estimate of what fraction of observations lie within k standard deviations of the mean. Specifically, it says that at a minimum the fraction $(1 - \frac{1}{k^2})$ of total observations lie within k standard deviations of the mean. The unusual feature of this theorem is that the distribution of the observations does not need to be known. If we roll a fair die ten times (or ten all at once), what is the chance that the total of the upfaces of the dice would exceed 46?

Example 3.21. Another result in the spirit of Chebyshev's theorem is Markov's theorem, first proved by another Russian mathematician Andrei Markov. It states that if X is a random variable that can only take on nonnegative values and has mean μ, then $P(X \geq a) \leq \frac{\mu}{a}$ for $a > 0$. The dice roll in Example 3.20 satisfies the condition on X. What is the probability that a ten dice roll exceeds 50?

Example 3.22. Repeat Example 3.20 for only two dice. What fraction of rolls should not exceed 2σ?

Example 3.23. Prove Chebyshev's theorem.

Example 3.24. Explain why the sample variance is computed as $s^2 = \frac{\sum_{k=1}^{n}(x_k-\mu)^2}{n}$ when the population mean μ is known, but as

$$s^2 = \frac{\sum_{k=1}^{n}(x_k - \overline{x})^2}{n-1}$$

when the sample mean $\overline{x}$ is used because μ is unavailable.

3.1.4 Grouped data

In Section 2.6 we introduced discrete probability distributions and showed how we could use a distribution to find the probability of a specific value or range of values of the random variable it represented. In this section we want to turn that problem around and go from a data set to an approximation of the discrete distribution that describes it. The obvious way to approach this is to order the numerical data from smallest to largest and partition the overall range interval into equally-sized subintervals of $\mathbb{R}$ called **classes** or **bins**. Every data point will be in some subinterval and the size of the subintervals, or **class width**, can be adjusted for our purposes. Class width equals the **upper class boundary** minus the **lower class boundary**. We do not want any data points falling exactly on a class boundary so, to ensure that this never happens, we express the class boundaries to one more decimal than the data. So if successive data points, recorded to the nearest integer, are 89 and 90, and we want to separate them into two different classes, we would set the class boundary at 89.5. The **upper class limit** is the largest data point that does not exceed its upper class boundary. Continuing with the example, 89 would be largest data point less than the upper class boundary at 89.5, and therefore would be the upper class limit of the class to which it belongs. Similarly, the **lower class limit** is the smallest data point that still exceeds its lower class boundary. So 90 would be the lower class limit relative to the lower class boundary at

89.5. The **class interval** is the upper limit minus the lower limit. Obviously, the upper class boundary of one class is also the lower class boundary of the next greater class. It is customary to call the midpoint between two adjacent class boundaries the **mark** of the class. Finally, the whole point of partitioning is to count the number of data points that fall into various classes. The number of data points in a given class will be the **frequency** of that class, and we will use this to construct a distribution.

Example 3.25. Here is a raw listing of 40 times to the nearest tenth of a second recorded by prospective football players for the 40-yard dash:

4.5	4.3	4.9	5.1	4.4	4.5	4.6	4.6
5.0	4.8	4.7	5.1	4.4	5.3	5.4	4.9
4.5	4.9	4.8	5.0	5.3	5.2	4.5	4.6
4.3	4.4	4.5	4.8	4.7	4.5	5.0	4.8
4.4	4.6	4.4	4.5	4.9	5.0	4.3	4.7

Prepare a table of class frequencies showing class intervals, class boundaries, marks, and frequencies.

The frequency data presented in Example 3.25 can be given more visual impact by constructing a **histogram** or **frequency polygon**. A histogram is a bar chart where the width of each bar is the width of a class boundary (so there is no open space between bars) and the height of each bar is the frequency recorded for that class, either as a raw number or percentage. A frequency polygon is an xy-plot where the class marks are plotted in the x-direction and the corresponding frequencies are plotted in the y-direction. Adjacent points on the graph are joined to create the polygonal profile. Examples of each follow.

Another style of presenting frequency data is to construct a cumulative frequency polygon, where the class marks are plotted in the x-direction and the cumulative frequency is plotted in the y-direction. The overall height of the polygon corresponds to the total number of data points. Because of the characteristic shape of a cumulative frequency curve, it is often called an **ogive**. The histogram and frequency polygons for the data in Example 3.25 appear in the figures below.

With the frequency information in Example 3.25, we can estimate the mean of the data. We say estimate because by grouping our data we have lost granular information on the specific data elements. Here is our workaround. We will assume that the mean of the data points in a particular class is the class mark. What would belie this assumption? If for some reason we set up the classes in such a way that the data points clustered around the class boundaries, then the mark might not be representative. But with enough data this danger subsides. So let x_k be the class mark of the kth class out of n classes and ω_k be the corresponding class frequency. Then the **grouped mean** of the sample is $\overline{x} = \frac{\sum_{k=1}^{n} x_k \omega_k}{N}$. This is just the frequency-weighted average of the class marks. Note that $N = \sum_{k=1}^{n} \omega_k$ is the total sample size.

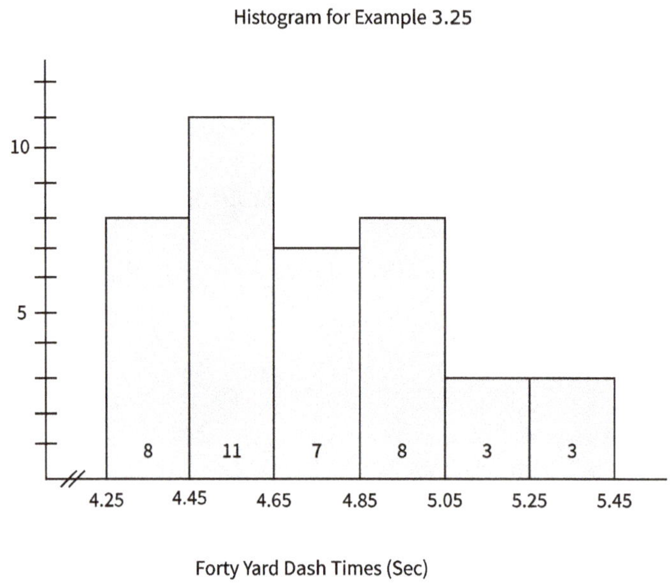

Figure 3.1: A frequency histogram (bar chart) based on the data given in Example 3.25.

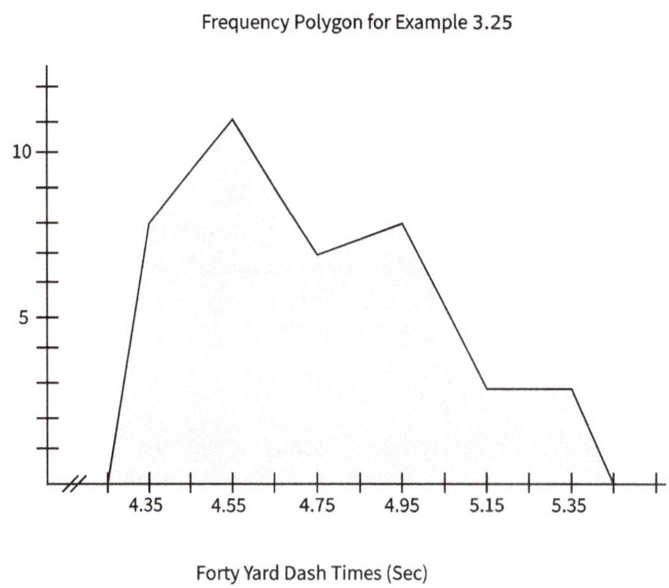

Figure 3.2: A frequency polygon based on the data given in Example 3.25.

We can also estimate the median of grouped data. Looking at a running total of frequencies, we can easily determine the class where the halfway point of the data resides. This class is the **median class**. We assume that the data in the median class is uniformly

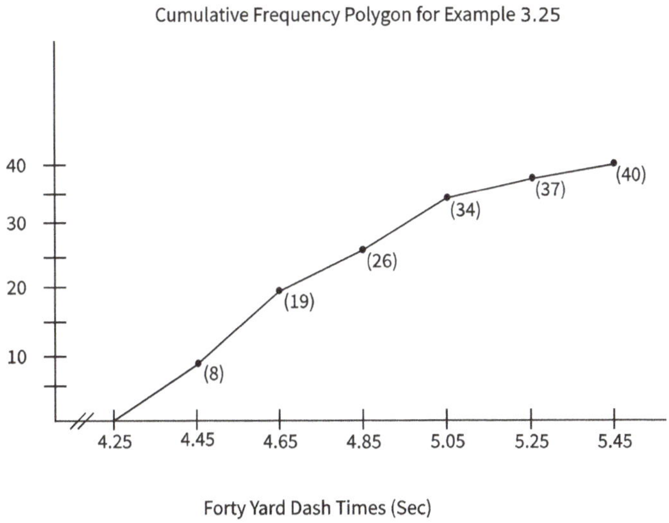

Figure 3.3: A cumulative frequency polygon based on the data given in Example 3.25.

spaced throughout the class. If the sample size is N, the cumulative frequency below the median class is M, the lower boundary of the median class is L, the median class has width r and frequency ω_{med}, we define the **grouped median** as $\tilde{x} = L + \frac{(\frac{N}{2}-M)r}{\omega_{\text{med}}}$.

If a grouped data set has one frequency greater than the rest, we call the corresponding class the **modal class** and define the **grouped mode** as the mark of the modal class. As we have noted, there is some information cost in grouping data, but the payoff is a reconstruction of a probability distribution reflecting the data as random variables. Even though we had to know what the true range of the data was to set up the frequency classes, the **grouped range** is defined a little differently. It is the difference between the overall highest class boundary and the lowest class boundary

Continuing with our adjustment of statistical formulas where the data is grouped, let us define the mean deviation $\overline{d}$. As usual let x_k be the class mark of the kth class out of n classes and ω_k be the corresponding class frequency. Then the **grouped mean deviation** is $\overline{d} = \frac{\sum_{k=1}^{n} |x_k - \overline{x}|\omega_k}{N}$. Here $\overline{x}$ is the grouped mean for the data set. Absolute value is still inconvenient to work with, so we do the same workaround as before and quantify dispersion with the **grouped variance**. This is defined as $s^2 = \frac{N\sum_{k=1}^{n}\omega_k x_k^2 - (\sum_{k=1}^{n}\omega_k x_k)^2}{N(N-1)}$, where again $N = \sum_{k=1}^{n}\omega_k$ is the sample size, x_k is the class mark of the kth class out of n classes and ω_k is the corresponding class frequency. This formula is patterned after the similar expression in Section 3.1.2 Example 3.10. Then s is the **grouped standard deviation**.

Example 3.26. Find the grouped mean of the data in Example 3.25.

Example 3.27. Find the grouped median of the data in Example 3.25.

Example 3.28. Find the grouped range of the data in Example 3.25.

Example 3.29. Find the grouped variance and standard deviation of the data in Example 3.25.

3.1.5 Percentiles & related

Given a large numerical data set organized from smallest to largest, we can identify points called **percentiles**, which tell us what fraction, expressed as a percentage, is below a particular percentile point. Percentile points depend on the data sample, so they are random variables. There are 99 percentile points that divide the range of the data set into 100 equal parts or zones. They are typically designated as P_k for $k = 1$ to 99.

Example 3.30. Find the 60th percentile point for the data in Example 3.25.

Another way to slice the data is by **deciles**, which partition the data range into ten parts. There are 9 decile points labeled D_k for $k = 1$ to 9. They are random variables and interpreted just like percentiles except they jump 10 % at a time.

Example 3.31. Find the 8th decile point for the data in Example 3.25.

A third way to sort data is into **quartiles**, which partition the range into just four parts. There are three quartile points Q_k for $k = 1$ to 3. Again, they are random variables depending on the sample and each quartile covers 25 % of the data range.

Example 3.32. Find the third quartile point for the data in Example 3.25.

3.2 Statistics of distributions

In Section 2.7 we introduced the notion of probability distributions for random variables. In Table 3.3 below, we summarize the salient facts regarding the most important of these distributions, as well as some new ones that we will need for inferential statistics problems.

We will explain the function Γ in our discussion of the t-distribution below.

Table 3.3: Common probability distribution facts (see Section 2.7 for details).

Distribution	Symbol	Range & parameters	PDF/PMF	Mean μ	Variance σ^2
Uniform Discrete	$f(x)$	$\{x_1, \dots, x_N\}$	$f(x_k) = \frac{1}{N}$	$\frac{\sum_{k=1}^{N} x_k}{N}$	$\frac{\sum_{k=1}^{N}(x_k-\mu)^2}{N}$
Bernoulli	$\tilde{b}(x;p)$	$x \in \{0,1\}, 0 < p < 1$	$p^x(1-p)^{1-x}$	p	$p(1-p)$
Binomial	$b(x;n,p)$	$x \in \{0,\dots,n\}, 0 < p < 1$	$\binom{n}{x}p^x(1-p)^{n-x}$	np	$np(1-p)$
Binomial proportion	$b(\frac{x}{n};n,p)$	$x \in \{0,\dots,n\}, 0 < p < 1$	$* * *$	p	$\frac{p(1-p)}{n}$
Pascal	$b^*(x;k,p)$	$k \in \mathbb{N}, x \in \{k, k+1, \dots\}, 0 < p < 1$	$\binom{x-1}{k-1}p^k(1-p)^{x-k}$	$\frac{k}{p}$	$\frac{k}{p}(\frac{1}{p}-1)$
Geometric	$g(x;p)$	$x \in \mathbb{N}, 0 < p < 1$	$p(1-p)^{x-1}$	$\frac{1}{p}$	$\frac{1}{p}(\frac{1}{p}-1)$
Hypergeometric	$h(x;n,N,M)$	$x \in \{0,\dots,n\}, x \le M, n-x \le N-M$	$\frac{\binom{M}{x}\binom{N-M}{n-x}}{\binom{N}{n}}$	$\frac{nM}{N}$	$\frac{nM(N-M)(N-n)}{N^2(N-1)}$
Poisson	$p(x;\lambda)$	$x \in \mathbb{N}$	$\frac{\lambda^x e^{-\lambda}}{x!}$	λ	λ
Multinomial	$f(x_i;n,p_i)$	$\sum_i^k x_i = n, 0 < p_i < 1, \sum_i^k p_i = 1$	$\binom{n}{x_1,\dots,x_k}p_1^{x_1}\cdots p_k^{x_k}$	$* * *$	$* * *$
Uniform continuous	$u_c(x;a,b)$	$a < x \le b, a, b, x \in \mathbb{R}$	$\frac{1}{b-a}$	$\frac{a+b}{2}$	$\frac{(b-a)^2}{12}$
Exponential	$f(t;\nu)$	$x, \nu > 0$	$\nu e^{-\nu t}$	$\frac{1}{\nu}$	$\frac{1}{\nu^2}$
Normal	$n(x;\mu,\sigma)$	$x \in \mathbb{R}$	$\frac{1}{\sigma\sqrt{2\pi}}\exp[-\frac{1}{2}\frac{(x-\mu)^2}{\sigma^2}]$	μ	σ^2
t-distribution	$f(t;\nu)$	$t \in \mathbb{R}$	$\frac{\Gamma(\frac{\nu+1}{2})}{\sqrt{\pi\nu}\Gamma(\frac{\nu}{2})}(1+\frac{t^2}{\nu})^{-\frac{\nu+1}{2}}$	0 for $\upsilon > 1$	$\frac{\nu}{\nu-2}$ for $\nu > 2$
χ^2 distribution	$f(x;\nu)$	$x > 0$	$\frac{1}{2^{(\frac{\nu}{2})}\Gamma(\frac{\nu}{2})}x^{\frac{\nu-2}{2}}\exp(-\frac{x}{2})$	ν	2ν
F distribution	$g(f)$	$f > 0$	$\frac{\Gamma(\frac{\nu_1+\nu_2}{2})}{\Gamma(\frac{\nu_1}{2})\Gamma(\frac{\nu_2}{2})}(\frac{\nu_1}{\nu_2})^{\frac{\nu_1}{2}}f^{(\frac{\nu_1}{2}-1)}(1+\frac{\nu_1}{\nu_2})^{-\frac{\nu_1+\nu_2}{2}}$	$\frac{\nu_2}{\nu_2-2}$ for $\nu > 2$	$\frac{2\nu_2^2(\nu_1+\nu_2-2)}{\nu_1(\nu_2-2)^2(\nu_2-4)}$ for $\nu_2 > 4$

Example 3.33. Justify the entries for mean and variance of the uniform discrete distribution in Table 3.3.

Example 3.34. Justify the entries for mean and variance of the Bernoulli distribution in Table 3.3.

Example 3.35. Justify the entries for mean and variance of the binomial distribution in Table 3.3.

Example 3.36. Justify the entries for mean and variance of the binomial proportion distribution in Table 3.3.

There is an interesting corollary to this if we frame it in terms of Chebyshev's theorem (Section 3.1.3 Example 3.22). Suppose $k\sigma = c > 0$. Then $\frac{1}{k^2} = \frac{\sigma^2}{c^2}$ and $(1 - \frac{1}{k^2}) = (1 - \frac{\sigma^2}{c^2})$. For the binomial proportion Y specifically, $(1 - \frac{1}{k^2}) = (1 - \frac{p(1-p)}{nc^2})$. You can see that as n grows without bound, the fraction $\frac{p(1-p)}{nc^2}$ approaches 0. In this context Chebyshev's result says $1 - \frac{p(1-p)}{nc^2} \le P(|y - p| < c)$, or, as n tends to infinity, the probability that y is within $\pm c$ of the mean p approaches 1. So if you run a binomial process long enough, the proportion of successes is going to approximate p as closely as you would like. This is called the **law of large numbers**. As an asymptotic claim, it is relative and not absolute. The *relative* proportion of successes to trials can approach p arbitrarily closely but the *absolute* difference between the number of successes and np could be still be huge.

Example 3.37. Justify the entries for mean and variance of the Poisson distribution in Table 3.3.

The remaining entries in our table of distributions require some calculus to derive, so we will just accept them as given. The calculus-oriented reader is encouraged to review the general method of moment-generating functions. The idea is to create a generating function (see Chapter 8) $M_X(t) = E[\exp(tX)]$ which is expanded in a power series in the variable t and where the various coefficients contain the so-called moments of the distribution of X about the origin. The moments are analogs of physical moments, where a force is multiplied by a distance from a fulcrum. Instead of a force, we use a probability mass or density function and, instead of a distance from a fulcrum, we use the value of a random variable referenced to the origin. This is exactly what we did when we defined $E[X]$. It represents the total moment of the first power of X about the origin. We have also used $E[X^2]$ in our formulas. This represents the total moment of the second power of X relative to the origin. Moments about the origin are often called "raw" moments to distinguish them from "central" moments, which are computed with respect to the mean. The variance $\sigma^2 = E[(X - \mu)^2]$ is an example of that. There are higher raw moments embedded in the moment-generating function and these are useful for characterizing the details of the shape of a distribution, such as symmetry and flatness, or more accurately skewness and kurtosis.

Here is some guidance for understanding the notation in advanced texts: moments are all denoted by μ, raw moments have primes, central moments have no primes. The order of a moment is indicated by a subscript. So the first moment about the origin is μ_1', better known to us as $E[X]$. The second moment about the mean is μ_2, which we have defined as $E[(X - E[X])^2]$. For the sake of tradition, or perhaps just familiarity, most texts quickly revert to calling $\mu_1' = \mu$, the mean, and $\mu_2 = \sigma^2$, the variance. The method of moment-generating functions allows us to acquire all of this information in an efficient and standardized manner. It also provides a platform for the justification of some important theoretical results, such as Laplace's approximation theorem and the central limit theorem. But it does require some calculus.

Example 3.38. If a fair penny is flipped 100 times, what is the mean number of heads (hence also tails), and what is the chance that the outcome of the experiment equals the mean exactly?

Example 3.39. About 15 % of a certain breed of dog is born hearing impaired. In a litter of ten puppies of this breed, what is the mean proportion of those with hearing problems and what is the standard deviation of this proportion?

Example 3.40. If you throw four fair coins on a table repeatedly, what is the probability that they will all come up heads for the third time on the seventh try? What is the average number of tries to get all heads for the third time?

Example 3.41. What is the probability of having the fourth boy born as the sixth child to a couple? What is the mean number of births needed to have the fourth boy?

Example 3.42. What is the probability of drawing (with replacement) 15 cards at random from a standard deck and have the last card drawn be the first ace? What is the average number of draws for this to occur?

Example 3.43. Five pieces of candy in a box of 20 chocolates are caramels. What is the probability of getting exactly three caramels in ten selections? What would be the average number of caramels picked in ten selections, and what would be the standard deviation? Apply Chebyshev's theorem for two standard deviations.

Example 3.44. Fifty tax returns are in a group subject to audit. Seven of them have fraudulent expense claims. If an auditor examines ten returns, what is the probability that none of the fraudulent returns will be audited? What would be the average number of fraudulent returns discovered by this process? What is the standard deviation and apply Chebyshev's theorem for two standard deviations.

Example 3.45. Over the last 100 years Sicily has had on average one magnitude 7.0 or greater earthquake every 14 years. Assuming earthquake events are independent (they are not completely) what is the probability of having five Richter 7+ earthquakes in the next half-century? How many should we expect during that time period? What is the standard deviation? Apply Chebyshev's theorem for two standard deviations.

Example 3.46. In Section 2.7 Example 2.47, we posed the following problem: Given the long-term average frequency of phone calls to the complaint desk at the department store is 30 per day. What is the probability that there is an hour that goes by with no complaint calls? The phone calls are Poisson events and the exponential distribution models the time between successive such events. The process is memoryless in the sense that estimating the time to the next event only depends on the start time and not on anything prior. Since the call frequency was 30 per ten-hour day, we have $\nu = 3$ calls per hour. The relevant distribution is $f(t;\nu) = 3e^{-3t}$ where t is in hours. We used a little calculus to answer the original question and determined the probability of a one hour gap between successive calls was 5 %. What would be the average length of time between successive calls with the given frequency and what would be the standard deviation? Apply Chebyshev's theorem for two standard deviations.

3.3 The normal distribution

As we mentioned in Chapter 2 the **normal** or **Gaussian distribution** is fundamental to statistics and is generally regarded as the most important probability distribution. It is a continuous distribution, so our methods for handling discrete distributions will not apply. Ironically, calculus-based methods do not offer much help either. Usually, we expect calculus to make short work of finding areas under probability density curves. Unfortunately, this is famously not the case with the particular function that generates the normal distribution curve. The silver lining to this cloud is that we may continue to sidestep any serious calculus and switch to the use of printed tables (historical approach) or computer algebra systems/calculators (preferred). Note that the family of exponential distributions is characterized by some sort of exponential function that describes each member. The exponential distribution of Section 3.2 Example 3.46 belongs to this family. We discovered it was very manageable with a little calculus. The last four distributions in Table 3.3 (normal, t-distribution, χ^2 distribution, and F distribution) belong to this family as well, but they are much more difficult to work with mathematically. We will have to rely on printed or online tables of one kind or another to solve problems.

If the random variable X is normally distributed, its distribution is given by the formula $n(x;\mu,\sigma) = \frac{1}{\sigma\sqrt{2\pi}}\exp(-\frac{1}{2}(\frac{x-\mu}{\sigma})^2)$. Here $x \in \mathbb{R}$, μ is the mean, and $\sigma > 0$ is the standard deviation. We can always "standardize" any distribution by creating a new random variable. This is achieved by subtracting the mean from the original random variable and dividing the result by its standard deviation. Define $Z = \frac{X-\mu}{\sigma}$, then the normal distribution simplifies to the **standard normal distribution** $n(z;0,1) = \frac{1}{\sqrt{2\pi}}\exp(-\frac{z^2}{2})$. Of course, $z \in \mathbb{R}$. This is essential for allowing us to construct tables for cumulative areas under the normal distribution curve. It would be completely impractical to generate tables that accommodated arbitrary μ and σ.

Here is a brief digression regarding how tabular values for cumulative area under the standard normal curve might be found. The function that defines the standard normal pdf is $n(z) = \frac{1}{\sqrt{2\pi}}\exp[-\frac{1}{2}z^2]$. It is unfortunate that the usual way of finding probabilities from a continuous distribution, namely integrating the pdf over an interval, just does not work for the normal distribution. The root of the problem is that $\int_a^b \exp[-\frac{1}{2}z^2]dz$ cannot be done the usual way in terms of elementary functions. But statisticians need to be able to compute probabilities one way or another, so somehow it has to be done. We can estimate integrals of this type by using the well-known power series $e^x = \sum_{n=0}^{\infty}\frac{x^n}{n!} = 1 + x + \frac{x^2}{2} + \frac{x^3}{6} + \cdots$, in which we replace x with $-\frac{z^2}{2}$ and integrate term-by-term from 0 to b. This integration is justified by the properties of the series. Doing the substitution for the first six terms, we have $\exp[-\frac{1}{2}z^2] \approx 1 - \frac{z^2}{2} + \frac{z^4}{8} - \frac{z^6}{48} + \frac{z^8}{384} - \frac{z^{10}}{3840}$. After integration the estimate of $P(0 \le Z \le b) = \frac{1}{\sqrt{2\pi}}\int_0^b \exp[-\frac{1}{2}z^2]dz$ becomes $\frac{1}{\sqrt{2\pi}}[b - \frac{b^3}{6} + \frac{b^5}{40} - \frac{b^7}{336} + \frac{b^9}{3456} - \frac{b^{11}}{42240}]$. Note that $n(z)$ is an even function, so we only need to find areas under the density function for $z > 0$. If $z < 0$, then $\int_a^0 n(z)dz = \int_0^{-a} n(z)dz$. Also, by the properties of integrals $P(0 \le a \le Z \le b) = P(0 \le Z \le b) - P(0 \le Z \le a)$. The factor $\frac{1}{\sqrt{2\pi}}$ may seem gratuitous, but it is needed for the purpose of **normalization**. Recall that for any pdf $f(x)$ the integral $\int_{-\infty}^{\infty} f(x)dx = 1$. It was shown in the eighteenth century by the Swiss mathematician Leonhard Euler that $\int_{-\infty}^{\infty}\exp[-\frac{1}{2}x^2]dx = \sqrt{2\pi}$, so we need to divide $\exp[-\frac{1}{2}x^2]$ by $\sqrt{2\pi}$ to make it a proper distribution.

Let us test this out and find the probability that a random variable with a standard normal distribution lies between zero and one using our estimating formula. Our result for $\int_0^1 n(x)dx$ is 0.3413. This agrees exactly with the tabular value to four decimals. The result for $\int_0^2 n(x)dx$ is 0.4729, which differs from the tabular value 0.4772 by less than one percent, so we could (almost) justifiably use our formula for $[-2 \le z \le 2]$. However, the estimate for $\int_0^3 n(x)dx$ is completely unusable as it differs considerably from the tabular value. This could be improved by adding more terms to the series estimation. At any rate, the tables and calculator/computer routines that report more precise cumulative areas are based on refinements of this basic method. The standard normal curve is the symmetric and instantly recognizable "bell curve."

As we know, the standard normal curve represents a probability density, so to find the probability that the random variable Z takes on a value between, say z_0 and z_1, we have to find the area under the curve between those two z values. For example, the area that is shaded in Fig. 3.4 lies between $z_0 = 0$ and $z_1 = 1$. We see that the area under the standard normal curve from $z_0 = 0$ to $z_1 = 1$ is 0.3413 either by reference to a table or our preceding calculation. This is the probability $P(0 \le Z \le 1)$. Our worked examples are essentially repetitions of this procedure under different conditions.

A ***Z*-score** for an arbitrary x in a general normal distribution is the transformed value $z = \frac{x-\mu}{\sigma}$. It is the signed fractional number of standard deviations that x is away

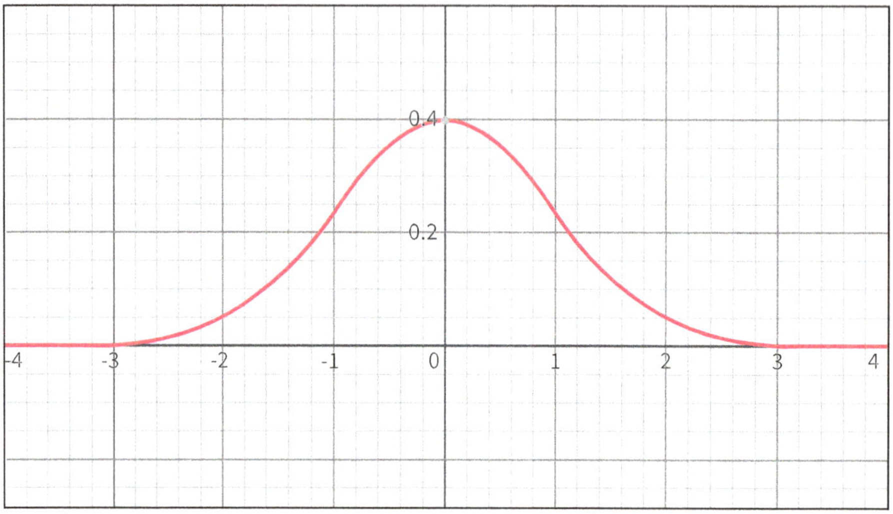

Figure 3.4: A graph of the standard normal distribution.

from the mean. The standard normal distribution is already transformed, so a particular value z is its own Z-score. Tables for cumulative area from the origin to a given z are typically organized according to increasing z from 0 to 3 with increments depending on desired precision. There is no need for negative z values since the bell curve is symmetrical and areas can be gotten by reflection through $z = 0$. Here is a table of cumulative areas under the standard normal curve between 0 and various Z-scores from 0 to 3. This table can be easily, if tediously, constructed with precision to any reasonable number of decimals by using an implementation of the error function, either by a computer algebra system or higher level calculator.

Example 3.47. Justify the previous statement that if X has mean μ and standard deviation σ, then $Z = \frac{X-\mu}{\sigma}$ has mean 0 and standard deviation 1.

Example 3.48. What is $P(-1 \leq Z \leq 1)$?

Example 3.49. Find $P(-2 \leq Z \leq 2)$ and $P(-3 \leq Z \leq 3)$.

Example 3.50. What is $P(-0.62 \leq Z \leq 1.73)$?

Table 3.4: Standard normal cumulative probabilities $P(0 \leq Z \leq z)$.

$z\downarrow$	0.00	0.01	0.02	0.03	0.04	0.05	0.06	0.07	0.08	0.09
0.0	0.0000	0.0040	0.0080	0.0120	0.0160	0.0199	0.0239	0.0279	0.0319	0.0359
0.1	0.0398	0.0438	0.0478	0.0517	0.0557	0.0596	0.0636	0.0675	0.0714	0.0753
0.2	0.0793	0.0832	0.0871	0.0910	0.0948	0.0987	0.1026	0.1064	0.1103	0.1141
0.3	0.1179	0.1217	0.1255	0.1293	0.1331	0.1368	0.1406	0.1443	0.1480	0.1517
0.4	0.1554	0.1591	0.1628	0.1664	0.1700	0.1736	0.1772	0.1808	0.1844	0.1879
0.5	0.1915	0.1950	0.1985	0.2019	0.2054	0.2088	0.2123	0.2157	0.2190	0.2224
0.6	0.2257	0.2291	0.2324	0.2357	0.2389	0.2422	0.2454	0.2486	0.2517	0.2549
0.7	0.2580	0.2611	0.2642	0.2673	0.2704	0.2734	0.2764	0.2794	0.2823	0.2852
0.8	0.2881	0.2910	0.2939	0.2967	0.2995	0.3023	0.3051	0.3078	0.3106	0.3133
0.9	0.3159	0.3186	0.3212	0.3238	0.3264	0.3289	0.3315	0.3340	0.3365	0.3389
1.0	0.3413	0.3438	0.3461	0.3485	0.3508	0.3531	0.3554	0.3577	0.3599	0.3621
1.1	0.3643	0.3665	0.3686	0.3708	0.3729	0.3749	0.3770	0.3790	0.3810	0.3830
1.2	0.3849	0.3869	0.3888	0.3907	0.3925	0.3944	0.3962	0.3980	0.3997	0.4015
1.3	0.4032	0.4049	0.4066	0.4082	0.4099	0.4115	0.4131	0.4147	0.4162	0.4177
1.4	0.4192	0.4207	0.4222	0.4236	0.4251	0.4265	0.4279	0.4292	0.4306	0.4319
1.5	0.4332	0.4345	0.4357	0.4370	0.4382	0.4394	0.4406	0.4418	0.4429	0.4441
1.6	0.4452	0.4463	0.4474	0.4484	0.4495	0.4505	0.4515	0.4525	0.4535	0.4545
1.7	0.4554	0.4564	0.4573	0.4582	0.4591	0.4599	0.4608	0.4616	0.4625	0.4633
1.8	0.4641	0.4649	0.4656	0.4664	0.4671	0.4678	0.4686	0.4693	0.4699	0.4706
1.9	0.4713	0.4719	0.4726	0.4732	0.4738	0.4744	0.4750	0.4756	0.4761	0.4767
2.0	0.4772	0.4778	0.4783	0.4788	0.4793	0.4798	0.4803	0.4808	0.4812	0.4817
2.1	0.4821	0.4826	0.4830	0.4834	0.4838	0.4842	0.4846	0.4850	0.4854	0.4857
2.2	0.4861	0.4864	0.4868	0.4871	0.4875	0.4878	0.4881	0.4884	0.4887	0.4890
2.3	0.4893	0.4896	0.4898	0.4901	0.4904	0.4906	0.4609	0.4911	0.4913	0.4916
2.4	0.4918	0.4920	0.4922	0.4925	0.4927	0.4929	0.4931	0.4932	0.4934	0.4936
2.5	0.4938	0.4940	0.4941	0.4943	0.4945	0.4946	0.4948	0.4949	0.4951	0.4952
2.6	0.4953	0.4955	0.4956	0.4957	0.4959	0.4960	0.4961	0.4962	0.4963	0.4964
2.7	0.4965	0.4966	0.4967	0.4968	0.4969	0.4970	0.4971	0.4972	0.4973	0.4974
2.8	0.4974	0.4975	0.4976	0.4977	0.4977	0.4978	0.4979	0.4979	0.4980	0.4981
2.9	0.4981	0.4982	0.4982	0.4983	0.4984	0.4984	0.4985	0.4985	0.4986	0.4988
3.0	0.4987	0.4987	0.4987	0.4988	0.4988	0.4989	0.4989	0.4989	0.4990	0.4990

Example 3.51. What is $P(Z < 0.75)$?

Example 3.52. What is $P(0.47 \leq Z \leq 2.06)$?

Example 3.53. Suppose X is normally distributed with mean 2 and standard deviation 1.5. Find $P(1 \leq X \leq 4)$.

Example 3.54. A coffee dispenser is adjusted so that its average fill amount is 5 ounces per cup. The amount dispensed X is a normally distributed random variable with standard deviation 0.4 ounces.
(i) What is the probability that the machine will dispense less than 5.5 ounces?
(ii) What is the probability that the machine will dispense between 4.5 and 5.5 ounces?
(iii) How big should the cups be to ensure that there is less than a 1 in 100 chance of overflowing?

Example 3.55. A certain electronic component has an average service life of eight years with standard deviation two years. The manufacturer has decided to offer a guarantee that if the component fails within the guarantee period he will replace it free. If his budget for covering this guarantee is equal to the cost of 5 % of the components, how long should he make the guarantee period?

Example 3.56. Psychometricians have confirmed that intelligence quotients (IQs) are normally distributed with mean 100 and standard deviation 16. It has been suggested that a minimum IQ of 110 is necessary for a college student to be successful in a rigorous STEM (Science/Technology/Engineering/Math) program. For various reasons, colleges and universities generally admit students with IQs of 90 and above. What fraction of the freshman college population is likely to succeed in an STEM discipline?

Example 3.57. Three-quarter inch bolts are produced by a machine that maintains an average diameter of 0.75 inches with a standard deviation of 0.003 inch. The bolts are advertised to be 0.75 ± 0.006 inch in diameter. What is the likelihood that a given bolt is out of specification?

Example 3.58. A random variable X has a normal distribution with $\sigma = 5$. If $P(X < 32) = 0.60$, what is $P(X > 40)$?

Before calculators were available, mathematicians needed a way to estimate binomial probabilities $b(x; n, p)$ for large n. Two workarounds were developed, the **Laplace approximation** for $p \approx \frac{1}{2}$ and the **Poisson approximation** for $p \approx 0$ or 1 (so $1-p \approx 0$). The Laplace approximation is based on the fact that as n increases, the binomial distribution begins to resemble the normal distribution. So for suitable n and p, we can set the binomial distribution mean equal to μ and the binomial standard deviation equal to σ in a normal distribution. Then the binomial random variable X becomes the standard normal random variable $Z = \frac{X-\mu}{\sigma}$. Binomial problems with large n can then be treated as standard normal problems. What are suitable n and p? A rule of thumb is that if np and $n(1-p)$ are both greater than 5, the approximation will be pretty good.

Example 3.59. In Section 3.2 Example 3.38, we flipped a fair coin 100 times and asked for the probability of a specific outcome. What is the probability that flipping a fair penny 1000 times results in the total number of heads between 495 and 505?

Example 3.60. Compare the result in Section 3.2 Example 3.38 to the corresponding Laplace approximation.

As mentioned earlier, another way to estimate binomial probabilities for large n but small p is the Poisson approximation. We include it here for contrast with the normal approximation. Another rule of thumb is that if $n \geq 20$ and $p \leq 0.05$ the Poisson approximation will be good. If $n \geq 100$ and $np < 10$ it will be very good. To apply the approximation, we take the binomial mean np and set it equal to the Poisson parameter λ. Then $b(x; n, p) \approx \frac{(np)^x e^{-np}}{x!}$.

Example 3.61. Estimate $b(2; 100, 0.05)$ using the Poisson approximation and compare with the exact value.

Example 3.62. Six sigma is a statistical quality control philosophy introduced by Motorola in the 1980s. The basic idea is to set upper and lower specification limits for some quantitative measure of a manufacturing process at $\pm 6\sigma$ from the ideal mean. Then if the actual mean drifts $\pm 1.5\sigma$ from the ideal, the probability that the process is operating out of spec is still acceptably low. Of course, $z = 6.0$ is way beyond our Z-score table, but we can use a CAS to calculate $P(Z > 6.0)$ and also $P(Z > 4.5)$ to quantify out of specification probabilities for a six sigma process.

3.4 Sampling distributions

Generally speaking, a **sampling distribution** is a probability distribution for a particular statistic. Sampling problems typically involve making a judgment whether a particular sample is consistent with the parameters of the population from which it came. This can lead us to question the representativeness of the sample if the population parameters are definitely known, or conversely to question the parameters if they have only been estimated. In Section 3.3 Example 3.57, we looked at a sample of size one and asked for the likelihood that its key measurement was within an acceptable range assuming a normal distribution for the population. In this case we decided the sample was representative of the population with given μ and σ. We would now like to extend this line of inquiry to arbitrary samples taken from populations where the parameters may or may not be precisely known.

3.4.1 Normal sampling distribution

In this section we consider the probability distribution of the mean of samples taken from the same normal population. We also look at the distributions for the difference of means and the mean proportion.

Sample mean

Suppose we have an infinite population of data points. The selection of any data point in the population is governed by the same probability distribution and all such selections are independent. Selecting n data points in this way creates a set of n independent, identically distributed random variables $\{X_1, X_2, \ldots, X_n\}$. This is a **random sample** drawn from the given population. As long as our sampling is "with replacement," a finite population is effectively an infinite one. As we noted in Section 3.1, statistics are random variables. We define two important **sampling statistics**, the sample mean and the sample variance:

- (sample mean) $\overline{X} = \frac{\sum_{k=1}^{n} X_k}{n}$,
- (sample variance) $S^2 = \frac{\sum_{k=1}^{n}(X_k - \overline{X})^2}{n-1}$.

Note that these two statistics are defined in terms of random variables and so are random variables themselves with their own probability distributions. The formula in Section 3.1.2 for the sample mean $\overline{x} = \frac{\sum_{k=1}^{n} x_k}{n}$ and in Section 3.1.3 for the sample standard deviation $s^2 = \frac{\sum_{k=1}^{n}(x_k - \overline{x})^2}{n-1}$ are numerical computations, so they are evaluations of random variables and not the variables themselves. We recognize that this distinction gets lost colloquially and people refer to either version as the sample mean or sample standard deviation, but to be mathematically consistent we make the distinction. It should be clear from context whether we are talking about the mean and variance as functions (capital letters) or as computations from specific sample data (lower case letters).

The probability distribution of $\overline{X}$ is called the **sampling distribution of the mean.** The standard deviation of any sampling statistic is called the **standard error** of that statistic, so the standard deviation of the sampling distribution for $\overline{X}$ is precisely the **standard error of the mean**. In what sense is it an "error"? The population from which the sample is drawn has fixed parameters. Sample statistics are analogs of these parameters, but because the n data points in a sample vary from sample to sample, there is unavoidably some dispersion in the values of the sample statistics. The standard error measures this spread.

Example 3.63. If a random sample of size n is drawn from an (effectively) infinite population with mean μ, show that the expected value of the sample mean $\mu_{\overline{X}}$ is the same as μ.

Example 3.64. If a random sample of size n is drawn from an (effectively) infinite population with variance σ^2, show that the expected value of the sample variance $V(\overline{X})$ is $\frac{\sigma^2}{n}$.

The **central limit theorem** is a key result that states if $\{X_1, X_2, \ldots, X_n\}$ is a random sample drawn from a population with known mean μ and variance σ^2, then the limiting distribution of

$$Z = \frac{\overline{X} - \mu}{\left(\frac{\sigma}{\sqrt{n}}\right)},$$

as n approaches infinity, is standard normal. A proof of this is far beyond the scope of our discussion, but we can still make practical use of it. This does not mean that in the limit we have a standard normal distribution, since in the limit we would have zero variance and Z itself would be infinite. What it does mean is that for reasonably sized n, say $n \geq 30$, it is valid to use a normal distribution with mean μ and variance $\frac{\sigma^2}{n}$ to approximate the sampling distribution of the mean, no matter how the population is distributed. In practice the central limit theorem is applied to samples from population distributions that are nonnormal. If n is small, this is inadvisable. But if the population from which the samples are being taken is itself normal with mean μ and variance σ^2, then the approximation is excellent with mean μ and variance $\frac{\sigma^2}{n}$.

Example 3.65. A company that manufactures heavy laminated wooden beams from smaller pieces of lumber may laminate 64 pieces to make one beam. The company purchases a specific grade of lumber and what is generally available are boards with an average strength modulus of 2.0×10^6 psi with standard deviation 4.0×10^5 psi. The company relies on the "law of averages" and sometimes luck to avoid assembling a laminated beam where all the component pieces are at the low end of the structural strength spectrum. What is the probability that a beam is assembled where the average strength of the component pieces is greater than 1.8×10^6 psi?

A complication arises if the population is finite and samples of size n are each picked without replacement. The sample mean is still the population mean, but the sample variance needs an adjustment. If the population size is N and its mean μ then $E(\overline{X}) = \mu$, but if the population has variance σ^2 then $V(\overline{X}) = \frac{\sigma^2}{n} \cdot (\frac{N-n}{N-1})$, where $(\frac{N-n}{N-1})$ is the **finite correction factor**. You can see that for fixed n, $\lim_{M\to\infty}(\frac{N-n}{N-1}) = 1$, so the correction factor has less effect on larger populations. A rule of thumb is that the correction factor can be ignored if samples are less than 5 % of the population. The difficulty with this distribution is the fact that the successive choices used to construct the sample have different probabilities as the prior choices are not replaced. The criterion for a random sample taken from a finite population is that each of the $\binom{N}{n}$ possible samples has the same probability of being selected. The justification of the finite population correction factor is similar to the arguments in Examples 3.63 and 3.64, but there are some technical details that will take us far afield, so we will just accept the result as given.

Example 3.66. Given the population $\{1, 2, 3, 4, 5, 6, 7, 8, 9, 10\}$, we would like to select samples of size 4 without replacement. There are, of course, $\binom{10}{4} = 210$ such samples. What is the sample mean and its standard error?

Difference of means

Suppose we have two populations, the first with mean μ_1 and variance σ_1^2 and the second with mean μ_2 and variance σ_2^2. Define $\overline{X}_1$ to be the mean of samples of size n_1 taken from the first population. Define $\overline{X}_2$ to be the mean of samples of size n_2 taken from the second population. Selections from one population are independent of those from the other. If the populations are not infinite, we assume all selections are with replacement. The statistic $\overline{X}_1 - \overline{X}_2$ has the distribution of the difference of two sample means.

Example 3.67. Find $E(\overline{X}_1 - \overline{X}_2) = \mu_{\overline{X}_1 - \overline{X}_2}$.

Example 3.68. Find $V(\overline{X}_1 - \overline{X}_2)$ and $\sigma_{\overline{X}_1 - \overline{X}_2}$.

We can invoke the central limit theorem again here. If n_1 and n_2 both exceed 30, the distribution for the difference of means is well approximated by the standard normal variable

$$Z = \frac{(\overline{X}_1 - \overline{X}_2) - (\mu_1 - \mu_2)}{\sqrt{\frac{\sigma_1^2}{n_1} + \frac{\sigma_2^2}{n_2}}}.$$

Example 3.69. LED spotlights made by Company A have a mean lifetime of ten years with standard deviation 1.5 year. Similar lamps made by Company B have a mean lifetime of nine years with a standard deviation of 2.5 years. What is the likelihood that a lot of 30 lamps from Company A will have a mean lifetime at least one year longer than a similar lot of 40 lamps from Company B?

Mean proportion

In Section 3.2 Example 3.36, we have already worked out the mean and variance of the random variable that represents the proportion of successes in a binomial experiment with n independent trials and probability of success p in each trial. We let $Y = \frac{X}{n}$ where X is binomially distributed with mean $\mu = np$ and variance $np(1-p)$. The quantity $y = \frac{x}{n}$ depends on the particular sample of n trials selected, so we may regard Y as a proportion statistic with sampling distribution given by $\mu_Y = p$ and $\sigma_Y^2 = \frac{p(1-p)}{n}$. Accordingly for suitably large n, by the central limit theorem this sampling distribution may be approximated by the standard normal distribution with

$$Z = \frac{Y - p}{\sqrt{\frac{p(1-p)}{n}}}.$$

Example 3.70. A typical baseball player might have 400 at bats (chances to hit) in a season and bat 0.250 (average number of hits per at bat). In a sequence of 200 at bats, what is the probability that the player would record between 35 and 60 hits. Assume performance is independent for each at bat (this is questionable in real life).

3.4.2 The t-distribution

This distribution was developed to handle smaller sample sizes ($n < 30$) where the central limit theorem does not support using a normal sampling distribution. The motivation for the t-distribution was the recognition that for smaller samples, if the population mean μ was only estimated and population variance σ^2 was unknown, the sample variance s^2 did not estimate it well. The problem was that s^2 varied too much between samples and using it as a proxy in

$$Z = \frac{\overline{X} - \mu}{(\frac{S}{\sqrt{n}})}$$

did not closely approximate a standard normal distribution. So a new statistic

$$T = \frac{\overline{X} - \mu}{(\frac{S}{\sqrt{n}})}$$

was invented and a formula derived for its probability distribution.

Recall from Chapter 3 that the details of the t-distribution were originally worked out by William S. Gosset about 1908, who has the distinction of being known to history as both a mathematician and brewer. He was employed by Guiness at the time and due to another employee having published a paper revealing some of Guiness' trade secrets, he was advised not to publish his mathematics under his own name or mention Guiness or even beer. Gosset used the alias "A Student" for his paper, hence the name Student's t-distribution. It did not hurt his career – he spent the rest of it with Guiness and rose to be Head Brewer. His mathematics was not done in a vacuum, as he managed to collaborate with both Karl Pearson and Ronald Fisher, who despised each other but got along with Gosset.

The probability density function for the t-distribution is

$$f(t;\nu) = \frac{\Gamma(\frac{\nu+1}{2})}{\sqrt{\pi\nu} \cdot \Gamma(\frac{\nu}{2})}\left(1 + \frac{t^2}{\nu}\right)^{-\frac{\nu+1}{2}},$$

for $t \in \mathbb{R}$ and shape parameter $\nu \in \mathbb{N}$. The distribution is symmetric about $t = 0$ due to the factor $(1 + \frac{t^2}{\nu})$, which is an even function of t, so the distributional mean is 0 for any ν. The graph of a typical t-distribution is shown below in Fig. 3.5. It has the bell shape of a standard normal distribution, but it has "fatter tails." Intuitively, we might expect this because for the Z statistic with σ known the variability of Z depends only on the variability of the sample means $\overline{X}$, but for the T statistic both $\overline{X}$ and S are contributing variability. This causes more area under the t-curve to be displaced from the center of the bell to the tails. As n increases the stability of s^2 improves, so it becomes a better proxy for σ^2, and a t-curve starts to resemble a z-curve. In fact, when $n = 30$, the

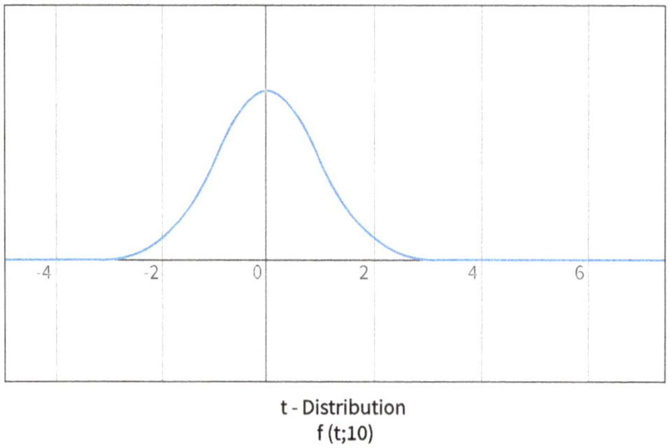

Figure 3.5: A graph of the t-distribution with $\nu = 10$, which looks similar to the standard normal distribution.

two distributions are nearly identical and as a practical matter we shift from using the t-distribution to the standard normal. Gosset made the seemingly restrictive assumption that these small samples were to be drawn from a normally distributed population. Ultimately, that assumption was shown to be unnecessary as long as the population distribution was roughly bell-shaped.

There is no need to deal with the details of the probability density function for T unless we are planning to create an accurate graph of the distribution. The T statistic is used much like the Z statistic and we will be interested in areas under the curve above or below certain t values. Note that a t-distribution depends on the parameter ν which changes the profile of the curve and hence the areas under it for fixed intervals of t. If our small sample size is n, we set $\nu = n - 1$ and call ν the number of **degrees of freedom** corresponding to that sample size. Here is why ν measures degrees of freedom. If you have a sample of size n with known $\overline{x}$, only $n-1$ data points can be freely specified, since after that the nth data point is implicitly determined by the need for the n data points to average out to $\overline{x}$. The formula $\frac{\sum_{k=1}^{n}(x_k - \overline{x})^2}{n-1}$ for sample variance s^2 has $n - 1$ and not n in the denominator for a similar reason.

The function $\Gamma(\cdot)$ that appears in the pdf for the t-distribution determines a constant for fixed ν. It is needed to make the area under the overall t curve equal to 1, as it must be for a proper distribution. Although it will not enter in to our practical application of the T statistic, we should at least have some idea of what it means. The symbol Γ stands for the gamma function, or generalized factorial function. The function $\Gamma(x)$ is defined for all x other than the negative integers by a complicated formula. If we restrict x to positive integer and half-integer values, things become much simpler. We have $\Gamma(x + 1) = x\Gamma(x)$ and $\Gamma(1) = 1$, so with a simple induction argument you can show $\Gamma(n) = (n - 1)!$. Do not forget $0! = 1$. Starting with $\Gamma(\frac{1}{2}) = \sqrt{\pi}$, we can recursively find Γ for half-integer arguments. For example, $\Gamma(\frac{\nu+1}{2})$ for $\nu = 2$ reduces to $\Gamma(\frac{3}{2}) = \frac{1}{2}\Gamma(\frac{1}{2}) = \frac{\sqrt{\pi}}{2}$. Finally, we should mention that the variance of the t-distribution is $\frac{\nu}{\nu-2}$, obviously for $\nu > 2$.

The t-distribution table is organized differently than the Z-score table. Degrees of freedom are listed in the leftmost column. For fixed ν, the number α in the top row is the percentage of area under the t curve that is to the right of the tabular entry $t_{\alpha,\nu}$ or in symbols $\alpha = P(T \geq t_{\alpha,\nu})$ For example, if we have a sample of size 16, ν would be 15. Then if $\alpha = 1\,\%$ we find the corresponding entry $t_{0.01,15} = 2.602$. So 1 % of the area under the curve is to the right of $t = 2.602$. By symmetry, another 1 % of the area is to the left of $t = -2.602$. It follows that $P(-2.602 \leq T \leq 2.602) = 98\,\%$. Equivalently, the computed t-value for a random sample of size 16 has a 98 % chance of being in the interval $[-2.602, 2.602]$. That would give us confidence that the μ we used to compute t is close to the true μ for the population.

We typically use t-values to test the reasonableness of estimating a particular value of the population mean μ. In the formula $t = \frac{\overline{x}-\mu}{(\frac{s}{\sqrt{n}})}$, if the sample mean is far from the assumed population mean, the numerator is large, all else being equal. This can mean one or the other (or both) of two things. Either the sample is very nonrepresentative and $\overline{x}$ is either too high or too low, or the assumed value of μ is unrealistic. This is detected by the computed sample t-score lying outside the interval containing the bulk of probability for T, as specified by the various area percentage points under the t-curve.

Here is a table of $t_{\alpha,\nu}$ for common area percentage points. This table can easily be constructed with the help of a computer algebra system or higher level calculator.

Example 3.71. A manufacturer of a particular airplane tire advertises that it lasts for an average of 750 landings. Thankfully, in the context of aviation safety, lasting is measured by maximum tread wear, not failure. Tread wear is assumed proportional to the number of landings. Each month the company tracks the performance of a different group of 20 tires from the service records of various airlines. One month the mean lifetime of a group of 20 tires was 765 landings with a standard deviation of 35. Assuming that tire lifetime measured by number of landings is roughly normal, how reasonable is the claim of a 750 landing average in view of this sample result?

Example 3.72. A Rubik's cube tournament featured two solvers competing simultaneously. After ten trials, ten time differences in seconds (Player 1 minus Player 2) were recorded as follows:

Game #	Player 1	Player 2	ΔScore
1	24.1	21.5	2.6
2	23.0	21.6	1.4
3	25.4	25.0	0.4
4	26.1	22.5	3.6
5	22.2	21.4	0.8
6	20.8	20.0	0.8
7	23.8	24.8	−1.0
8	25.5	25.6	−0.1
9	24.4	24.5	−0.2
10	24.0	21.2	2.8
Total	*	*	11.1

How reasonable is it to say that the two players are equally skilled?

Table 3.5: Area percentage points for T.

$\nu \downarrow \alpha \rightarrow$	10 %	5 %	2.5 %	1 %	0.5 %
1	3.078	6.314	12.706	31.821	63.657
2	1.886	2.920	4.303	6.965	9.925
3	1.638	2.353	3.182	4.541	5.841
4	1.533	2.132	2.776	3.747	4.604
5	1.476	2.015	2.571	3.365	4.032
6	1.440	1.943	2.447	3.143	3.707
7	1.415	1.895	2.365	2.998	3.499
8	1.397	1.860	2.306	2.896	3.355
9	1.383	1.833	2.262	2.821	3.250
10	1.372	1.812	2.228	2.764	3.169
11	1.363	1.796	2.201	2.718	3.106
12	1.356	1.782	2.179	2.681	3.055
13	1.350	1.771	2.160	2.650	3.012
14	1.345	1.761	2.145	2.624	2.977
15	1.341	1.753	2.131	2.602	2.947
16	1.337	1.746	2.120	2.583	2.921
17	1.333	1.740	2.110	2.567	2.898
18	1.330	1.734	2.101	2.552	2.878
19	1.328	1.729	2.093	2.539	2.861
20	1.325	1.725	2.086	2.528	2.845
21	1.323	1.721	2.080	2.518	2.831
22	1.321	1.717	2.074	2.508	2.819
23	1.319	1.714	2.069	2.500	2.807
24	1.318	1.711	2.064	2.492	2.797
25	1.316	1.708	2.060	2.485	2.787
26	1.315	1.706	2.056	2.479	2.779
27	1.314	1.703	2.052	2.473	2.771
28	1.313	1.701	2.048	2.467	2.763
29	1.311	1.699	2.045	2.462	2.756
∞	1.282	1.645	1.960	2.326	2.576

3.4.3 The χ^2 distribution

Now that we have distributions that allow us to assess $\overline{X}$ for a complete range of sample sizes, we would like to have a comparable distribution for sample standard deviation S^2. More specifically, we want to consider the distribution of the random variable X^2, called "chi squared." Note that χ (chi) is a lower case Greek X and we will use the symbol χ^2 both as a specific value of $X^2 = \frac{(n-1)S^2}{\sigma^2}$ and as the name of its probability distribution. Recalling that $\nu = n - 1$, any random variable with probability density function

$$f(x) = \frac{1}{2^{(\frac{\nu}{2})}\Gamma(\frac{\nu}{2})} x^{\frac{\nu-2}{2}} \exp\left(-\frac{\nu}{2}\right)$$

is said to have a χ^2 **distribution with ν degrees of freedom**. There is potential confusion if we are talking about a random variable X and we do not know if that is the Latin X, where a specific value is x, or the Greek X, where a specific value is χ. We will be sure to make this clear from context. Some typical graphs of the χ^2 distribution for various ν are shown below. Since (Greek) $X^2 \geq 0$, the graphs are entirely in the right half-plane and for $\nu > 2$ the graph starts at the origin due to the factor $x^{\frac{\nu-2}{2}}$.

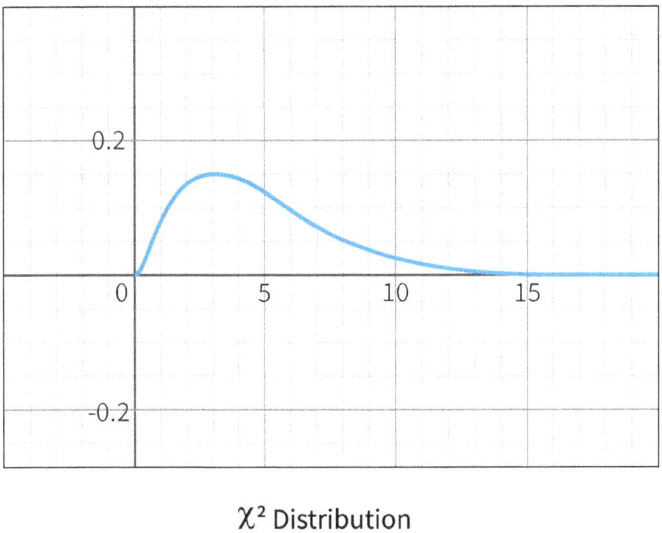

Figure 3.6: A graph of the chi-squared distribution with ten degrees of freedom also depicting a 5 % level of significance.

The χ^2 distribution is typically used to test the reasonableness of assumed values of population variance, hence standard deviation. The table of χ^2 values is laid out similarly to the table of t-values. Degrees of freedom are again listed in the leftmost column. For fixed ν, the number α in the top row is the percentage of area under the χ^2 curve that is to the right of the tabular entry $\chi^2_{\alpha,\nu}$ or in symbols $\alpha = P(T \geq \chi^2_{\alpha,\nu})$. As noted, we do not have the convenience of symmetry about the origin for the curve, so the scheme of percentages has both high and low α's. For example, a sample of size 21 would have $\nu = 20$. Choosing $\alpha = 99\,\%$ corresponds to $\chi^2_{0.99,20} = 8.260$ and choosing $\alpha = 1\,\%$ corresponds to $\chi^2_{0.01,20} = 37.566$. Evidently 98 % of the total area under the χ^2 curve is between these two points. So if we compute a representative sample χ^2 value, it should have a 98 % chance of being in the interval $[8.260, 37.566]$. If that occurs, we would have some confidence in our assumed value of σ^2. Similar to the analysis for the t-score, if s^2 is unrepresentative or if σ^2 is unrealistic, χ^2 being extreme in terms of probability is a confirmation that one or both of these conditions have occurred.

Here is a table of $\chi^2_{\alpha,\nu}$ for common area percentage points. This table can easily be constructed with the help of a computer algebra system or higher level calculator.

Table 3.6: Area percentage points for $\chi^2_{\alpha,\nu}$.

$\nu \downarrow \alpha \rightarrow$	99.5 %	99 %	97.5 %	95 %	5 %	2.5 %	1 %	0.5 %
1	0.0000393	0.000157	0.000982	0.00393	3.841	5.024	6.635	7.879
2	0.0100	0.0201	0.0506	0.103	5.991	7.378	9.210	10.597
3	0.0717	0.115	0.216	0.352	7.815	9.348	11.345	12.838
4	0.207	0.297	0.484	0.711	9.488	11.143	13.277	14.860
5	0.412	0.554	0.831	1.145	11.070	12.832	15.086	16.750
6	0.676	0.872	1.237	1.635	12.592	14.449	16.812	18.548
7	0.989	1.239	1.690	2.167	14.067	16.013	18.475	20.278
8	1.344	1.646	2.180	2.733	15.507	17.535	20.090	21.955
9	1.735	2.088	2.700	3.325	16.919	19.023	21.666	23.589
10	2.156	2.558	3.247	3.940	18.307	20.483	23.209	25.188
11	2.603	3.053	3.816	4.575	19.675	21.920	24.725	26.757
12	3.074	3.571	4.404	5.226	21.026	23.337	26.217	28.300
13	3.565	4.107	5.009	5.892	22.362	24.736	27.688	29.819
14	4.075	4.660	5.629	6.571	23.685	26.119	29.141	31.319
15	4.601	5.229	6.262	7.261	24.996	27.488	30.578	32.801
16	5.142	5.812	6.908	7.962	26.296	28.845	32.000	34.267
17	5.697	6.408	7.564	8.672	27.587	30.191	33.409	35.718
18	6.265	7.015	8.231	9.390	28.869	31.526	34.805	37.156
19	6.844	7.633	8.907	10.117	30.144	32.852	36.191	38.582
20	7.434	8.260	9.591	10.851	31.410	34.170	37.566	39.997
21	8.034	8.897	10.283	11.591	32.671	35.479	38.932	41.401
22	8.643	9.542	10.982	12.338	33.924	36.781	40.289	42.796
23	9.260	10.196	11.689	13.091	35.172	38.076	41.638	44.181
24	9.886	10.856	12.401	13.848	36.415	39.364	42.980	45.558
25	10.520	11.524	13.120	14.611	37.652	40.646	44.314	46.928
26	11.160	12.198	13.844	15.379	38.885	41.923	45.642	48.290
27	11.808	12.879	14.573	16.151	40.113	43.194	46.963	49.645
28	12.461	13.565	15.308	16.928	41.337	44.461	48.278	50.993
29	13.121	14.256	16.047	17.708	42.557	45.722	49.588	52.336
30	13.787	14.953	16.791	18.493	43.773	46.979	50.892	53.672

Example 3.73. A certain weed control chemical is sold as a 50 % concentrate with standard deviation of 2 %. Laboratory analysis of six different lots gives these concentrations in %: 50.1, 50.2, 49.6, 49.4, 50.8, and 51.0. Is the claim of $S = 2\,\%$ reasonable?

Example 3.74. A weight loss medication claims an average weight loss of 20 % after six months with standard deviation 3 %. A cohort of 25 people took the drug for six months and the result was an average of 18 % bodyweight loss with a 7 % standard deviation. Is this claim a fraud?

3.4.4 The *F* distribution

This distribution, named after Ronald Fisher, is the ratio of two independent χ^2 distributions, each divided by its respective degree of freedom. So

$$f(\nu_1, \nu_2) = \frac{\frac{\chi_1^2}{\nu_1}}{\frac{\chi_2^2}{\nu_2}}.$$

Since $\frac{\chi^2}{\nu} = \frac{s^2}{\sigma^2}$, this can also be written as

$$f = \frac{\frac{s_1^2}{\sigma_1^2}}{\frac{s_2^2}{\sigma_2^2}} = \frac{s_1^2\sigma_2^2}{s_2^2\sigma_1^2}.$$

In a subsequent section, we will use it to make inferences about the ratio of variances between two normal populations. The probability density function, as one might imagine, is complicated (see Table 3.3); however, we can at least say that $f \geq 0$ and has a shape similar to the χ^2 distribution. Tables for the *F* distribution are set up like the χ^2 table, but each area percentage point α takes a full page, since it must accommodate a matrix of (ν_1, ν_2)-values. We recommend that values of $f_\alpha(\nu_1, \nu_2)$ for a given α be gotten from a computer or calculator source. One convenient identity for working with the *F* distribution is $f_{1-\alpha}(\nu_1, \nu_2) = \frac{1}{f_\alpha(\nu_2, \nu_1)}$.

3.5 Estimation

We would like to develop a method for estimating unknown population parameters by calculating statistics of samples from the population. There are two choices for framing an estimate, either as a **point estimate** or as an **interval estimate**. There are sophisticated methods for estimating parameters based on the concepts of loss functions and risk functions, but often these approaches are difficult to apply in practical situations. We are going to rely on classical estimation methods assuming loss and risk functions are unavailable or impractical.

Let the statistic $\widehat{\Theta}$ be chosen to estimate the parameter θ. A specific value of $\widehat{\Theta}$ is denoted by $\hat{\theta}$. Our first criterion for a good estimator is that it is **unbiased**. As noted earlier, we say that $\widehat{\Theta}$ is an **unbiased estimator** of θ whenever $E(\widehat{\Theta}) = \theta$. Note that the graph of an unbiased estimator of θ has to be centered at θ. Our second criterion is that between two unbiased estimators of θ, we prefer the one whose probability distribution has the least variance. So given $\widehat{\Theta}_1$ and $\widehat{\Theta}_2$, if $V(\widehat{\Theta}_1) < V(\widehat{\Theta}_2)$ then we would choose $\widehat{\Theta}_1$ as the **more efficient estimator**. Among all possible estimators of θ, that with the least variance is called the **most efficient estimator**. It can be shown that for samples from a normal population both the mean $\overline{x}$ and the median $\tilde{x}$ are unbiased estimators of μ, but the mean has a smaller variance and hence is more efficient. Each important parameter θ has a most efficient point estimator and we will review these in the following discussion.

Another way to estimate θ is by locating it in an interval (a, b). The endpoints of the interval will depend on a point estimate $\widehat{\theta}$ together with the sampling distribution for $\widehat{\Theta}$. For example, if we are estimating μ by using $\overline{X}$, we already know that the variance $V(\overline{X}) = \frac{\sigma^2}{n}$ decreases as the sample size increases. For a given level of probability, or **confidence**, that the true μ was in a given interval, we would expect larger n to allow a shorter interval. In general, based on the sampling distribution of $\widehat{\Theta}$, we can find a and b such that for any $0 < \alpha < 1$ there is a probability of $1 - \alpha$ that $\theta \in (a, b)$. We say (a, b) is a $(1 - \alpha) \cdot 100\,\%$ **confidence interval** for θ. The interval endpoints are called the **confidence limits** of the estimate and $(1 - \alpha)$ itself is called the **confidence coefficient**.

3.5.1 Estimating the mean μ

The statistic $\overline{X}$ is the most efficient estimator of the parameter μ. So a sample value $\overline{x}$ is the best candidate for a point estimate of μ. As noted above, the larger the sample size, the better the estimate. Let us develop an interval estimate with a predetermined confidence level. We know that if sample size $n \geq 30$ the sampling distribution for the mean is approximately normal with mean $\mu_{\overline{X}} = \mu$ and standard deviation $\sigma_{\overline{X}} = \frac{\sigma}{\sqrt{n}}$. Set the confidence level at 90 %. From Table 3.4, we have $P(0 \leq Z \leq 1.65) = 0.4505$, so $P(-1.65 \leq Z \leq 1.65) = 90\,\%$. Then setting

$$Z = \frac{\overline{X} - \mu}{\frac{\sigma}{\sqrt{n}}}$$

we get

$$P\left(-1.65 \leq \frac{\overline{X} - \mu}{\frac{\sigma}{\sqrt{n}}} \leq 1.65\right) = 90\,\%.$$

Simplifying this somewhat, we have

$$P\left(-1.65\frac{\sigma}{\sqrt{n}} \leq \overline{X} - \mu \leq 1.65\frac{\sigma}{\sqrt{n}}\right) = P\left(-1.65\frac{\sigma}{\sqrt{n}} - \overline{X} \leq -\mu \leq 1.65\frac{\sigma}{\sqrt{n}} - \overline{X}\right)$$

and, multiplying through by -1 to reverse the inequalities,

$$P\left(\overline{X} - 1.65\frac{\sigma}{\sqrt{n}} \le \mu \le \overline{X} + 1.65\frac{\sigma}{\sqrt{n}}\right) = 90\,\%.$$

Plugging in a specific $\overline{x}$, we have found that the random interval

$$\left(\overline{x} - 1.65\frac{\sigma}{\sqrt{n}}, \overline{x} + 1.65\frac{\sigma}{\sqrt{n}}\right)$$

has a 90 % chance of containing the population mean μ. If we had chosen a higher confidence level, the value of Z would have been greater and then the confidence interval would have been larger.

This specific example is easily generalized. If we set the confidence level at $(1 - \alpha)$, then we are setting $P(-z_{\alpha/2} \le Z \le z_{\alpha/2}) = 1 - \alpha$. Mimicking the steps in the example, we finally get the interval estimate

$$P\left(\mu \in \left(\overline{x} - z_{\alpha/2}\frac{\sigma}{\sqrt{n}}, \overline{x} + z_{\alpha/2}\frac{\sigma}{\sqrt{n}}\right)\right) = (1 - \alpha) \cdot 100\,\%.$$

Example 3.75. Suppose we select 100 data points from a normally distributed population with variance 2.25 and compute the sample mean $\overline{x} = 9.3$. Find a 95 % confidence interval for the population mean μ.

Example 3.75 is not typical because we knew the population variance but not the population mean. A more realistic scenario is that we know neither parameter. As long as $n \ge 30$, we can use the sample variance as a proxy for population variance and proceed as in Example 3.75.

Example 3.76. Four hundred students take a common final exam in College Algebra. A sample of 49 student scores had an average of 73 with a standard deviation of 15. Find 90 % and 99 % confidence intervals for the entire cohort mean μ.

A legitimate concern with these estimates is how well μ is being estimated by $\overline{x}$. Since

$$P\left(\mu \in \left(\overline{x} - z_{\alpha/2}\frac{\sigma}{\sqrt{n}}, \overline{x} + z_{\alpha/2}\frac{\sigma}{\sqrt{n}}\right)\right) = (1 - \alpha) \cdot 100\,\%,$$

it is apparent that we may be $(1 - \alpha) \cdot 100\,\%$ confident that $|\mu - \overline{x}| < z_{\alpha/2}\frac{\sigma}{\sqrt{n}}$. If we want to determine a sample size that gives us $(1 - \alpha) \cdot 100\,\%$ confidence that the error $|\mu - \overline{x}|$ is no greater than Δ, we can solve $z_{\alpha/2}\frac{\sigma}{\sqrt{n}} = \Delta$ to find the minimum $n = (z_{\alpha/2}\frac{\sigma}{\Delta})^2$.

Example 3.77. Revisiting Example 3.76, how big would a sample of scores have to be to give us 95 % confidence that the sample mean was within two points of the cohort mean μ?

If sample sizes are under 30, we again have to use the t-distribution instead of the z-distribution in making these confidence interval estimates. We can derive the confidence limits in exactly the same way we did for $n \geq 30$ except we use the area percentage points of the T table. The expression

$$P\left(\mu \in \left(\overline{x} - z_{\alpha/2}\frac{\sigma}{\sqrt{n}}, \overline{x} + z_{\alpha/2}\frac{\sigma}{\sqrt{n}}\right)\right) = (1-\alpha)\cdot 100\,\%$$

becomes

$$P\left(\mu \in \left(\overline{x} - t_{\alpha/2}\frac{s}{\sqrt{n}}, \overline{x} + t_{\alpha/2}\frac{s}{\sqrt{n}}\right)\right) = (1-\alpha)\cdot 100\,\%.$$

Notice we have already substituted s for σ. We assume the samples are taken from a roughly normal population. As usual, $t_{\alpha/2}$ is the t-value that leaves an area of $\frac{\alpha}{2}$ area to its right under the t-distribution curve with $\nu = n-1$.

Example 3.78. Suppose we select 15 data points from a normally distributed population with unknown population variance but computed sample mean $\overline{x} = 10$ and sample variance $s = 1.3$. Find a 95 % confidence interval for the population mean μ.

We can extend this approach further to estimating a confidence interval for the difference of two means $\mu_1 - \mu_2$ where samples of size $n_1 > 30$ and $n_2 > 30$ are drawn from normal populations with known variances σ_1^2 and σ_2^2. If $\overline{x}_1$ and $\overline{x}_2$ are the respective sample means, then

$$P\left([\mu_1-\mu_2] \in \left([\overline{x}_1-\overline{x}_2] - z_{\alpha/2}\sqrt{\frac{\sigma_1^2}{n_1}+\frac{\sigma_2^2}{n_2}}, [\overline{x}_1-\overline{x}_2] + z_{\alpha/2}\sqrt{\frac{\sigma_1^2}{n_1}+\frac{\sigma_2^2}{n_2}}\right)\right) = (1-\alpha)\cdot 100\,\%.$$

Other more specialized cases, including small sample sizes, missing population variance information, and lack of independence in sample selection, are covered in the references for this chapter.

3.5.2 Estimating the variance σ^2

The most efficient point estimator of σ^2 is the sample variance S^2. We know that the statistic X^2 (chi squared) defined as $\frac{(n-1)S^2}{\sigma^2}$ has a χ^2 distribution with $(n-1)$ degrees of freedom. We assume that the samples are taken from a normally distributed population. Recall the χ^2 distribution is asymmetric, so if we intend to have an interval estimate, once we choose a confidence factor α we will need to find an upper χ^2 value where $\frac{\alpha}{2}$ of the area under the χ^2 curve is to the right and a lower χ^2 value where $\frac{\alpha}{2}$ of the area under the curve is to left. Denoting the upper limit by $\chi^2_{\alpha/2}$ and the lower by $\chi^2_{1-\alpha/2}$, we have $P(\chi^2_{1-\alpha/2} < X^2 < \chi^2_{\alpha/2}) = (1-\alpha)\cdot 100\,\%$. Then mimicking the calculation we made for estimating μ, we set

$$P(\chi^2_{1-\alpha/2} < X^2 < \chi^2_{\alpha/2}) = P\left(\chi^2_{1-\alpha/2} < \frac{(n-1)S^2}{\sigma^2} < \chi^2_{\alpha/2}\right)$$
$$= P\left(\frac{(n-1)S^2}{\chi^2_{\alpha/2}} < \sigma^2 < \frac{(n-1)S^2}{\chi^2_{1-\alpha/2}}\right)$$
$$= (1-\alpha) \cdot 100\,\%.$$

We have used the fact that if $0 < a < b$ then $0 < \frac{1}{b} < \frac{1}{a}$. So given a random sample of size n and variance s^2, a $(1-\alpha) \cdot 100\,\%$ confidence interval for population variance σ^2 is $(\frac{(n-1)s^2}{\chi^2_{\alpha/2}}, \frac{(n-1)s^2}{\chi^2_{1-\alpha/2}})$.

Example 3.79. Eight boxes of the same cereal are selected at random from a production line and weighed. The weights in ounces are 20.1, 20.4, 20.3, 20.5, 20.2, 19.9, 20.4 and 20.3. Find a 95 % confidence interval for the variance of the weights for the entire production run.

3.5.3 Estimating the ratio of variances $\frac{\sigma_1^2}{\sigma_2^2}$

We already know the most efficient point estimator of σ^2 is S^2, so the statistic $\frac{S_1^2}{S_2^2}$ is the estimator of $\frac{\sigma_1^2}{\sigma_2^2}$. Again we assume σ_1^2 and σ_2^2 represent normal population variances. The statistic that will allow us to construct confidence intervals for $\frac{\sigma_1^2}{\sigma_2^2}$ is the F statistic. Recall from Section 3.4.4 that

$$F = \frac{\frac{S_1^2}{\sigma_1^2}}{\frac{S_2^2}{\sigma_2^2}} = \frac{\sigma_2^2 S_1^2}{\sigma_1^2 S_2^2}.$$

If the sample from population 1 has size n_1 and the sample from population 2 has size n_2, then $v_1 = n_1 - 1$ and $v_2 = n_2 - 1$. It follows that the distribution we want is $f(v_1, v_2)$ and $f_{1-\alpha/2}(v_1, v_2) < F < f_{\alpha/2}(v_1, v_2)$ will correspond to a $(1-\alpha) \cdot 100\,\%$ confidence interval. Replacing F, we have

$$P\left(f_{1-\alpha/2}(v_1, v_2) < \frac{\sigma_2^2 S_1^2}{\sigma_1^2 S_2^2} < f_{\alpha/2}(v_1, v_2)\right) = (1-\alpha) \cdot 100\,\%.$$

We can isolate $\frac{\sigma_1^2}{\sigma_2^2}$ by multiplying across-the-board by $\frac{S_2^2}{S_1^2}$ to get

$$P\left(\frac{S_2^2}{S_1^2} f_{1-\alpha/2}(v_1, v_2) < \frac{\sigma_2^2}{\sigma_1^2} < \frac{S_2^2}{S_1^2} f_{\alpha/2}(v_1, v_2)\right)$$

and then we flip the fractions and reverse the sense of the inequalities as before. This results in

$$P\left(\frac{S_1^2}{S_2^2}\frac{1}{f_{\alpha/2}(\nu_1,\nu_2)} < \frac{\sigma_1^2}{\sigma_2^2} < \frac{S_1^2}{S_2^2}\frac{1}{f_{1-\alpha/2}(\nu_1,\nu_2)}\right) = (1-\alpha)\cdot 100\,\%.$$

Using the identity $f_{1-\alpha}(\nu_1,\nu_2) = \frac{1}{f_\alpha(\nu_2,\nu_1)}$ from Section 3.4.4 and computed sample variances, we can finally write

$$\left(\frac{s_1^2}{s_2^2}\frac{1}{f_{\alpha/2}(\nu_1,\nu_2)}, \frac{s_1^2}{s_2^2}f_{\alpha/2}(\nu_2,\nu_1)\right)$$

as the $(1-\alpha)\cdot 100\,\%$ confidence interval for the population variance ratio $\frac{\sigma_1^2}{\sigma_2^2}$.

Example 3.80. Rifle ammunition of a certain caliber can be rated according to its consistency in following the same path to a target under identical conditions with a fixed aim. Ideally, all shots would hit the bullseye, but in practice there is a spread around the bullseye caused by slight manufacturing variations. The data points making up the spread are the distances that various bullets hit the target measured in inches from the exact center. If a box of 10 rounds from manufacturer A is fired at the target resulting in a spread with standard deviation of 2 inches and a box of 25 rounds from manufacturer B is fired with a spread with standard deviation of 3 inches, find a 98 % confidence interval for the ratio of variances for all ammunition of this caliber between the two manufacturers.

3.6 Hypothesis testing

Decision theory addresses the problem of characterizing data populations based on examining their samples. Estimation theory is the branch of decision theory devoted to quantifying population parameters, subject to some predetermined tolerance, using sample statistics. The other branch is statistical inference, which attempts to assign probabilistic truth values to general statements about populations based on the behavior of samples. A **statistical hypothesis** is such a statement, and whether it is true or false can never be known with certainty unless the sample grows to be the entire population. Since this is usually impractical for various reasons, such as infeasibility or cost, we have to content ourselves with making judgments based on the incomplete information contained in smaller samples. A hypothesis may be **accepted** if the information from a sample supports it, or it may be **rejected** if it does not. Bear in mind that there is a spectrum of acceptance and rejection, and it is more meaningful if we can attach a probability to the level of support or rejection. Being 98 % sure that such-and-such is true about a population is better than being 55 % sure.

Statisticians endow the words accept and reject with meaning deeper than the colloquial. If evidence from a sample does not probabilistically support a hypothesis and it is rejected, it may not actually be false, but it is treated as operationally false. But if that evidence does support the hypothesis, we accept it only in the sense that we do not reject it. That has created the appearance of philosophical double-talk in statistics for 100 years. If we accept a hypothesis, it only suggests we may regard it as true because no evidence has been adduced to show it is questionable. It has also been said (tongue not entirely in cheek) that when you accept a hypothesis you do not really have to believe it is true, you just act as though it is true. Acceptance is, therefore, a very loaded word.

So a hypothesis can be weakly true or false in these technical senses. This is important for the way we may choose to frame arguments. It is considered stronger to reject (the denial of) a hypothesis than to accept (not reject) the hypothesis directly. Accordingly, our arguments are often logically indirect in the same sense as a proof by contradiction. Some modern statisticians have taken issue with the whole scheme of classical hypothesis testing and what confidence is really justified in its conclusions. The concern is that unknown and unknowable confounding factors may significantly influence the probability of accepting or rejecting a given hypothesis. Any conclusions one way or the other are therefore weakened to an unknown extent. Controlling for as many of these factors as possible would improve the quality of the statistical test, but since the problem is not epistemologically reducible to zero, the basic issue does not entirely go away. Be that as it may, we present the traditional Neyman–Pearson protocol as follows:

(1) Formulate the **null hypothesis**, denoted by H_0. This is a clear statement about some property of a population that can be expressed as a parameter θ and therefore related to a sample statistic $\widehat{\theta}$, the so-called **test statistic**. Testing the null hypothesis amounts to applying rules to decide whether to accept H_0 or reject it. Often we are trying to show statistical support for the effectiveness of some new treatment or procedure. The word "null" implies that there is no difference between the new treatment and the old treatment. So ordinarily we hope to reject H_0 to make our case.
(2) Formulate the **alternative hypothesis**, denoted by H_1, that is logically complementary to H_0. This H_1 typically takes the form of some comparison of θ with a particular value θ_0. For example, if H_0 claims $\theta = \theta_0$, then H_1 might be $\theta \neq \theta_0$, $\theta > \theta_0$, or $\theta = \theta_1$.
(3) Decide on the size of the **rejection region** or critical region for H_0. This size is a probability, denoted by α (often 5 % as originally championed by Fisher), that determines a region under the sampling distribution curve for $\widehat{\theta}$.
(4) Compute the test statistic $\widehat{\theta}$ from the sample data.
(5) Conclude that if the sample statistic $\widehat{\theta}$ falls in the rejection region, we discard H_0 and accept H_1.

The possible values of a test statistic can fall into two zones: the **acceptance region** for H_0 or the **critical region** for H_0. The boundary value between the acceptance and critical regions is called the **critical value** for the test. The chance of having a value of the test statistic inside the critical region when H_0 is true is α. It is important to specify α before computing $\widehat{\theta}$. There is a lurking temptation (imagine a billion dollar drug approval trial) to first find $\hat{\theta}$ and subsequently tweak α to justify acceptance or rejection of a hypothesis that favors a particular decision.

Recognizing that all of our decisions to accept or reject a hypothesis are ultimately subject to probability, there are several types of mistakes we can make unwittingly. If we accept H_0 and in fact H_0 is true – no problem. But if we reject H_0 and H_0 is true, then we have committed a **Type I error**. If we reject H_0 and H_0 is false – no problem. But if we accept H_0 and H_0 is false, then we have committed a **Type II error**. In the US justice system, the accused is innocent until proven guilty. Let H_0 be the statement that the accused is innocent. Then convicting him is a Type I error. If he is truly guilty and the court acquits him, that is a Type II error. Here is a little diagram that should clarify things:

Testing Errors	H_0 true	H_0 false
accept H_0	OK	Type II
reject H_0	Type I	OK

The probability of making a Type I error is the same as the critical region size α above. It is also called the **level of significance** of the test or maximum **false positive** rate. The smaller the level of significance, the more extreme the test statistic has to be to warrant rejecting H_0. The probability of making a Type II error is denoted by β. Unfortunately, α and β are inversely related, so reduction in both simultaneously is not possible.

A supplementary piece of information for hypothesis testing called the ***p*-value** was formally defined by Karl Pearson and popularized by Ronald Fisher, although a concept equivalent to it was in use 200 years prior. By definition, a p-value is the lowest level of significance (α) at which the null hypothesis would be rejected based on an observed test statistic. Equivalently, it is the probability that, assuming the validity of the null hypothesis, the test statistic computed for an additional sample would be more extreme than for the original sample. A more descriptive term for the p-value is the **observed level of significance**. In the precomputer era, α would be set in advance based on tabular area percentage points, usually at 5 % or 1 % if the test were particularly important, and the sample test statistic would be calculated and located inside or outside of the critical region. Then H_0 would be rejected or not. The observed α (p-value) requires calculating an area under a distribution curve beyond an arbitrary point determined by the test statistic. This means a customized area calculation requiring some computer power. The accept/reject decision for H_0 still has to be made (perhaps relative to nominal $\alpha = 0.05$), but a miniscule p-value would give extra confidence in rejecting H_0. Currently, the professional statistics community is not entirely happy with the proliferation of papers and reports using the p-value in a misleading way. Accordingly, the American Statistical Association has published guidance on the proper use and interpretation of p-values. Some statisticians accept the mathematical aspects of p-values but question their true significance. The issue of contention seems to be the valid framing of a testable hypothesis and hence what a small p-value ("wee p") really means.

Example 3.81. A manufacturer of nylon tow ropes claims that one of his products has a mean breaking strength of 5,000 lbs with a standard deviation of 250 lbs. Forty ropes were tested to destruction with the result that the sample mean was 4900 lbs. Test the validity of his claim using a level of significance $\alpha = 2.5\,\%$.

Example 3.82. Repeat Example 3.81 if the sample size is reduced to 16 and $\alpha = 2.0\,\%$.

Example 3.83. A random sample of 36 identical delivery trucks were retired from service after reaching an average of 100,000 miles with a standard deviation of 20,000 miles. Does this support the manufacturer's claim that the average lifetime mileage of one of these trucks is greater than 90,000 miles? Use a 1 % level of significance. What is the p-value?

Example 3.84. A sample of 80 English students took an exam on the history of the United Kingdom. They scored an average $\mu_1 = 81\,\%$ with a standard deviation $s_1 = 5\,\%$. A sample of 40 Scottish students took the same exam and scored an average $\mu_2 = 79\,\%$ with a standard deviation $s_2 = 7\,\%$. Is it reasonable to say that there is no difference between the two groups in terms of performance on the exam? Use a 5 % level of significance.

Example 3.85. A certain English lady with exquisitely refined taste claimed that she could tell whether her tea was prepared by first adding milk to the tea or by first adding tea to the milk. Ronald Fisher put her to the test and gave her eight cups of tea to distinguish, four of each mode of preparation. All she was told was that there were four of each type. The number of four cup subsets of the eight was, of course, $\binom{8}{4} = 70$, so the probability of selecting by chance the unique correct subset was 1.43 %. The null hypothesis H_0 in this experiment was that her ability to distinguish the two preparation modes was due entirely to chance (luck). In fact, she identified all eight cups correctly. With a p-value of 1.43 %, Fisher rejected H_0 and accepted her claim that she had this odd talent.

There are a number of additional specialized tests including those designed to test several means simultaneously (ANOVA). These are somewhat more involved procedurally than the hypothesis testing we have done above, so we refer the reader to the chapter references below which discuss them in detail.

3.7 Correlation

Perhaps you have heard the expression "correlation is not causation" usually explained by noting that wet streets do not imply it is raining. The fire hydrant may have been turned on. However, lack of correlation really does imply lack of causation. After all, we detect causation by noting the degree of correlation. If every other time it rained the streets were not wet, we would probably wonder about the exact connection between rain and wet streets. The direction of causality is the thing.

We would like to have a formal expression that captures the degree of correlation between two random variables X and Y. Specifically, we want to quantify the **linear**

correlation between the variables. If we select a random sample from the Cartesian product $X \times Y$ consisting of n point pairs (x_k, y_k) and plot them in the xy-plane we get so-called scatter plots as shown below.

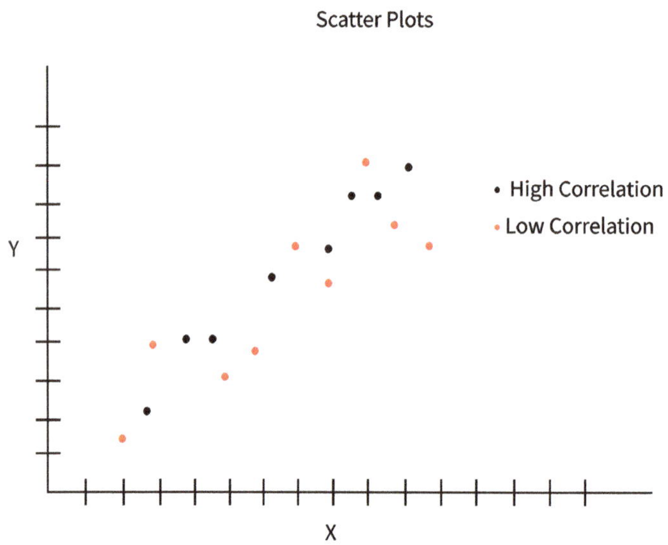

Figure 3.7: A scatter plot showing two cases, high positive correlation and low positive correlation.

Subjectively, we can see if the points tend to cluster along a line of positive slope (**positive correlation**) or negative slope (**negative correlation**). If the pattern of points shows neither tendency then we say there is no linear correlation. Another possibility is that the scatter plot shows a pattern of association that is recognizable but not linear. The (Pearson) correlation coefficient ρ measures linear correlation. It is defined by an unwieldy formula that amounts to dividing the covariance of X and Y by the product of their standard deviations.

$$\begin{aligned}\rho &= \frac{E[(X-\mu_X)(Y-\mu_Y)]}{\sigma_X\sigma_Y} \\ &= \frac{n\sum_{k=1}^n x_k y_k - (\sum_{k=1}^n x_k)(\sum_{k=1}^n y_k)}{\sqrt{(n\sum_{k=1}^n x_k^2 - (\sum_{k=1}^n x_k)^2)\cdot(n\sum_{k=1}^n y_k^2 - (\sum_{k=1}^n y_k)^2)}}.\end{aligned}$$

We can show that $-1 \leq \rho \leq 1$. If the pattern of point pairs lies in a straight line with positive slope, then $\rho = 1$. If the slope is negative, then $\rho = -1$. As the pattern of points deviates slightly from a straight line, $|\rho| < 1$. Finally, when the pattern decays to the point where no linear relation is evident, we say there is no correlation and $\rho = 0$.

Example 3.86. The following is a table of height and weight for eight female gymnasts:

X = Height (in)	57	63	63	64	60	61	62	65
Y = Weight (lbs)	102	110	107	115	105	105	106	120

Find the correlation coefficient between height and weight.

3.7.1 Linear regression

Suppose we have collected data in the form of n point pairs (x_k, y_k) and suspect an underlying linear relationship between the two random variables X and Y. We propose the formula $Y = mX + b$ and check how well it predicts y_k if we plug in x_k. The prediction will not be perfect, so (unless we are really lucky) $y_k \neq mx_k + b = \tilde{y}_k$, where y_k is the actual value and $\tilde{y}_k$ is the corresponding predicted value. The best formula, or the best choice of m and b, would be the one that minimizes the overall discrepancies between y_k and $\tilde{y}_k$. How should we measure these discrepancies? We could try to minimize the function $\sum_{k=1}^{n} |y_k - \tilde{y}_k|$, but as we have repeatedly seen, working with absolute values is awkward. A better measure from an algebraic standpoint would be $\sum_{k=1}^{n}(y_k - \tilde{y}_k)^2$ and minimizing this function by selecting optimal m and b is the essence of the **method of least squares**. This idea was originally developed by the French mathematician A.-M. Legendre at the turn of the nineteenth century.

Define

$$\phi(m,b) = \sum_{k=1}^{n}(y_k - \tilde{y}_k)^2 = \sum_{k=1}^{n}(y_k - (mx_k + b))^2.$$

There is a calculus method that finds the values of m and b so that $\phi(m,b)$ is a minimum. We will skip over those details, but a calculus-tolerant reader may easily reconstruct the result using two dimensional optimization. Denoting the values of m and b that minimize $\phi(m,b)$ as m_{best} and b_{best}, we have

$$m_{\text{best}} = \frac{n\sum_{k=1}^{n} x_k y_k - (\sum_{k=1}^{n} x_k)(\sum_{k=1}^{n} y_k)}{n\sum_{k=1}^{n} x_k^2 - (\sum_{k=1}^{n} x_k)^2}.$$

Writing $\sum_x$ for $\sum_{k=1}^{n} x_k$ and so forth, we can render the formula more memory friendly as

$$m_{\text{best}} = \frac{n\sum_{xy} - \sum_x \sum_y}{n\sum_{x^2} - \sum_x^2}.$$

Then $b_{\text{best}} = \bar{y} - m_{\text{best}}\bar{x}$ in terms of the means $\bar{x}$ and $\bar{y}$. The line $y = m_{\text{best}}x + b_{\text{best}}$ is a **regression line** where the independent variable x is the **regressor**. There are many generalizations of this concept.

Example 3.87. Find the regression line for the data in Chapter 6 Example 6.1 where height is the regressor.

Example 3.88. Suppose we are attending a jewelry auction and six diamond sales (same grade) have been recorded:

X = Weight (ct)	0.7	1.0	1.1	0.9	1.2	0.6
Y = price ($)	1000	2100	2650	1650	3100	750

A comparable 0.8 carat stone is being offered and we do not want to pay too much relative to the rest of the stones. We know that the price of a rough diamond varies with the square of its weight. What is a fair price for the stone?

References & suggested further reading

1) *John E. Freund's Mathematical Statistics with Applications* (7th ed.) by I. Miller and M. Miller (2004), Pearson – *a go-to reference for a more sophisticated presentation*
2) *Applied Statistics and Probability for Engineers* (4th ed.) by D. Montgomery and G. Runger (2007), Wiley
3) *An Introduction to Probability Theory and Its Applications (Vol. 1)* by W. Feller (1950), Wiley – *classic multivolume treatise*
4) *Statistical Methods for Research Workers* by R. A. Fisher (1925), Oliver & Boyd, London – *presentation of hypothesis testing by an originator*
5) *Introduction to Statistics* by R. Walpole (1968), Macmillan – *very readable textbook that is lighter on math*
6) *Probability and Statistics* by M. H. DeGroot (1975), Addison-Wesley
7) *Statistics for the Behavioral Sciences* (8th ed.) by F. Gravetter and L. Wallnau (2009), Cengage/Wadsworth

Solutions to Chapter 3 examples

Example 3.1

$$E[aX+b] = \sum_x (ax+b)\cdot f(x) = \sum_x ax\cdot f(x) + \sum_x b\cdot f(x) = a\sum_x x\cdot f(x) + b\sum_x f(x)$$
$$= aE[X] + b(1) = aE[X] + b.$$

Note $\sum_x f(x) = 1$.

Example 3.2

The probability of $g(x)$ as a data point is $f(x)$, so this is a definition.

Example 3.3

$$E[(X-E[X])^2] = E[X^2 - 2X\cdot E[X] + (E[X])^2] = E[X^2] - 2E[X]\cdot E[X] + (E[X])^2$$

by Example 3.1. This reduces to $E[X^2] - (E[X])^2$.

Example 3.4

By Example 3.2,

$$\begin{aligned} E\left[\sum_{i=1}^{n} c_i g_i(X)\right] &= \sum_x \left[\sum_{i=1}^{n} c_i g_i(X)\right] f(x) = \sum_{i=1}^{n}\left[\sum_x c_i g_i(X)\cdot f(x)\right] \\ &= \sum_{i=1}^{n} c_i \left[\sum_x g_i(X)\cdot f(x)\right] = \sum_{i=1}^{n} c_i E[g_i(X)]. \end{aligned}$$

Example 3.5

Let $f(x) = P(X = x)$ and $g(y) = P(Y = y)$. We also set $\phi(x,y) = P(X = x) \cap Y = y)$. This ranges over all pairs (x,y) which implies that $\sum_y \phi(x,y) = f(x)$ and $\sum_x \phi(x,y) = g(y)$. So

$$E[X \pm Y] = \sum_x \sum_y (x \pm y)\phi(x,y) = \sum_x \sum_y x\phi(x,y) \pm \sum_x \sum_y y\phi(x,y).$$

This in turn is the same as

$$\sum_x x\left(\sum_y \phi(x,y)\right) \pm \sum_y y \sum_x \phi(x,y) = \sum_x xf(x) \pm \sum_y yg(y)$$

by the preceding. It follows that

$$E[X \pm Y] = \sum_x xf(x) \pm \sum_y yg(y) = E[X] \pm E[Y].$$

Note that we need not assume X and Y are statistically independent.

Example 3.6

With $\phi(x,y) = P(X = x) \cap Y = y)$ as in Example 3.5,

$$E[XY] = \sum_x \sum_y (xy)\phi(x,y) = \sum_x \sum_y (xy)f(x)g(y)$$

by the independence of X and Y. Then

$$\sum_x xf(x) \cdot \sum_y yg(y) = E[X]E[Y].$$

Note that the converse of this statement is false. There are cases where $E[XY] = E[X]E[Y]$ but X and Y are not statistically independent.

Example 3.7

Putting the data in increasing order we have $\{4.9, 5.0, 5.2, 5.3, 5.4, 5.5, 5.5, 5.8\}$. The mean $\overline{x} = \frac{4.9+5.0+5.2+5.3+5.4+5.5+5.5+5.8}{8} = 5.325$. The median $\widetilde{x} = \frac{5.3+5.4}{2} = 5.35$. The mode appears to be $\widehat{x} = 5.5$. Here is a tip for calculating the mean easily. "Eyeball" the data and pick a value that looks easy to compare to all the rest. In this case we take 5.0. Now add the signed deviations of each data element from 5.0. We get 2.6, so find $\frac{2.6}{8} = 0.325$ and add it to 5.0. Even when the data are big numbers, if the deviations are small we can usually do this with mental arithmetic.

Example 3.8

$\overline{x} = \frac{1+2+3+4+5+6}{6} = 3.5$. The median happens to be the same and there is no mode.

Example 3.9

$\frac{76+71+79+84+x}{5} = 80$, so $76 + 71 + 79 + 84 + x = 400$. This implies $x = 90$. Maybe a long shot.

Example 3.10

The fair price to play would be the expected value of your winnings $E[X] = \sum_{k=1}^{N} x_k \cdot f(x_k)$, where $x_k = 2^k$ and $f(x_k) = \frac{1}{2^k}$. Recall the geometric distribution from Section 2.7. The probability that the head first appears on the kth flip is the same as the probability of getting $k-1$ tails in a row then the head. The probability of $k-1$ tails is $\frac{1}{2^{k-1}}$ and the probability of the head is $\frac{1}{2}$, hence $f(x_k) = \frac{1}{2^k}$. Finally, since there is no natural limit to how long the game can go on, we must consider an infinite sum: $E[X] = \lim_{N\to\infty} \sum_{k=1}^{N} 2^k \cdot \frac{1}{2^k} = \lim_{N\to\infty} N = \infty$. Apparently the expected winnings from this game are infinite. Hence there is no fair price to play. This example is called the **St. Petersburg paradox**. It was first posed by Nicolaus Bernoulli in 1713.

Example 3.11

Recognizing that the sample in Section 3.1.2 Example 3.8 above is the entire population, we have

$$\sigma^2 = \sigma_X^2 = \frac{\sum_{k=1}^{N}(x_k - 3.5)^2}{N} = \frac{1}{6} \cdot 2(6.25 + 2.25 + 0.25) = 2.92.$$

Hence $\sigma = 1.71$.

Example 3.12

Let $f(x)$ be the probability that the dice total x,

x	$f(x)$
2	$\frac{1}{36}$
3	$\frac{2}{36}$
4	$\frac{3}{36}$
5	$\frac{4}{36}$
6	$\frac{5}{36}$
7	$\frac{6}{36}$
8	$\frac{5}{36}$
9	$\frac{4}{36}$
10	$\frac{3}{36}$
11	$\frac{2}{36}$
12	$\frac{1}{36}$

Then $\mu = \frac{252}{36} = 7$. Then

$$\sigma_X^2 = E[(X-7)^2] = \sum_{k=1}^{N}(x_k - \mu)^2 \cdot f(x_k)$$
$$= 2 \cdot \left[\frac{(-5)^2}{36} + \frac{2(-4)^2}{36} + \frac{3(-3)^2}{36} + \frac{4(-2)^2}{36} + \frac{5(-1)^2}{36}\right] = \frac{105}{18} = 5.83.$$

And $\sigma = 2.42$. These are the answers based on appealing to the basic definitions. Knowing the mean and variance for one die and the formulas below greatly simplifies this calculation; see Example 3.19.

Example 3.13

$(1 + 29) + (2 + 28) + \cdots (14 + 16) + 15 = 14 \cdot 30 + 15 = 435$, so $\mu = \frac{435}{29} = 15$. Then $\sigma^2 = \frac{2\cdot(14^2+13^2+\cdots+1^2)}{29} = \frac{2\cdot 1015}{29} = 70$. So $\sigma = 8.37$.

Example 3.14

$$s^2 = \frac{\sum_{k=1}^{n}(x_k - \overline{x})^2}{n-1} = \frac{\sum_{k=1}^{n}(x_k^2 - 2x_k\overline{x} + \overline{x}^2)}{n-1}.$$

Note $\overline{x}$ is constant, so by Example 3.14,

$$s^2 = \frac{(\sum_{k=1}^{n} x_k^2) - 2\overline{x}(\sum_{k=1}^{n} x_k) + n\overline{x}^2}{n-1}.$$

Substituting $\frac{\sum_{k=1}^{n} x_k}{n}$ for $\overline{x}$ gives

$$s^2 = \frac{\sum_{k=1}^n x_k^2 - 2(\frac{\sum_{k=1}^n x_k}{n})(\sum_{k=1}^n x_k) + n(\frac{\sum_{k=1}^n x_k}{n})^2}{n-1},$$

which after multiplying by $\frac{n}{n}$ reduces to

$$\frac{n\sum_{k=1}^n x_k^2 - 2(\sum_{k=1}^n x_k)^2 + n^2(\frac{\sum_{k=1}^n x_k}{n})^2}{n(n-1)} = \frac{n\sum_{k=1}^n x_k^2 - (\sum_{k=1}^n x_k)^2}{n(n-1)}.$$

Example 3.15

$E[(X-\mu)^2] = E[X^2 - 2\mu X + \mu^2] = E[X^2] - 2\mu E[X] + \mu^2$ by Example 3.3. Since $E[X] = \mu$, this reduces to $E[X^2] - 2\mu^2 + \mu^2 = E[X^2] - \mu^2$.

Example 3.16

First, by Example 3.14, we have $\mu_{X+a} = E[X+a] = E[X] + a = \mu_X + a$. Then $\sigma_{X+a}^2 = E[((X+a) - (\mu_X + a))^2] = E[(X+a) - (\mu_X - a)^2] = E[(X-\mu_X)^2] = \sigma_X^2$.

Example 3.17

Evidently, $\mu_{bX} = E[bX] = bE[X] = b\mu_X$ again by Table 3.1(4). Then $\sigma_{bX}^2 = E[(bX - b\mu_X)^2] = E[b^2(X-\mu_X)^2] = b^2E[(X-\mu_X)^2]$. This establishes that $\sigma_{bX}^2 = b^2\sigma_X^2$. So multiplying each data element by the same constant increases the variance of the population by the square of that constant.

Example 3.18

We have $\mu_{X+Y} = E[X+Y] = E[X] + E[Y] = \mu_X + \mu_Y$. Then

$$\begin{aligned}\sigma_{X+Y}^2 &= E\big[(X + Y - (\mu_X + \mu_Y))^2\big] = E\big[((X-\mu_X) + (Y-\mu_Y))^2\big] \\ &= E\big[((X-\mu_X)^2 + 2(X-\mu_X)(Y-\mu_Y) + (Y-\mu_Y)^2)\big].\end{aligned}$$

This is equivalent to

$$E[((X-\mu_X)^2 + (Y-\mu_Y)^2)] + E[2(X-\mu_X)(Y-\mu_Y)] = \sigma_X^2 + \sigma_Y^2 + E[2(X-\mu_X)(Y-\mu_Y)].$$

We claim $E[(X-\mu_X)(Y-\mu_Y)] = 0$. Certainly, $E[(X-\mu_X)(Y-\mu_Y)] = E[XY - \mu_X Y - \mu_Y X + \mu_X\mu_Y]$. This further reduces to $E[XY] - \mu_X E[Y] - \mu_Y E[X] + \mu_X\mu_Y$. From Table 3.1(6) $E[XY] = E[X]E[Y] = \mu_X\mu_Y$ and both $\mu_X E[Y]$ and $\mu_Y E[X] = \mu_X\mu_Y$, so $E[(X-\mu_X)(Y-\mu_Y)] = 0$, establishing the claim.

Example 3.19

For one die, the variance is 2.92. By Table 3.2(3), for two dice (certainly statistically independent from each other), $\sigma_{\text{pair}}^2 = 2\cdot\sigma_1^2$. So $\sigma_{\text{pair}}^2 = 2(2.915) = 5.83$. This agrees with Example 3.12.

Example 3.20

The minimum count is 10 and the maximum is 60. We do know in advance by the nature of the experiment that the distribution of observations is symmetrical about the mean, which is $10 \cdot 3.5 = 35$. Using Example 3.19 revised for ten dice, we get $\sigma_{10}^2 = 10\sigma_1^2 = 10 \cdot 2.92 = 29.2$, so one standard deviation is $\sqrt{29.2} = 5.4$. Take $k = 2$ in Chebyshev's theorem, then we would expect at least $(1 - \frac{1}{4}) = \frac{3}{4}$ of the observations to be between $35 - 10.8 \approx 24$ and $35 + 10.8 \approx 46$. So $\frac{1}{4}$ would be out of that range with equal fractions above 46 and below 24. We conclude that at most $\frac{1}{8}$ of the rolls should exceed 46.

Example 3.21

Here $\mu = 35$, so with $a = 50$, $P(X \geq 50) \leq \frac{35}{50} = 70\,\%$. True, but not a very sharp result.

Example 3.22

Using Chebyshev's theorem with $\sigma = 2.42$, hence $2\sigma = 4.84$ and $\mu = 7.0$, we should have $(1 - \frac{1}{4}) = \frac{3}{4}$ of the observations between $7.0 - 4.84 = 2.16$ and $7.0 + 4.84 = 11.84$. The only possible outcomes not in this range are 2 and 12, which have a combined probability of $\frac{1}{18} = 5.56\,\%$. So the actual probability that an observation (roll of two dice) is within two standard deviations of the mean is 94.4 %. The Chebyshev estimate of at least 75 % is certainly true, but far from the actual.

Example 3.23

We start with

$$\sigma^2 = E[(X - \mu)^2] = \sum_{k=1}^{N} (x_k - \mu)^2 \cdot f(x_k).$$

We break the sum into three parts, below $\mu - k\sigma$, between $\mu - k\sigma$ and $\mu + k\sigma$, and above $\mu + k\sigma$. So we have

$$\sigma^2 = \sum_{x \leq \mu - k\sigma} (x_k - \mu)^2 \cdot f(x_k) + \sum_{x \geq \mu - k\sigma}^{x \leq \mu + k\sigma} (x_k - \mu)^2 \cdot f(x_k) + \sum_{x \geq \mu + k\sigma} (x_k - \mu)^2 \cdot f(x_k).$$

Since the middle sum $\sum_{x \geq \mu - k\sigma}^{x \leq \mu + k\sigma} (x_k - \mu)^2 \cdot f(x_k) \geq 0$, we can remove it to create an inequality,

$$\sigma^2 \geq \sum_{x \leq \mu - k\sigma} (x_k - \mu)^2 \cdot f(x_k) + \sum_{x \geq \mu + k\sigma} (x_k - \mu)^2 \cdot f(x_k).$$

Now for $x \leq \mu - k\sigma$, it is true that $x_k - \mu \leq -k\sigma$, or $-(x_k - \mu) \geq k\sigma$, so it follows that $(x_k - \mu)^2 \geq k^2\sigma^2$. Likewise, for $x \geq \mu + k\sigma$, one gets $(x_k - \mu)^2 \geq k^2\sigma^2$. So rewriting the inequality gives us

$$\sigma^2 \geq \sum_{x \leq \mu - k\sigma} k^2\sigma^2 \cdot f(x_k) + \sum_{x \geq \mu + k\sigma} k^2\sigma^2 f(x_k).$$

Dividing by σ^2 yields

$$1 \geq \sum_{x \leq \mu - k\sigma} k^2 f(x_k) + \sum_{x \geq \mu + k\sigma} k^2 f(x_k) = k^2\Big(\sum_{x \leq \mu - k\sigma} f(x_k) + \sum_{x \geq \mu + k\sigma} f(x_k)\Big).$$

What is $\sum_{x \leq \mu - k\sigma} f(x_k) + \sum_{x \geq \mu + k\sigma} f(x_k)$? It is the total probability that $x \leq \mu - k\sigma$ or $x \geq \mu + k\sigma$. That is the same as $P(|x - \mu| \geq k\sigma)$. So $\frac{1}{k^2} \geq P(|x - \mu| \geq k\sigma)$ and hence $1 - \frac{1}{k^2} \leq P(|x - \mu| < k\sigma)$.

Example 3.24

Recall that the population mean μ is a fixed parameter, but the sample mean $\bar{x}$ is itself a random variable with some uncertainty built into it. Using the sample mean in place of the population mean in the first formula does not account properly for the additional variability of $\bar{x}$ and therefore understates the sample variance. Statisticians recognize that using sample statistics to estimate population parameters may introduce a degree of bias into calculations. On average this bias is minimized if the expected value of the estimating statistic equals the corresponding parameter. So, in general, if we use $\hat{\theta}$ to estimate θ, we would like to have $E[\hat{\theta}] = \theta$. This condition defines an **unbiased estimator**. In the context of our immediate problem, we want to show that $E[S^2] = \sigma^2$, or that the expected value of the sample variance equals the population variance. Starting with

$$\begin{aligned} E\left[\frac{\sum_{k=1}^{n}(X_k - \overline{X})^2}{n-1}\right] &= E\left[\frac{\sum_{k=1}^{n}[(X_k - \mu) - (\overline{X} - \mu)]^2}{n-1}\right] \\ &= E\left[\frac{\sum_{k=1}^{n}[(X_k - \mu)^2 - 2(X_k - \mu)(\overline{X} - \mu) + (\overline{X} - \mu)^2]}{n-1}\right], \end{aligned}$$

we can distribute the expectation operator, which gives us

$$\frac{\sum_{k=1}^{n} E[(X_k - \mu)^2] - 2\sum_{k=1}^{n} E[(X_k - \mu)(\overline{X} - \mu)] + \sum_{k=1}^{n} E[(\overline{X} - \mu)^2]}{n-1}.$$

Note that $-2\sum_{k=1}^{n} E[(X_k - \mu)(\overline{X} - \mu)] = -2\sum_{k=1}^{n} E[(\overline{X} - \mu)^2]$, so after cancellation we have $\frac{\sum_{k=1}^{n} E[(X_k - \mu)^2] - \sum_{k=1}^{n} E[(\overline{X} - \mu)^2]}{n-1}$. Also $\sum_{k=1}^{n} E[(\overline{X} - \mu)^2] = n \cdot E[(\overline{X} - \mu)^2]$. Summarizing our equality we get

$$E\left[\frac{\sum_{k=1}^{n}(X_k - \overline{X})^2}{n-1}\right] = \frac{(\sum_{k=1}^{n} E[(X_k - \mu)^2]) - n \cdot E[(\overline{X} - \mu)^2]}{n-1}.$$

Now $E[(X_k - \mu)^2] = \sigma^2$ and $E[(\overline{X} - \mu)^2] = \frac{\sigma^2}{n}$, so finally,

$$\frac{\sum_{k=1}^{n} E[(X_k - \mu)^2] - n \cdot E[(\overline{X} - \mu)^2]}{n-1} = \frac{n \cdot \sigma^2 - n \cdot \frac{\sigma^2}{n}}{n-1} = \frac{(n-1)\sigma^2}{n-1} = \sigma^2.$$

The expected value of the adjusted formula for sample variance is exactly the population variance. This verifies that S^2 computed using the sample mean and denominator $n - 1$ is an unbiased estimator of the population variance. We have also used the fact that the variance of the sample mean is $E[(\overline{X} - \mu)^2] = \frac{\sigma^2}{n}$ (see Section 3.4.1, Example 3.64).

Example 3.25

We scan the data and determine the range is from 4.3 to 5.4. If we set the class widths equal to 0.2 seconds we will have six classes.

class interval	class boundaries	mark	frequency
4.3–4.4	4.25–4.45	4.35	8
4.5–4.6	4.45–4.65	4.55	11
4.7–4.8	4.65–4.85	4.75	7
4.9–5.0	4.85–5.05	4.95	8
5.1–5.2	5.05–5.25	5.15	3
5.3–5.4	5.25–5.45	5.35	3

Example 3.26

We want

$$\begin{aligned}\overline{x} &= \frac{\sum_{k=1}^{n} x_k \omega_k}{N} \\ &= \frac{(4.35)(8) + (4.55)(11) + (4.75)(7) + (4.95)(8) + (5.15)(3) + (5.35)(3)}{40} \\ &= \frac{189.2}{40} = 4.73.\end{aligned}$$

Example 3.27

There are 40 observations and the 20th observation happens in the third class, which is the median class and has lower boundary $L = 4.65$. Also, $M = 19$, $\omega_{\text{med}} = 7$, and $r = 0.20$. Plugging this data into $\tilde{x} = L + \frac{(\frac{N}{2} - M)r}{\omega_{\text{med}}}$, we have $\tilde{x} = 4.65 + \frac{(20-19)\cdot 0.20}{7} = 4.68$.

Example 3.28

$5.35 - 4.25 = 1.10$. The actual data range is $5.4 - 4.3 = 1.10$ as well.

Example 3.29

We want

$$\begin{aligned} s^2 &= \frac{N\sum_{k=1}^{n}\omega_k x_k^2 - \left(\sum_{k=1}^{n}\omega_k x_k\right)^2}{N(N-1)} \\ &= \frac{1}{40\cdot 39}\Big[(40)\,[(18.92)(8) + (20.70)(11) + (22.56)(7) \\ &\quad + (24.50)(8) + (26.52)(3) + (28.62)(3)] - (189.2)^2\Big] \\ &= \frac{35936 - 35797}{1560} = 0.0891. \end{aligned}$$

So $s = 0.299$. For the sake of comparison, if we calculate the mean, variance, and standard deviation from the actual individual data points instead of grouping, we get $\overline{x} = 4.728$, $s^2 = 0.0912$, and $s = 0.302$.

Example 3.30

There are 40 data points in the sample, so we want to find a value for which 24 data points are below. There is a lot of duplication in the data, so we will have to be content with finding an ideal percentile point based on the assumption that data are uniformly distributed within a class. There are 19 observations falling below the lower boundary of the third frequency class, so we will need to add five observations to that by adding $\frac{5}{7}$ of the width of the third frequency class which has 7 assumed uniformly distributed data points in it. The lower class boundary is 4.65, so we set P_{60} at $4.65 + \frac{5}{7}(0.20) = 4.79$. So the 60th percentile for the 40-yard dash times is 4.79 seconds, and we can then say that 60 % of the runners had times lower than 4.79 seconds. Since having a lower time means a runner is faster, we could also say that 60 % of the runners were faster than 25.1 feet per second.

Example 3.31

First, we need to find the class containing the 32nd entry out of 40. That would be the fourth frequency class with 8 data points. The lower class boundary is 4.85 and there are 26 data points below it, so we will need to add 6 more, again assuming the 8 in the class are uniformly distributed. So $D_8 = 4.85 + \frac{6}{8}(0.20) = 5.0$. We conclude that 80 % of the runners had elapsed times lower than 5.0 seconds or ran faster than 24.0 feet per second. Note that being in a higher percentile/decile for this particular data set is less desirable from an athletic standpoint.

Example 3.32

The 30th data point is again in the fourth frequency class, so we can use the information in Example 3.31 to expedite our calculation. We need to add four data points to the 26 data points below 4.85. With the usual assumption of uniform distribution in the class, we have $Q_3 = 4.85 + \frac{4}{8}(0.20) = 4.95$. This is also P_{75}, where 75 % of the runners had elapsed times less than 4.95 seconds or ran faster than 24.2 feet per second.

Example 3.33

We have already done this in Section 3.1.2 and Section 3.1.3 in connection with defining the mean and variance of a population data set. Recall that for the mean $\mu = E[X] = \sum_{k=1}^{N} x_k \cdot \frac{1}{N} = \frac{\sum_{k=1}^{N} x_k}{N}$ and for the variance $\sigma^2 = E[(X-\mu)^2] = \frac{\sum_{k=1}^{N}(x_k-\mu)^2}{N}$.

Example 3.34

A Bernoulli trial is a single experiment with either success or failure. It is useful to us because we can represent a binomial random variable as a sum of identical Bernoulli random variables. The calculation of mean and variance is immediate from the probability mass function. If X is Bernoulli distributed, then $\mu_X = E[X] = 0 \cdot p^0(1-p)^{1-0} + 1 \cdot p^1(1-p)^{1-1} = p$. Likewise, $\sigma_X^2 = E[(X-p)^2] = E[X^2] - p^2$ by Table 3.1(3). But $E[X^2] = 0^2 \cdot p^0(1-p)^{1-0} + 1^2 \cdot p^1(1-p)^{1-1} = p$. Then $\sigma_X^2 = p - p^2 = p(1-p)$.

Example 3.35

A binomial experiment with probability of success p can be viewed as n independent repetitions of a Bernoulli experiment with the same probability of success. So let the binomial random variable be $X = \sum_{i=1}^{n} Y_i$, where each Y_i is a Bernoulli distributed random variable for $i = 1$ to n. By a simple inductive generalization of Table 3.1(5),

$$\mu_X = E[X] = E\left[\sum_{i=1}^{n} Y_i\right] = \sum_{i=1}^{n} E[Y_i] = \sum_{i=1}^{n} p = np.$$

Likewise,

$$\sigma_X^2 = \sum_{i=1}^{n} \sigma_{Y_i}^2 = \sum_{i=1}^{n}(p(1-p)) = np(1-p),$$

again by a simple generalization of Section 3.1.3 Example 3.18. Notice that the only difference between the Bernoulli and binomial pmf's is the presence of the factor $\binom{n}{x}$ in $b(x; n, p)$. This binomial coefficient simply counts how the various cases of x successes can appear in a sequence of n Bernoulli trials.

Example 3.36

If X is binomially distributed with mean p and variance $np(1-p)$, we can easily find the mean and variance of $Y = \frac{X}{n}$. For the mean, $\mu_Y = E[\frac{X}{n}] = \frac{1}{n}E[X] = \frac{np}{n} = p$. For the variance, $\sigma_Y^2 = \frac{np(1-p)}{n^2}$ by Section 3.1.3 Example 3.17. This simplifies to $\sigma_Y^2 = \frac{p(1-p)}{n}$.

The random variable Y represents the proportion of successes out of n repetitions in a binomial process.

Example 3.37

For this distribution,

$$\mu_X = E[X] = \sum_{x=0}^{\infty} x \cdot \frac{\lambda^x e^{-\lambda}}{x!} = \lambda \sum_{x=1}^{\infty} \frac{\lambda^{x-1} e^{-\lambda}}{(x-1)!},$$

where the $x = 0$ term is zero and we have cancelled x against $x!$. Reindexing with $k = x - 1$, we have $\lambda \sum_{k=0}^{\infty} \frac{\lambda^k e^{-\lambda}}{k!}$. But $\sum_{k=0}^{\infty} \frac{\lambda^k e^{-\lambda}}{k!} = 1$ since it is the sum of the probabilities of all possible discrete values of the Poisson distribution. Hence $\mu_X = \lambda$. Now for the variance, consider $E[(X - E[X])^2] = E[X^2] - (E[X])^2$ from Table 3.1(3). In turn, we have $E[X^2] = E[X(X-1)] + E[X]$. So let us find $E[X(X-1)]$. This is

$$\sum_{x=0}^{\infty} x(x-1) \cdot \frac{\lambda^x e^{-\lambda}}{x!} = \sum_{x=2}^{\infty} \frac{\lambda^x e^{-\lambda}}{(x-2)!}$$

because both the $x = 0$ and $x = 1$ terms are zero. However,

$$\sum_{x=2}^{\infty} \frac{\lambda^x e^{-\lambda}}{(x-2)!} = \lambda^2 \sum_{x=2}^{\infty} \frac{\lambda^{x-2} e^{-\lambda}}{(x-2)!} = \lambda^2 \sum_{k=0}^{\infty} \frac{\lambda^k e^{-\lambda}}{k!}$$

after we reindex $k = x - 2$. Again the sum is 1, so we conclude $E[X(X-1)] = \lambda^2$. Now since $E[X] = \lambda$, $E[X^2] = E[X(X-1)] + E[X] = \lambda^2 + \lambda$. Finally, $\sigma^2 = E[X^2] - (E[X])^2 = \lambda^2 + \lambda - \lambda^2 = \lambda$. Oddly, the mean and variance of the Poisson distribution are the same.

Example 3.38

This is a binomial experiment, so the distribution is

$$\begin{aligned} b(x; 100, 0.5) &= \binom{100}{x} (0.5)^x (0.5)^{100-x} = \binom{100}{x} (0.5)^{100} \\ &= \binom{100}{x} (7.89 \times 10^{-31}). \end{aligned}$$

The mean is given by Table 3.3 as $\mu = (100)(0.5) = 50$. To calculate the number

$$b(50; 100, 0.5) = \binom{100}{50} (7.89 \times 10^{-31}),$$

we rely on CAS. It turns out to be about 7.96 %.

Example 3.39

This is a binomial proportion problem with distribution

$$b\left(\frac{x}{n};n,p\right)=b\left(\frac{x}{10};10,0.15\right).$$

From Table 3.3, we have $\mu = 0.15$ and $\sigma^2 = \frac{(0.15)(0.85)}{10} = 0.01275$. So $\sigma = 0.113$.

Example 3.40

This is a Pascal distribution problem where we are waiting for the third success to occur on the seventh try. The probability of an all heads toss is $\frac{1}{2^4} = \frac{1}{16}$. Recall from Table 3.3 that

$$b^*(x;k,p)=\binom{x-1}{k-1}p^k(1-p)^{x-k},$$

so we have

$$b^*\left(7;3,\frac{1}{16}\right)=\binom{7-1}{3-1}\left(\frac{1}{16}\right)^3\left(\frac{15}{16}\right)^{7-3}=\binom{6}{2}\left(\frac{15^4}{16^7}\right)=0.282\,\%.$$

For the second part, $\mu = \frac{k}{p} = \frac{3}{0.0625} = 48$.

Example 3.41

Assuming the probability of a boy being born is 50 %, we have another Pascal problem,

$$b^*\left(6;4,\frac{1}{2}\right)=\binom{5}{3}\left(\frac{1}{2}\right)^6=15.6\,\%.$$

For the mean $\mu = \frac{k}{p}$, it would be $\frac{4}{0.5} = 8$.

Example 3.42

This is a geometric distribution problem, since we are looking to succeed for the first time on the 15th try. We can assume the deck is shuffled between draws to rerandomize the cards previously drawn. The probability of drawing an ace is $\frac{1}{13}$. From Table 3.3,

$$g(x;p)=p(1-p)^{x-1}=g\left(15;\frac{1}{13}\right)=\frac{1}{13}\left(\frac{12}{13}\right)^{14}=2.51\,\%.$$

For the second part, $\mu = (\frac{1}{13})^{-1} = 13$.

Example 3.43

This is a hypergeometric problem where the number of selections made is $n = 10$, the number of caramels wanted is $x = 3$, the number of selections available is $N = 20$, and the total number of caramels among the available selections is $M = 5$. From Table 3.3,

$$h(x; n, N, M) = \frac{\binom{M}{x}\binom{N-M}{n-x}}{\binom{N}{n}} = h(3; 10, 20, 5) = \frac{\binom{5}{3}\binom{20-5}{10-3}}{\binom{20}{10}} = 34.8\,\%.$$

Then $\mu = \frac{nM}{N} = \frac{10\cdot 5}{20} = 2.5$. Finally, $\sigma^2 = \frac{nM(N-M)(N-n)}{N^2(N-1)}$ or $\sigma^2 = \frac{(10)(5)(15)(10)}{(400)(19)} = 0.987$, so $\sigma = 0.993$. Rounding off σ to 1, Chebyshev theorem implies that 75 % of the time you would expect to get between 0.5 and 4.5 caramels, or practically speaking, one to four.

Example 3.44

This is hypergeometric with

$$h(x; n, N, M) = h(0; 10, 50, 7) = \frac{\binom{7}{0}\binom{43}{10}}{\binom{50}{10}} = 18.7\,\%.$$

Then $\mu = \frac{10\cdot 7}{50} = 1.4$. Also $\sigma^2 = \frac{(10)(7)(43)(40)}{(2500)(49)} = 0.983$, so $\sigma = 0.991$. Again rounding σ to 1, we would expect that 75 % of the time, ten audits would expose between zero and 3.4 fraudulent returns, or in practical terms, three or fewer.

Example 3.45

These kinds of phenomena are modeled as Poisson processes. The Poisson distribution is $p(x; \lambda) = \frac{\lambda^x e^{-\lambda}}{x!}$, where λ is the average occurrence of an event during a specific amount of time. If the rate of occurrence α is stated per unit of time, then multiply by the overall time t to get λ. In this case there is $\frac{1}{14}$ of a 7+ earthquake per year, so $\alpha = 0.0714$. Then the average for 50 years is $(0.0714)(50) = 3.57$. So $p(5; 3.57) = \frac{(3.57)^5 e^{-3.57}}{5!} = 13.6\,\%$. The Poisson mean $\mu = \lambda = 3.57$ and the variance $\sigma^2 = \lambda = 3.57$, so $\sigma = 1.89$. The interval 3.57 ± 3.78 is $[0, 7.35]$, so 75 % of the time Sicily should experience no more than 7 earthquakes of this magnitude per half-century. For perspective, Sicily has an earthquake that is noticeable to the inhabitants every 22 days on average.

Example 3.46

From Table 3.2, $\mu = \frac{1}{\nu} = \frac{1}{3} = 20$ minutes between calls. Also, $\sigma^2 = \frac{1}{\nu^2}$, so $\sigma = \mu = 20$. In minutes, the $\pm 2\sigma$ interval around 20 is $[0, 60]$, so 75 % of the time, the gap between calls will be less than an hour. We found a more accurate estimate in Section 2.7 Example 2.47. Actually, 95 % of the time the gap will be under an hour.

Example 3.47

$$\mu_Z = E\left[\frac{X-\mu}{\sigma}\right] = \frac{1}{\sigma}E[X-\mu] = \frac{1}{\sigma}(E[X]-\mu) = \frac{1}{\sigma}(\mu-\mu) = 0.$$

Similarly,

$$\sigma_Z^2 = E\left[\left(\frac{X-\mu}{\sigma}\right)^2\right] - \left(E\left[\frac{X-\mu}{\sigma}\right]\right)^2 = E\left[\left(\frac{X-\mu}{\sigma}\right)^2\right],$$

since $E[\frac{X-\mu}{\sigma}] = 0$. Then $E[(\frac{X-\mu}{\sigma})^2] = \frac{1}{\sigma^2}E[(X-\mu)^2]$, which in turn is $\frac{1}{\sigma^2}E[X^2 - 2\mu X + \mu^2] = \frac{1}{\sigma^2}(E[X^2] - \mu^2) = \frac{\sigma^2}{\sigma^2} = 1$.

Example 3.48

The bell curve is symmetric, so from the discussion above, $P(-1 \le Z \le 1) = 2 \cdot (0.3413) = 0.6826$. Roughly speaking, two-thirds of the area under the bell curve is within one standard deviation of zero.

Example 3.49

Again by the symmetry of the bell curve, we will imitate what we did in Example 3.48. From the table, $P(0 \le Z \le 2) = 0.4772$, so $P(-2 \le Z \le 2) = 2 \cdot (0.4772) = 0.9544$. Also $P(0 \le Z \le 3) = 0.4987$, so $P(-3 \le Z \le 3) = 0.9974$.

Example 3.50

$P(-0.62 \le Z \le 1.73) = P(0 \le Z \le 0.62) + P(0 \le Z \le 1.73)$ by symmetry. From the table, $P(0 \le Z \le 0.62) = 0.2324$, and $P(0 \le Z \le 1.73) = 0.4582$. So $P(-0.62 \le Z \le 1.73) = 0.2324 + 0.4582 = 0.6906$.

Example 3.51

By symmetry, $P(Z \le 0.0) = 0.50$. Then from the table $P(0 < Z < 0.75) = 0.2734$. So $P(Z < 0.75) = 0.50 + 0.2734 = 0.7734$.

Example 3.52

$P(0.47 \le Z \le 2.06) = P(0 \le Z \le 2.06) - P(0 \le Z \le 0.47) = 0.4803 - 0.1808 = 0.2995$.

Example 3.53

Transform $X \to Z$. So $Z = \frac{X-2}{1.5}$. Then $x = 1$ corresponds to $z_1 = -\frac{1}{1.5} = -0.667$ and $x = 4$ corresponds to $z_2 = \frac{2}{1.5} = 1.333$. We need $P(-0.667 \le Z \le 1.333)$. This is similar to Example 3.50. We have $P(-0.667 \le Z \le 1.333) = P(0 \le Z \le 0.667) + P(0 \le Z \le 1.333) = 0.2474 + 0.4987 = 0.7461$. So $P(1 \le X \le 4) = 0.7461$.

Example 3.54

(i) $\mu = 5$ and $\sigma = 0.4$, so $Z = \frac{X-5}{0.4}$. Hence $x = 5.5$ corresponds to $z = \frac{5.5-5}{0.4} = 1.25$. So $P(Z \le 1.25) = 0.500 + 0.3944 = 0.8944$ or 89.4 %.

(ii) $P(4.5 \le X \le 5.5) = P(\frac{4.5-5}{0.4} \le Z \le \frac{5.5-5}{0.4}) = P(-1.25 \le Z \le 1.25) = 2 \cdot P(0 \le Z \le 1.25) = 2 \cdot 0.3944 = 0.7888 = 78.9\,\%$.

(iii) Let c be the volume of the cup. Then $P(X > c) \le 0.01$. This is equivalent to $P(Z > \frac{c-5}{0.4}) \le 0.01$. From the table, $P(0 \le Z \le 2.33) = 0.4901$, so $P(2.33 < Z) = 0.500 - 0.4901 = 0.01$. Evidently, $\frac{c-5}{0.4} = 2.33$, or $c = 5.93$ ounces, so a 6-ounce cup will do. Real coffee machines are better than this.

Note that in this problem as well as others, to determine $P(Z > \alpha)$ for $\alpha > 0$, it is only necessary to find $P(0 \le Z \le \alpha)$ and subtract it from $P(0 \le Z \le \infty) = 0.500$. For $\alpha > 0$, we have $P(Z \le \alpha) = P(-\infty \le Z \le 0) + P(0 \le Z \le \alpha) = 0.500 + P(0 \le Z \le \alpha)$. Then $P(Z > \alpha) = 1 - P(Z \le \alpha) = 1 - [0.500 + P(0 \le Z \le \alpha)] = 0.500 - P(0 \le Z \le \alpha)$.

Example 3.55

We assume the component life X is normally distributed. Restating the deal, the manufacturer wants a component to have a 95 % chance of surviving the guarantee period because he has budgeted to furnish free replacements for those that do not. Let g be the guarantee period in years. Then $P(X > g) \ge 0.95$. This is equivalent to $P(Z > \frac{g-8}{2}) \ge 0.95$ or $P(Z \le \frac{g-8}{2}) \le 0.05$. From the table, $P(0 \le Z \le 1.65) = 0.500 - 0.4505 = 0.05$. So $\frac{g-8}{2} = 1.65$ implies $g = 11.3$ years. We see that 95 % of the components should last at least 11.3 years.

Example 3.56

We want the ratio of the 110+ IQ population to the 90+ IQ population. Let X be the IQ random variable. Then $Z = \frac{X-100}{16}$. So $x = 90$ corresponds to $z = \frac{-10}{16} = -0.625$ and $x = 110$ corresponds to $z = \frac{10}{16} = 0.625$. The fraction of the general population with IQs above 90 is $P(-0.625 \le Z) = P(-0.625 \le Z \le 0) + P(0 \le Z) = P(0 \le Z \le 0.625) + 0.500$ by symmetry of the bell curve. This works out to $0.2341 + 0.500 = 0.7341$. The fraction with IQs above 110 is $P(Z \ge 0.625) = 0.500 - 0.2341 = 0.2659$. Finally, the population ratio is $\frac{110+}{90+} = \frac{0.2659}{0.7341} = 36.2\,\%$.

Example 3.57

Let the diameter of a manufactured bolt be D. Variability in a process of this type is usually modeled by a normal distribution. We have $Z = \frac{D-0.75}{0.003}$. A bolt will be out of spec if $Z > \frac{0.006}{0.003} = 2.0$ or $Z < \frac{-0.006}{0.003} = -2.0$. The probability of a bolt to be in spec is $P(-2.0 \le Z \le 2.0) = 2 \cdot P(0 \le Z \le 2.0) = 2 \cdot 0.4772 = 95.4\,\%$, so the probability that a random bolt is out of specification is 4.6 %.

Example 3.58

Let $Z = \frac{X-\mu}{5}$. Then $P(Z < \frac{32-\mu}{5}) = 0.60$. Since $0.60 > 0.50 = P(-\infty \leq Z \leq 0)$, it must be the case that $\frac{32-\mu}{5} > 0$. Also, from the table $0.60 = 0.50 + 0.10 = 0.50 + P(0 \leq Z \leq 0.255)$. We conclude $\frac{32-\mu}{5} = 0.255$ or $\mu = 32 - 1.275 = 30.725$. Then $P(X > 40) = P(Z > \frac{40-30.725}{5}) = P(Z > 1.855) = 0.500 - P(0 \leq Z \leq 1.855) = 0.500 - .4682 = 0.0318 = 3.2\,\%$.

Example 3.59

Let X be the number of heads. Then $Z = \frac{X-\mu}{\sigma}$ where we substitute $\mu = np$ and $\sigma = \sqrt{np(1-p)}$. So $Z = \frac{X-np}{\sqrt{np(1-p)}}$ and for this problem $Z = \frac{X-500}{5\sqrt{10}}$. We seek $P(\frac{495-500}{5\sqrt{10}} \leq Z \leq \frac{505-500}{5\sqrt{10}})$, which simplifies to $P(-0.310 \leq Z \leq 0.310)$. This is $2 \cdot P(0 \leq Z \leq 0.310) = 2 \cdot (0.1217) = .2434 = 24.3\,\%$. So if this experiment were done repeatedly, almost a quarter of the outcomes would be within ± 5 heads of the mean. If you repeat the calculation with ± 20 heads, essentially all the outcomes are likely to be in this range.

Example 3.60

Recall we determined that $b(50; 100, 0.5) = 7.96\,\%$. Letting $Z = \frac{X-np}{\sqrt{np(1-p)}} = \frac{X-50}{5}$, we force the point estimation by finding $P(\frac{49.5-50}{5} \leq Z \leq \frac{50.5-50}{5})$. This is equivalent to $P(\frac{-0.5}{5} \leq Z \leq \frac{0.5}{5})$ or $P(-0.10 \leq Z \leq 0.10)$ which is $2 \cdot P(0 \leq Z \leq 0.10) = 2 \cdot 0.0398 = .0796 = 7.96\,\%$. In this example we used the so-called **continuity correction**, which allows us to approximate a discrete distribution (binomial) with a continuous distribution (normal). The correction amounts to treating k in the binomial distribution as the interval $[k - \frac{1}{2}, k + \frac{1}{2}]$ in the normal distribution. So for $k = 50$ in the binomial, we used $z \in [50 - \frac{1}{2}, 50 + \frac{1}{2}]$ in the normal.

Example 3.61

$b(2; 100, 0.05) \approx \frac{(5)^2 e^{-5}}{2!} = 0.0842$. The actual value is $\binom{100}{2}(0.05)^2(0.95)^{98} = 0.081$.

Example 3.62

Maple's error function (erf) command gives

$$P(Z > 6.0) = 1.00 - 0.999999999013412 \approx 10^{-9} \quad \text{and}$$

$$P(Z > 4.5) = 1 - 0.999996602326875 = 0.0000033977 \approx 3.4 \times 10^{-6}.$$

So for the ideal 6σ process, the probability of exceeding specification limits either high or low is an infinitesimal $2 \times 10^{-7}\%$. If the mean drifts to $+1.5\sigma$, the probability of exceeding the upper specification limit (USL) becomes $3.4 \times 10^{-4}\%$. In this case the drift favors not exceeding the lower specification limit (LSL). In fact, the probability of that happening becomes $P(Z < 7.5) \approx 3 \times 10^{-14} \approx 0$. Adding these two probabilities, we conclude that the chance of exceeding the overall specification limits is $3.4 \times 10^{-4}\%$. If the process consisted of making electronic components, this would mean we could expect about 7 out-of-spec components in two million. A similar argument applies to the mean drifting to -1.5σ.

Example 3.63

We want

$$E(\overline{X}) = E\left(\frac{\sum_{k=1}^{n} X_k}{n}\right) = \frac{1}{n}\sum_{k=1}^{n} E(X_k),$$

by Table 3.1(4). Now the X_k are identically distributed with $E(X_k) = \mu$, so

$$\frac{1}{n}\sum_{k=1}^{n} E(X_k) = \frac{1}{n}\sum_{k=1}^{n} \mu = \frac{n\mu}{n} = \mu = \mu_{\overline{X}}.$$

Example 3.64

Recall (Section 3.1.3 Example 3.18, generalized) that for independent random variables, the variance of the sum is equal to the sum of the variances. Since $\overline{X} = \frac{\sum_{k=1}^{n} X_k}{n} = \sum_{k=1}^{n} \frac{X_k}{n}$, we have $V(\overline{X}) = \sum_{k=1}^{n} V(\frac{X_k}{n})$. Now $V(\frac{X_k}{n}) = \frac{1}{n^2}V(X_k)$ by Section 3.1.3 Example 3.17. Also, since the X_k are identically distributed with $V(X_k) = \sigma^2$, we have

$$\sum_{k=1}^{n} V\left(\frac{X_k}{n}\right) = \sum_{k=1}^{n} \frac{1}{n^2}V(X_k) = \frac{n\sigma^2}{n^2} = \frac{\sigma^2}{n} = V(\overline{X}).$$

Unlike the sample mean, the sample variance depends on the size of the sample. We might add that this dependence is "in a good way", as all else being equal a bigger sample implies a smaller standard error, $\sigma_{\overline{X}} = \frac{\sigma}{\sqrt{n}}$.

Example 3.65

We are considering a sample of size $n = 64$ taken from a population of component pieces with average strength $\mu = 2.0 \times 10^6$ and standard deviation $\sigma = 4.0 \times 10^5$. The central limit theorem is warranted (the lumber strength is probably normally distributed anyway). We look at the standard normal variable

$$z = \frac{\overline{x} - 2.0 \times 10^6}{\left(\frac{4.0\times10^5}{\sqrt{64}}\right)} = \frac{\overline{x} - 2.0 \times 10^6}{50{,}000}.$$

The Z value corresponding to the weak beam is $\frac{1.8\times10^6 - 2.0\times10^6}{50{,}000} = -4$. Substantially all of the area under the bell curve is to the right of $z = -4$, so the probability $\approx 100\,\%$ that the beam is not composed of lumber pieces with average strength less than 1.8×10^6 psi. The company need not worry.

Example 3.66

Sample size is 40 % of the population, so we will need the finite population correction factor. We know $\mu = 5.5$ and $\sigma^2 = 8.25$. So $\mu_{\overline{X}} = 5.5$ and $V(\overline{x}) = \frac{8.25}{4} \cdot \frac{10-4}{10-1} = 1.375$, so $\sigma_{\overline{X}} = 1.173$. This example is way below the threshold where we can use the central limit theorem, but in the upcoming discussion of the t-distribution we will have a fix for small samples.

Example 3.67

From Table 3.1(5), $E(\overline{X}_1 - \overline{X}_2) = E(\overline{X}_1) - E(\overline{X}_2) = \mu_1 - \mu_2 = \mu_{\overline{X}_1 - \overline{X}_2}$.

Example 3.68

From the squaring argument in Section 3.1.3 Example 3.18, we see that we can change the sign on the left in $\sigma^2_{X-Y} = \sigma^2_X + \sigma^2_Y$ without changing the right-hand side. So $V(\overline{X}_1 - \overline{X}_2) = V(\overline{X}_1) + V(\overline{X}_2)$. Using their equivalents from Example 3.64, we have $V(\overline{X}_1) + V(\overline{X}_2) = \frac{\sigma_1^2}{n_1} + \frac{\sigma_2^2}{n_2}$. So $V(\overline{X}_1 - \overline{X}_2) = \frac{\sigma_1^2}{n_1} + \frac{\sigma_2^2}{n_2}$ and $\sigma_{\overline{X}_1 - \overline{X}_2} = \sqrt{\frac{\sigma_1^2}{n_1} + \frac{\sigma_2^2}{n_2}}$. If the samples are taken from finite populations without replacement, the variances in the formula can be modified with an appropriate finite correction factor for each.

Example 3.69

Let $\overline{x}_1$ be the mean lifetime of a Company A lamp and $\overline{x}_2$ be the mean lifetime of a Company B lamp. We want to know $P(\overline{x}_1 - \overline{x}_2 > 1)$. Let

$$z = \frac{(\overline{x}_1 - \overline{x}_2) - (10 - 9)}{\sqrt{\frac{(1.5)^2}{30} + \frac{(2.5)^2}{40}}} = \frac{(\overline{x}_1 - \overline{x}_2) - 1}{0.481}.$$

The Z value corresponding to $(\overline{x}_1 - \overline{x}_2) = 1$ is $z = 0$, so $P(\overline{x}_1 - \overline{x}_2 > 1) = P(z > 0) = 0.500$.

Example 3.70

From a binomial proportion standpoint, we are asking what is the probability that the player is successful in between 0.175 and 0.300 of his at bats. Since we are modeling a discrete process with a continuous distribution, we should use the continuity correction and restate the success interval as [34.5, 60.5]. Then $z_{\text{low}} = \frac{0.1725-0.250}{0.0306} = -2.53$ and $z_{\text{hi}} = \frac{0.300-0.250}{0.0306} = 1.63$. So we want $P(-2.53 \leq Z \leq 1.63)$, which is the same as $P(0 \leq Z \leq 2.53) + P(0 \leq Z \leq 1.63) = 0.4943 + 0.4484 = 0.9427$. So the player has a 94.3 % chance of achieving between 35 and 60 hits.

Example 3.71

The sample size is below the threshold for using the normal distribution, so we use the t-distribution. We have

$$t = \frac{765 - 750}{\frac{35}{\sqrt{20}}} = \frac{15}{7.83} = 1.92.$$

Note that $t_{0.025,19} = 2.093$, so $-t_{0.025,19} \le t \le t_{0.025,19}$. If the claimed mean $\mu = 750$ is correct, then there is a 95 % probability that a t-value based on a random sample of size 20 would fall in this interval. Since the sample t-value did, we conclude that the claim of an average lifetime of 750 landings is reasonable.

Example 3.72

We assume the score differential random variable D is normally distributed from a population with mean μ_D and unknown variance σ_D^2. The mean sample score difference $\overline{d} = 1.11$ and sample variance $s_d^2 = 2.16$. We form

$$t = \frac{\overline{d} - \mu_D}{\frac{s_d}{\sqrt{n}}}$$

and evaluate it with the given data. Assuming the two players are equal, $\mu_D = 0$. So

$$t = \frac{1.11 - 0}{\frac{1.47}{3.16}} = 2.39.$$

For $v = 9$, we have from Table 3.4 that $P(-2.26 \le t \le 2.26) = 95\,\%$. Since our value $t = 2.39$ is outside this range, it appears that the assumption $\mu_D = 0$ is not warranted and hence the players are probably not equal.

This test was set up to determine if the computed t value was outside the range $(-2.26, 2.26)$. The areas under the T distribution curve below $t = -2.26$ and above $t = 2.26$ are called, respectively, the left and right **tails** of the distribution corresponding to the 95 % area within the interval $(-2.26, 2.26)$. Since the distribution is symmetric, each tail represents 2.5 % of the total area under the curve, and the test is called a **two-tailed test**. This type of test is used whenever we are dealing with a two-sided alternative, such as this case where either player could have been more skilled. Although a brief glance at the data suggests that Player 1 is noticeably better than Player 2, we purposely suppress this observation and set the test up as if either player could be dominant. The actual data shows that they are probably not equally skilled, so our suspicion is justified. Whenever we want to test a one-sided alternative we use a **one-tailed test** and use either the right or left hand tail as appropriate. We will discuss this in more detail in Section 3.6, Hypothesis Testing.

Example 3.73

For this sample, $\overline{x} = 50.18$, $s^2 = 0.402$. So $\chi^2 = \frac{5\cdot(0.402)}{(2.0)^2} = 0.503$. Checking the table, $\chi^2_{0.99,5} = 0.554$. So $\chi^2 < \chi^2_{0.99,5}$ means this χ^2 value falls below 99 % of all χ^2 values with 5 degrees of freedom. This is unlikely, but if the quoted standard deviation were less, say 1 %, $\chi^2 = \frac{5\cdot(0.402)}{(1.0)^2} = 2.01$. Again from the table there is a 90 % probability that χ^2 values with 5 degrees of freedom are between 1.145 and 11.070. So quoting a 1 % standard deviation is very reasonable and would allow the manufacturer to improve his quality claim.

Example 3.74

$\chi^2 = \frac{24\cdot 7^2}{3^2} = 130.7$. From the table, 99.5 % of χ^2 values with 24 degrees of freedom are less than 45.558. So, yes, the claim is unreasonable. To fix it, the marketing department might claim $s = 6$, then $\chi^2 = \frac{24\cdot 7^2}{6^2} = 32.67 < 36.415 = \chi^2_{0.95,24}$. Much more believable.

Example 3.75

Here $(1 - \alpha) = 0.95$, so $\frac{\alpha}{2} = 0.025$ and $z_{\alpha/2} = 1.96$. We also have $\frac{\sigma}{\sqrt{n}} = \frac{1.5}{10} = 0.15$. So

$$P(\mu \in (9.3 - 1.96(0.15), 9.3 + 1.96(0.15))) = P(\mu \in (9.006, 9.594)) = 95\,\%.$$

Example 3.76

Using s for σ, for 90 % confidence we have

$$P\left(\mu \in \left(73 - 1.65\left(\frac{15}{7}\right), 73 + 1.65\left(\frac{15}{7}\right)\right)\right) = P(\mu \in (69.46, 76.54)) = 90\,\%.$$

For 99 %, one gets

$$P\left(\mu \in \left(73 - 2.575\left(\frac{15}{7}\right), 73 - 2.575\left(\frac{15}{7}\right)\right)\right) = P(\mu \in (67.48, 78.52)) = 99\,\%.$$

Example 3.77

We do not have σ, so we will use s instead, but only accept an answer that exceeds 30. Since $\Delta = 2$, we get

$$n = \left(1.96\frac{s}{\Delta}\right)^2 = (1.96)^2\left(\frac{15}{2}\right)^2 = (56.25)(3.84) = 216.$$

We have met the $n \geq 30$ threshold, but this sample would be over half the population!

Example 3.78

Since $n = 15$ we need to use the t-distribution with $\nu = 14$. For $\alpha = 0.05$, we find from Table 3.5 that $t_{0.025,14} = 2.145$. So

$$P\left(\mu \in \left(10 - 2.145\left(\frac{1.3}{\sqrt{15}}\right), 10 + 2.145\left(\frac{1.3}{\sqrt{15}}\right)\right)\right)$$

simplifies to $P(\mu \in (9.28, 10.72)) = 95\,\%$. Had we mistakenly used the standard normal distribution with $z_{0.025} = 1.96$ our 95 % confidence interval would have contracted to $(9.34, 10.66)$, which gives the illusion of more precision.

Example 3.79

Calculating s^2, we get 0.0369. From Table 3.6, for $\alpha = 0.05$ and 7 degrees of freedom, $\chi^2_{\alpha/2} = 16.013$ and $\chi^2_{1-\alpha/2} = 1.69$. So

$$\left(\frac{(n-1)s^2}{\chi^2_{\alpha/2}}, \frac{(n-1)s^2}{\chi^2_{1-\alpha/2}}\right) = \left(\frac{(7)(0.0369)}{16.013}, \frac{(7)(0.0369)}{1.69}\right) = (0.016, 0.153).$$

Example 3.80

We have $s_1 = 2$, $n_1 = 10$, and $v_1 = 9$. Likewise, $s_2 = 3$, $n_2 = 25$, and $v_2 = 24$. Moreover, $\alpha = 0.02$, so $\frac{\alpha}{2} = 0.01$. We verify that $f_{0.01}(9, 24) = 3.22$ and $f_{0.01}(24, 9) = 4.73$. Plugging these values into our formula for the 98 % confidence interval for $\frac{\sigma_1^2}{\sigma_2^2}$, we get $(\frac{2^2}{3^2}(\frac{1}{3.22}), \frac{2^2}{3^2}(4.73)) = (0.138, 2.10)$. In terms of the standard deviation ratio $\frac{\sigma_1}{\sigma_2}$, this is $(0.371, 1.45)$. If 1.0 were in the center of this interval, we would say the ammunition from both manufacturers has about the same consistency. However, it looks like manufacturer A makes a slightly better product.

Example 3.81

Based on the discrepancy between the claim and the sample mean, we are skeptical that the claim is true. Let H_0 = "the mean breaking strength is 5000 lbs." The complementary statement is H_1 = "the mean breaking strength is not 5000 lbs." Since $\alpha = 0.025$, and the sampling distribution is normal ($n = 40$), the acceptance region is $(-2.25 < Z < 2.25)$. Then the critical region is **two-tailed** $Z < -2.25$ or $Z > 2.25$. This is a consequence of the formulation of the statement of H_1. Calculating z using $\overline{x} = 4900$, we get

$$z = \frac{4900 - 5000}{\frac{250}{\sqrt{40}}} = -2.53 < -2.25.$$

So we reject H_0 and the manufacturer's claim is unsupported. This calculation would also reject a revised H_0 = "the mean breaking strength is at least 5000 lbs."

Example 3.82

We have to use T-scores because $n < 30$. We have the benefit of knowing σ, so we use it instead of s in computing

$$t = \frac{4900 - 5000}{\frac{250}{\sqrt{16}}} = 1.6.$$

The acceptance region for $\alpha = 0.02$ and $\nu = 15$ is $(-2.602 \le T \le 2.602)$. In this case the null hypothesis should be accepted (fail to be rejected) and the manufacturer's claim appears reasonable.

Example 3.83

Let μ_0 be 90,000 and μ be the manufacturer's claimed mileage. Then H_0 = "$\mu = \mu_0$" and H_1 = "$\mu > \mu_0$." The sample size $n \ge 30$, so we compute

$$z = \frac{\overline{x} - \mu_0}{\frac{s}{\sqrt{n}}} = \frac{100,000 - 90,000}{\frac{20,000}{\sqrt{36}}} = 3.0.$$

The critical region consists of the area to the right of $z = 2.33$. The statement of H_1 makes this a **one-tailed** critical region. Our z computed from the sample is well above 2.33, so we reject H_0 and accept H_1. The manufacturer's claim looks very reasonable. For $z = 3.0$, the observed level of significance $p = 0.0013 \ll 0.01$. Another vote for rejection.

Example 3.84

We are testing the null hypothesis H_0 = "$\mu_1 = \mu_2$" with H_1 = "$\mu_1 \neq \mu_2$." This will be a two-tailed test because the inequality between groups could go either way. We compute

$$z = \frac{(\overline{x}_1 - \overline{x}_2) - (\mu_1 - \mu_2)}{\sqrt{\frac{s_1^2}{n_1} + \frac{s_2^2}{n_2}}} = \frac{(81 - 79) - 0}{\sqrt{\frac{25}{80} + \frac{49}{40}}} = \frac{2}{1.24} = 1.61.$$

The critical region is $Z < -1.96$ or $Z > 1.96$. Based on the calculated z, we accept H_0, so it is reasonable to assume there is no difference in performance between English and Scottish students in general on this exam.

Example 3.86

The pieces of the formula are: $n = 8$,

$$\sum_{k=1}^{8} x_k = 495, \quad \sum_{k=1}^{8} y_k = 870, \quad \sum_{k=1}^{8} x_k y_k = 53,922,$$
$$\sum_{k=1}^{8} x_k^2 = 30,673, \quad \text{and} \quad \sum_{k=1}^{8} y_k^2 = 94,864.$$

So

$$\rho = \frac{(8)(53922) - (495)(870)}{\sqrt{((8)(30673) - (495)^2) \cdot ((8)(94864) - (870)^2)}}.$$

This works out to $\rho = \frac{726}{849.9} = 0.854$. This is a fairly high positive correlation, which might be expected due to the similar levels of conditioning required for the sport. The general population would probably be somewhat lower.

Example 3.87

Using the sums from Chapter 6 Example 6.1, we have

$$m_{\text{best}} = \frac{(8)(53922) - (495)(870)}{(8)(30673) - (495)^2} = \frac{726}{359} = 2.02.$$

Now $\overline{x} = 61.9$ and $\overline{y} = 108.8$, so $b_{\text{best}} = 108.8 - (2.02)(61.9) = -16.2$. So the regression line has equation $y = 2.02x - 16.2$. We would predict that a four-foot-ten gymnast weighs $y = 2.02(58) - 16.2 = 101.0$ lbs and a five-foot-six gymnast $y = 2.02(66) - 16.2 = 117.1$ lbs. A rule of thumb, which we have conveniently broken here, is that regression formulas should not be used for extrapolation outside the regressor data set. We have no assurance that our linearized formula remains valid outside that region.

Example 3.88

The governing relationship is $Y = AX^2$ for some constant A. This does not work directly as a linear regression, but we can rewrite it as $\sqrt{Y} = \sqrt{A}X$. Now we do have a linear relationship between $\sqrt{Y}$ and X. So let us redo the table as

X	0.7	1.0	1.1	0.9	1.2	0.6
$V = \sqrt{Y}$	31.62	45.83	51.48	40.62	55.68	27.39

where we set $V = \sqrt{Y}$ to avoid confusion. We have

$$\sum_x = 5.5, \quad \sum_v = 252.62, \quad \sum_{xv} = 244.4, \quad \sum_{x^2} = 5.31.$$

Plugging in to our formula,

$$m_{\text{best}} = \frac{(6)(244.4) - (5.5)(252.62)}{(6)(5.31) - (5.5)^2} = \frac{76.99}{1.61} = 47.82.$$

Then $b_{\text{best}} = \overline{v} - m\overline{x} = 42.10 - (47.82)(0.917) = -1.75$. So the linear regression line has equation $\sqrt{y} = 47.82x - 1.75$. Recovering y, we have $y = (47.82x - 1.75)^2$. A 0.8 carat diamond should sell at this auction for about $(47.82(0.8) - 1.75)^2 = \$1333$. That seems consistent with the table. Yes, ideally b_{best} should be zero, but the nonzero value is an artifact of the approximation process.

4 Vector methods

The primary focus of linear algebra is on vector spaces and their transformations, which can be represented by matrices. We will cover vector methods in this chapter and discuss matrix methods separately. Vector space theory in its present form is a relatively recent mathematical invention. There are glimmers of the notion of vector in the works of the Greek geometers, but the core idea seems to have been dormant until it reappeared in a form that we would recognize today about the time of Isaac Newton. Although Newton worked with quantities having both magnitude and direction like force, displacement, and velocity, he never took the step of regarding them as examples of the broader concept we now call "vector." Several prominent mathematicians including C. F. Gauss and W. R. Hamilton moved the concept closer to our current formulation in the first half of the nineteenth century. The German high school teacher Hermann Grassmann published an amazingly prescient theory in the 1840s touching on n-dimensional spaces and the use of matrix transformations, but his work was largely and unjustly ignored. Grassmann had no stature in the scientific community and his book was written with a cryptic notation, but he was vindicated 30 years later when the significance of his contributions was finally appreciated. The major impetus for the adoption of vector methods came from the physicists James Clerk Maxwell and Oliver Heaviside, who framed their theories of electromagnetism in terms of vectors. The American physicist J. Willard Gibbs had studied Maxwell's work and adapted it to thermodynamics. Gibbs turned out to be the prime mover in codifying and popularizing modern vector space theory. He was a professor at Yale and in the early 1880s began to circulate printed notes explaining vector analysis, first to his students and ultimately to scientists in the United States and Europe. In 1901 he published the first text on modern vector space theory entitled *Vector Analysis*. Acceptance of Gibbs' theory was wide but not unanimous. There were still holdouts who championed Hamilton's old theory of quaternions, but that view eventually faded as the vector concept proved superior.

A typical vector theory course taught today to engineering and physics students would have a practical focus probably very similar to a course Gibbs might have offered. This is understandable in view of how the subject developed. More recently, mathematicians have pursued abstractions of vector space theory that would be barely recognizable to an engineer. Our discussion will emphasize the down-to-earth problem-solving aspects of vector methods and leave the abstruse points to the theorists. We now turn to a presentation of the basic definitions.

4.1 Basic definitions

A **vector space (linear space)** is a pair of sets $\langle V, F\rangle$, where V consists of elements called **vectors** equipped with the operation of **vector addition** and F is a so-called **scalar field**,

https://doi.org/10.1515/9783111506654-005

typically $\mathbb{R}$ or $\mathbb{C}$. Elements of the scalar field may be combined with vectors in accordance with **scalar multiplication**. We will write vectors in boldface and scalars unbolded. Vector addition and scalar (field) addition will be denoted by "+". There should be no ambiguity based on what elements are being added, but if it helps with clarity the symbol $\oplus$ can be used for vector addition. Scalar multiplication will be indicated by juxtaposition with the scalar always to the left. Operations within the field have been covered in Section 0.2.1. When the scalar field is clear from context, we often abuse the notation and refer to the vector space as just V. The operations specific to a vector space behave as follows:

Table 4.1: Axioms for vector spaces. $\mathbf{x}, \mathbf{y}, \mathbf{z} \in V$, $a, b, 0, 1 \in F$.

1) Associativity of addition	$\mathbf{x} + (\mathbf{y} + \mathbf{z}) = (\mathbf{x} + \mathbf{y}) + \mathbf{z}$
2) Commutativity of addition	$\mathbf{x} + \mathbf{y} = \mathbf{y} + \mathbf{x}$
3) Existence of an additive identity (the zero vector)	$\mathbf{x} + \mathbf{0} = \mathbf{x}$ (note $\mathbf{0} \neq 0$)
4) Existence of additive inverses	$\mathbf{x} + (-\mathbf{x}) = \mathbf{0}$
5) Unitarity	$1\mathbf{x} = \mathbf{x}$
6) Annihilation	$0\mathbf{x} = \mathbf{0}$
7) Distributivity of scalar multiplication over vector addition	$a(\mathbf{x} + \mathbf{y}) = a\mathbf{x}+a\mathbf{y}$
8) Distributivity of scalar multiplication over scalar addition	$(a + b)\mathbf{x} = a\mathbf{x} + b\mathbf{x}$
9) Associativity of scalar multiplication	$a(b\mathbf{x}) = (ab)\mathbf{x}$

Axioms 1) through 4) involve only vector addition; V with these four properties forms an **abelian group**.

Example 4.1. Show that $\mathbb{R}^2$ is a vector space over the real field ($\mathbb{R}$ again).

Example 4.2. Show that $\mathbb{R}^n$ is a vector space over $\mathbb{R}$. Of course, this will supersede Example 4.1 and separately show that $\langle \mathbb{R}^3, \mathbb{R} \rangle$ is a vector space.

There is a subtlety in Examples 4.1 and 4.2 that needs to be clarified. We have shown that the various n-tuples equipped with their relevant operations fulfill the nominal requirements for a vector space. Both $\mathbb{R}^2$ and $\mathbb{R}^3$ as Cartesian products are technically just collections of points or geometric addresses. To develop a geometry on these spaces, we typically impose the so-called **Euclidean metric** ρ, which gives us the ability to determine the distance between points. We remark that this distance function is available for any $\mathbb{R}^n$, and occasionally you will see the notation $\mathbb{E}^n = \langle \mathbb{R}^n, \rho \rangle$ to emphasize the presence of the metric. The metric boils down to a generalization of the Pythagorean formula in n dimensions. Given $\mathbf{x} = (x_1, x_2, \ldots, x_n)$ and $\mathbf{y} = (y_1, y_2, \ldots, y_n)$, $\rho(\mathbf{x}, \mathbf{y}) = \sqrt{\sum_{i=1}^{n}(x_i - y_i)^2}$.

So far we have not seen an element that explicitly has magnitude and direction – properties that we are expecting from our naïve understanding of vector spaces. We can easily remedy that. Using $\mathbb{R}^2$ as an example, let us define the vector $\langle a, b\rangle$ to be the position vector of the point (a, b). It is a **directed line segment** connecting the origin $(0, 0)$ of $\mathbb{R}^2$ as the **initial point (tail)** to the point (a, b) as the **terminal point (head)**. This scheme clearly gives us a bijective map between points (a, b) and vectors $\langle a, b\rangle$. For the former, we view $\mathbb{R}^2$ as just a set of locations in the *xy*-plane, and for the latter we view $\mathbb{R}^2$ as a collection of directed line segments emanating from the origin. From context and notation, it will be easy to distinguish a location from its position vector. The general vector space axioms are clearly satisfied as well for the vector interpretation, so we are now justified in calling it a vector space. The applicable metric is just the familiar Pythagorean formula in $\mathbb{R}^2$ for distance between points $P_1 = (x_1, y_1)$ and $P_2(x_2, y_2)$, namely $d(\overline{P_1P_2}) = \sqrt{(x_1 - x_2)^2 + (y_1 - y_2)^2}$. This immediately gives us the length of $\langle a, b\rangle$ as $\sqrt{a^2 + b^2}$. So now the vector $\langle a, b\rangle$ is finally shown to have both magnitude and direction, as we were anticipating.

In general, the **length** (or **norm**) of an arbitrary vector $\mathbf{v}$ is customarily written as $|\mathbf{v}|$, or, if there may be confusion (as in Table 4.2) with absolute value, as $\|\mathbf{v}\|$. For a vector $\mathbf{x} = (x_1, x_2, \ldots, x_n)$ in $\mathbb{R}^n$, we have immediately $|\mathbf{x}| = \sqrt{\sum_{i=1}^{n} x_i^2}$. Vector spaces where such a measure of vector length is available are called **normed vector spaces**. There are abstract spaces which are not normable.

Table 4.2: Axioms for vector space norms. $\mathbf{x}, \mathbf{y} \in V$, $a \in F \|\cdot\|$ is norm, $|\cdot|$ is absolute value or modulus if $F = \mathbb{C}$.

1) Faithfulness	$\|\mathbf{x}\| = 0$ implies $\mathbf{x} = \mathbf{0}$
2) Absolute (or positive) homogeneity	$\|a\mathbf{x}\| = \|a\| \cdot \|\mathbf{x}\|$
3) Subadditivity (triangle inequality)	$\|\mathbf{x} + \mathbf{y}\| \leq \|\mathbf{x}\| + \|\mathbf{y}\|$

These axioms immediately imply that:

(i) $\|\mathbf{0}\| = 0$, since for any $\mathbf{x} \in V$, $\|\mathbf{0}\| = \|0\mathbf{x}\| = |0|\|\mathbf{x}\| = 0$. This seemingly obvious property is the converse of faithfulness, so now we know that the only vector with zero length is $\mathbf{0}$.

(ii) for any $\mathbf{x} \in V$, $\|\mathbf{x}\| \geq 0$, since $\|\mathbf{x} - \mathbf{x}\| \leq \|\mathbf{x}\| + \|\mathbf{x}\|$ implies $2\|\mathbf{x}\| \geq \|\mathbf{0}\| = 0$. Hence every nonzero vector has positive length.

Faithfulness deserves an explanation. It ensures that the norm function recognizes, or separates, different vectors, since if $\|\mathbf{x} - \mathbf{y}\| = 0$, then $\mathbf{x} - \mathbf{y} = \mathbf{0}$ and $\mathbf{x} = \mathbf{y}$. A norm-like function $\|\cdot\|_s$ that satisfies axioms 2) and 3), but not 1), is called a **seminorm**. A seminorm may assign zero length to a nonzero vector, so if two vectors $\mathbf{x}$ and $\mathbf{y}$ are not equal, it may still be that $\|\mathbf{x} - \mathbf{y}\|_s$ is zero. We say in that case that the seminorm does not distinguish or separate the vectors $\mathbf{x}$ and $\mathbf{y}$.

Some vector spaces come equipped with a function called a **bilinear form**. If (V, F) is a vector space, then $f : V \times V \to F$ is a bilinear form if the following axioms are satisfied.

Table 4.3: Axioms for bilinear forms. $\mathbf{x}, \mathbf{y}, \mathbf{z} \in V$, $a \in F$.

1) Additivity in the first argument	$f(\mathbf{x}+\mathbf{y}, \mathbf{z}) = f(\mathbf{x}, \mathbf{z}) + f(\mathbf{y}, \mathbf{z})$
2) Additivity in the second argument	$f(\mathbf{x}, \mathbf{y}+\mathbf{z}) = f(\mathbf{x}, \mathbf{y}) + f(\mathbf{x}, \mathbf{z})$
3) Homogeneity in either argument	$f(a\mathbf{x}, \mathbf{y}) = f(\mathbf{x}, a\mathbf{y}) = af(\mathbf{x}, \mathbf{y})$

Mathematicians generally call a vector space equipped with a bilinear form an **inner product space**. This seems to be too general for physicists. They have a more specific concept of an inner product space that meshes with their formalism for quantum mechanics. Physicists define an inner product space as a vector space over $F = \mathbb{R}$ or $\mathbb{C}$ with a function called an inner product $\phi : V \times V \to F$ that satisfies the following axioms:

Table 4.4: Axioms for inner product spaces (applied version). $\mathbf{x}, \mathbf{y}, \mathbf{z} \in V$, $a \in F$.

1) Linearity in the first argument	$\phi(a\mathbf{x}+\mathbf{y}, \mathbf{z}) = a\phi(\mathbf{x}, \mathbf{z}) + a\phi(\mathbf{y}, \mathbf{z})$
2) Conjugate symmetry	$\phi(\mathbf{x}, \mathbf{y}) = \overline{\phi(\mathbf{y}, \mathbf{x})}$
3) Positivity	$\phi(\mathbf{x}, \mathbf{x}) \geq 0$
4) Definiteness	$\phi(\mathbf{x}, \mathbf{x}) = 0$ implies $\mathbf{x} = 0$

Note Axiom 2) implies $\phi(\mathbf{u}, \mathbf{u}) \in \mathbb{R}$, since $\phi(\mathbf{u}, \mathbf{u}) = \overline{\phi(\mathbf{u}, \mathbf{u})}$. Also Axiom 4) implies $\phi(\mathbf{0}, \mathbf{0}) = \phi(0\mathbf{x}, \mathbf{0}) = 0\phi(0\mathbf{x}, \mathbf{0}) = 0$, then Axiom 1) gives the converse.

These axioms immediately imply that

$$\phi(\mathbf{x}, a\mathbf{y} + \mathbf{z}) = \overline{\phi(a\mathbf{y} + \mathbf{z}, \mathbf{x})} = \overline{a}\overline{\phi(\mathbf{y}, \mathbf{x})} + \overline{a}\overline{\phi(\mathbf{z}, \mathbf{x})} = \overline{k}\phi(\mathbf{x}, \mathbf{y}) + \overline{k}\phi(\mathbf{x}, \mathbf{z}).$$

This is antilinearity in the second argument. If $F = \mathbb{R}$, the conjugations are irrelevant, and the inner product is both symmetric and linear in the second argument. This recovers all of the axioms for the bilinear form but also imposes symmetry, which the bilinear form does not have definitionally. A mapping which is linear in the first argument and antilinear in the second argument is often called sesquilinear (one and a half times linear) by mathematicians, although physicists often reverse the argument properties. *Bottom line*: an inner product for a real vector space is a symmetric bilinear form which is positive definite, and an inner product for a complex vector space is a conjugate symmetric sesquilinear form which is positive definite. Another notation for inner product is $\phi(\mathbf{u}, \mathbf{v}) = \langle \mathbf{u}, \mathbf{v} \rangle$, which steps on the toes of our vector component notation.

The inner product defines a norm by the formula $\|\mathbf{v}\| = \sqrt{\phi(\mathbf{u},\mathbf{u})}$. Shortly, we will introduce a version of vector multiplication called the "dot product," which is a real inner product defined for the space $\mathbb{R}^n$.

A point of slight confusion: note that it is required for any vector in $\mathbb{R}^n$ to have the origin as its initial point to conform to the axioms in Table 4.1. For example, a directed line segment constructed between the points $(1,1)$ and $(2,3)$ has the same length and direction as the vector $\langle 1,2\rangle$, but since it is not a position vector it is not directly expressible in the form $\langle a,b\rangle$. Hence it cannot legitimately participate in the operations defined for the vector space. Moreover, if we were to call it a vector, there would not be a unique vector corresponding to a specified length and direction. But there is a strong practical temptation to identify all directed line segments, no matter where their location, as vectors or at least vector-like. The resolution of this issue is to consider all directed line segments with the same length and direction as members of an equivalence class whose representative is always chosen as that whose initial point is the origin, namely a bona fide vector. Then adding two arbitrary directed line segments becomes the same as adding their vector representatives.

The expressions **bound vector** and **free vector** are sometimes encountered in physics and engineering literature. Our definition of a vector in $\mathbb{R}^2$ or $\mathbb{R}^3$ corresponds to a bound vector, namely one with a specific initial point at the origin. Arbitrary directed line segments are then free vectors which are elements of a so-called **affine space**. A not-entirely-facetious description of an affine space is "a vector space which has forgotten its origin." Every affine space has an associated vector space, so the free versus bound vector issue is perhaps a distinction without a difference. The specification of a directed line segment by its initial point (x_0,y_0,z_0) and its final point (x_1,y_1,z_1) automatically generates the corresponding bound vector $\langle x_1-x_0,y_1-y_0,z_1-z_0\rangle$, so all operations on directed line segments may be viewed as operations on bound equivalence class vector representatives. With the understanding that we are abusing our terminology, if we refer to a directed line segment as a vector we mean the bound vector that represents it.

Example 4.3. Since the vector spaces $\mathbb{R}^2$ and $\mathbb{R}^3$ are models of our physical world, they are called **spatial**. Give an example of a vector space that is not spatial.

Example 4.4. Show that the set of all real functions on an arbitrary nonvoid set is a vector space over $\mathbb{R}$.

Under some circumstances, a subset of a vector space $\langle V,F\rangle$ will satisfy the axioms in Table 4.1 and be a vector space on its own, or a **vector subspace**. So if $W \subset V$ as a set, we would like to identify the conditions under which W is a vector space. Note that we are forming a subset of the vectors V and not a subfield of the scalar field F. First of all, as a subgroup of V, it must be the case that W is closed under vector addition, that is, $\mathbf{x},\mathbf{y} \in W$ must imply $\mathbf{x}+\mathbf{y} \in W$. Likewise, W must be closed under scalar multiplication, so if $c \in F$, we must have $c\mathbf{x} \in W$ for all $\mathbf{x} \in W$. All of the operations for elements of V are inherited by W, so associativity and commutativity of vector addition are preserved in W. Zero belongs to every field F so by closure of scalar multiplication in W, the element $0\mathbf{x} \in W$. Now certainly $(0 + 0)\mathbf{x} = 0\mathbf{x}$. Viewing $\mathbf{x}$ as an element of V (so the axioms apply) by Axiom 7 we can write $(0 + 0)\mathbf{x} = 0\mathbf{x} + 0\mathbf{x} = 0\mathbf{x}$. Then by Axiom 4 we add the additive inverse of $0\mathbf{x}$ to both sides and, applying Axiom 3, we obtain $0\mathbf{x} + \mathbf{0} = \mathbf{0}$. Finally, another application of Axiom 3 yields $0\mathbf{x} = \mathbf{0}$. So closure of the two operations implies that $\mathbf{0} \in W$ and clearly the additive inverse of the 1 in the scalar field can be multiplied by any $\mathbf{x} \in W$ to produce the additive inverse of $\mathbf{x}$. Now we have Axioms 1 through 4 satisfied for W and again the remaining axioms are valid for W due to the properties of the scalar field. We conclude that closure of vector addition and scalar multiplication in the set W is sufficient for $\langle W, F\rangle$ to be a subspace of $\langle V, F\rangle$.

We can distill the sufficient condition for a subset of a vector space to be a subspace into one condition:

Subspace test for vector spaces

Given the vector space $\langle V, F\rangle$ and $W \subset V$ as a subset,
for every $\mathbf{w}_1$ and $\mathbf{w}_2$ in W and every $c \in F$, if $c\mathbf{w}_1 + \mathbf{w}_2 \in W$, then $\langle W, F\rangle$ is a subspace of $\langle V, F\rangle$.

This condition implies closure for vector addition and scalar multiplication in W. For closure of addition, set $c = 1$ to show $1\mathbf{w}_1 + \mathbf{w}_2 = \mathbf{w}_1 + \mathbf{w}_2 \in W$. We know $1\mathbf{w}_1 = \mathbf{w}_1$ as an element of V. For closure of scalar multiplication, set $\mathbf{w}_1 = \mathbf{w}_2$ and let $c = -1$, then $(-1)\mathbf{w}_2 + \mathbf{w}_2 = \mathbf{0}$ as an element of V, so evidently $\mathbf{0} \in W$. Now set $\mathbf{w}_2 = \mathbf{0}$ and we see $c\mathbf{w}_1 + \mathbf{0} = c\mathbf{w}_1 \in W$. Both operations are hence closed for the subset and conformance with all the axioms is verified. We conclude the subspace test is a sufficient condition for a subset to be a subspace.

Example 4.5. From Example 4.2 and the subsequent discussion, we know that $\mathbb{R}^3$ is a vector space over $\mathbb{R}$. Vectors in $\mathbb{R}^3$ are specified by ordered triples $\langle x, y, z\rangle$ where $x, y, z \in \mathbb{R}$. Let us choose a subset W of these ordered triples where the third entry is zero. So $W = \{\langle x, y, 0\rangle : x, y\} \in \mathbb{R}$. We claim $\langle W, \mathbb{R}\rangle$ is a vector space. Take $w_1 = \langle x_1, y_1, 0\rangle$ and $w_2 = \langle x_2, y_2, 0\rangle$ and any $c \in \mathbb{R}$. Form $cw_1 + w_2 = c\langle x_1, y_1, 0\rangle + \langle x_2, y_2, 0\rangle$. This can be rewritten as $\langle cx_1 + x_2, cy_1 + y_2, 0\rangle$. But this is the form of an element in W, so by the subspace test we conclude $\langle W, \mathbb{R}\rangle$ is a subspace of $\mathbb{R}^3$. You may recognize W as a plane in the three-dimensional space, in fact, it is precisely the xy-plane. If we add a nonzero constant vector $\mathbf{z} \in \mathbb{R}^3$ to every element of the subspace W, we are **translating** W by $\mathbf{z}$. In general, if M is a subspace of the vector space V and $\mathbf{x} \in V$, the translated subspace $\mathbf{x} + M$ is an **affine subspace** (or **linear manifold**) with the same dimension as M. It has "forgotten" its origin, so it is no longer technically a subspace of V, but vector operations still work (recall free vectors above). A maximal proper affine subspace of V is called a **hyperplane**. Any hyperplane in V necessarily has dimension one less than the dimension of V, or **codimension** 1. In our example, $W + \mathbf{z}$ is a hyperplane in $\mathbb{R}^3$.

4.2 Linear combinations & spans

A linear combination of vectors from the space $\langle V, F\rangle$ is simply a sum of vectors with various scalar coefficients applied. So $a\mathbf{x} + b\mathbf{y} + c\mathbf{z} = \mathbf{u}$ is a typical linear combination. If we choose $V = \mathbb{R}^3$, this combination would look like $a\langle x_1, x_2, x_3\rangle + b\langle y_1, y_2, y_3\rangle + c\langle z_1, z_2, z_3\rangle = \langle u_1, u_2, u_3\rangle$. Applying the scalar multiplications and then performing the vector additions, we get

$$\langle ax_1 + bx_1 + cx_1, ax_2 + bx_2 + cx_2, ax_3 + bx_3 + cx_3\rangle = \langle u_1, u_2, u_3\rangle.$$

Matching the respective entries in the ordered triples, we see that $u_1 = ax_1 + bx_1 + cx_1$, $u_2 = ax_2 + bx_2 + cx_2$, and $u_3 = ax_3 + bx_3 + cx_3$.

Example 4.6. In $\mathbb{R}^2$ we have the linear combination $a\langle 1, 2\rangle + b\langle 1, 3\rangle = \langle -2, 5\rangle$. Determine a and b.

Example 4.7. In $\langle \mathfrak{P}_x, \mathbb{Q}\rangle$ (see Example 5.3 in Section 5.1) we have the linear combination $a(x^3 - 3x^2 + 1) + b(2x^3 + 4x^2 + 3x - 5) + c(3x^3 + 2x^2 - 2) + d(x^2 - x + 1) = (11x^3 + 8x^2 + 7x - 14)$. Determine a, b, c, and d.

If we have a set of vectors, say $\{\mathbf{x}_1, \mathbf{x}_2, \dots, \mathbf{x}_n\} \in \langle V, F\rangle$, we might ask what other vectors could be generated as linear combinations of these. The answer is any vector that can be expressed as $a_1\mathbf{x}_1 + a_2\mathbf{x}_2 + \cdots + a_n\mathbf{x}_n$ where $a_i \in F$ for $1 \le i \le n$. We say the **linear span** (or just the **span**) of the set $S = \{\mathbf{x}_1, \mathbf{x}_2, \dots, \mathbf{x}_n\} \subset V$ is the set $sp(S) = \{\sum_{i=1}^{n} a_i\mathbf{x}_i : a_i \in F\}$. We do not rule out the possibility that $S = \emptyset$, in which case $sp(S) = \{\mathbf{0}\}$. If S spans W, we say S is a **spanning set** of W. Note that the span of a set of vectors from V is always a subspace of V. It should be clear that if $W \subset V$ and $S \subset W$, then $sp(S) \subset W$. In other words, no vector generated as a linear combination of vectors already in W can escape from being just an element of W.

4.3 Linear independence & bases

We define a set of *nonzero* vectors $\{\mathbf{x}_1, \mathbf{x}_2, \dots, \mathbf{x}_n\}$ to be **linearly independent** (or simply **independent**) if every linear combination that equals the zero vector forces every coefficient in the combination to be zero. So if $a_1\mathbf{x}_1 + a_2\mathbf{x}_2 + \cdots + a_n\mathbf{x}_n = \mathbf{0}$, then $a_i = 0$ for $1 \le i \le n$. This means that there is no redundancy in the set with regard to generating their linear span. Suppose that $\{\mathbf{x}_1, \mathbf{x}_2, \dots, \mathbf{x}_n\}$ is linearly independent and $sp(\{\mathbf{x}_1, \mathbf{x}_2, \dots, \mathbf{x}_n\}) = sp(\{\mathbf{x}_2, \dots, \mathbf{x}_n\})$. This would mean that $\mathbf{x}_1$ could be written as a linear combination of the vectors in the set $\{\mathbf{x}_2, \dots, \mathbf{x}_n\}$. So $\mathbf{x}_1 = b_2\mathbf{x}_2 + \cdots + b_n\mathbf{x}_n$. This means $\mathbf{x}_1 - b_2\mathbf{x}_2 + \cdots + b_n\mathbf{x}_n = \mathbf{0}$, and by linear independence all the coefficients (including 1) would have to be zero. This is impossible, so every member of the original set is essential for generating the span. This is obviously true for any member of the spanning set. A set of vectors that is not linearly independent is said to be **dependent**.

Given a set of vectors S in the space V, it may be the case that $sp(S) = V$ and also the vectors in S are linearly independent. In that case we call S a **basis** for V. It is beyond the scope of our discussion, but it can be shown that every vector space has a spanning set that is maximal with regard to linearly independence. A basis is not unique, but as we will show there is a property shared by all bases of a particular vector space and that is the number of vectors in the basis. This number is called the **dimension** of the vector space. A vector space V with a finite basis consisting of n **basis vectors** is said to have dimension n. This can be abbreviated to $\dim V = n$. If the space does not have a finite basis, we say it is infinite dimensional. Examples of finite-dimensional vector spaces are the $\mathbb{R}^n$ spaces in Section 4.1, Example 4.2. An example of an infinite dimensional space is $\langle \mathfrak{P}_X, \mathbb{Q} \rangle$ in Section 4.1, Example 4.3. Sometimes we qualify a basis as being a **Hamel basis** or **algebraic basis** to distinguish it from other types of bases which are defined in terms of different properties.

Example 4.8. Show that if we are given a vector space V with finite basis $B = \{\mathbf{e}_1, \mathbf{e}_2, \ldots, \mathbf{e}_n\}$, there is only one way to write an arbitrary $\mathbf{v} \in V$ in terms of the basis vectors.

Example 4.9. Show that if the set $\{\mathbf{v}_1, \mathbf{v}_2, \ldots, \mathbf{v}_n\}$ spans the vector space V, then the set $\{\mathbf{w}, \mathbf{v}_1, \mathbf{v}_2, \ldots, \mathbf{v}_n\}$ is linearly dependent, where $\mathbf{w} \neq 0$.

Example 4.10. Show that any set of $n+1$ vectors in a vector space V of dimension n is linearly dependent.

Example 4.11. Show that the dimension of a finite-dimensional vector space V is well defined.

4.4 Vectors in component form

One of the many advantages of identifying a basis for a vector space is that arbitrary vectors can be expressed in a uniform way. For example, we can show that a basis for the vector space $\mathbb{R}^2$ is $\{\langle 1, 0\rangle, \langle 0, 1\rangle\}$. It is traditional to denote the basis vectors $\langle 1, 0\rangle$ by $\hat{\mathbf{i}}$ and $\langle 0, 1\rangle$ by $\hat{\mathbf{j}}$. An arbitrary vector $\langle x, y\rangle$ can then be written $x\hat{\mathbf{i}} + y\hat{\mathbf{j}}$. This is yet another way to distinguish points in $\mathbb{R}^2$ from vectors in $\mathbb{R}^2$. Section 4.3 Example 4.8 assures us that writing a vector in this or any other manner relative to a fixed basis is unique. We call $\hat{\mathbf{i}}$ the **unit vector** in the x-direction and $\hat{\mathbf{j}}$ the corresponding unit vector in the y-direction. The vector space $\mathbb{R}^3$ is an easy extension of this. We add $\hat{\mathbf{k}}$, the unit vector in the z-direction, and $\langle x, y, z\rangle$ becomes $x\hat{\mathbf{i}} + y\hat{\mathbf{j}} + z\hat{\mathbf{k}}$. The set $\{\hat{\mathbf{i}}, \hat{\mathbf{j}}, \hat{\mathbf{k}}\}$ is called the **standard basis** for $\mathbb{R}^3$ to distinguish it from other possible bases. We assume the standard basis is in use for $\mathbb{R}^2$ and $\mathbb{R}^3$ unless otherwise noted. The **components** of a vector relative to a given basis are the coefficients of the linear combination of basis vectors expressing the vector. So the x-component of $\langle 3, -5, 6\rangle$ is 3, and so forth. The space $\langle \mathfrak{P}_x, \mathbb{Q}\rangle$ has basis $B = \{x^n : n \in \mathbb{N}_0\}$. This has to be an infinite spanning set in order to express all possible rational polynomials of arbitrary but finite degree. The condition of linear independence is applied just to finite subsets of B. Relative to B, the polynomial $3x^{10} - 2x^8 + 5x^7 - 4x^6 + 7x^4 + 11x^3 - x^2 + x + 6$ would be written $\langle 6, 1, -1, 11, 7, 0, -4, 5, -2, 0, 2, 0, \dots, 0, \dots\rangle$. Some advanced books on algebra start with this expression to define a polynomial.

Vector addition, subtraction, and scalar multiplication can all be done componentwise. So in $\mathbb{R}^3$ we would have $\langle 8, -2, 7\rangle + \langle 4, 5, 11\rangle = \langle 12, 3, 18\rangle$ and $3\langle 7, -5, 9\rangle = \langle 21, -15, 27\rangle$. Extension to other vector spaces is immediately obvious.

Example 4.12. A cruise ship leaves port due east and travels for 10 hours at 15 knots, then turns south and steams for 5 hours at 20 knots, then turns west and goes for 3 hours at 10 knots. How far from port is it and which direction should it take to return?

Example 4.13. A 50 N force is acting in the direction of the line segment running from (1, 2, 3) to (4, 7, 10) Express this force as a vector.

Unitizing a vector

Given vector $\mathbf{v} \neq \mathbf{0}$, a parallel unit vector $\mathbf{u}_\mathbf{v}$ is $\frac{\mathbf{v}}{|\mathbf{v}|}$
Given $\mathbf{v} = \langle a, b, c\rangle$, $\mathbf{u}_\mathbf{v} = \frac{\langle a,b,c\rangle}{\sqrt{a^2+b^2+c^2}}$
Similarly for $\mathbf{v} \in \mathbb{R}^n$

Example 4.14. Three forces in a common plane are acting at the same point on a structural connection. Force #1 is 10 tons pulling due north. Force #2 is 5 tons pulling due west. Force #3 is 15 tons pulling southeast. What additional force and in what direction will balance the existing forces?

4.5 Dot product

As we mentioned above in connection with general inner product spaces, there is a type of multiplication between vectors in $\mathbb{R}^n$ called the **dot product**. The dot product combines two vectors in $\mathbb{R}^n$ and returns us to $\mathbb{R}$. Viewed as the mapping "$\cdot$" we have "$\cdot$": $\mathbb{R}^n \times \mathbb{R}^n \to \mathbb{R}$, where "$\cdot$" $(\langle x_i\rangle, \langle y_j\rangle) = \sum_{i=1}^n \sum_{j=1}^n x_i y_j \delta_{ij}$. The bookkeeping symbol δ_{ij} is known as the **Kronecker delta** and is defined to be 1 if $i = j$ and 0 otherwise. Practically speaking, all we are doing is multiplying the corresponding components of two n-dimensional vectors and adding up all the individual products. So in $\mathbb{R}^3$, for example, $\langle 2, 5, 8\rangle \cdot \langle -1, 5, 7\rangle = (2)(-1) + (5)(5) + (8)(7) = -2 + 25 + 56 = 79$. All of our applications will be in $\mathbb{R}^2$ or $\mathbb{R}^3$, but it is worth knowing that the construction is valid for arbitrary n. The dot product is a well-behaved operation. The following computational rules are simply an adaptation of the inner product axioms given in Table 4.4.

Table 4.5: Dot product rules for $\mathbf{x}, \mathbf{y}, \mathbf{z} \in \mathbb{R}^n$.

1) Commutativity	$\mathbf{x} \cdot \mathbf{y} = \mathbf{y} \cdot \mathbf{x}$
2) Distributivity over vector addition	$\mathbf{x} \cdot (\mathbf{y} + \mathbf{z}) = \mathbf{x} \cdot \mathbf{y} + \mathbf{x} \cdot \mathbf{z}$
3) Associativity of scalar multiplication	$c(\mathbf{x} \cdot \mathbf{y}) = (c\mathbf{x}) \cdot \mathbf{y} = \mathbf{x} \cdot (c\mathbf{y})$
4) Annihilation	$\mathbf{0} \cdot \mathbf{x} = 0$
5) Norm	$\sqrt{\mathbf{x} \cdot \mathbf{x}} = \lvert\mathbf{x}\rvert$

There are some particularly convenient formulas involving the dot product. They help reduce some sticky problems in 2- and 3-dimensional geometry to almost routine calculations.

Example 4.15. Show that the angle θ between two vectors $v, w \in \mathbb{R}^3$ is given by $\theta = \arccos(\frac{\mathbf{v} \cdot \mathbf{w}}{|\mathbf{v}||\mathbf{w}|})$.

Example 4.16. A 100-foot tall building with a square 50×50 foot base is oriented so that one wall faces due east. A drain pipe is to be run from the northeast corner of the roof to the southwest corner of the base. The pipe will descend along the east wall 50 feet to the southeast corner of the building, then turn and descend another 50 feet along the south wall to its destination. An angled fitting must be manufactured to accommodate the turn at the southeast corner. What should be the angle?

Example 4.17. Show that $\mathbf{v} \cdot \mathbf{w} = |\mathbf{v}||\mathbf{w}| \cos\theta$ and therefore the dot product detects **perpendicularity (orthogonality)**.

Example 4.18. Show that the scalar **projection** of the vector **v** in the direction of vector **w** is $\frac{\mathbf{v} \cdot \mathbf{w}}{|\mathbf{w}|}$.

4.6 Cross product

In the vector space $\mathbb{R}^3$, there is another type of multiplication between vectors called the **cross product**. Viewed as a mapping "$\times$" we have "$\times$": $\mathbb{R}^3 \times \mathbb{R}^3 \to \mathbb{R}^3$, so we are combining two vectors and the result is another vector. It may seem odd, but this version of vector multiplication only works in $\mathbb{R}^3$ for reasons that are beyond the scope of our discussion. However, this will not prevent us from using it effectively for practical problems. The definition of the cross product may seem unduly complicated at first, but once we appreciate its utility that feeling should give way to respect. To economize on notation, we are going to give the standard unit vectors in $\mathbb{R}^3$ temporary new names: $\hat{\mathbf{i}} = \hat{\mathbf{e}}_1, \hat{\mathbf{j}} = \hat{\mathbf{e}}_2$, and $\hat{\mathbf{k}} = \hat{\mathbf{e}}_3$. Then we define $\mathbf{x} \times \mathbf{y} = \sum_{i=1}^{3} \sum_{j=1}^{3} \sum_{k=1}^{3} \varepsilon_{ijk} \hat{\mathbf{e}}_i x_j y_k$ The bookkeeping symbol ε_{ijk} is known as the **Levi-Civita symbol** and is defined to be 1 if ijk is an even permutation of 123, –1 if ijk is an odd permutation of 123, and 0 for any other case. The even permutations of 123 are 123, 312, and 231. These retain the original order with only "wraparounds" permitted. The odd permutations are 132, 213, and 321. So apparently only six of the 27 possible terms of the triple sum survive. We are thankful for that. More explicitly, if $\mathbf{x} = \langle x_1, x_2, x_3 \rangle$ and $\mathbf{y} = \langle y_1, y_2, y_3 \rangle$ then $\mathbf{x} \times \mathbf{y} = \hat{\mathbf{i}}(x_2y_3 - x_3y_2) + \hat{\mathbf{j}}(x_3y_1 - x_1y_3) + \hat{\mathbf{k}}(x_1y_2 - x_2y_1)$. After we discuss matrices and determinants, we will have a scheme using formal determinants that organizes this calculation for us. Conveniently enough, it is possible to extend this formula to vectors in $\mathbb{R}^2$ by artificially giving each vector a zero component in the $\hat{\mathbf{k}}$ direction. Of course, the result will turn out to be a vector in $\mathbb{R}^3$ with only a $\hat{\mathbf{k}}$ component.

The cross product is less well-behaved than the dot product.

Table 4.6: Cross product rules for $\mathbf{x}, \mathbf{y}, \mathbf{z} \in \mathbb{R}^3$.

1) Anticommutativity	$\mathbf{x} \times \mathbf{y} = -\mathbf{y} \times \mathbf{x}$
2) Left distributivity over vector addition	$\mathbf{x} \times (\mathbf{y} + \mathbf{z}) = \mathbf{x} \times \mathbf{y} + \mathbf{x} \times \mathbf{z}$
3) Right distributivity over vector addition	$(\mathbf{y} + \mathbf{z}) \times \mathbf{x} = \mathbf{y} \times \mathbf{x} + \mathbf{z} \times \mathbf{x}$
4) Associativity of scalar multiplication	$c(\mathbf{x} \times \mathbf{y}) = (c\mathbf{x}) \times \mathbf{y} = \mathbf{x} \times (c\mathbf{y})$
5) Annihilation	$\mathbf{0} \times \mathbf{x} = \mathbf{0}$

Example 4.19. An alternative definition of the cross product is $\mathbf{x} \times \mathbf{y} = |\mathbf{x}||\mathbf{y}| \sin\theta \, \hat{\mathbf{e}}_\perp$, where θ is the angle between $\mathbf{x}$ and $\mathbf{y}$ and $\hat{\mathbf{e}}_\perp$ is a unit vector perpendicular to the common plane of $\mathbf{x}$ and $\mathbf{y}$ and points in the direction determined by the **right-hand rule**. Show that this is equivalent to the definition above.

Example 4.20. If two vectors form adjacent sides of a parallelogram, find the area of the figure.

Example 4.21. Find the area of a triangle in $\mathbb{R}^3$ with vertices at $(1, 2, 3)$, $(3, 5, 7)$, and $(-1, -1, 4)$.

Example 4.22. Show that the cross product detects parallelism and antiparallelism.

Example 4.23. ($\int$) With just the triangle area formula and some very basic differential calculus, we can prove **Kepler's second law**, namely that the planets travel in their orbits in such a way that a line connecting the sun to a planet sweeps out equal areas in equal times. This was cutting-edge astronomy in the 1590s. The practical effect is that when a planet is near the sun (near perihelion) it travels faster than when it is far from the sun (near aphelion). Kepler's first law states that the planets travel around the sun in elliptical orbits with the sun at one focus. He deduced these laws from extensive naked-eye astrolabe observations prior to 1600 (the telescope was invented in 1608 in Holland) without any prior valid quantitative theory of planetary motion. We present this calculation as a testament to the power of vector methods.

Suppose the vector connecting the sun to a planet is $\mathbf{r}(t)$, where t is time. After an increment Δt of time passes, the vector changes to $\mathbf{r}(t + \Delta t)$ and the planet travels $\Delta\mathbf{r}(t) = \mathbf{r}(t + \Delta t) - \mathbf{r}(t)$. Since Δt is assumed small, we can regard the area swept out by the radius vector during Δt to be a (very thin) triangle. Now $\mathbf{r}(t)$ and $\Delta\mathbf{r}(t)$ are two sides of this triangle, so its area $A(t) = \frac{1}{2}|\mathbf{r}(t) \times \Delta\mathbf{r}(t)|$. We would like to develop an expression for $\frac{dA}{dt}$, the rate of sweeping out area. Specifically, we want to show it is a constant. To do that we will need to look at $\frac{d^2A}{dt^2}$ and show it is zero. So

$$\frac{dA}{dt} = \lim_{\Delta t \to 0} \frac{\Delta A}{\Delta t} = \lim_{\Delta t \to 0} \frac{1}{2}\left|\mathbf{r}(t) \times \frac{\Delta\mathbf{r}(t)}{\Delta t}\right|.$$

Evidently

$$\frac{dA}{dt} = \frac{1}{2}\left|\mathbf{r}(t) \times \lim_{\Delta t \to 0} \frac{\Delta\mathbf{r}(t)}{\Delta t}\right| = \frac{1}{2}\left|\mathbf{r}(t) \times \frac{d\mathbf{r}(t)}{dt}\right|.$$

Now we use the product rule to differentiate a cross product (see Table 4.7 below). It works the same way as for a regular scalar product as long as the factors are kept in the same order (recall the cross product does not commute). The result is

$$\frac{d^2A}{dt^2} = \frac{1}{2}\frac{d}{dt}\left|\mathbf{r}(t) \times \frac{d\mathbf{r}(t)}{dt}\right| = \frac{1}{2}\left|\frac{d\mathbf{r}(t)}{dt} \times \frac{d\mathbf{r}(t)}{dt} + \mathbf{r}(t) \times \frac{d^2\mathbf{r}(t)}{dt^2}\right|.$$

The first cross product vanishes because a vector is parallel to itself (see Example 4.22). In the second cross product, the term $\frac{d^2\mathbf{r}(t)}{dt^2}$ is the acceleration of the planet along the radius. The force of gravity causes this acceleration and it is given by Newton's gravity law $\mathbf{F}(r) = -\frac{GM_s m_p}{r^2}\hat{e}_r$. In this formula the force of gravity acting on the planet at distance $|\mathbf{r}| = r$ from the sun is $\mathbf{F}(r)$, G is the gravitational constant, and M_s and m_p are the masses of the sun and planet, respectively. The unit vector $\hat{e}_r = \frac{\mathbf{r}(t)}{|\mathbf{r}(t)|}$ points from the sun towards the planet. Since $\mathbf{F}(r) = m_p \frac{d^2\mathbf{r}(t)}{dt^2}$ (**Newton's second law of motion**), we see that $\hat{e}_r$ and $\frac{d^2\mathbf{r}(t)}{dt^2}$ are parallel. Hence the second cross product also vanishes and we conclude $\frac{d^2A}{dt^2} = 0$. Since this is the derivative of $\frac{dA}{dt}$, it must be that $\frac{dA}{dt}$ is a constant. But this is precisely Kepler's second law. So our little formula leads to the validation of a major historical advance in astronomy.

Example 4.24. There is a three-dimensional volume analogue of the cross product formula for parallelogram area. A parallelepiped is a six-sided figure in three dimensions bounded by opposite pairs of congruent parallelograms. Imagine a rectangular solid like a shoebox that has been subjected to shear in two directions. Any corner of the figure has three vectors emanating from it. Two define a parallelogram base and the third extends out of that plane to give the figure volume. Let the vectors forming the base be **B** and **C**. The area of the base is then $|\mathbf{B} \times \mathbf{C}|$. But before we take the norm, note that the vector $\mathbf{B} \times \mathbf{C}$ is perpendicular to the parallelogram base and that is exactly the direction in which the altitude of the figure is measured. In fact, the altitude is the projection of the out-of-plane third vector **A** onto $\mathbf{B} \times \mathbf{C}$. The scalar projection of **A** onto $\mathbf{B} \times \mathbf{C}$ is precisely $\mathbf{A} \cdot \mathbf{B} \times \mathbf{C}$. Since the volume V of the parallelepiped is the area of its base times its altitude, we have $V = |\mathbf{A} \cdot \mathbf{B} \times \mathbf{C}|$. The vertical bars here signify absolute value.

$\mathbf{A}\cdot\mathbf{B}\times\mathbf{C}$ is called the **scalar triple product** of **A**, **B**, and **C**. It is an unambiguous expression as the cross product must be executed first. Computationally, if $\mathbf{A} = \langle a_1, a_2, a_3\rangle$, $\mathbf{B} = \langle b_1, b_2, b_3\rangle$, and $\mathbf{C} = \langle c_1, c_2, c_3\rangle$, we have $\mathbf{A}\cdot\mathbf{B}\times\mathbf{C} = a_1(b_2c_3 - b_3c_2) + a_2(b_3c_1 - b_1c_3) + a_3(b_1c_2 - b_2c_1)$. When we discuss determinants, we will present a scheme that self-organizes this calculation. Interestingly, any even permutation of the vectors while holding the operation symbols in place does not change the value of the scalar triple product. So $\mathbf{A}\cdot\mathbf{B}\times\mathbf{C} = \mathbf{B}\cdot\mathbf{C}\times\mathbf{A} = \mathbf{C}\cdot\mathbf{A}\times\mathbf{B}$. An odd permutation changes the sign.

Example 4.25. A three-dimensional lattice is described by a repeating unit in the shape of a parallelepiped with corner vectors $\langle 0, 1, 0\rangle$, $\langle -2, 2, 0\rangle$, and $\langle -1, 2, 3\rangle$. Units are in millimeters. How many lattice units would be in a cubic centimeter?

Example 4.26. It makes sense to form $\mathbf{A}\times\mathbf{B}\times\mathbf{C}$, but as written the expression is ambiguous because the cross product is not associative. In other words, $\mathbf{A}\times(\mathbf{B}\times\mathbf{C}) \neq (\mathbf{A}\times\mathbf{B})\times\mathbf{C}$. We can develop a convenient expression for $\mathbf{A}\times(\mathbf{B}\times\mathbf{C})$ two different ways. The first involves grinding out all the components by brute calculation and then seeking simplifications. This will work but it is not as pleasant a derivation as one based on our bookkeeping symbols δ_{ij} and ε_{ijk}. We need two things – a fact and a notational convention:

(i) The bookkeeping symbols satisfy the following identity: $\varepsilon_{kij}\varepsilon_{kpq} = \delta_{ip}\delta_{jq} - \delta_{iq}\delta_{jp}$, as may be established by checking the nonvanishing terms (almost all do).

(ii) Let us use the so-called Einstein summation convention, where if two indices are repeated in an expression there is an implied summation. The range of the sum is clear from context. For example, we would interpret the expression $a_{ij}b_{jk}$ as $\sum_{j=1}^{3} a_{ij}b_{jk}$ where $1 \le j \le 3$. Another example is the dot product definition in $\mathbb{R}^3$: $\mathbf{x}\cdot\mathbf{y} = \sum_{i=1}^{3}\sum_{j=1}^{3} x_iy_j\delta_{ij}$ reduces to just $x_iy_j\delta_{ij}$.

So, with the vector components as in Example 4.24,

$$\mathbf{A}\times(\mathbf{B}\times\mathbf{C}) = \varepsilon_{ijk}\hat{\mathbf{e}}_i a_j(\mathbf{B}\times\mathbf{C})_k = \varepsilon_{ijk}\hat{\mathbf{e}}_i a_j(\varepsilon_{kpq}b_pc_q) = \varepsilon_{ijk}\varepsilon_{kpq}\hat{\mathbf{e}}_i a_j b_p c_q.$$

Now $\varepsilon_{ijk} = \varepsilon_{kij}$ and $\varepsilon_{kij}\varepsilon_{kpq} = \delta_{ip}\delta_{jq} - \delta_{iq}\delta_{jp}$, so

$$\begin{aligned}\mathbf{A}\times(\mathbf{B}\times\mathbf{C}) &= (\delta_{ip}\delta_{jq} - \delta_{iq}\delta_{jp})\hat{\mathbf{e}}_i a_j b_p c_q = \hat{\mathbf{e}}_i a_j b_i c_j - \hat{\mathbf{e}}_i a_j b_j c_i \\ &= \mathbf{B}(a_jc_j) - \mathbf{C}(a_jb_j) = \mathbf{B}(\mathbf{A}\cdot\mathbf{C}) - \mathbf{C}(\mathbf{A}\cdot\mathbf{B}).\end{aligned}$$

That is it. The beauty of the bookkeeping symbols is that they act like scalars and can be shifted around in expressions without worrying about commutativity. In summary, $\mathbf{A}\times(\mathbf{B}\times\mathbf{C}) = \mathbf{B}(\mathbf{A}\cdot\mathbf{C}) - \mathbf{C}(\mathbf{A}\cdot\mathbf{B})$. This is universally called the **BAC–CAB identity** for the **vector triple product**. It is useful in characterizing certain electromagnetic phenomena, but the physics gets pretty deep, so we will just illustrate an identity due to Jacobi that comes up in nonassociative (like the cross product) algebras.

Example 4.27. With **A**, **B**, and **C** as above, show that $\mathbf{A}\times(\mathbf{B}\times\mathbf{C}) + \mathbf{B}\times(\mathbf{C}\times\mathbf{A}) + \mathbf{C}\times(\mathbf{A}\times\mathbf{B}) = \mathbf{0}$.

4.7 Vector geometry

We can harness the tools developed thus far to solve additional two- and three-dimensional geometric problems involving space curves.

4.7.1 Straight lines

Restricting our discussion to $\mathbb{R}^2$ or the xy-plane for the moment, we ask what is the bare minimum information required to specify a straight line graph. We need either two points (according to Euclid) or a point along with a direction. Given two points (x_1,y_1) and (x_2,y_2), we can determine the relative inclination (change in y divided by change in x) of the line running through them by calculating $\frac{y_2-y_1}{x_2-x_1} = m$. Reasoning that this should be the same as the inclination using an arbitrary point (x,y) on the line, we have $\frac{y-y_1}{x-x_1} = m$ or $y = mx + (y_1 - mx_1)$. This is the **point–slope form** of the straight line equation. Calling m the **slope** and $b = y_1 - mx_1$ the y-**intercept**, we have recovered the formula $y = mx + b$, which is the **slope–intercept form** of the straight line equation. Given a generic point (x,y) on a line and a direction specified as the slope m, we can easily calculate $b = y - mx$ and reconstruct $y = mx + b$ for that case. In the event that a line is vertical ($m = \infty$) and cuts the x-axis at $x = c$, that expression will serve as the equation of the line. We now have a complete picture of graphing straight lines in $\mathbb{R}^2$. What about $\mathbb{R}^3$? Suddenly, the planar experience in $\mathbb{R}^2$ seems inadequate.

We will develop an analogue of the slope–intercept formula for three dimensions. Once again, we start off with two points (a_1, a_2, a_3) and (b_1, b_2, b_3). How shall we define the slope of the line through these points? There are three reference planes to choose from, the xy-plane, xz-plane, and yz-plane. A better idea is to let the **direction vector** $\mathbf{v} = \langle b_1 - a_1, b_2 - a_2, b_3 - a_3\rangle$ express the orientation of the line in space. This is effectively the "slope" in $\mathbb{R}^3$. Now we can write the **equation of a line** as $\mathbf{r}(t) = \mathbf{r}_0 + \mathbf{v}t$, where $\mathbf{r}(t)$ is the position vector of a point on the line, $\mathbf{r}_0$ is the position vector of a fixed point on the line, $\mathbf{v}$ is the direction vector, and t is a **parameter**, typically $(-\infty < t < \infty)$. Thinking of t as time does no harm and will make practical sense when we view the line as the path of a moving object and need to compute velocity and acceleration. You can imagine starting at the origin $(0, 0, 0)$ and moving along the vector $\mathbf{r}_0 = \langle a_1, a_2, a_3\rangle$ to get to the line (the bus stop), then moving along the line (riding the bus) by taking various scalar multiples of $\mathbf{v}$. In effect you are adding the constant vector $\mathbf{r}_0$ to a multiple t of the direction vector $\mathbf{v}$ to generate the vector $\mathbf{r}(t)$, which is the position vector of a generic point on the line.

Example 4.28. Write the vector equation for the straight line through the point $(1, 2, 3)$ in the direction of the vector $\langle -1, 1, 4\rangle$. Does the line go through $(-2, 5, 15)$ and if so what is t?

Example 4.29. Find the equation of the straight line that goes through $(2, 1, 4)$ and $(7, -7, 3)$.

Example 4.30. Express the vector equation in Example 4.29 in component form.

Example 4.31. Find the equation of a line parallel to the line in Example 4.28 but passing through (1, 1, 1).

Example 4.32. Do the lines $r_1(t) = \langle 1, 3, 5\rangle + t_1\langle 4, -5, 3\rangle$ and $r_2(t) = \langle 6, 2, 2\rangle + t_2\langle 4, -5, 3\rangle$ intersect?

Example 4.33. Find the equation of a line through (1, 2, 3) perpendicular to $r(t) = \langle 1, 2, 3\rangle + t\langle 2, 4, 3\rangle$.

A **point-to-line** distance is easily calculated using our vector methods. Suppose we are given the line $\mathbf{r}(t) = \mathbf{r}_0 + \mathbf{v}t$ and the point $Q = (a, b, c)$. We can pick a completely arbitrary point P on the line and construct the vector $\mathbf{w}$ from P to Q. Note that a unit vector parallel to the line is $\frac{\mathbf{v}}{|\mathbf{v}|}$. Then the scalar projection of $\mathbf{w}$ in the direction perpendicular to the line is $|\mathbf{w}| \sin\theta$, where θ is the angle between $\mathbf{w}$ and $\mathbf{v}$. Using our cross product formula, we see that the perpendicular point-to-line distance

$$d = |\mathbf{w}| \sin\theta = \frac{|\mathbf{w} \times \mathbf{v}|}{|\mathbf{v}|}.$$

Some examples will make this clear.

Example 4.34. What is the distance between the point (4, 3, 7) and the line $\mathbf{r}(t) = \langle 1, 2, 3\rangle + \langle 2, -3, 1\rangle t$?

Example 4.35. What is the distance from the point (−1, 5, 9) to the line $\mathbf{r}(t) = \langle -2, -2, 11\rangle + \langle 1, -5, 8\rangle t$?

There are three cases to consider in connection with **line-to-line** distances. Two lines can intersect, they can be parallel, or they can be neither. Lines that are not parallel and do not intersect are called **skew** lines. Lines that intersect obviously have zero distance between them. For lines that are parallel, the distance between them can be found by picking an arbitrary point on one line and finding its distance to the other line as we did above. Skew lines offer more of a challenge. The distance between skew lines, of course, means the distance perpendicular to both. Our plan is to create a vector $\mathbf{w}$ by connecting two arbitrary points, one on each line, and then, by the magic of vector methods, find a unit vector $\mathbf{v}$ that is simultaneously perpendicular to both lines. The scalar projection of $\mathbf{w}$ in the direction of $\mathbf{v}$ will be the required perpendicular distance d. So given $\mathbf{r}(t_1) = \mathbf{r}_0 + \boldsymbol{\alpha} t_1$ and $\mathbf{s}(t_2) = \mathbf{s}_0 + \boldsymbol{\beta} t_2$, we form the unit vector $\mathbf{v} = \boldsymbol{\alpha} \times \boldsymbol{\beta}$ and then unitize it. This unit vector is necessarily perpendicular to both lines and will serve as the "screen" upon which we project the vector $\mathbf{w}$. Now suppose the points on the two lines are $P = (a, b, c)$ and $Q = (d, e, f)$, so $\mathbf{w} = \langle a - d, b - e, c - f\rangle$. The distance between skew lines

$$d = \left| \frac{\mathbf{w} \cdot \boldsymbol{\alpha} \times \boldsymbol{\beta}}{\|\boldsymbol{\alpha} \times \boldsymbol{\beta}\|} \right|.$$

Here we have used the convention that if norms and absolute values appear in the same expression, norms are denoted by $\| \cdot \|$.

Example 4.36. What is the distance between the lines given by $\mathbf{r}(t_1) = \langle 1,2,3\rangle + \langle 1,0,3\rangle t_1$ and $\mathbf{s}(t_2) = \langle -1,5,7\rangle + \langle 4,4,6\rangle t_2$?

4.7.2 General curves ($\int$)

We can investigate more general space curves than just straight lines by considering the **curve** or **path** function $\mathbf{r}(t) = \langle x(t), y(t), z(t)\rangle$, where t is a parameter and $x(t)$, $y(t)$, and $z(t)$ are component functions. It is customary to think of $t \in [t_1, t_2]$ as time and then $\mathbf{r}(t)$ as the position vector of a point in space which we imagine has some velocity as it traces out the curve. There are several restrictions we impose on $\mathbf{r}(t)$ to ensure it is mathematically well behaved. First, we insist that $\mathbf{r}(t)$ must be a continuous function so the curve is never broken. Second, we require $\mathbf{r}(t)$ to be an injective function on (t_1, t_2), so the curve never crosses itself. We allow $\mathbf{r}(t_1) = \mathbf{r}(t_2)$ in the case of a **closed curve**. Third, we impose the condition that $\mathbf{r}'(t) \neq 0$, which ensures that the tracing process never pauses. It could never reverse itself either, since this would violate injectivity. Fourth and finally, we want $\mathbf{r}(t)$ to be a **smooth** function, which means $\mathbf{r}'(t)$ should be continuous. This is another guarantee that tracing could not reverse itself, since there would be a time t_0 when $\mathbf{r}'(t_0) = 0$. Many of the space curves encountered in real world settings also have continuous higher order path functions (orbits, trajectories, and so forth) and this allows us to characterize them more completely.

Example 4.37. Find the three-dimensional path function corresponding to a circle of radius 2 centered at (0, 0, 1) in the z-plane

Example 4.38. Find the path function in the xy-plane corresponding to an ellipse with equation $\frac{x^2}{9} + \frac{y^2}{25} = 1$.

Example 4.39. Find the path function for a circular helix of radius 3 and pitch 1 with axis coincident with the positive z-axis and extending from $z = 0$ to $z = 10$.

We can view a space curve as a completed entity where the tracing process implicit in the definition of the path function is considered to be a fait accompli as in the prior three examples. However, if the path function describes some dynamic process like an airplane flying or a satellite orbiting, we might be more interested in the local rather than global properties of the curve. Given $\mathbf{r}(t) = \langle x(t), y(t), z(t)\rangle$, we can easily get the **velocity** vector along the curve by calculating $\mathbf{r}'(t) = \lim_{\Delta t \to 0} \frac{\mathbf{r}(t+\Delta t) - \mathbf{r}(t)}{\Delta t}$, which is the usual definition of derivative for a function of one variable. Applying this to $\mathbf{r}(t)$ written in terms of components, we can easily confirm that $\mathbf{r}'(t) = \langle x'(t), y'(t), z'(t)\rangle$. Note that $|\mathbf{r}'(t)|$ is the scalar **speed**. This vector is never zero ($\mathbf{r}'(t) \neq 0$ was assumed) and represents the direction vector of a straight line tangent to the curve at a given point, much like the derivative $y'(x)$ represents the slope of a tangent to the graph of $y(x)$. If we want only the direction of the tangent, we can unitize $\mathbf{r}'(t)$ and define $\frac{\mathbf{r}'(t)}{|\mathbf{r}'(t)|} = \mathbf{T}(t)$ as the **unit tangent vector**. As the curve wends its way through space, the unit tangent vector will generally be changing its direction, unless we are on a piece of "straight track," so to speak. Assuming we can differentiate $\mathbf{r}(t)$ one more time, the derivative $\mathbf{T}'(t)$ measures this rate of change. There is no reason to suppose that the raw derivative $\mathbf{T}'(t)$ is a unit vector, so we unitize that and define $\mathbf{N}(t) = \frac{\mathbf{T}'(t)}{|\mathbf{T}'(t)|}$. This $\mathbf{N}(t)$ is called the **unit (principal) normal vector** for the curve $\mathbf{r}(t)$. It is perpendicular (normal) to $\mathbf{T}(t)$ and points in the direction that the unit tangent vector is turning towards. It will be convenient to have a table showing the interplay of differentiation and vector operations.

Table 4.7: Differentiation of vectors for $\mathbf{x}, \mathbf{y} \in \mathbb{R}^k$.

1) Constant multiple rule	$(c\mathbf{x})' = c(\mathbf{x}')$
2) Sum rule	$(\mathbf{x} + \mathbf{y})' = \mathbf{x}' + \mathbf{y}'$
3) Difference rule	$(\mathbf{x} - \mathbf{y})' = \mathbf{x}' - \mathbf{y}'$
4) Dot product rule	$(\mathbf{x} \cdot \mathbf{y})' = \mathbf{x}' \cdot \mathbf{y} + \mathbf{x} \cdot \mathbf{y}'$
5) Cross product rule	$(k = 3)$ $(\mathbf{x} \times \mathbf{y})' = \mathbf{x}' \times \mathbf{y} + \mathbf{x} \times \mathbf{y}'$ (maintain order!)

Example 4.40. Show that $\mathbf{T}(t)$ and $\mathbf{N}(t)$ are perpendicular.

Now that we have $\mathbf{T}(t)$ and $\mathbf{N}(t)$, we can define $\mathbf{B}(t) = \mathbf{T}(t) \times \mathbf{N}(t)$. Note $\mathbf{T}(t) \perp \mathbf{N}(t)$ means $|\mathbf{B}(t)| = |\mathbf{T}(t)||\mathbf{N}(t)| \sin \frac{\pi}{2} = 1$, so we are justified in calling $\mathbf{B}(t)$ the **unit binormal vector.** Vectors $\mathbf{T}$, $\mathbf{N}$, and $\mathbf{B}$ are mutually perpendicular and are attached to each point of a space curve as a miniature mobile version of the basis vectors for $\mathbb{R}^3$. Together they are called the "moving trihedron" and provide a localized Cartesian coordinate system to investigate motion along the path. In pairs they define three orthogonal planes: $\mathbf{T}$ and $\mathbf{N}$ determine the **osculating plane**, $\mathbf{T}$ and $\mathbf{B}$ determine the **rectifying plane**, and $\mathbf{N}$ and $\mathbf{B}$ together determine the **normal plane**. A good intuitive picture of what $\mathbf{T}$, $\mathbf{N}$, and $\mathbf{B}$ represent is this: imagine an airplane flying along the space curve with these three axes emanating from the center of the plane. A climb or dive would correspond to rotation around $\mathbf{N}$. A roll to either side would correspond to rotation around $\mathbf{T}$. And a yaw (turn without banking) would correspond to rotation around $\mathbf{B}$.

Two scalar quantities are also useful in characterizing the local behavior of space curves. The first is **curvature**, denoted by $\kappa(t)$. We discuss the second below. Curvature is, roughly speaking, the ratio of the tendency of the curve to change direction to the velocity along the curve. By definition, $\kappa(t) = \frac{|\mathbf{T}'(t)|}{|\mathbf{r}'(t)|}$. Intuitively, if the change in direction is small compared to velocity, the curvature of the path is small. Conversely, if the direction change is significant for a small velocity, the curvature is large. The **radius**

of curvature $\rho(t) = \frac{1}{\kappa(t)}$. This is the imputed radius of a circle having the same curvature as $\mathbf{r}(t)$. This circle lies in the osculating plane and is called the **osculating circle**. It represents the circle that most nearly conforms to the curve at some momentary t. If $\mathbf{r}(t)$ is not changing direction at a point, then $\kappa(t) = 0$ and $\rho(t) = \infty$. Curvature can be expressed directly in terms of $\mathbf{r}(t)$ as $\kappa(t) = \frac{|\mathbf{r}'(t)\times\mathbf{r}''(t)|}{|\mathbf{r}'(t)|^3}$.

Example 4.41. Find **T**, **N**, **B**, and κ for the circle in the xy-plane given by $\mathbf{r}(t) = \langle a\cos t, a\sin t\rangle$ for $0 \le t \le 2\pi$.

Example 4.42. Find **T**, **N**, **B**, and κ for the circular helix in Example 4.39 above.

By Example 4.41, a circle of radius 3 would have curvature $\frac{1}{3}$. The helix goes around the z-axis at a constant distance of 3 units, so in that respect it behaves like the circle, but it also climbs 1 unit every time it goes around. The extra distance that the helical path function traverses compared to the circular path function with the same nominal diameter "opens up" the helix and causes its curvature to be less than that for the circle. Also note that the **TNB** frame tracking the helix has a constant backwards tilt in order to follow **T** along the curve.

Example 4.43. The curve $\mathbf{r}(t) = \langle t, t^2, t^3\rangle$, known as the "twisted cubic," is often used as an example illustrating the features of space curve theory. Find **T**, **N**, and **B** when $t = 1$.

The second scalar quantity that gives insight into the local behavior of space curves is **torsion**, denoted by $\tau(t)$. With nonplanar curves, the osculating plane can rock back and forth as the curve is traced out. This means the direction of the unit normal (and the other two unit vectors, of course) may be changing. The tendency of the unit normal to change direction along the curve, measured by its projection onto the unit binormal, is called the torsion of the curve. We have thus far developed all of our space curve formulas in terms of $\mathbf{r}(t)$ and its derivatives. The corresponding expression for torsion is unfortunately rather complicated, but for the sake of completeness we present it here,

$$\tau(t) = \frac{[\mathbf{r}'(t) \times \mathbf{r}''(t)] \cdot \mathbf{r}'''(t)}{|\mathbf{r}'(t) \times \mathbf{r}''(t)|^2}.$$

We are now going to explore a different, but equivalent, way of characterizing space curves that has a more "natural" feel to it.

Up to this point, we have expressed space curves in terms of their intuitively appealing, but somewhat artificial, dynamic behavior with respect to the parameter time. Many applications in physics and engineering involve objects large and small following paths in space over time, so this viewpoint is certainly convenient and it makes sense to develop the theory around this parameter. In the mid-nineteenth century, mathematicians wished to develop a more intrinsic way of parametrizing curves that did not depend (explicitly at least) on the notion of time. The French mathematician J.-F. Frenet proposed using the distance along a curve from some fixed starting point as a more natural parameter (the **Frenet coordinate**). This allowed space curves to be expressed in terms of a parameter that was already contained, so to speak, in the curve itself, viewed as a static and completed geometric object. The first order of business in this approach is to find an expression that yields the length of a curve.

Let $s(t)$ be the length along the curve given by $\mathbf{r}(t) = \langle x(t), y(t), z(t)\rangle$ where we start measuring length at t_0 and stop at t_1. The goal is to get an expression for $\frac{ds}{dt}$ in terms of what we know and then form the integral

$$s(t_1) - s(t_0) = \int_{t_0}^{t_1} ds = \int_{t_0}^{t_1} \frac{ds}{dt} dt.$$

Differential distance ds in $\mathbb{R}^3$ obeys the Pythagorean theorem, $ds^2 = dx^2 + dy^2 + dz^2$, so we can write

$$\frac{ds}{dt} = \sqrt{\left(\frac{dx}{dt}\right)^2 + \left(\frac{dy}{dt}\right)^2 + \left(\frac{dz}{dt}\right)^2}.$$

Note that the velocity vector along the curve is

$$\mathbf{r}'(t) = \left\langle \frac{dx}{dt}, \frac{dy}{dt}, \frac{dz}{dt} \right\rangle.$$

Evidently,

$$\mathbf{r}'(t) \cdot \mathbf{r}'(t) = \left(\frac{dx}{dt}\right)^2 + \left(\frac{dy}{dt}\right)^2 + \left(\frac{dz}{dt}\right)^2,$$

so $\sqrt{\mathbf{r}'(t) \cdot \mathbf{r}'(t)} = |\mathbf{r}'(t)| = \frac{ds}{dt}$. Plugging this into our integral, we have $s(t_1) - s(t_0) = \int_{t_0}^{t_1} |\mathbf{r}'(t)| dt$. Setting $t_0 = 0$ and $s(t_0) = 0$, we can reexpress the integral for arc length as a function of time as $s(t) = \int_0^t |\mathbf{r}'(\tau)| d\tau$. We have used τ as the dummy integration variable time.

Example 4.44. Find the total length L along the helix of Example 4.39 above. The curve is given by $\mathbf{r}(t) = \langle 3\cos t, 3\sin t, \frac{t}{2\pi}\rangle$ and t ranges from 0 to 20π.

Example 4.45. Find the length L of the "twisted cubic" $\mathbf{r}(t) = \langle t, t^2, t^3\rangle$ in Example 4.43 from $t = 0$ to $t = 1$.

With our formula for the arc length $s(t)$, we are in a position to restate the formulas for $\mathbf{T}$, $\mathbf{N}$, $\mathbf{B}$, and κ relative to the Frenet coordinate s. Applying the fundamental theorem of calculus to the arc length formula $s(t) = \int_0^t |\mathbf{r}'(\tau)|d\tau$, we see that $s(t)$ is a strictly increasing function, since the norm $|\mathbf{r}'(t)| > 0$ (curve tracing never stops). It follows that $s(t)$ is injective and therefore invertible (at least in principle) as $t(s)$. We can also compose $\mathbf{r}(t)$ with $t(s)$ to form $\mathbf{r}(t(s)) = \mathbf{r}(s)$. The explicit appearance of the time parameter is now eliminated. Moreover, since $\frac{ds}{dt} = |\mathbf{r}'(t)|$ and $|\mathbf{r}'(t)|$ is never zero, we have $\frac{dt}{ds} = \frac{1}{|\mathbf{r}'(t)|}$. We will use this fact immediately below.

Consider

$$\mathbf{T}(t) = \frac{\mathbf{r}'(t)}{|\mathbf{r}'(t)|}$$

by definition. Note that

$$\frac{\mathbf{r}'(t)}{|\mathbf{r}'(t)|} = \frac{d\mathbf{r}}{dt}\frac{dt}{ds} = \frac{d\mathbf{r}}{ds},$$

so $\frac{d\mathbf{r}}{ds}$ is already the unit tangent vector. We denote $\frac{d\mathbf{r}}{ds}$ by $\mathbf{T}(s)$. Next we define the curvature κ in terms of the arc length parameter s, $\kappa(s) = |\frac{d\mathbf{T}(s)}{ds}|$. Simpler than using t. Now for the unit normal, we have $\mathbf{N}(s) = \frac{1}{\kappa(s)}\frac{d\mathbf{T}(s)}{ds}$. The curvature was exactly the factor needed to normalize $\frac{d\mathbf{T}(s)}{ds}$. Finally, the binormal $\mathbf{B}(s) = \mathbf{T}(s) \times \mathbf{N}(s)$. Our program of expressing the vectors of the **TNB** frame along with curvature in terms of arc length is now complete. One loose end remains, namely how we can express torsion τ in terms of s. Translating into symbols our qualitative description of torsion as the projection of the rate of change of the unit normal onto the unit binormal, we have $\tau(s) = \frac{d\mathbf{N}(s)}{ds} \cdot \mathbf{B}(s)$. This is mercifully simpler than the formula for torsion purely in terms of $\mathbf{r}(t)$ and its derivatives.

One might reasonably ask why we bother to develop the space curve formalism in terms of time when everything seems simpler in terms of arc length. The disappointing lesson here is that the formulas expressed in terms of arc length appear to be relatively simpler than those for the time parameter, but that may be an illusion, since finding $s(t)$ in order to eliminate t can easily become a nightmare – see Example 4.45 above for a fairly innocent but frustrating case. We note that $\mathbf{r}(t)$ does have its charm for practical problems, especially those involving motion along space curves.

4.7.3 Trajectories ($\int$)

Ballistic trajectories are those space curves followed by non-self-propelled objects launched from the surface of the Earth and subject to the returning force of gravity. These curves are assumed to happen in a plane, so we can frame our analysis in $\mathbb{R}^2$. It is customary to neglect air resistance or any other force other than the initial launch. Mass of the object does not enter into the calculation of the flight path since downward acceleration is a constant $g = 32.16$ feet per second^2 (9.81 meters per second^2) for any object under these assumptions.

Suppose the projectile is launched along a straight line elevated at an angle θ with respect to the ground or other parallel datum plane with initial velocity v_0. You can imagine a cannon with the barrel inclined at angle θ firing a cannonball with muzzle velocity v_0. The horizontal component of velocity $v_h = v_0 \cos\theta$ and the vertical component of velocity is $v_v = v_0 \sin\theta$. Gravity causes the cannonball to have a net upward velocity at time t of $v_0 \sin\theta - gt$. Expressing the trajectory as the vector function $\mathbf{r}(t)$, what we have at this point is $\mathbf{r}'(t) = \langle v_0 \cos\theta, v_0 \sin\theta - gt\rangle$. To recover $\mathbf{r}(t)$, we integrate each component with respect to time, so

$$\mathbf{r}(t) = \int_0^t \mathbf{r}'(\xi)d\xi = \left\langle v_0 t \cos\theta, v_0 t \sin\theta - \frac{1}{2}gt^2 \right\rangle.$$

Several important facts about the trajectory can be immediately deduced from this expression.

The **flight time** t_f for the projectile is the nonzero solution to the quadratic equation $v_0 t \sin\theta - \frac{1}{2}gt^2 = 0$. This gives $t_f = \frac{2v_0 \sin\theta}{g}$. Since we are assuming no air resistance, v_h is undiminished, and the total horizontal distance traveled by the projectile until it comes back to the Earth is

$$v_0 t_f \cos\theta = v_0 \cos\theta \cdot \frac{2v_0 \sin\theta}{g} = \frac{v_0^2 \sin 2\theta}{g}$$

by the trigonometric identity $\sin 2\theta = 2\sin\theta\cos\theta$. This distance is called the **range** of the projectile. Also under this assumption the path $\mathbf{r}(t)$ is a parabola whose vertex is the highest point reached by the projectile. By symmetry, this height is reached at time $\frac{1}{2}t_f = \frac{v_0 \sin\theta}{g}$ after launch. At this moment the vertical velocity of the projectile is momentarily zero before it starts to accelerate back to earth. Substituting $\frac{1}{2}t_f = \frac{v_0 \sin\theta}{g}$ into the $\hat{\mathbf{j}}$ component of $\mathbf{r}(t)$, we get the **maximum height** reached is

$$h_{\max} = (v_0 \sin\theta)\left(\frac{v_0 \sin\theta}{g}\right) - \frac{1}{2}g\left(\frac{v_0 \sin\theta}{g}\right)^2 = \frac{(v_0 \sin\theta)^2}{2g}.$$

In the event that the projectile is launched at the point (a, b) instead of the origin, the trajectory would be revised to

$$\mathbf{r}(t) = \left\langle a + v_0 t \cos\theta, b + v_0 t \sin\theta - \frac{1}{2}gt^2 \right\rangle.$$

The formulas for flight time, range, and maximum height would have to be adjusted accordingly. The presence of a headwind or tailwind in the plane of flight would result in the adjustment of the horizontal velocity component but not the vertical.

Table 4.8: Key trajectory formulas. Launch at origin with velocity v_0, inclination θ over level ground.

1) Flight time	$\frac{2v_0 \sin\theta}{g}$
2) Range	$\frac{v_0^2 \sin 2\theta}{g}$
3) Maximum height	$\frac{(v_0 \sin\theta)^2}{2g}$

In the examples below, unless otherwise noted, assume the projectile is launched at the origin and travels over level ground.

Example 4.46. A rifle shoots a bullet with velocity 1000 feet per second (fps) with the barrel elevated 30° above horizontal. Find the flight time, range, and maximum height attained.

Example 4.47. The "Green Monster" at Fenway Park in Boston is the left field fence at the baseball park. The fence is 310 feet from home plate down the left field line. Although this distance is smaller than that in most major league parks, the fact that the fence is 37 feet high makes it difficult to hit a home run down the left field line. If a baseball is hit down the left field line starting at 3 feet off the ground and traveling at 110 mph initial velocity (which is constant since we are neglecting air resistance), what is the minimum angle of inclination from horizontal for which the ball would just clear the fence and what is the maximum angle that would still result in a home run?

Example 4.48. For a cannon with given muzzle velocity v_0, find the launch angle yielding maximum range. Assume both launch and impact at ground level.

Example 4.49. An antiaircraft gun is aimed due east and fires projectiles traveling at 1200 [fps]. An enemy aircraft is flying due west towards the location of the gun at a constant altitude of 5,000 [ft]. If the gun is inclined 60° from horizontal, is it capable of shooting down the plane? If so, how many miles east of the gun location could this occur?

4.7.4 Planes

A **plane** in $\mathbb{R}^3$ is uniquely determined by three noncollinear points, say $P_1 = (a_1, b_1, c_1)$, $P_2 = (a_2, b_2, c_2)$, and $P_3 = (a_3, b_3, c_3)$. We would like to derive the equation of the plane they determine in the form $ax + by + cz = D$. This is the **canonical** or **standard** form of the equation of a plane in $\mathbb{R}^3$. Here is a simple plan to achieve this. Form the vectors $\mathbf{v}_1 = \overrightarrow{P_1P_2}$ and $\mathbf{v}_2 = \overrightarrow{P_1P_3}$ and take the cross product $\mathbf{v}_1 \times \mathbf{v}_2$. By the properties of the cross product, we know $\mathbf{v}_1 \times \mathbf{v}_2$ will be perpendicular to both $\mathbf{v}_1$ and $\mathbf{v}_2$, hence perpendicular to the common plane of P_1, P_2, and P_3. We will use this fact to construct a generic vector in that common plane. Let $\mathbf{v} = \langle x - a_1, y - b_1, z - c_1\rangle$. Now set $\mathbf{v} \cdot \mathbf{v}_1 \times \mathbf{v}_2 = 0$. By the properties of the dot product, the generic vector $\mathbf{v}$ is now perpendicular to the cross product which by design was already perpendicular to the original plane. In other words, the generic vector is back in the original plane. Let us carry out this construction in finer detail and actually derive the equation of the plane in canonical form.

We have $\mathbf{v}_1 = \langle a_2 - a_1, b_2 - b_1, c_2 - c_1\rangle$ and $\mathbf{v}_2 = \langle a_3 - a_1, b_3 - b_1, c_3 - c_1\rangle$. Before we apply our formula for the cross product, let us compress our notation. Set $\langle a_2 - a_1, b_2 - b_1, c_2 - c_1\rangle = \langle A_2, B_2, C_2\rangle$ and $\langle a_3 - a_1, b_3 - b_1, c_3 - c_1\rangle = \langle A_3, B_3, C_3\rangle$. Now $\mathbf{v}_1 \times \mathbf{v}_2 = \langle B_2C_3 - B_3C_2, -A_2C_3 + A_3C_2, A_2B_3 - A_3B_2\rangle$. Dotting this with $\mathbf{v} = \langle x - a_1, y - b_1, z - c_1\rangle$ we finally can write

$$(x - a_1)(B_2C_3 - B_3C_2) + (y - b_1)(-A_2C_3 + A_3C_2) + (z - c_1)(A_2B_3 - A_3B_2) = 0.$$

Putting this in canonical form,

$$\begin{aligned}&(B_2C_3 - B_3C_2)x + (-A_2C_3 + A_3C_2)y + (A_2B_3 - A_3B_2)z\\ &\quad = a_1(B_2C_3 - B_3C_2) + b_1(-A_2C_3 + A_3C_2) + c_1(A_2B_3 - A_3B_2).\end{aligned}$$

Some examples are in order. The first is very simple.

Example 4.50. Find the equation of the plane in standard form that contains the points $(0, 0, 0)$, $(1, 2, 3)$, and $(3, 5, 7)$.

Example 4.51. Find the equation of the plane in standard form that contains the points $(1, 2, 2)$, $(3, -1, 4)$, and $(6, 5, 8)$.

A **direction vector** for a plane is any vector perpendicular to the plane. We can use the second part of the preceding construction to recover the equation of the plane from a direction vector if we also know one point on the plane. Note that a vector in the plane is not sufficient to define the plane.

Example 4.52. A plane containing the point $(1, 1, 1)$ has direction vector $\mathbf{u} = \langle 2, -1, 3\rangle$. Find the equation of the plane in standard form.

Example 4.53. A plane containing the point $(4, 7, -13)$ has direction vector $\langle 11, -23, 8\rangle$. Find its equation in standard form.

Planes can either be parallel (possibly identical) or intersect in a line. The angle between intersecting planes is called a **dihedral angle**. The dihedral angle between planes is the same as the angle between their direction vectors. Recall (Section 4.5, Example 4.15) that this angle is given by the formula $\theta = \arccos(\frac{\mathbf{v}\cdot\mathbf{w}}{|\mathbf{v}||\mathbf{w}|})$. If two planes are parallel, the respective coefficients in the standard equation for one plane would be the same multiple as the coefficients in the other one. For example, the plane given by $6x - 15y + 3z = 13$ is parallel to the plane given by $2x - 5y + z = 11$, since the respective coefficients of the former are three times the coefficients of the latter. In the event that the constant terms are also in the same ratio, the two equations would describe the identical plane. So $6x - 15y + 3z = 33$ clearly gives the same plane as that given by $2x - 5y + z = 11$. Note that a direction vector for the line of intersection between two planes is simply the cross product of the direction vectors for the two planes.

Example 4.54. Determine if the planes given by $2x - 5y + z = 11$ and $3x + 7y - 4z = 9$ are parallel. If not, find the dihedral angle in which they intersect.

Example 4.55. Find the equation of the line of intersection of the two nonparallel planes given by $x - y + z = 6$ and $2x + y - z = 3$.

Example 4.56. Continuing Example 4.55, find a vector that lies in the plane given by $x - y + z = 6$ but has its initial point on the line $\mathbf{r}(t)$ and is perpendicular to that line.

It will be useful to have a formula that gives a **point-to-plane distance**. Our derivation is based on connecting a given point to any arbitrary point in the given plane and calculating the length of that line segment. Of course, it would be very unlikely that the line segment chosen so randomly would be perpendicular to the plane, which would be required to get the true distance of the point. The idea of projecting the line segment onto the direction vector of the plane comes to our rescue. The projected length becomes the perpendicular and therefore true distance point-to-plane. Suppose we are given the point (x_1, y_1, z_1) and the plane with equation $ax + by + cz = D$. Choose an arbitrary point (x_2, y_2, z_2) in the plane and construct the vector $\mathbf{v} = \langle x_1 - x_2, y_1 - y_2, z_1 - z_2 \rangle$. A unit vector parallel to the direction vector of the plane is

$$\mathbf{u} = \frac{\langle a, b, c \rangle}{\sqrt{a^2 + b^2 + c^2}}.$$

The scalar projection of $\mathbf{v}$ in the direction of $\mathbf{u}$ will be the (signed) perpendicular distance of (x_1, y_1, z_1) from the plane. From Section 4.5, Example 4.18, this projection is

$$\frac{\langle a, b, c \rangle}{\sqrt{a^2 + b^2 + c^2}} \cdot \langle x_1 - x_2, y_1 - y_2, z_1 - z_2 \rangle = \frac{a(x_1 - x_2) + b(y_1 - y_2) + c(z_1 - z_2)}{\sqrt{a^2 + b^2 + c^2}}.$$

So the point-to-plane distance

$$d = \left| \frac{a(x_1 - x_2) + b(y_1 - y_2) + c(z_1 - z_2)}{\sqrt{a^2 + b^2 + c^2}} \right|.$$

Example 4.57. Given the plane with equation $2x + 3y - 5z = 10$, find the distance from the origin to the plane.

Example 4.58. Given the plane with equation $x - 2y + 4z = 7$, find the distance from the point $(2, 1, 5)$ to the plane.

Regarding **plane-to-plane distance**, recall that two planes can have one of three relationships: (i) they can intersect, (ii) they can be parallel and separated by a positive distance, or (iii) they can be parallel and be identical. In cases (i) and (iii), the distance between planes is zero. To handle case (ii), select a point in either plane and find its distance to the other plane. The distance from a point to a solid figure bounded by planes can be determined by finding the equations of the planes bounding the solid and carrying out the calculation in Examples 4.57 or 4.58 with respect to each of the bounding planes. Then select the shortest distance.

References & suggested further reading

1) *Mathematical Methods for Physicists* (7th ed.) by George Arfken, Hans Weber, & Frank Harris (2013), Elsevier/Academic Press – *classic that presents the vector basics and much more*
2) *Calculus Early Transcendentals* (12 ed) by George Thomas, Maurice Weir, & Joel Hass (2010), Pearson/Addison-Wesley – *great book for calculus that covers all the vector background material*
3) *Advanced Calculus for Applications* by Francis Hildebrand (1962), Prentice-Hall

Solutions to Chapter 4 examples

Example 4.1

The elements of $\mathbb{R}^2$ are point pairs (x, y). Define vector addition as $(x_1, y_1) + (x_2, y_2) = (x_1 + x_2, y_1 + y_2)$. Define scalar multiplication by $c \in \mathbb{R}$ as $c(x, y) = (cx, cy)$. The associativity and commutativity are obvious from the properties of $\mathbb{R}$. The element $(0, 0)$ serves as the additive identity. The additive inverse of (x, y) is clearly $(-x, -y)$. Unitarity and annihilation are trivial. For distributivity of scalar multiplication over vector addition, we have $c[(x_1, y_1) + (x_2, y_2)] = c(x_1, y_1) + c(x_2, y_2)$ and for distributivity of scalar multiplication over scalar addition, this is immediate from the field properties of $\mathbb{R}$. Finally, $c_1(c_2(x, y)) = (c_1 c_2)(x, y)$. The axioms are satisfied, so we conclude $\langle \mathbb{R}^2, \mathbb{R} \rangle$ is a vector space.

Example 4.2

The elements of $\mathbb{R}^n$ *are n*-**tuples** $(x_1, x_2, \ldots, x_n)$. Define vector addition and scalar multiplication in complete analogy with Example 4.1. So $(x_1, x_2, \ldots, x_n) + (x_1', x_2', \ldots, x_n') = (x_1 + x_1', x_2 + x_2', \ldots, x_n + x_n')$ and $c(x_1, x_2, \ldots, x_n) = (cx_1, cx_2, \ldots, cx_n)$. Verification that elements of $\mathbb{R}^n$ conform to the axioms proceeds exactly parallel to Example 4.1, and we conclude that $\langle \mathbb{R}^n, \mathbb{R} \rangle$ and hence $\langle \mathbb{R}^3, \mathbb{R} \rangle$ are vector spaces.

Example 4.3

Consider the set of polynomials $\mathfrak{P}_x$ of any degree in the variable x with rational coefficients. A typical element is $a_n x^n + a_{n-1} x^{n-1} + \cdots + a_1 x + a_0$ where $a_i \in \mathbb{Q}$. Define vector addition as polynomial addition where the coefficients of monomial terms of equal degree are added. Adding two rational numbers produces another rational, so sums remain in the set. Define multiplication by $c \in \mathbb{Q}$ as $c(a_n x^n + a_{n-1} x^{n-1} + \cdots + a_1 x + a_0) = ca_n x^n + ca_{n-1} x^{n-1} + \cdots + ca_1 x + ca_0$. Multiplying two rational numbers results in another rational, so scalar multiples stay in the set. The additive identity is the zero polynomial. The additive inverse is $-a_n x^n - a_{n-1} x^{n-1} - \cdots - a_1 x - a_0$. The other axioms are satisfied since $\mathbb{Q}$ is a field. We conclude $\langle \mathfrak{P}_x, \mathbb{Q} \rangle$ is a vector space, decidedly not spatial. We could try to emulate the manner in which we wrote the elements of $\mathbb{R}^n$ in Example 4.2. The typical element might be written as $(a_0, a_1, a_2, \ldots, a_{n-1}, a_n)$, with the understanding that the coefficient in the ith position applies to x^i. A moment's reflection reveals that we may need arbitrarily many positions in the parenthetical expression because the degree of a polynomial, although finite, can be arbitrarily high. Instead of an n-tuple, we really need a ω-tuple with infinitely many positions. Of course, only finitely many entries in the ω-tuple could be nonzero to ensure that it represents a proper polynomial.

Example 4.4

This example gives us a hint as to why the theory of vector spaces is so strongly connected to functional analysis. Let $X \neq \varnothing$ be any set and $\mathfrak{F}(X) = \{f : X \to \mathbb{R}\}$. The vectors are simply real functions on X, such as $f(x)$ and $g(x)$. For all $x \in X$, we define vector addition as $(f+g)(x) = f(x) + g(x)$, and scalar multiplication by $c \in \mathbb{R}$ as $(cf)(x) = cf(x)$. The constant zero function acts as the additive identity and for each $f(x)$ its negative is clearly the additive inverse, since $f(x) + (-f(x)) = 0$. Axioms 1 through 4 are therefore satisfied and the remaining axioms depending on the field properties of $\mathbb{R}$ are verified as well.

Example 4.6

Matching entries, we obtain two equations: (i) $a + b = -2$ and (ii) $2a + 3b = 5$. Solving these simultaneously, we have $a = -11$ and $b = 9$.

Example 4.7

Matching entries, we obtain four equations: (i) $a + 2b + 3c = 11$, (ii) $-3a + 4b + 2c + d = 8$, (iii) $3b - d = 7$, and (iv) $a - 5b - 2c + d = -14$. Solving simultaneously, we have $a = 1$, $b = c = 2$, and $d = -1$.

Example 4.8

For the sake of contradiction, suppose $\mathbf{v} = a_1\mathbf{e}_1 + a_2\mathbf{e}_2 + \cdots + a_n\mathbf{e}_n$ and also that $\mathbf{v} = b_1\mathbf{e}_1 + b_2\mathbf{e}_2 + \cdots + b_n\mathbf{e}_n$. Subtracting the second from the first expression, we get $(a_1 - b_1)\mathbf{e}_1 + (a_2 - b_2)\mathbf{e}_2 + \cdots + (a_n - b_n)\mathbf{e}_n = \mathbf{0}$. Since the basis vectors are linearly independent, it must be the case that $a_i - b_i = 0$ for $1 \le i \le n$. But this means $a_i = b_i$ and we see that the expression of any vector with respect to the given basis is unique.

Example 4.9

We know $\mathbf{w}$ is some linear combination of the $\mathbf{v}_i$, say $\mathbf{w} = a_1\mathbf{v}_1 + a_2\mathbf{v}_2 + \cdots + a_n\mathbf{v}_n$. Then $\mathbf{w} - a_1\mathbf{v}_1 + a_2\mathbf{v}_2 + \cdots + a_n\mathbf{v}_n = \mathbf{0}$. At least one $a_i \neq 0$, so the condition for linear independence is violated.

Example 4.10

Consider an arbitrary set of vectors $S = \{\mathbf{v}_1, \mathbf{v}_2, \ldots, \mathbf{v}_n, \mathbf{v}_{n+1}\}$ in V which has a basis $B = \{\mathbf{e}_1, \mathbf{e}_2, \ldots, \mathbf{e}_n\}$. For the sake of contradiction, suppose the vectors in S are independent. This requires that the test expression $\beta_1\mathbf{v}_1 + \beta_2\mathbf{v}_2 + \cdots + \beta_{n+1}\mathbf{v}_{n+1} = \mathbf{0}$ forces each $\beta = 0$. First, express each $\mathbf{v}_i$ in terms of the basis vectors. We have $\mathbf{v}_1 = \sum_{j=1}^{n} a_{1j}\mathbf{e}_j$, and in general $\mathbf{v}_i = \sum_{j=1}^{n} a_{ij}\mathbf{e}_j$ for $1 \le i \le n+1$. Substituting into the test expression, we get

$$\beta_1 \sum_{j=1}^{n} a_{1j}\mathbf{e}_j + \beta_2 \sum_{j=1}^{n} a_{2j}\mathbf{e}_j + \cdots + \beta_{n+1} \sum_{j=1}^{n} a_{(n+1)j}\mathbf{e}_j = \mathbf{0}.$$

Rewriting this sum by collecting coefficients of the basis vectors, we obtain

$$\mathbf{e}_1 \sum_{k=1}^{n+1} a_{k1}\beta_k + \mathbf{e}_2 \sum_{k=1}^{n+1} a_{k2}\beta_k + \cdots + \mathbf{e}_n \sum_{k=1}^{n+1} a_{kn}\beta_k = \mathbf{0}.$$

Suppose we set each coefficient sum for an $\mathbf{e}_i$ to zero. This creates a system of n equations, one for each basis vector, in $n+1$ variables, namely the β_i. The underdetermined system then has one free variable, allowing us to construct a solution with some nonzero β_i. But this contradiction establishes that the $n+1$ vectors cannot be independent. This result clearly extends to any set of $n+k$ vectors, where $k \ge 1$.

Example 4.11

Consider all linearly independent spanning sets for $V \neq \varnothing$. There is at least one. By well-ordering, there is a minimum cardinality n among these sets. Since such a set would constitute a basis for V, and any set with $n+1$ or more vectors could not be a basis by Example 4.10, we conclude that $\dim V = n$.

Example 4.12

This is a case where if we drew the port and waypoints to scale on a map, some of the displacements would be directed line segments (free vectors). The task is to find the vectors in $\mathbb{R}^2$ that represent displacements along each leg of the journey. The answer to the question will be the vector that is the additive inverse of the sum of the three legs. We do not have displacements yet – the problem data is in terms of speed and time. Orient the xy-plane so the east aligns with positive values of x. Measuring distance in (nautical) miles the eastward leg is $10 \cdot 15 = 150$ miles, so the first leg displacement vector is $\langle 150, 0 \rangle$. The second is $5 \cdot 20 = 100$ miles, so that vector is $\langle 0, -100 \rangle$. The length of the third vector is $3 \cdot 10 = 30$ miles, so that vector is $\langle -30, 0 \rangle$. The sum of these three vectors represents where the ship is. So the vector describing its position is $\langle x, y \rangle = \langle 150, 0 \rangle + \langle 0, -100 \rangle + \langle -30, 0 \rangle = \langle 120, -100 \rangle$. The vector that leads from that position to port is then $\langle -120, 100 \rangle$. The norm of this vector is $\sqrt{(-120)^2 + 100^2} = 156.2$ miles. The direction in which to proceed back to port would be $\theta = 129.8°$ for an engineer or $N50.2°W$ for an actual sailor. Recall θ is the angle in the xy-plane measured counterclockwise from the positive x-axis.

Example 4.13

The force vector must have a magnitude of 50 newtons. The direction of the force will be the same as the direction of the line segment starting at $(1, 2, 3)$ and terminating at $(4, 7, 10)$. This line segment extends $4 - 1 = 3$ units in the x-direction, $7 - 2 = 5$ units in the y-direction, and $10 - 3 = 7$ units in the z-direction. So the corresponding vector is $\langle 3, 5, 7 \rangle$. The norm of $\langle 3, 5, 7 \rangle$ is $\sqrt{3^2 + 5^2 + 7^2} = \sqrt{83} = 9.11$, so it captures the direction but not the correct magnitude of the force. It would be very convenient to have a vector of length 1 in the direction of $\langle 3, 5, 7 \rangle$. The operation of **unitizing** a nonzero vector is always possible and easy to do. We simply divide any vector by its norm to produce a unit vector in the same direction. So $\frac{1}{9.11}\langle 3, 5, 7 \rangle$ saves the direction of $\langle 3, 5, 7 \rangle$ but has length 1. The force vector we want has this direction but norm 50, so our solution is $\mathbf{F} = \frac{50}{9.11}\langle 3, 5, 7 \rangle \mathrm{N} = \langle 16.47, 27.44, 38.42 \rangle$. Checking, $\sqrt{(16.47)^2 + (27.44)^2 + (38.42)^2} \approx 50\,\mathrm{N}$.

Example 4.14

The vectors corresponding to these forces are respectively $\langle 0, 10 \rangle$, $\langle -5, 0 \rangle$, and $\frac{\sqrt{2}}{2}\langle 15, -15 \rangle$. Adding componentwise, we get the net force $\mathbf{F} = \langle 5.61, -0.61 \rangle$ tons. So the balancing force is $-\mathbf{F} = \langle -5.61, 0.61 \rangle$ tons.

Example 4.15

The terminal points of the two vectors along with their common initial point determine a plane, so this problem immediately devolves to $\mathbb{R}^2$. The law of cosines is a generalization of the Pythagorean formula. It supplies a correction term that applies when the "right" angle fails to be right. Suppose a scalene triangle has sides a, b, and c where we want to calculate the length of c in terms of a and b but the angle θ opposite side c fails to be 90°. The law of cosines says $c^2 = a^2 + b^2 - 2ab\cos\theta$. Now let us use this in the context of vectors.

Suppose $\mathbf{v} = \langle v_1, v_2, v_3\rangle$ and $\mathbf{w} = \langle w_1, w_2, w_3\rangle$. The directed line segment that runs from the terminal point of $\mathbf{v}$ to the terminal point of $\mathbf{w}$ completes a triangle with $\mathbf{v}$ and $\mathbf{w}$ as the other sides. The vector $\mathbf{w} - \mathbf{v}$ has the same length and orientation as this line segment. That length is $|\mathbf{w} - \mathbf{v}| = \sqrt{(w_1 - v_1)^2 + (w_2 - v_2)^2 + (w_3 - v_3)^2}$. Now identify this expression as c in the law of cosines and likewise treat a as $|\mathbf{v}|$ and $|\mathbf{w}|$ as b. With these substitutions, we have $(w_1 - v_1)^2 + (w_2 - v_2)^2 + (w_3 - v_3)^2 = (v_1^2 + v_2^2 + v_3^2) + (w_1^2 + w_2^2 + w_3^2) - 2(|\mathbf{v}||\mathbf{w}|)\cos\theta$. Bit of a mess at this point, but wait – expanding the left-hand side, we get $(w_1^2 - 2w_1v_1 + v_1^2) + (w_2^2 - 2w_2v_2 + v_2^2) + (w_3^2 - 2w_3v_3 + v_3^2) = (v_1^2 + v_2^2 + v_3^2) + (w_1^2 + w_2^2 + w_3^2) - 2(w_1v_1 + w_2v_2 + w_3v_3)$. Now harvest the cancellations on both sides to get $-2(w_1v_1 + w_2v_2 + w_3v_3) = -2(|\mathbf{v}||\mathbf{w}|)\cos\theta$. Dividing by -2 and noticing that $w_1v_1 + w_2v_2 + w_3v_3 = \mathbf{v}\cdot\mathbf{w}$, we finally have $\mathbf{v}\cdot\mathbf{w} = |\mathbf{v}||\mathbf{w}|\cos\theta$ or $\cos\theta = \frac{\mathbf{v}\cdot\mathbf{w}}{|\mathbf{v}||\mathbf{w}|}$. The desired formula follows immediately.

Example 4.16

A 90° fitting will not work because the drain is turning through 90° at the same time as it is descending. We will impose a coordinate system on the building and locate the origin at the point of the fitting. Then the two runs of pipe will be represented by vectors, and our formula will give us the necessary angle. Letting the z-axis coincide with the southeast corner of the building extending towards the roof, we then locate the y-axis in the south face of the building. Now we can find the components of the two vectors. The pipe running from the origin to the roof has no extension in the y-direction, moves -50 feet in the x-direction, and 50 feet in the z-direction. So its vector is $\langle -50, 0, 50\rangle$. The pipe descending across the south face has no extension in the x-direction, moves -50 feet in the y-direction, and -50 feet in the z-direction. Then its vector is $\langle 0, -50, -50\rangle$. We compute the fitting angle

$$\theta = \arccos\left(\frac{\langle -50, 0, 50\rangle \cdot \langle 0, -50, -50\rangle}{|\langle -50, 0, 50\rangle||\langle 0, -50, -50\rangle|}\right) = \arccos\left(\frac{-2500}{(\sqrt{5000})^2}\right) = \arccos\left(-\frac{1}{2}\right) = 120^\circ.$$

Example 4.17

This expression for the dot product is often used as the fundamental definition, but it has the disadvantage that the angle between vectors needs to be known. It is an obvious consequence of the calculation in Example 4.15 that if the dot product of two nonzero vectors is zero, then the cosine of the angle between them must be zero. It follows that the acute angle must be $\frac{\pi}{2}$ or 90°.

Example 4.18

The scalar projection of a vector is the length of its “shadow” in some direction. Specifically, it is the amount of the vector’s length that aligns with a given direction. If that direction is given in the form of a unit vector $\mathbf{u}$, which makes an angle θ with $\mathbf{v}$, then the projection of $\mathbf{v}$ in that direction is $|\mathbf{v}|\cos\theta = \mathbf{v}\cdot\mathbf{u}$. Note that $\frac{\mathbf{w}}{|\mathbf{w}|}$ is a unit vector in the direction of $\mathbf{w}$, so $\mathbf{v}\cdot\frac{\mathbf{w}}{|\mathbf{w}|}$ is precisely the projection of $\mathbf{v}$ in the direction of vector $\mathbf{w}$.

Example 4.19

Consider the following eight equations which develop the puzzle pieces in an orderly way. This is a lot of algebra, but it is worth seeing the connection between the definition and a more practical formula. Much cancellation will be possible:

(i) By definition,

$$\begin{aligned}(\mathbf{x}\times\mathbf{y})\cdot(\mathbf{x}\times\mathbf{y}) &= (\hat{\mathbf{i}}(x_2y_3 - x_3y_2) + \hat{\mathbf{j}}(x_3y_1 - x_1y_3) + \hat{\mathbf{k}}(x_1y_2 - x_2y_1)) \\ &\quad\cdot(\hat{\mathbf{i}}(x_2y_3 - x_3y_2) + \hat{\mathbf{j}}(x_3y_1 - x_1y_3) + \hat{\mathbf{k}}(x_1y_2 - x_2y_1)) \\ &= (x_2y_3 - x_3y_2)^2 + (x_3y_1 - x_1y_3)^2 + (x_1y_2 - x_2y_1)^2.\end{aligned}$$

(ii) Expanding the squares,

$$\begin{aligned}&(x_2y_3 - x_3y_2)^2 + (x_3y_1 - x_1y_3)^2 + (x_1y_2 - x_2y_1)^2 \\ &\quad = (x_2y_3)^2 - 2(x_2y_2x_3y_3) + (x_3y_2)^2 + (x_3y_1)^2 - 2(x_3y_1x_1y_3) \\ &\qquad + (x_1y_3)^2 + (x_1y_2)^2 - 2(x_1y_2x_2y_1) + (x_2y_1)^2.\end{aligned}$$

(iii) Then regrouping,

$$\begin{aligned}&(x_2y_3)^2 - 2(x_2y_2x_3y_3) + (x_3y_2)^2 + (x_3y_1)^2 - 2(x_3y_1x_1y_3) \\ &\quad + (x_1y_3)^2 + (x_1y_2)^2 - 2(x_1y_2x_2y_1) + (x_2y_1)^2 \\ &= \left[(x_2y_3)^2 + (x_3y_2)^2 + (x_3y_1)^2 + (x_1y_3)^2 + (x_1y_2)^2 + (x_2y_1)^2\right] \\ &\quad - 2[x_2y_2x_3y_3 + x_3y_1x_1y_3 + x_1y_2x_2y_1].\end{aligned}$$

(iv) Generating similar terms,

$$\begin{aligned}|\mathbf{x}|^2|\mathbf{y}|^2 &= (x_1^2 + x_2^2 + x_3^2)(y_1^2 + y_2^2 + y_3^2) \\ &= (x_1^2y_1^2 + x_1^2y_2^2 + x_1^2y_3^2) + (x_2^2y_1^2 + x_2^2y_2^2 + x_2^2y_3^2) + (x_3^2y_1^2 + x_3^2y_2^2 + x_3^2y_3^2).\end{aligned}$$

(v) More similar terms

$$\begin{aligned}(\mathbf{x}\cdot\mathbf{y})^2 &= (x_1y_1 + x_2y_2 + x_3y_3)^2 \\ &= (x_1y_1)^2 + (x_2y_2)^2 + (x_3y_3)^2 + x_1y_1x_2y_2 + x_1y_1x_3y_3 + x_2y_2x_1y_1 \\ &\quad + x_2y_2x_3y_3 + x_3y_3x_1y_1 + x_3y_3x_2y_2.\end{aligned}$$

(vi) Regrouping (v),

$$(\mathbf{x}\cdot\mathbf{y})^2 = (x_1y_1)^2 + (x_2y_2)^2 + (x_3y_3)^2 + 2[x_1y_1x_2y_2 + x_1y_1x_3y_3 + x_2y_2x_3y_3].$$

(vii) Setting up cancellations,

$$\begin{aligned}|\mathbf{x}|^2|\mathbf{y}|^2 - (\mathbf{x}\cdot\mathbf{y})^2 \\ &= [(x_1^2y_1^2 + x_1^2y_2^2 + x_1^2y_3^2) + (x_2^2y_1^2 + x_2^2y_2^2 + x_2^2y_3^2) + (x_3^2y_1^2 + x_3^2y_2^2 + x_3^2y_3^2)] \\ &\quad - [(x_1y_1)^2 + (x_2y_2)^2 + (x_3y_3)^2 + 2(x_1y_1x_2y_2 + x_1y_1x_3y_3 + x_2y_2x_3y_3)].\end{aligned}$$

(viii) Harvesting cancellations,

$$\begin{aligned}|\mathbf{x}|^2|\mathbf{y}|^2 - (\mathbf{x}\cdot\mathbf{y})^2 &= [(x_2^2y_1^2 + x_2^2y_2^2 + x_2^2y_3^2) + (x_3^2y_1^2 + x_3^2y_2^2 + x_3^2y_3^2)] \\ &\quad - 2[x_2y_2x_3y_3 + x_3y_1x_1y_3 + x_1y_2x_2y_1].\end{aligned}$$

Comparing equations (iii) and (viii), assemble the puzzle pieces to get

$$(\mathbf{x}\times\mathbf{y})\cdot(\mathbf{x}\times\mathbf{y}) = |\mathbf{x}|^2|\mathbf{y}|^2 - (\mathbf{x}\cdot\mathbf{y})^2.$$

Since $(\mathbf{x}\cdot\mathbf{y})^2 = |\mathbf{x}|^2|\mathbf{y}|^2\cos^2\theta$, evidently $|\mathbf{x}\times\mathbf{y}|^2 = |\mathbf{x}|^2|\mathbf{y}|^2(1-\cos^2\theta) = |\mathbf{x}|^2|\mathbf{y}|^2\sin^2\theta$. Upon taking square roots, we finally get $|\mathbf{x}\times\mathbf{y}| = |\mathbf{x}||\mathbf{y}|\sin\theta$.

Now for the matter of $\widehat{\mathbf{e}}_\perp$. Notice that $\mathbf{x}\cdot(\mathbf{x}\times\mathbf{y}) = x_1(x_2y_3 - x_3y_2) + x_2(x_3y_1 - x_1y_3) + x_3(x_1y_2 - x_2y_1) = 0$. Likewise, $\mathbf{y}\cdot(\mathbf{x}\times\mathbf{y}) = 0$. We know that a zero dot product is an indicator of orthogonality, so we can conclude that $\mathbf{x}\times\mathbf{y}$ is perpendicular to the common plane of $\mathbf{x}$ and $\mathbf{y}$. Using the convention of a right-handed coordinate system, we will choose the direction of $\widehat{\mathbf{e}}_\perp$ to be given by the right-hand rule. If the fingers of the right hand are extended in the direction of $\mathbf{x}$ and then curled in the direction of $\mathbf{y}$, the thumb will point in the direction of $\widehat{\mathbf{e}}_\perp$. This completes the verification of the equivalence of the two definitions. As with the dot product, the trade-off for using the simpler definition is having to know the angle between vectors in advance.

Example 4.20

Let $\mathbf{x}$ and $\mathbf{y}$ be the two vectors. The length of an altitude of the parallelogram drawn from the terminal point of $\mathbf{x}$ and intersecting $\mathbf{y}$, or if necessary a parallel extension of $\mathbf{y}$, would be $|\mathbf{x}|\sin\theta$, where θ is the angle between $\mathbf{x}$ and $\mathbf{y}$. The base of the figure is $|\mathbf{y}|$. So the area of the parallelogram is base times altitude or $|y||x|\sin\theta$. This happens to be $|\mathbf{x}\times\mathbf{y}|$. For a triangle, take half, so the area would be $\frac{1}{2}|\mathbf{x}\times\mathbf{y}|$.

Example 4.21

Label the points P_1, P_2, and P_3, respectively. The line segment connecting P_1 and P_2 will be written $\overline{P_1P_2}$, and so forth. Pick any point – P_1 will do – and find the vectors that represent the line segments $\overline{P_1P_2}$ and $\overline{P_1P_3}$. For $\overline{P_1P_2}$, we have $\langle 3-1, 5-2, 7-3\rangle$ or $\mathbf{x} = \langle 2, 3, 4\rangle$. For $\overline{P_1P_3}$, we have $\langle -1-1, -1-2, 4-3\rangle$ or $\mathbf{y} = \langle -2, -3, 1\rangle$. We do not know the angle between these vectors, so we resort to the component definition of cross product $\mathbf{x}\times\mathbf{y} = \widehat{\mathbf{i}}(x_2y_3 - x_3y_2) + \widehat{\mathbf{j}}(x_3y_1 - x_1y_3) + \widehat{\mathbf{k}}(x_1y_2 - x_2y_1)$. Then in this case, $\mathbf{x}\times\mathbf{y} = \widehat{\mathbf{i}}(3\cdot 1 - 4(-3)) + \widehat{\mathbf{j}}(4(-2) - 2\cdot 1) + \widehat{\mathbf{k}}(2(-3) - 3(-1))$. Simplified, this is the vector $\langle 15, -10, 3\rangle$. The area is $\frac{1}{2}|\langle 15, -10, 3\rangle| = \frac{1}{2}\sqrt{225 + 100 + 9} = 9.14$ units. This method is easier to use in practice than a purely analytical geometry approach.

Example 4.22

For nonzero vectors, if $\mathbf{x}\times\mathbf{y} = |\mathbf{x}||\mathbf{y}|\sin\theta\,\widehat{\mathbf{e}}_\perp = \mathbf{0}$, it must be the case that $\sin\theta = 0$, or $\theta = 0^\circ$ for parallelism or 180° for antiparallelism.

Example 4.25

Here $V = |\langle 0, 1, 0\rangle\cdot\langle -2, 2, 0\rangle\times\langle -1, 2, 3\rangle| = 6\times 10^{-9}\,\mathrm{m}^3$, so $\frac{1\times 10^{-6}}{6\times 10^{-9}} = 1.67\times 10^2$ units.

Example 4.27

We have

$$\begin{aligned}\mathbf{A}\times(\mathbf{B}\times\mathbf{C}) &= \mathbf{B}(\mathbf{A}\cdot\mathbf{C}) - \mathbf{C}(\mathbf{A}\cdot\mathbf{B}),\\ \mathbf{B}\times(\mathbf{C}\times\mathbf{A}) &= \mathbf{C}(\mathbf{B}\cdot\mathbf{A}) - \mathbf{A}(\mathbf{B}\cdot\mathbf{C}),\\ \mathbf{C}\times(\mathbf{A}\times\mathbf{B}) &= \mathbf{A}(\mathbf{C}\cdot\mathbf{B}) - \mathbf{B}(\mathbf{C}\cdot\mathbf{A}).\end{aligned}$$

Adding the six terms on the right and cancelling in pairs (recall the dot product is commutative), we obtain the zero vector.

Example 4.28

Here $\mathbf{r}_0 = \langle 1, 2, 3\rangle$ and $\mathbf{v} = \langle -1, 1, 4\rangle$, so $\mathbf{r}(t) = \langle 1, 2, 3\rangle + t\langle -1, 1, 4\rangle$. Note $\langle 1, 2, 3\rangle + 3\langle -1, 1, 4\rangle = \langle -2, 5, 15\rangle$, so "yes" when $t = 3$.

Example 4.29

Here $\mathbf{v} = \langle 7-2, -7-1, 3-4\rangle = \langle 5, -8, -1\rangle$. Use either given point for $\mathbf{r}_0$, say $\mathbf{r}_0 = \langle 2, 1, 4\rangle$, then $\mathbf{r}(t) = \langle 2, 1, 4\rangle + t\langle 5, -8, -1\rangle$. Note that the order of subtraction in calculating $\mathbf{v}$ from the two given points could be reversed. Then reversing the sign of t generates the same line in the same direction.

Example 4.30

Separating the components of $\langle x(t), y(t), z(t)\rangle$, we get $x(t) = 2 + 5t$, $y(t) = 1 - 8t$, and $z(t) = 4 - t$.

Example 4.31

We use the same direction vector as in Example 4.28 but adjust $\mathbf{r}_0$ to $\langle 1,1,1\rangle$. So $\mathbf{r}(t) = \langle 1,1,1\rangle + t\langle -1,1,4\rangle$.

Example 4.32

The direction vector for both lines is the same, so they are parallel. If they intersect they would be the same line, so if $t_1 = 0$ there must be some t_2 such that $\langle 6,2,2\rangle + t_2\langle 4,-5,3\rangle = \langle 1,3,5\rangle$. Check the first two component equations, (i) $6 + 4t_2 = 1$ and (ii) $2 - 5t_2 = 3$. Equation (i) gives $t_2 = -\frac{5}{4}$ and (ii) gives $t_2 = -\frac{1}{5}$. Evidently, there is no consistent parameter value that would allow $\mathbf{r}_2(t)$ to equal $\langle 1,3,5\rangle$, so the lines cannot be the same, hence they do not intersect.

Example 4.33

Suppose the direction vector of the perpendicular line is $\langle x,y,z\rangle$, then the dot product of the two direction vectors should be zero, so $\langle 2,4,3\rangle \cdot \langle x,y,z\rangle = 0$. All we need is a solution to $2x + 4y + 3z = 0$. We have two degrees of freedom to work with, so we can easily find $z = 0$, $x = 2$, and then $y = -1$ as a solution. Then $\mathbf{r}_1(t) = \langle 1,2,3\rangle + t\langle 2,-1,0\rangle$ satisfies the criteria. Recall that in $\mathbb{R}^2$ a line that is perpendicular to that given by $y = mx + b$ is $y = m'x + b'$, where $mm' = -1$. Restating these formulas in our vector notation, $y = mx + b$ becomes $\mathbf{r}(t) = \langle 0,b\rangle + x\langle 1,m\rangle$ and $\mathbf{r}'(t) = \langle 0,b'\rangle + x\langle 1,m'\rangle$. Then the dot product condition yields $\langle 1,m\rangle \cdot \langle 1,m'\rangle = mm' + 1 = 0$ which returns the perpendicular slope formula $mm' = -1$.

Example 4.34

We first pick the point P on the line. The simplest case usually is to let $t = 0$, so $P = (1,2,3)$. The vector $\mathbf{w}$ from P to $Q = (4,3,7)$ is then $\langle 3,1,4\rangle$. Finding $\frac{\mathbf{v}}{|\mathbf{v}|}$, we have $\frac{\langle 2,-3,1\rangle}{\sqrt{14}}$. Finally, using our cross product formula, we get

$$\langle 3,1,4\rangle \times \frac{\langle 2,-3,1\rangle}{\sqrt{14}} = \frac{\langle 13,5,-11\rangle}{\sqrt{14}},$$

so the distance d is $\frac{1}{\sqrt{14}}|\langle 13,5,-11\rangle| = 4.74$.

Example 4.35

With $t = 0$, $P = (-2,-2,11)$ and $Q = (-1,5,9)$. Then $\mathbf{w} = \langle 1,7,-2\rangle$ and $\frac{\mathbf{v}}{|\mathbf{v}|} = \frac{\langle 1,-5,8\rangle}{3\sqrt{10}}$. The cross product is

$$\langle 1,7,-2\rangle \times \frac{\langle 1,-5,8\rangle}{3\sqrt{10}} = \frac{|\langle 46,-10,-12\rangle|}{3\sqrt{10}}.$$

Then $d = |\frac{48.58}{9.49}| = 5.12$.

Example 4.36

First, $\mathbf{v} = \langle 1,0,3\rangle \times \langle 4,4,6\rangle = \langle -12,6,4\rangle$. Next, set $t_1 = t_2 = 0$ so that $P = (1,2,3)$ and $Q = (-1,5,7)$. Then $\mathbf{w} = \langle 2,-3,-4\rangle$. Finally,

$$d = \left|\frac{\langle 2,-3,-4\rangle \cdot \langle 1,0,3\rangle}{\|\langle 1,0,3\rangle\|}\right| = \left|\frac{-10}{\sqrt{10}}\right| = 3.162.$$

We wanted to be precise in representing the two line equations so we distinguished the two parameters. This is not strictly necessary since any value in either line equation still yields a point on the line (provided $-\infty < t < \infty$).

Example 4.37

Here $\mathbf{r}(t) = \langle 2\cos t, 2\sin t, 1\rangle$ where $0 \le t \le 2\pi$.

Example 4.38

Let $x = a\cos t$ and $y = b\sin t$. Then $\frac{a^2\cos^2 t}{9} + \frac{b^2\sin^2 t}{25} = 1$. Regrouping, $(\frac{a}{3})^2\cos^2 t + (\frac{b}{5})^2\sin^2 t = 1$. We see $(\frac{a}{3})^2 = (\frac{b}{5})^2 = 1$, so $a = \pm 3$ and $b = \pm 5$. Choosing the positive values, we can finally write $\mathbf{r}(t) = \langle 3\cos t, 5\sin t\rangle$.

Example 4.39

Looking down at the helix along the z-axis from a height above $z = 10$, we would see a circle of radius 3. Following Example 4.37, the path function would be $\mathbf{r}(t) = \langle 3\cos t, 3\sin t, z\rangle$, where we have initialized the curve at the point $(3,0,0)$ and the dependence in the z-direction still needs to be determined. Since the pitch is one unit, whenever the parameter t increases by 2π units, the position vector has the same x and y coordinates, but the z-coordinate has increased by exactly one. We can make this happen by setting $z = \frac{t}{2\pi}$. Now the complete path function is $\mathbf{r}(t) = \langle 3\cos t, 3\sin t, \frac{t}{2\pi}\rangle$, and we allow t to range over the interval $[0, 20\pi]$ to ensure that the helix completes ten revolutions and terminates at $(3,0,10)$.

Example 4.40

This is a consequence of the more general fact that if a vector $\mathbf{v}(t)$ has constant length, then $\mathbf{v}'(t)$ is perpendicular to $\mathbf{v}(t)$. If $\mathbf{v}(t)$ has constant length, then $(\mathbf{v}(t)\cdot\mathbf{v}(t))' = [|\mathbf{v}(t)|^2]' = 0$. Then by the dot product rule, $\mathbf{v}'(t)\cdot\mathbf{v}(t) + \mathbf{v}(t)\cdot\mathbf{v}(t)' = 2(\mathbf{v}(t)\cdot\mathbf{v}(t)') = 0$. Evidently, $\mathbf{v}'(t) \perp \mathbf{v}(t)$. Now $\mathbf{T}(t)$ is a unit vector so it has constant length one. Then

$\mathbf{T}(t) \cdot \mathbf{T}'(t) = 0$ by the preceding and therefore $\mathbf{T}(t) \perp \mathbf{T}'(t)$. A constant scalar multiple of one of these vectors does not influence perpendicularity, so

$$\mathbf{T}(t) \perp \frac{\mathbf{T}'(t)}{|\mathbf{T}'(t)|} \quad \text{or} \quad \mathbf{T}(t) \perp \mathbf{N}(t).$$

Example 4.41

First, we need $\mathbf{r}'(t) = \langle -a \sin t, a \cos t \rangle$, so

$$\mathbf{T}(t) = \frac{\mathbf{r}'(t)}{|\mathbf{r}'(t)|} = \frac{\langle -a \sin t, a \cos t \rangle}{\sqrt{a^2 \cos^2 t + a^2 \sin^2 t}} = \langle -\sin t, \cos t \rangle.$$

Then

$$\mathbf{N}(t) = \frac{\mathbf{T}'(t)}{|\mathbf{T}'(t)|} = \frac{\langle -\cos t, -\sin t \rangle}{\sqrt{\cos^2 t + \sin^2 t}} = \langle -\cos t, -\sin t \rangle.$$

Recall from Section 4.6 that the cross product is a three-dimensional construction, but we can apply it to two-dimensional vectors by embedding them in $\mathbb{R}^3$, each with a zero component in the z-direction. The cross product formula then reduces to $\mathbf{x} \times \mathbf{y} = \widehat{\mathbf{k}}(x_1 y_2 - x_2 y_1)$. It follows that $\mathbf{B}(t) = \mathbf{T}(t) \times \mathbf{N}(t) = \langle -\sin t, \cos t, 0 \rangle \times \langle -\cos t, -\sin t, 0 \rangle = \widehat{\mathbf{k}}[(-\sin t)(-\sin t) - (-\cos t)(\cos t)] = \widehat{\mathbf{k}}$. The alternate formula $\mathbf{x} \times \mathbf{y} = |\mathbf{x}||\mathbf{y}| \sin \theta \widehat{\mathbf{e}}_\perp$ (Section 4.6 Example 4.19) yields the same result with $\widehat{\mathbf{e}}_\perp$ given by the right-hand rule as $\widehat{\mathbf{k}}$. Now for the curvature,

$$\kappa(t) = \frac{|\langle -\cos t, -\sin t \rangle|}{|\langle -a \sin t, a \cos t \rangle|} = \frac{1}{a}.$$

Since $\rho(t) = \frac{1}{\kappa(t)}$, the radius of curvature is the radius of the original circle a. In fact, the osculating circle is the given circle itself and the osculating plane is the xy-plane. Although $\mathbf{T}(t)$ and $\mathbf{N}(t)$ vary with t, $\mathbf{B}(t) = \widehat{\mathbf{k}}$ is constant.

Example 4.42

We have

$$\mathbf{r}(t) = \left\langle 3 \cos t, 3 \sin t, \frac{t}{2\pi} \right\rangle$$

for the helix, where $0 \le t \le 20\pi$. Then

$$\mathbf{r}'(t) = \left\langle -3 \sin t, 3 \cos t, \frac{1}{2\pi} \right\rangle \quad \text{and} \quad |\mathbf{r}'(t)| = \sqrt{9 + \frac{1}{4\pi^2}} = \frac{\sqrt{36\pi^2 + 1}}{2\pi}.$$

So

$$\mathbf{T}(t) = \frac{2\pi}{\sqrt{36\pi^2+1}}\left\langle -3\sin t, 3\cos t, \frac{1}{2\pi}\right\rangle.$$

Let us set $\alpha = \frac{2\pi}{\sqrt{36\pi^2+1}}$ to simplify notation. Next,

$$\mathbf{N}(t) = \frac{\mathbf{T}'(t)}{|\mathbf{T}'(t)|} = \frac{\alpha\langle -3\cos t, -3\sin t, 0\rangle}{\alpha\sqrt{9\cos^2 t + 9\sin^2 t}} = \langle -\cos t, -\sin t, 0\rangle.$$

So $\mathbf{N}(t)$ is antiparallel to the projection of $\mathbf{r}(t)$ onto the xy-plane. In other words, as the helix winds up the z axis, $\mathbf{N}(t)$ constantly points to that axis. Getting the unit binormal will not be as easy as in Example 4.41:

$$\mathbf{B}(t) = \mathbf{T}(t)\times\mathbf{N}(t) = \alpha\left\langle -3\sin t, 3\cos t, \frac{1}{2\pi}\right\rangle \times \langle -\cos t, -\sin t, 0\rangle.$$

This boils down to

$$\mathbf{B}(t) = \alpha\left\langle \frac{\sin t}{2\pi}, \frac{-\cos t}{2\pi}, 3\right\rangle.$$

As a check we compute

$$|\mathbf{B}(t)| = \frac{2\pi}{\sqrt{36\pi^2+1}}\left|\left\langle \frac{\sin t}{2\pi}, \frac{-\cos t}{2\pi}, 3\right\rangle\right| = \frac{2\pi}{\sqrt{36\pi^2+1}}\cdot\frac{\sqrt{36\pi^2+1}}{2\pi} = 1.$$

Finally,

$$\kappa(t) = \frac{|\mathbf{T}'(t)|}{|\mathbf{r}'(t)|} = \frac{3\alpha}{1/\alpha} = 3\alpha^2 = 3\cdot\frac{4\pi^2}{36\pi^2+1} = \frac{12\pi^2}{36\pi^2+1}.$$

Note that

$$\frac{12\pi^2}{36\pi^2+1} < \frac{12\pi^2}{36\pi^2} = \frac{1}{3}.$$

Example 4.43

For the unit vectors, $\mathbf{T}(t) = \frac{\langle 1,2t,3t^2\rangle}{\sqrt{1+4t^2+9t^4}}$ in general, and $\mathbf{T}(1) = \frac{\langle 1,2,3\rangle}{\sqrt{14}}$.

Now for $\mathbf{N}$,

$$\frac{d}{dt}\mathbf{T}(t) = \frac{d}{dt}\left(\frac{1}{\sqrt{1+4t^2+9t^4}}\right)\hat{\mathbf{i}} + \frac{d}{dt}\left(\frac{2t}{\sqrt{1+4t^2+9t^4}}\right)\hat{\mathbf{j}} + \frac{d}{dt}\left(\frac{3t^2}{\sqrt{1+4t^2+9t^4}}\right)\hat{\mathbf{k}},$$

where

$$\frac{d}{dt}\left(\frac{1}{\sqrt{1+4t^2+9t^4}}\right) = -\frac{1}{2}\left(\frac{36t^3+8t}{(1+4t^2+9t^4)^{3/2}}\right),$$

$$\frac{d}{dt}\left(\frac{2t}{\sqrt{1+4t^2+9t^4}}\right) = \frac{2}{\sqrt{1+4t^2+9t^4}} - \frac{t(36t^3+8t)}{(1+4t^2+9t^4)^{3/2}}, \text{ and}$$
$$\frac{d}{dt}\left(\frac{3t^2}{\sqrt{1+4t^2+9t^4}}\right) = \frac{6t}{\sqrt{1+4t^2+9t^4}} - \frac{3}{2}\left(\frac{t^2(36t^3+8t)}{(1+4t^2+9t^4)^{3/2}}\right).$$

With these substitutions, this is $\mathbf{T}'(t)$. Since $\mathbf{N}(1) = \frac{\mathbf{T}'(1)}{|\mathbf{T}'(1)|}$, we have

$$\mathbf{N}(1) = \frac{\frac{\sqrt{14}}{98}\langle 11, 8, -9\rangle}{\frac{\sqrt{14}}{98}\sqrt{121+64+81}} = \frac{1}{\sqrt{266}}\langle 11, 8, -9\rangle,$$

which is the unit principal normal at $t = 1$.

Then $\mathbf{B}(1) = \mathbf{T}(1) \times \mathbf{N}(1) = \frac{1}{61.02}\langle -42, 42, -14\rangle = 0.2294\langle -3, 3, -1\rangle$, which is the unit binormal at $t = 1$. Lots of calculation for a seemingly simple curve.

Example 4.44

Here $L = s(20\pi) = \int_0^{20\pi} |\mathbf{r}'(t)| dt$. We know

$$\mathbf{r}'(t) = \left\langle \frac{dx}{dt}, \frac{dy}{dt}, \frac{dz}{dt} \right\rangle = \left\langle -3\sin t, 3\cos t, \frac{1}{2\pi} \right\rangle,$$

so

$$|\mathbf{r}'(t)| = \sqrt{9\sin^2 + 9\cos^2 + \frac{1}{4\pi^2}} = \sqrt{9 + \frac{1}{4\pi^2}} \approx 3.0042,$$

then

$$L = (3.0042)(20\pi) = 188.76 \text{ units}.$$

Example 4.45

Here $\mathbf{r}'(t) = \langle 1, 2t, 3t^2\rangle$, so $|\mathbf{r}'(t)| = \sqrt{1+4t^2+9t^4}$. Then $L = s(1) = \int_0^1 \sqrt{1+4t^2+9t^4}dt \approx$ 1.863 (by numerical integration).

Example 4.46

Here $v_0 = 1000$ [fps], $\theta = 30^\circ$, $g = 32.16$ [fps^2]. Flight time is

$$\frac{2(1000)\sin 30^\circ}{32.16} = 31.09 \text{ [s]}.$$

Range is

$$\frac{(1000)^2 \sin 60^\circ}{32.16} = 26928.7 \text{ [ft]}$$

or about 5.1 [mi]. Maximum height is

$$\frac{(1000 \sin 30^\circ)^2}{2(32.16)} = 3886.8\ [\text{ft}].$$

Example 4.47

The trajectory equation is

$$\mathbf{r}(t) = (v_0 \cos\theta)t\,\hat{\mathbf{i}} + ((v_0 \sin\theta)t - 16t^2)\hat{\mathbf{j}}.$$

We do the problem in feet per second and note that 110 miles per hour is 161.333 [fps]. Each component in the trajectory equation gives a relation between t and θ, but although we have two equations in two unknowns, it is a nonlinear situation. We solve by eliminating t and obtain a trigonometric relation for θ. So, since the fence is 310 ft from the point of launch, the time to reach the wall is given by $t_{\text{wall}} = \frac{310}{161.333\cos\theta} = \frac{1.921}{\cos\theta}$. Plugging this into the equation for height and setting it equal to 34 ft (we start 3 ft off the ground), we get $(161.333 \sin\theta)(\frac{1.921}{\cos\theta}) - 16(\frac{1.921}{\cos\theta})^2 = 34$. This simplifies to $310\frac{\sin\theta}{\cos\theta} - \frac{59.044}{\cos^2\theta} = 34$, which we write as $310\tan\theta - 59.044\sec^2\theta = 34$, since $\frac{1}{\cos\theta} = \sec\theta$. Using the trigonometric formula $\sec^2\theta = 1 + \tan^2\theta$, we recover a quadratic equation in $\tan\theta$, namely $310\tan\theta - 59.044(1 + \tan^2\theta) = 34$, or $59.044\tan^2\theta - 310\tan\theta + 93.044 = 0$. Solving this (quadratic formula), we get the roots $\tan\theta = 0.3196$ and 4.931. These yield the angles 17.72° and 78.54°. The smaller angle corresponds to a flat trajectory where the ball just makes it over the fence and the larger angle corresponds to a high parabolic arc that, if it were any higher, would come down and hit the fence or the field before 310 feet. Any launch angle in between these angles would result in the ball clearing the fence.

Example 4.48

Let $R(\theta)$ be the range as a function of launch angle for a given v_0. We know $R(\theta) = \frac{v_0^2 \sin 2\theta}{g}$, so the answer to the question involves finding the angle θ that maximizes $\sin 2\theta$. The sine function attains its maximum at 90°, so $2\theta = 90^\circ$ gives $\theta = 45^\circ$ as the angle yielding maximum range for any muzzle velocity.

Example 4.49

The trajectory equation is

$$r(t) = (1200 \cos 60^\circ)t\,\hat{\mathbf{i}} + ((1200 \sin 60^\circ)t - 16t^2)\hat{\mathbf{j}}.$$

The antiaircraft shell reaches the plane altitude whenever $(1200 \sin 60^\circ)t - 16t^2 = 5000$. Rewriting this as a quadratic, $16t^2 - 1039.2t + 5000 = 0$. This has solutions 5.233 and 59.717 [s]. Obviously, the shorter time corresponds to hitting the enemy plane as the shell rises rather than falls. After 5.233 seconds, the shell is (1200)(0.500)(5.233) = 3139.8 feet (0.595 [mi]) east of the gun at the point of interception.

Example 4.50

Let the origin be the common initial point for the two vectors. Then $\mathbf{v}_1 = \langle 1, 2, 3 \rangle$ and $\mathbf{v}_2 = \langle 3, 5, 7 \rangle$. We have $\mathbf{v}_1 \times \mathbf{v}_2 = \langle -1, 2, -1 \rangle$. Now let $\mathbf{v} = \langle x - 0, y - 0, z - 0 \rangle$. Finally, $\mathbf{v} \cdot \mathbf{v}_1 \times \mathbf{v}_2 = 0$ yields $-x + 2y - z = 0$. Done. We can easily check that the three points belong to the plane.

Example 4.51

Select (1, 2, 2) as the common initial point. Then $\mathbf{v}_1 = \langle 2, -3, 2 \rangle$ and $\mathbf{v}_2 = \langle 5, 3, 6 \rangle$. We have $\mathbf{v}_1 \times \mathbf{v}_2 = \langle -24, -2, 21 \rangle$. Also $\mathbf{v} = \langle x - 1, y - 2, z - 2 \rangle$, so $\mathbf{v} \cdot \mathbf{v}_1 \times \mathbf{v}_2 = 0$ yields $-24(x - 1) - 2(y - 2) + 21(z - 2) = 0$. Putting this in the required form, we have $-24x - 2y + 21z = 14$.

Example 4.52

A generic vector in the plane is $\mathbf{v} = \langle x - 1, y - 1, z - 1 \rangle$. Since $\mathbf{u}$ is perpendicular to $\mathbf{v}$, we have $\mathbf{u} \cdot \mathbf{v} = 0$. So $2(x - 1) - (y - 1) + 3(z - 1) = 0$. Putting in standard form, we get $2x - y + 3z = 4$. Notice that the components of the direction vector always reappear as the respective coefficients in the plane equation. This is a very convenient fact that we will exploit regularly. If the direction vector had been reversed (antiparallel), we would still have $(-\mathbf{u}) \cdot \mathbf{v} = -(\mathbf{u} \cdot \mathbf{v}) = \mathbf{u} \cdot \mathbf{v} = 0$ with no change to the plane equation.

Example 4.53

From the remark in Example 4.52, we can immediately write $11(x-4)-23(y-7)+8(z+13) = 0$. Moving the constants to the right-hand side, we have $11x - 23y + 8z = 44 - 161 - 104 = -221$.

Example 4.54

Checking the respective coefficients, we confirm they are not related by a fixed multiple, so they are not parallel. To get the angle of intersection we construct direction vectors based on the respective coefficients. Let $\mathbf{v} = \langle 2, -5, 1 \rangle$ and $\mathbf{w} = \langle 3, 7, -4 \rangle$. Then

$$\theta = \arccos\left(\frac{\langle 2, -5, 1 \rangle \cdot \langle 3, 7, -4 \rangle}{|\langle 2, -5, 1 \rangle||\langle 3, 7, -4 \rangle|} \right) = \arccos\left(\frac{-33}{\sqrt{30}\sqrt{74}} \right) = \arccos(-0.701) = 134.5^\circ.$$

The corresponding acute angle is 45.5°.

Example 4.55

We want the points (x, y, z) that belong to both planes, so we solve $x - y + z = 6$ and $2x + y - z = 3$ simultaneously. Since we have two equations in three unknowns, we will have one variable free as a parameter. The solution with z as the parameter is $x = 3$, $y = z - 3$, and $z = z$. Expressing this in the manner of Section 4.7.1, we find the line of intersection is $\mathbf{r}(t) = \langle 3, -3, 0\rangle + t\langle 0, 1, 1\rangle$.

Example 4.56

We are free to pick the initial point on the line, so to keep it simple we select $(3, -3, 0)$. The proposed vector will be $\mathbf{v} = \langle x - 3, y + 3, z\rangle$. It must satisfy the condition of perpendicularity with respect to the line $\mathbf{r}(t)$ and also have its components satisfy $x - y + z = 6$. For perpendicularity, we have $\mathbf{v}\cdot\langle 0, 1, 1\rangle = 0$ or $(y + 3) + z = 0$. Solving this simultaneously with $x - y + z = 6$, we find that $x = 3 - 2z$, $y = -3 - z$, and $z = z$ give a set of solutions with parameter z. We may choose z arbitrarily, so for simplicity take $z = 1$. Now $(x, y, z) = (1, -4, 1)$. Plugging these values back into $\mathbf{v} = \langle x - 3, y + 3, z\rangle$, we find $\mathbf{v} = \langle -2, -1, 1\rangle$. This vector answers the question. Checking that it is perpendicular to the direction vector of the line, we get $\langle -2, -1, 1\rangle \cdot \langle 0, 1, 1\rangle = 0$. Likewise, with respect to the direction vector of the plane we have $\langle -2, -1, 1\rangle \cdot \langle 1, -1, 1\rangle = 0$, so it lies in the plane.

Example 4.57

To get an arbitrary point in the plane, the simplest approach is to set $x = y = 0$ so that $z = -2$. Then the given point is $(0, 0, 0)$ and the arbitrary point in the plane is $(0, 0, -2)$. The vector $\mathbf{v}$ defined above is $\langle 0, 0, 2\rangle$, and the direction vector of the plane is $\langle 2, 3, -5\rangle$. Using our formula,

$$d = \left|\frac{2(0) + 3(0) - 5(2)}{\sqrt{2^2 + 3^2 + (-5)^2}}\right| = \frac{10}{\sqrt{38}} \approx 1.62.$$

Example 4.58

Keeping things as simple as possible, let the point in the plane be $(7, 0, 0)$. Then

$$\begin{aligned}\left|\frac{a(x_1 - x_2) + b(y_1 - y_2) + c(z_1 - z_2)}{\sqrt{a^2 + b^2 + c^2}}\right| &= \left|\frac{(2-7) - 2(1-0) + 4(5-0)}{\sqrt{1^2 + (-2)^2 + 4^2}}\right| \\ &= \frac{13}{\sqrt{33}} \approx 2.26.\end{aligned}$$

5 Matrix methods

Rectangular arrays of numbers were used in the first millennium BCE by Chinese mathematicians to solve simultaneous linear equations. The method was introduced to Europe by Geralamo Cardano in his landmark 1645 treatise *Ars Magna* (which also presented a solution to the cubic equation). The British (and briefly American) mathematician J. J. Sylvester first used the word matrix about 1850 to refer to these arrays. The word matrix means "womb" in Latin, and he envisioned a matrix as containing smaller matrices formed by deleting rows and columns of the original array. Another British mathematician, Arthur Cayley, was the first to view matrices as objects in their own right with special algebraic operations defined on them. This moved matrices beyond their traditional role in solving systems of linear equations and in calculating determinants towards a more comprehensive role in linear algebra and abstract algebra in general. A determinant is a special type of mapping of a square matrix over a given field back to the field. Cayley discovered many foundational properties of matrices, such as the facts that matrix multiplication is not commutative and every square matrix satisfies its own so-called characteristic equation, which is derived from its determinant. We will clarify these terms and concepts in the following discussion. More recently, mathematicians have expressed linear transformations and operators on vector spaces in terms of matrices. Engineers and physicists have found that many of their concepts can be framed in terms of matrix methods. One of the original formulations of quantum mechanics by the German physicist Werner Heisenberg was actually called matrix mechanics.

5.1 Linear systems

A system of m linear equations in n variables can be written in the following standard way. We assume in the sequel that $m \leq n, 1 \leq j \leq m, 1 \leq k \leq n$, $a_{jk} \in \mathbb{R}$ and $c_j \in \mathbb{R}$,

$$\begin{aligned}
a_{11}x_1 + a_{12}x_2 + \cdots + a_{1n}x_n &= c_1, \\
a_{21}x_1 + a_{22}x_2 + \cdots + a_{2n}x_n &= c_2, \\
&\vdots \\
a_{j1}x_1 + a_{j2}x_2 + \cdots + a_{jn}x_n &= c_j, \\
&\vdots \\
a_{m1}x_1 + a_{m2}x_2 + \cdots + a_{mn}x_n &= c_m.
\end{aligned}$$

The coefficient a_{jk} appears in the jth row multiplied by the kth variable x_k. Another system of m linear equations in n variables with the same solution as the first is called an **equivalent system**. It may be that solving the second system is easier than the first, so if we can perform a sequence of incremental operations to systematically change the first

https://doi.org/10.1515/9783111506654-006

system into the second one, making sure that after each step equivalence is maintained, we can get the solution with less effort. What sort of operations would they be? Here is a list of operations that maintain equivalence.

Table 5.1: Elementary row operations (ERO).

1) Interchange two rows
2) Multiply a row by a nonzero scalar β
3) Add a nonzero multiple β of a row to some other row

We will show that these operations maintain equivalence and allow us to simplify a given system so that is in **triangular** (if $m = n$) or **echelon** (if $m < n$) form. If $m > n$, the system is **overdetermined** and has no solution unless there are $m - n$ redundancies in the system. A triangular system has $a_{jk} = 0$ if $j > k$. An echelon system has the property that no row is **degenerate** (all coefficients zero but the constant is not) and the first nonzero term in each row is to the right of the first nonzero term of the row above. The **row reduction algorithm** consists of applying the operations in Table 5.1 to a given system in order to achieve a triangular or echelon system where a solution is obtainable by the process of **back substitution** from the bottom to top. If an equation in the given system is determined to be a multiple of another, it may be removed from the system during execution of the algorithm. If an equation is degenerate, there is no solution to the system. If the row reduction algorithm produces a triangular form, the system has a unique solution. If the algorithm produces a nontriangular echelon form, the system has infinitely many solutions. If the algorithm encounters a degenerate equation, the system has no solution.

Example 5.1. Show the **elementary row operations** in Table 5.1 preserve system equivalence.

Example 5.2. Given a linear system with equations (i) $a_{k1}x_1 + a_{k2}x_2 + \cdots + a_{kn}x_n = c_k$ and (ii) $a_{j1}x_1 + a_{j2}x_2 + \cdots + a_{jn}x_n = c_j$, show how to eliminate the x_1 term in (ii) by using (i) and elementary row operations.

Given m linear equations in n variables by ERO 1 we could relist them all (by ERO 1) by putting the ones with nonzero x_1 coefficients in the top rows. There must be at least one or it is not a system with n variables. Then by the scheme in Example 5.2, we could use the equation in the topmost row to eliminate the x_1 coefficients in all the successive rows (some may already be zero). Now leaving the topmost row fixed, we could relist the second to mth equations to put the ones with nonzero x_2 coefficients in the rows just under the topmost. Repeating the procedure in Example 5.2, we could use the equation in row two to eliminate the x_2 coefficient in all subsequent rows. If there were no equation with a nonzero x_2 coefficient, we would move to the next variable with a nonzero coefficient and repeat the whole process again. Eventually, this scheme terminates in the mth row subject to the caveats above regarding dependence and degeneracy. Some examples should clarify the details.

Example 5.3. Solve the system

$$\begin{aligned} 2x + 5y &= 19, \\ 6x + 3y &= 21. \end{aligned}$$

Example 5.4. Solve the system

$$\begin{aligned} 3x + y - 2z &= 0, \\ 9x - 10y + 4z &= -2, \\ 12x + 20y - z &= 99. \end{aligned}$$

Example 5.5. Solve the system

$$\begin{aligned} x + 2y - 3z &= 8, \\ 2x + 3y + 6z &= 4. \end{aligned}$$

5.2 Matrices

A **matrix** M is a rectangular array of entries that are elements from an algebraic field. For our discussion, we assume that the field is $\mathbb{R}$, unless noted otherwise. With a broader perspective, the entries could be variables or functions of them. We write M as (a_{ij}) or

$$\begin{bmatrix} a_{11} & a_{12} & a_{13} & \dots & a_{1n} \\ a_{21} & a_{22} & a_{23} & \dots & a_{2n} \\ a_{31} & a_{32} & a_{33} & \dots & a_{3n} \\ \vdots & \vdots & \vdots & a_{ij} & \vdots \\ a_{m1} & a_{m2} & a_{m3} & \dots & a_{mn} \end{bmatrix},$$

where $1 \le i \le m$ and $1 \le j \le n$, depending on whether we need a shorthand or detailed representation in brackets. A typical **entry** $a_{ij} \in \mathbb{R}$ appears in the ith row and jth column, and we say that M is a **matrix over** $\mathbb{R}$, and that the **shape** of a matrix with m rows and n columns is $m \times n$ or "m by n." If $m = n$, which turns out to be a very important case, we call it a **square matrix**. Since matrices were inspired by arrays of coefficients for linear systems, we have some parallel terminology. If M is square and $a_{ij} = 0$ if $i > j$, then we say M is **upper triangular**. This corresponds to the triangular form of a linear system. If $a_{ij} = 0$ if $i < j$, then we say M is **lower triangular**. If $a_{ij} = 0$ if $i \neq j$, then we say M is a **diagonal matrix**. If we relax the condition that $m = n$, then we call M an **echelon**

matrix by analogy with an echelon system if (i) all rows with only zeroes are relegated to the bottom, and (ii) the "echelon property" holds, where the first nonzero entry in a row is to the right of the first nonzero entry in the row just above. A more specialized echelon matrix is said to be in **row canonical form** if two other conditions are imposed, namely (iii) the leading nonzero entry in each row is a 1, and (ii) each leading nonzero entry (a 1) is the only nonzero entry in its column. Canonical is a fancy word for "standard" in some sense. It can be shown that every matrix is **row equivalent** to a unique matrix in row canonical form.

Matrix

$$M_1 = \begin{bmatrix} 1 & 2 & 4 & 0 \\ 0 & 3 & 2 & 4 \\ 0 & 0 & 0 & 0 \end{bmatrix}$$

is in echelon but not row canonical form. Matrix

$$M_2 = \begin{bmatrix} 1 & 0 & 4 & 0 \\ 0 & 1 & 1 & 4 \\ 0 & 0 & 0 & 0 \end{bmatrix}$$

is in row canonical form.

The elementary row operations in Table 5.1 carry over directly to matrices. The reason for this is easy to see, since a matrix is basically an abbreviation of a system of equations where the explicit variables are replaced by column positions with the rightmost column corresponding to the constants in the system. We would like to convert an arbitrary matrix to row canonical form. This can be accomplished in two steps, first by converting the arbitrary matrix to echelon form by **Gaussian elimination** and then converting the echelon form to row canonical form by continuing with the **Gauss–Jordan algorithm**. The sequence of steps is certainly not unique, but the motivation is to use row operations appropriately in sequence to drive the matrix into its final desired form. Immediately following is a detailed example of the process. In Section 5.5 we use it to find the multiplicative inverse of a square matrix.

Example 5.6. Convert the following matrix to echelon form: $\begin{bmatrix} 2 & 10 & 4 & 6 \\ 1 & 3 & 2 & 4 \\ 2 & 0 & 1 & 0 \end{bmatrix}$.

Example 5.7. Convert the echelon matrix of Example 5.6 to row canonical form.

Example 5.8. Solve the linear system

$$
\begin{aligned}
2x + 10y + 4z &= 6,\\
x + 3y + 2z &= 4,\\
2x + z &= 0
\end{aligned}
$$

using matrices.

Example 5.9. For comparison with the matrix method, solve the system

$$
\begin{aligned}
2x + 10y + 4z &= 6,\\
x + 3y + 2z &= 4,\\
2x + z &= 0
\end{aligned}
$$

using successive substitutions.

5.3 Matrix algebra

As mentioned in Section 5, it was realized about 1850 that operations could be defined on sets of matrices that were analogous to addition and multiplication of real numbers. Matrices offer a special challenge in this regard since they may have different shapes. In order for two matrices to be equal, they must have the same shape and corresponding entries must to be equal. So

$$
\begin{bmatrix} a_{11} & a_{12} & \dots & a_{1n} \\ a_{21} & a_{22} & \dots & a_{2n} \\ \vdots & \vdots & \dots & \vdots \\ a_{m1} & a_{m2} & \dots & a_{mn} \end{bmatrix} = \begin{bmatrix} a'_{11} & a'_{12} & \dots & a'_{1s} \\ a'_{21} & a'_{22} & \dots & a'_{2s} \\ \vdots & \vdots & \ddots & \vdots \\ a'_{r1} & a'_{r2} & \dots & a'_{rs} \end{bmatrix}
$$

if and only if $m = r, n = s$, and $a_{ij} = a'_{ij}$, for all (i, j) pairs. A $1 \times n$ matrix $[a_{11} \quad a_{12} \quad \dots \quad a_{1n}]$ is called a **row vector**, and an $m \times 1$ matrix $\begin{bmatrix} a_{11} \\ a_{21} \\ \vdots \\ a_{m1} \end{bmatrix}$ is called a **column vector**. Matrices may be multiplied by scalars, so if $c \in \mathbb{R}$, then

$$
c \begin{bmatrix} a_{11} & a_{12} & \dots & a_{1n} \\ a_{21} & a_{22} & \dots & a_{2n} \\ \vdots & \vdots & \ddots & \vdots \\ a_{m1} & a_{m2} & \dots & a_{mn} \end{bmatrix} = \begin{bmatrix} ca_{11} & ca_{12} & \dots & ca_{1n} \\ ca_{21} & ca_{22} & \dots & ca_{2n} \\ \vdots & \vdots & \ddots & \vdots \\ ca_{m1} & ca_{m2} & \dots & ca_{mn} \end{bmatrix}.
$$

So for matrix M, $1 \cdot M = M$ and $0 \cdot M = [0]$, the all-zero matrix. Two matrices of the same shape may be added entrywise,

$$\begin{bmatrix} a_{11} & a_{12} & \dots & a_{1n} \\ a_{21} & a_{22} & \dots & a_{2n} \\ \vdots & \vdots & \ddots & \vdots \\ a_{m1} & a_{m2} & \dots & a_{mn} \end{bmatrix} \oplus \begin{bmatrix} a'_{11} & a'_{12} & \dots & a'_{1n} \\ a'_{21} & a'_{22} & \dots & a'_{2n} \\ \vdots & \vdots & \ddots & \vdots \\ a'_{m1} & a'_{m2} & \dots & a'_{mn} \end{bmatrix}$$
$$= \begin{bmatrix} a_{11}+a'_{11} & a_{12}+a'_{12} & \ddots & a_{1n}+a'_{1n} \\ a_{21}+a'_{21} & a_{22}+a'_{22} & \dots & a_{2n}+a'_{2n} \\ \vdots & \vdots & \ddots & \vdots \\ a_{m1}+a'_{m1} & a_{m2}+a'_{m2} & \dots & a_{mn}+a'_{mn} \end{bmatrix}.$$

We have used the symbol $\oplus$, defined by this rule, to distinguish matrix addition from ordinary addition in $\mathbb{R}$.

The general rules for matrix addition and scalar multiplication are parallel to those for the real number system.

Table 5.2: Matrix addition & scalar multiplication. $[0], M_1, M_2, M_3$ are matrices with same shape, $c_1, c_2 \in \mathbb{R}$.

1) $M_1 \oplus (M_2 \oplus M_3) = (M_1 \oplus M_2) \oplus M_3$	associativity
2) $M_1 \oplus [0] = M_1$	identity
3) $M_1 \oplus (-1)M_1 = [0]$	inverse
4) $M_1 \oplus M_2 = M_2 \oplus M_1$	commutativity
5) $c_1(M_1 \oplus M_2) = c_1M_1 \oplus c_1M_2$	distributivity
6) $(c_1 + c_2)M_1 = c_1M_1 \oplus c_2M_1$	distributivity
7) $(c_1c_2)M_1 = c_1(c_2M_1)$	associativity

These rules follow immediately from the observation that we are applying the same rules for real numbers simultaneously in every entry position.

Matrix multiplication offers a new challenge by the way it is defined. For same size matrices $A = (a_{ij})$ and $B = (b_{ij})$, we defined $A \oplus B$ as $(a_{ij} + b_{ij})$. It is tempting, but incorrect, to define the matrix product of A and B to be $(a_{ij} \cdot b_{ij})$. For reasons that will become apparent, we define it as follows. Notice first of all that to be multiplied together, two matrices must have shapes that are **conformable**. This means that the number of columns of the left factor must equal the number of rows of the right factor. We will do a specific example and then generalize later. The symbol $\otimes$ means matrix multiplication. Consider

$$A \otimes B = \begin{bmatrix} a_{11} & a_{12} \\ a_{21} & a_{22} \\ a_{31} & a_{32} \end{bmatrix} \otimes \begin{bmatrix} b_{11} & b_{12} & b_{13} & b_{14} \\ b_{21} & b_{22} & b_{23} & b_{24} \end{bmatrix}$$
$$= \begin{bmatrix} a_{11}b_{11}+a_{12}b_{21} & a_{11}b_{12}+a_{12}b_{22} & a_{11}b_{13}+a_{12}b_{23} & a_{11}b_{14}+a_{12}b_{24} \\ a_{21}b_{11}+a_{22}b_{21} & a_{21}b_{12}+a_{22}b_{22} & a_{21}b_{13}+a_{22}b_{23} & a_{21}b_{14}+a_{22}b_{24} \\ a_{31}b_{11}+a_{32}b_{21} & a_{31}b_{12}+a_{32}b_{22} & a_{31}b_{13}+a_{32}b_{23} & a_{31}b_{14}+a_{32}b_{24} \end{bmatrix}.$$

The pattern may not be obvious, but what we have done is to take the two entries in the first row of A and multiply them in order by the two entries in the first column of B, that gives $a_{11}b_{11}$ and $a_{12}b_{21}$, then added these together and put the sum in the first row/first column entry of the product matrix. We do this for the three rows of A and the four columns of B to get 12 sums that populate the entries in the 3×4 product matrix. The ith row of A is matched with the jth column of B to generate the (i,j) entry of $A \otimes B$. Now it should be clear why conformability is essential. The product of an $m \times n$ matrix on the left by an $n \times p$ matrix on the right yields an $m \times p$ product matrix. You could not reverse the order of factors unless $m = p$. This is another reason why square matrices are important. There is never a conformability issue for multiplication in any order.

Given the $m \times n$ matrix

$$A = \begin{bmatrix} a_{11} & a_{12} & \dots & a_{1n} \\ a_{21} & a_{22} & \dots & a_{2n} \\ \vdots & \vdots & \ddots & \vdots \\ a_{m1} & a_{m2} & \dots & a_{mn} \end{bmatrix}$$

and the $n \times p$ matrix

$$B = \begin{bmatrix} b_{11} & b_{12} & \dots & b_{1p} \\ b_{21} & b_{22} & \dots & b_{2p} \\ \vdots & \vdots & \ddots & \vdots \\ b_{n1} & b_{n2} & \dots & b_{np} \end{bmatrix},$$

the (i,j) entry of $A \otimes B$, which is an $m \times p$ matrix is

$$[a_{i1} \quad a_{i2} \quad \dots \quad a_{in}] \otimes \begin{bmatrix} b_{1j} \\ b_{2j} \\ \vdots \\ b_{nj} \end{bmatrix} = (A \otimes B)_{ij} = \sum_{k=1}^{n} a_{ik} b_{kj}.$$

So we can think of the matrix product as the result of all the row vectors from A pairwise multiplied by all the column vectors of B.

Table 5.3: Matrix multiplication properties. $[0], I_n, M_1, M_2, M_3$ are matrices, all products defined, $c \in \mathbb{R}$.

1) $M_1 \otimes (M_2 \otimes M_3) = (M_1 \otimes M_2) \otimes M_3$	associativity
2) $M_1 \otimes (M_2 \oplus M_3) = (M_1 \otimes M_2) \oplus (M_1 \otimes M_3)$	left distributivity
3) $(M_1 \oplus M_2) \otimes M_3 = (M_1 \otimes M_3) \oplus (M_2 \otimes M_3)$	right distributivity
4) $c(M_1 \otimes M_2) = cM_1 \otimes M_2 = M_1 \otimes cM_2$	scalar distributivity
5) $[0] \otimes M_1 = M_1 \otimes [0] = [0]$	annihilation
6) If M_1 is $n \times n$, then $I_n \otimes M_1 = M_1 \otimes I_n = M_1$	identity

Example 5.10. Given

$$A = \begin{bmatrix} 2 & 5 & 7 \\ 3 & 8 & 4 \\ 9 & 0 & 1 \end{bmatrix} \quad \text{and} \quad B = \begin{bmatrix} 3 & 6 & 6 \\ 1 & 9 & 8 \\ 4 & 5 & 7 \end{bmatrix},$$

find $A \oplus B$, $A \otimes B$, and $B \otimes A$.

Example 5.11. Find

$$\begin{bmatrix} 3 & 4 & -8 \\ 9 & 5 & 1 \\ 6 & 7 & -6 \end{bmatrix} \otimes \begin{bmatrix} x \\ y \\ z \end{bmatrix}.$$

Example 5.12. Express the linear system

$$\begin{aligned} 5x + 3y - 2z &= 0, \\ 8x - 9y + 4z &= -2, \\ 6x + y - 7z &= 14 \end{aligned}$$

as a matrix equation.

5.4 Square matrices

Square matrices expand what we can do with the theory. Two $n \times n$ matrices can be added as well as multiplied in either order and the result is an $n \times n$ matrix. This means that the set of such matrices with real entries, denoted by $M_n(\mathbb{R})$, works essentially like the integers except for the lack of multiplicative commutativity. The square matrix $I_n = (a_{ij})$ where $a_{ij} = 1$ if $i = j$ and $a_{ij} = 0$ if $i \neq j$, is called the $n \times n$ identity matrix. More specifically, the set constitutes what algebraists call a noncommutative ring with identity. So we need to keep this in mind during calculations and pay attention to the order of factors. In some cases $A \otimes B$ may equal $B \otimes A$, but unless these are matrices of a special type, this is coincidental. So the door is now open to taking successive powers of an $n \times n$ matrix, which will be well defined because a matrix always commutes with itself. Then we can assemble various powers of such a matrix into a matrix-valued polynomial or even an infinite series. As unnatural as it sounds, this will make it possible to use a matrix as an exponent.

Polynomials are covered in detail in Section 6.1. A polynomial of degree k in the indeterminate x over the reals is an expression such as $P(x) = a_k x^k + a_{k-1}x^{k-1} + \cdots + a_1 x + a_0$. Omitting the trivial case of degree zero, k is a positive integer and the coefficients $a_0, \ldots, a_k \in \mathbb{R}$. The **evaluation** of $P(x)$ at $x = c \in \mathbb{R}$ is $P(c) = a_k c^k + a_{k-1}c^{k-1} + \cdots + a_1 c + a_0$. This $P(c)$ would be a real number, but it is legitimate to let $x = A \in M_n(\mathbb{R})$, in which case $P(A) = a_k A^k + a_{k-1}A^{k-1} + \cdots + a_1 A + a_0 I_n$ would be a matrix of real numbers. We multiply a_0

by the I_n to give the constant matrix character. The usual rules for adding, multiplying, or forming compositions of polynomial functions are compatible with using a matrix argument.

Example 5.13. Given $P(x) = 2x^3 + x^2 - x + 3$, and

$$A = \begin{bmatrix} 2 & 3 \\ 1 & 4 \end{bmatrix},$$

calculate $P(A)$.

Example 5.14. Given

$$A = \begin{bmatrix} \frac{1}{2} & \frac{1}{4} \\ \frac{1}{3} & \frac{1}{2} \end{bmatrix},$$

find the first four terms of the series for

$$\exp\left(\begin{bmatrix} \frac{1}{2} & \frac{1}{4} \\ \frac{1}{3} & \frac{1}{2} \end{bmatrix}\right).$$

An $n \times n$ matrix A is defined to be **invertible** or **nonsingular** if there exists a matrix, denoted by A^{-1} such that $A \otimes A^{-1} = A^{-1} \otimes A = I_n$. When it exists, we call the unique $n \times n$ matrix A^{-1} the **inverse of** A. In the special case $n = 2$, if $A = \begin{bmatrix} a & b \\ c & d \end{bmatrix}$ it is possible to immediately write down

$$A^{-1} = \frac{1}{ad - bc} \begin{bmatrix} d & -b \\ -c & a \end{bmatrix}.$$

We note that A^{-1} exists if and only if $ad - bc \neq 0$.

Example 5.15. Show

$$\begin{bmatrix} a & b \\ c & d \end{bmatrix} \otimes \frac{1}{ad - bc} \begin{bmatrix} d & -b \\ -c & a \end{bmatrix} = \begin{bmatrix} 1 & 0 \\ 0 & 1 \end{bmatrix}.$$

Example 5.16. Show that for $n \times n$ matrices $A_1, A_2, \ldots, A_n$ it is true that $(A_1 \otimes A_2 \otimes \cdots \otimes A_n)^{-1} = A_n^{-1} \otimes \cdots \otimes A_2^{-1} \otimes A_1^{-1}$.

The elementary row operations in Table 5.1 have elementary column operation analogs, specifically (i) interchange, (ii) scaling, and (iii) addition. If an $n \times n$ identity matrix is subjected to an elementary row or column operation, the result is an **elementary matrix.** Multiplying an arbitrary $n \times n$ matrix on the left (premultiplying) by a row elementary matrix performs the corresponding elementary row operation. Multiplying it on the right (postmultiplying) by an elementary column matrix performs the corresponding elementary column operation. Matrix M_1 is row equivalent to matrix M_2 if and only if there is an invertible matrix A such that $M_1 = A \otimes M_2$. This A is a product of elementary row matrices. Similarly, M_1 is column equivalent to M_2 if and only if there is an invertible matrix B such that $M_1 = M_2 \otimes B$. This B is a product of elementary column matrices. Matrix M_1 is **equivalent** to M_2 if and only if there are invertible matrices A and B such that $M_1 = A \otimes M_2 \otimes B$. Stated differently, two matrices are equivalent if each can be obtained from the other by a suitable sequence of elementary row and column operations. Because A^{-1} and B^{-1} exist, we have $A^{-1} \otimes M_1 \otimes B^{-1} = M_2$ and it is clear that matrix equivalence is an equivalence relation in the broader sense of Section 0.3.1.

Example 5.17. Find the 3×3 elementary matrix that interchanges the first and second rows of $\begin{bmatrix} 1 & 4 & 7 \\ 2 & 0 & 9 \\ 3 & 6 & 6 \end{bmatrix}$.

Interchanging the rows and columns of any matrix results in its **transpose**. For a nonsquare matrix, the operation changes the shape of the matrix. We restrict ourselves to the square matrix case, which preserves the ability to combine matrices and their transposes algebraically. Given a matrix M, we denote its transpose by M^T. If $M = M^T$, or equivalently $(m_{ij}) = (m_{ji})$, we say M is a **symmetric matrix**. If $M = -M^T$, we say M is antisymmetric. For an **antisymmetric matrix**, it must be the case that $(m_{ii}) = (-m_{ii})$, so the diagonal elements are forced to be zero. A matrix is **normal** if it commutes with its transpose, or $M \otimes M^T = M^T \otimes M$. If $M^T = M^{-1}$ then it is **orthogonal**, and necessarily $M \otimes M^T = M^T \otimes M = I_n$. Here are the rules for working with transposes.

Table 5.4: Rules for transposes. M, N are $n \times n$ matrices, $\alpha \in \mathbb{R}$.

1) $(M^T)^T = M$
2) $(M \oplus N)^T = M^T \oplus N^T$
3) $(\alpha M)^T = \alpha M^T$
4) $(M \otimes N)^T = N^T \otimes M^T$ (reversed)

With Rule 4, we can write $(M^{-1})^T \otimes M^T = (M \otimes M^{-1})^T = I_n^T = I_n$, so conveniently enough, the inverse of the transpose is the transpose of the inverse. With that fact in mind, we can examine another type of equivalence relation for matrices. If there is an invertible matrix P such that $M_1 = P^T \otimes M_2 \otimes P$, we say that M_1 is **congruent** to M_2. Congruence is certainly reflexive because $M_1 = I^T \otimes M_1 \otimes I$, and it is symmetric because if $M_1 = P^T \otimes M_2 \otimes P$, then $(P^{-1})^T \otimes M_1 \otimes P^{-1} = (P^{-1})^T \otimes P^T \otimes M_2 \otimes P \otimes P^{-1} = M_2$. Finally, it is transitive because $M_1 = P^T \otimes M_2 \otimes P$ and $M_2 = Q^T \otimes M_3 \otimes Q$ imply $M_1 = P^T \otimes Q^T \otimes M_3 \otimes Q \otimes P = (Q \otimes P)^T \otimes M_3 \otimes (Q \otimes P)$.

Congruence preserves symmetric matrices, so if $M_2 = M_2^T$, we have

$$M_1^T = (P^T \otimes M_2 \otimes P)^T = P^T \otimes M_2^T \otimes (P^T)^T = P^T \otimes M_2 \otimes P = M_1.$$

An old result due to J. J. Sylvester states that if M is a real symmetric matrix, then there exists an invertible matrix P such that $D = P^T \otimes M \otimes P$ is a diagonal matrix. This is called the **law of inertia**. The diagonalization algorithm is much like the matrix inversion algorithm in Section 5.6, but it operates on both rows and columns. Since it is a bit lengthy (and tedious), we refer the reader to the chapter references for a detailed explanation. The algorithm can be extended (even more tediously) to allow the matrix P to be chosen as orthogonal. We present a simple example of diagonalization with an orthogonal matrix as part of our discussion in Section 5.9.

Example 5.18. Suppose M is an arbitrary real symmetric matrix, $D = P^T \otimes M \otimes P$ is diagonal, and P is orthogonal. Show how to find M^k.

5.5 Determinants

Every square matrix A has a **determinant** $\det A$ defined on it. Determinants can reveal a lot of useful information about the underlying matrix. Rather than immediately writing down a formula that requires a bit of parsing at the start, it may be clearer to build the expression with explanation as we go along. Given the matrix

$$\begin{bmatrix} a_{11} & a_{12} & \dots & a_{1n} \\ a_{21} & a_{22} & \dots & a_{2n} \\ \vdots & \vdots & \ddots & \vdots \\ a_{n1} & a_{n2} & \dots & a_{nn} \end{bmatrix},$$

we are going to form products of n of the entries at a time in a special way. Then we will sum up all of these products with appropriate **parity** ($\pm$) **factors** and produce a final single number that will be the determinant. The products of n of the elements are constructed as follows. We pick any element from the first row, say a_{1j}, then we choose any element from the second row, say a_{2k}, where we make sure a_{2k} is not in the same column as a_{1j}, or $k \neq j$. Next we move to row three and pick a_{3s} so that $j \neq s$ and $k \neq s$. You can see the pattern developing. Working our way down the n rows we choose one entry from each row subject to the condition that no two chosen entries share the same column. One way to make sure this happens is to permute the row index numbers to get the paired column index numbers. Why does this work? A permutation (see Section 1.1.1) is a bijective mapping, so if we permute the numbers 1 through n, we will get the same numbers back only shuffled. The important thing is we will get each number back only once. For example, if $n = 5$, a typical permutation of $\{1, 2, 3, 4, 5\}$ might be $\{4, 3, 5, 1, 2\}$.

Denoting the permutation function by π, this means $\pi(1) = 4, \pi(2) = 3, \pi(3) = 5, \pi(4) = 1$, and $\pi(5) = 2$. Then $a_{1\pi(1)}, a_{2\pi(2)}, a_{3\pi(3)}, a_{4\pi(4)}$, and $a_{5\pi(5)}$ are entries from the five different rows, none of which appear in the same column. This is exactly what we want to form a product. For the current example, we can write the product of all five terms as $\prod_{i=1}^{5} a_{i\pi(i)}$. Now that we have expressed the product, we want for a particular permutation π, let us do the same thing for *all* permutations of 1 through 5 and then add them up. We are almost there. Returning to the general case where we have an $n \times n$ matrix, we denote by S_n the set of all possible permutations of 1 through n. In Section 1.1.1 we found this set had $n!$ elements. We use the signum function to assign +1 to an even permutation and −1 to an odd permutation. Every element of S_n can be factored into cycles of length two, or transpositions. While the factorization of a permutation into transpositions may not be unique, the evenness or oddness of the number of transpositions is invariant, so parity is well defined. If a permutation has fixed points (elements not moved by the mapping) these are customarily omitted when writing a permutation in cycle notation. Finally, we can write down the Leibnitz formula for the determinant of the $n \times n$ matrix A,

$$\det A = \sum_{\pi \in S_n} \operatorname{sgn}(\pi) \prod_{i=1}^{n} a_{i\pi(i)}.$$

It may seem complicated, but in practice for small n it can be done by hand.

Example 5.19. Find the determinant of

$$A = \begin{bmatrix} a_{11} & a_{12} \\ a_{21} & a_{22} \end{bmatrix}.$$

Example 5.20. Find the determinant of

$$A = \begin{bmatrix} a_{11} & a_{12} & a_{13} \\ a_{21} & a_{22} & a_{23} \\ a_{31} & a_{32} & a_{33} \end{bmatrix}.$$

Example 5.21. Find

$$\det \begin{bmatrix} 2 & 0 & 4 \\ 1 & 3 & 5 \\ 6 & 3 & 8 \end{bmatrix}.$$

The formula

$$\det A = \sum_{\pi \in S_n} \operatorname{sgn}(\pi) \prod_{i=1}^{n} a_{i\pi(i)}$$

requires the sum of $n!$ permuted products. Unfortunately, the factorial grows very fast, so the determinant of a 4×4 matrix requires 24 products in general. In the precomputer days, this was a computational headache. We do have a workaround known as the **Laplace expansion**. Let us use the matrix in Example 5.21 to demonstrate. We need to pick a row or column of the matrix that has small entries, if possible, and certainly any with zeroes. So for the given matrix choose row one. Now take the first entry in that row and suppress the first row and first column. What is left? The 2×2 submatrix $\begin{bmatrix} 3 & 5 \\ 3 & 8 \end{bmatrix}$. The determinant of this submatrix is called a **minor**. We calculate the minor to be $24 - 15 = 9$. Next we assign a parity to the minor based on the position of the suppressed row and column. Since 2 was the $(1,1)$ entry in the matrix, the **parity** factor is $(-1)^{1+1} = 1$. Finally, we call the minor with its parity a **cofactor**, or **signed minor**, of 2. We are going to multiply each element in row one by its cofactor and add them up to get the determinant. That is why the zero in the second column is convenient, since the cofactor is 0. The cofactor of 4 is

$$(-1)^{1+3} \det \begin{bmatrix} 1 & 3 \\ 6 & 3 \end{bmatrix} = 1 \cdot (3 - 18) = -15.$$

The determinant is then $2 \cdot 9 - 4 \cdot 15 = 18 - 60 = -42$ as before.

Example 5.22. Find

$$\det \begin{bmatrix} 1 & 0 & -2 & 0 \\ 3 & 4 & 0 & 1 \\ 4 & 1 & 3 & 3 \\ 2 & 5 & 0 & 1 \end{bmatrix}.$$

It can be shown that the determinant of a product of matrices is the same as the product of their individual determinants, or in symbols $\det(M_1 \otimes M_2) = \det(M_1) \cdot \det(M_2)$. If either M_1 or M_2 had a zero determinant, then $\det(M_1 \otimes M_2) = 0 \neq \det(I_n) = 1$. This makes it clear that a matrix with zero determinant cannot have an inverse. So we now have a criterion for invertible matrices, namely those with nonzero determinants.

For some types of matrices, calculation of the determinant is greatly simplified and the arithmetic contortions above are unnecessary. Upper triangular, lower triangular, and diagonal matrices have the property that their determinants are just the product of the factors along the main diagonal.

Example 5.23. Find the determinants for

$$A = \begin{bmatrix} 1 & 4 & 6 & -2 \\ 0 & 3 & 5 & 9 \\ 0 & 0 & -1 & 7 \\ 0 & 0 & 0 & -3 \end{bmatrix} \quad \text{and} \quad B = \begin{bmatrix} 2 & 0 & 0 & 0 \\ 0 & 4 & 0 & 0 \\ 0 & 0 & 7 & 0 \\ 0 & 0 & 0 & -1 \end{bmatrix}.$$

5.6 Matrix inverses

We would like to have an algorithm for inverting a matrix (finding its multiplicative inverse). Suppose we want to invert

$$M = \begin{bmatrix} 2 & 0 & -1 \\ 2 & -1 & -2 \\ 0 & 3 & 2 \end{bmatrix}.$$

First, we check that it has nonzero determinant, or is **nonsingular**. The idea is to write down M and I_3 side-by-side and perform row operations on both simultaneously. We will choose operations that convert M into I_3 and those same operations will convert I_3 into M^{-1}. So to start,

$$\begin{bmatrix} 2 & 0 & -1 \\ 2 & -1 & -2 \\ 0 & 3 & 2 \end{bmatrix} \begin{bmatrix} 1 & 0 & 0 \\ 0 & 1 & 0 \\ 0 & 0 & 1 \end{bmatrix}.$$

Adding (−1) times the top row to the second row gives

$$\begin{bmatrix} 2 & 0 & -1 \\ 0 & -1 & -1 \\ 0 & 3 & 2 \end{bmatrix} \begin{bmatrix} 1 & 0 & 0 \\ -1 & 1 & 0 \\ 0 & 0 & 1 \end{bmatrix}.$$

Multiplying the second row by 3 and adding it to the bottom row gives

$$\begin{bmatrix} 2 & 0 & -1 \\ 0 & -1 & -1 \\ 0 & 0 & -1 \end{bmatrix} \begin{bmatrix} 1 & 0 & 0 \\ -1 & 1 & 0 \\ -3 & 3 & 1 \end{bmatrix}.$$

We now have an echelon matrix on the left. Continuing with the Gauss–Jordan algorithm we multiply the bottom row by (−1) and add it to the second row to get

$$\begin{bmatrix} 2 & 0 & -1 \\ 0 & -1 & 0 \\ 0 & 0 & -1 \end{bmatrix} \begin{bmatrix} 1 & 0 & 0 \\ 2 & -2 & -1 \\ -3 & 3 & 1 \end{bmatrix}.$$

We repeat that operation but add to the top row, giving

$$\begin{bmatrix} 2 & 0 & 0 \\ 0 & -1 & 0 \\ 0 & 0 & -1 \end{bmatrix} \begin{bmatrix} 4 & -3 & -1 \\ 2 & -2 & -1 \\ -3 & 3 & 1 \end{bmatrix}.$$

Now multiply the top row by $\frac{1}{2}$ to produce

$$\begin{bmatrix} 1 & 0 & 0 \\ 0 & -1 & 0 \\ 0 & 0 & -1 \end{bmatrix} \begin{bmatrix} 2 & -\frac{3}{2} & -\frac{1}{2} \\ 2 & -2 & -1 \\ -3 & 3 & 1 \end{bmatrix}.$$

The end is in sight. We can do two operations at once. Multiplying the second and third rows by (−1), we finally have

$$\begin{bmatrix} 1 & 0 & 0 \\ 0 & 1 & 0 \\ 0 & 0 & 1 \end{bmatrix} \begin{bmatrix} 2 & -\frac{3}{2} & -\frac{1}{2} \\ -2 & 2 & 1 \\ 3 & -3 & -1 \end{bmatrix}.$$

Evidently,

$$\begin{bmatrix} 2 & 0 & -1 \\ 2 & -1 & -2 \\ 0 & 3 & 2 \end{bmatrix}^{-1} = \begin{bmatrix} 2 & -\frac{3}{2} & -\frac{1}{2} \\ -2 & 2 & 1 \\ 3 & -3 & -1 \end{bmatrix}.$$

Although Gauss and Jordan were fairly recent mathematicians in the grand scheme of mathematical history, the basic method has been in use for two millennia.

Why exactly does this algorithm work? Recall each row operation on a square matrix can be implemented by premultiplication by an elementary (row) matrix. Suppose we have a sequence of elementary matrices operating on the given matrix M that converts it into the identity matrix. This can be symbolized as $(E_r \otimes E_{r-1} \otimes \cdots \otimes E_2 \otimes E_1 \otimes I) \otimes M = I$. Remember that as factors are applied they go on the left. Apparently, $(E_r \otimes E_{r-1} \otimes \cdots \otimes E_2 \otimes E_1 \otimes I) = M^{-1}$. This notation is technically ambiguous at the moment as it indicates either a left and right inverse of M. Although it is not an obvious proof, thankfully the left and right inverses of a matrix (if it has one at all) are equal.

5.7 Cramer's rule

We have discussed solving a system of linear equations by reducing the system with row operations to either echelon form or row canonical form. There is an elegant algorithm called **Cramer's rule** for solving linear systems based instead on determinants. The Swiss mathematician Gabriel Cramer published it about 1750. It works by taking ratios of various determinants that come from the coefficient matrix of a linear system. Let us examine the simplest general case where the system is

$$ax + by = s,$$
$$cx + dy = t.$$

We can solve for x by multiplying the top equation by d and the bottom by b to get

$$adx + bdy = sd,$$
$$bcx + bdy = bt$$

and now subtract the bottom from the top. This gives $adx - bcx = sd - bt$. Observe that this equation can be expressed by determinants as

$$\begin{vmatrix} a & b \\ c & d \end{vmatrix} x = \begin{vmatrix} s & b \\ t & d \end{vmatrix}.$$

So

$$x = \frac{\begin{vmatrix} s & b \\ t & d \end{vmatrix}}{\begin{vmatrix} a & b \\ c & d \end{vmatrix}}.$$

If we let the determinant of the coefficient matrix be D and the modified determinant where first column (the x column) has been replaced by the constants, we can write $x = \frac{D_x}{D}$. A similar argument shows $y = \frac{D_y}{D}$. This is the essence of Cramer's rule and it can be generalized with some tedious algebra to systems with n equations in n variables provided the coefficient matrix is nonsingular.

Example 5.24. Solve the system

$$5x + 3y - 2z = 0,$$
$$8x - 9y + 4z = -2,$$
$$6x + y - 7z = 14$$

from Section 5.3 Example 5.12 using Cramer's rule to four decimals.

5.8 Changing bases

The background in Section 4.3 and Section 4.4 will be helpful for this section. Let us frame our discussion in terms of $\mathbb{R}^2$ and $\mathbb{R}^3$, which are the most familiar spaces to us. Matrices provide a natural way to express linear transformations. In Section 4.4 we described the component or coordinate representation of a vector relative to a fixed basis. Suppose we have the vector v in a given space expressed relative to the basis B_1 and we would like to express it relative to a different basis B_2. We will distinguish these cases by writing them as $[v]_{B_1}$ and $[v]_{B_2}$. Our immediate goal is to construct a matrix that transforms $[v]_{B_1}$ into $[v]_{B_2}$.

Example 5.25. Assuming our vector space is $\mathbb{R}^2$, find the matrix that converts $[v]_{B_1} = [\alpha_1, \alpha_2]$ into $[v]_{B_2} = [\beta_1, \beta_2]$.

Example 5.26. Using the method in Example 5.25 find the change-of-basis matrix connecting $B_1 = \{e_1, e_2, e_3\}$ and $B_2 = \{f_1, f_2, f_3\}$.

Example 5.27. Use the result of Example 5.25 to find the change-of-basis matrix for

$$B_1 = \left\{ \begin{bmatrix} 1 \\ 0 \end{bmatrix}, \begin{bmatrix} 0 \\ 1 \end{bmatrix} \right\} \quad \text{and} \quad B_2 = \left\{ \begin{bmatrix} 1 \\ 1 \end{bmatrix}, \begin{bmatrix} -1 \\ 1 \end{bmatrix} \right\}.$$

This corresponds to rotating the xy-plane counterclockwise 45°.

Example 5.28. Use the matrix in Example 5.27 to find the components of the vector $\begin{bmatrix} 2 \\ 4 \end{bmatrix}_{B^1}$ relative to B_2.

Example 5.29. Show that any change-of-basis matrix is invertible.

The change-of-basis matrix can also be used to find the matrix of a linear operator relative to a new basis by applying it to the matrix of the operator relative to the original basis.

5.9 Eigenvectors & eigenvalues

As in Section 5.4, we let M be an $n \times n$ matrix over $\mathbb{R}$. We define a function of λ by forming $\lambda I_n - M$ where I_n is the $n \times n$ identity matrix. Expressing M as (m_{ij}) we can write down a more illuminating version of this matrix,

$$\begin{bmatrix} \lambda - m_{11} & m_{12} & m_{13} & \dots & m_{1n} \\ m_{21} & \lambda - m_{22} & m_{23} & \dots & m_{1n} \\ m_{31} & m_{32} & \lambda - m_{33} & \dots & m_{1n} \\ \vdots & \vdots & \vdots & \ddots & \vdots \\ m_{n1} & m_{n2} & m_{n3} & \dots & \lambda - m_{nn} \end{bmatrix}.$$

This is called the **characteristic matrix** of M. Now we build on this and define $\Delta_M(\lambda)$ to be the determinant of $\lambda I_n - M$, or $\Delta_M(\lambda) = \det(\lambda I_n - M)$. Of course, this will be an nth degree polynomial in λ. We call this the **characteristic polynomial** of M and the equation $\Delta_M(\lambda) = 0$ is called the **characteristic equation** of M. The **Cayley–Hamilton theorem** states that every matrix satisfies its characteristic equation. This is fairly involved to prove, so we will just accept it as given. Here is an example to see it in action.

Example 5.30. Apply the Cayley–Hamilton theorem to

$$M = \begin{bmatrix} 1 & 3 \\ 2 & 4 \end{bmatrix}.$$

Example 5.31. Show that the Cayley–Hamilton theorem can be used to find the inverse of an $n \times n$ matrix for small (convenient) n.

Example 5.32. Find the eigenvalues and eigenvectors corresponding to the symmetric matrix

$$M = \begin{bmatrix} 5 & 2\sqrt{2} \\ 2\sqrt{2} & 3 \end{bmatrix}$$

and show the eigenvectors are independent.

Now we turn our attention to diagonalizing M from Example 5.32. Let the matrix P consist of the two orthonormal eigenvectors as columns, so

$$P = \begin{bmatrix} \frac{1}{\sqrt{3}} & \sqrt{\frac{2}{3}} \\ -\sqrt{\frac{2}{3}} & \frac{1}{\sqrt{3}} \end{bmatrix}.$$

From Section 5.4 we know that

$$P^{-1} = \frac{1}{\frac{1}{3} + \frac{2}{3}} \begin{bmatrix} \frac{1}{\sqrt{3}} & -\sqrt{\frac{2}{3}} \\ \sqrt{\frac{2}{3}} & \frac{1}{\sqrt{3}} \end{bmatrix} = \begin{bmatrix} \frac{1}{\sqrt{3}} & -\sqrt{\frac{2}{3}} \\ \sqrt{\frac{2}{3}} & \frac{1}{\sqrt{3}} \end{bmatrix}.$$

Evidently $P^T = P^{-1}$, so P is an orthogonal matrix according to the definition in Section 5.4 Example 5.13. Finally, we assemble the matrix product

$$P^{-1} \otimes M \otimes P = \left(\frac{1}{\sqrt{3}}\right)^2 \begin{bmatrix} 1 & -\sqrt{2} \\ \sqrt{2} & 1 \end{bmatrix} \otimes \begin{bmatrix} 5 & 2\sqrt{2} \\ 2\sqrt{2} & 3 \end{bmatrix} \otimes \begin{bmatrix} 1 & \sqrt{2} \\ -\sqrt{2} & 1 \end{bmatrix}.$$

Note that

$$\begin{bmatrix} 1 & -\sqrt{2} \\ \sqrt{2} & 1 \end{bmatrix} \otimes \begin{bmatrix} 5 & 2\sqrt{2} \\ 2\sqrt{2} & 3 \end{bmatrix} = \begin{bmatrix} 1 & -\sqrt{2} \\ 7\sqrt{2} & 7 \end{bmatrix}$$

and then

$$\begin{bmatrix} 1 & -\sqrt{2} \\ 7\sqrt{2} & 7 \end{bmatrix} \otimes \begin{bmatrix} 1 & \sqrt{2} \\ -\sqrt{2} & 1 \end{bmatrix} = \begin{bmatrix} 3 & 0 \\ 0 & 21 \end{bmatrix},$$

so

$$\frac{1}{3}\begin{bmatrix} 3 & 0 \\ 0 & 21 \end{bmatrix} = \begin{bmatrix} 1 & 0 \\ 0 & 7 \end{bmatrix}.$$

This calculation has produced a diagonal matrix D with the same determinant as M consisting of the eigenvalues of M as diagonal entries. If we wanted an efficient way to compute high powers of M, we could use the scheme described in Section 5.4 Example 5.18. In that example we showed that $M^k = P \otimes D^k \otimes P^{-1}$. So, for example,

$$M^{10} = \begin{bmatrix} \frac{1}{\sqrt{3}} & \sqrt{\frac{2}{3}} \\ -\sqrt{\frac{2}{3}} & \frac{1}{\sqrt{3}} \end{bmatrix} \otimes \begin{bmatrix} 1 & 0 \\ 0 & 7^{10} \end{bmatrix} \otimes \begin{bmatrix} \frac{1}{\sqrt{3}} & -\sqrt{\frac{2}{3}} \\ \sqrt{\frac{2}{3}} & \frac{1}{\sqrt{3}} \end{bmatrix},$$

where we have used the property that the kth power of a diagonal matrix is simply the same matrix with the diagonal entries each raised to the kth power. This sort of calculation arises when there is a system consisting of entities in various states with probabilities that entities in one state can transition to another state. These transition probabilities make up the entries in an overall **transition matrix** for the process. The initial state of the system is represented as a vector and each time the process is repeated the state vector changes via multiplication by the transition matrix. The typical question asks what the state vector looks like after many iterations of the process or even in principle infinitely many iterations. Our method is particularly useful in computing arbitrary powers of the transition matrix.

References & suggested further reading

1) *Introduction to Matrix Analysis* (2nd ed.) by Richard Bellman (1997), Society for Industrial & Applied Mathematics (SIAM) – *higher level, widely regarded as a classic*
2) *Matrices and Linear Transformations* (2nd ed.) by Charles Cullen (1990), Dover (Addison Wesley) – *readable basic introduction*
3) *Linear Algebra* (2nd ed.) by Seymour Lipschutz (1991), Schaum – *outline series*

Solutions to Chapter 5 examples

Example 5.1

1) If $(x_1', x_2', \ldots, x_k', \ldots, x_n')$ solves the given system, it also solves the system with two rows interchanged. The x_k are not permuted by an interchange.
2) If $(x_1', x_2', \ldots, x_k', \ldots, x_n')$ solves $a_{j1}x_1 + a_{j2}x_2 + \cdots + a_{jn}x_n = c_j$ for all j, then it solves the system where the kth equation is

$$\beta(a_{k1}x_1 + a_{k2}x_2 + \cdots + a_{kn}x_n) = \beta c_k$$

for some k.

3) If $(x_1', x_2', \ldots, x_k', \ldots, x_n')$ solves $a_{j1}x_1 + a_{j2}x_2 + \cdots + a_{jn}x_n = c_j$ for all j, then it solves the system where the jth equation is $(a_{j1} + \beta a_{k1})x_1 + (a_{kj} + \beta a_{k2})x_2 + \cdots + (a_{jn} + \beta a_{kn})x_n = c_j + \beta c_k$ for any j. If $k = j$, this operation is the same as 2) as long as $\beta \neq -1$. Obviously, we may also use ERO 3 in the form of subtracting a nonzero multiple β of a row from some other row since it is the same as adding $-\beta$.

Example 5.2

If $a_{j1} = 0$, we are done, so assume a_{j1} and a_{k1} are nonzero. Use ERO 2 to multiply row/equation (i) by $\frac{a_{j1}}{a_{k1}}$ and then ERO 3 to subtract 1 times that to row/equation (ii). The result is

$$(a_{j1} - a_{j1})x_1 + \left(a_{j2} - \frac{a_{j1}a_{k2}}{a_{k1}}\right)x_2 + \cdots + \left(a_{jn} - \frac{a_{j1}a_{kn}}{a_{k1}}\right)x_n = c_j - \frac{a_{j1}}{a_{k1}}c_k.$$

The coefficient of x_1 in (ii) is now zero and the system is equivalent to the original.

Example 5.3

We use ERO 3 and multiply row 1 by (3) to get $6x + 15y = 57$ which we subtract from row 2 to get row $2'$, $0 - 12y = -36$. Now the system becomes

$$\begin{aligned} 2x + 5y &= 19, \\ -12y &= -36. \end{aligned}$$

We have reduced the original system of two equations in two variables to an equivalent triangular system where the second equation has only one variable and can be solved immediately. We find $y = 3$. Now we backsubstitute this value of y into the first equation to get $2x + 5(3) = 19$, or $2x = 4$. So the solution set for the system is $\{x = 2, y = 3\}$.

Example 5.4

We expect three equations in three unknowns to yield a triangular system. If there is some dependence among the equations this may fail to be true. For example, one equation might be a multiple of another. We will shortly have a test to sort these cases out. Use ERO 3 and multiply row 1 by (3) to get $9x + 3y - 6z = 0$, then subtract that from row 2 to get $0 - 13y + 10z = -2$. The transformed system is now

$$\begin{aligned} 3x + y - 2z &= 0, \\ -13y + 10z &= -2, \\ 12x + 20y - z &= 99. \end{aligned}$$

Now use ERO 3 again, this time multiplying equation 1 by (4) to get $12x + 4y - 8z = 0$, then subtract that from row 3 to get $0 + 16y + 7z = 99$. Our system is now

$$\begin{aligned} 3x + y - 2z &= 0, \\ -13y + 10z &= -2, \\ 16y + 7z &= 99. \end{aligned}$$

If we can eliminate $16y$ from row $3'$, we will have reached the triangular configuration. So we use ERO 3 yet again, multiplying row 2 this time by $(\frac{16}{13})$ to get $-16y + \frac{160}{13}z = -\frac{32}{13}$, then add that to row 3 to get $\frac{251}{13}z = \frac{1255}{13}$ or by ERO 2, $251z = 1255$. Our final equivalent system is

$$\begin{aligned} 3x + y - 2z &= 0, \\ -13y + 10z &= -2, \\ 251z &= 1255. \end{aligned}$$

Backsubstituting in row 3 gives us $z = 5$, then row 2 becomes $-13y + 50 = -2$, which implies $y = 4$. Row 1 becomes $3x + 4 - 10 = 0$ giving $x = 2$, so the solution set is $\{x = 2, y = 4, z = 5\}$.

Example 5.5

This is an underdetermined system with two equations in three unknowns, so the best we can do is transform it into an echelon system. One of the variables will be a **free variable** which can take on arbitrary values. Use ERO 3 by multiplying row 1 by (2) to get $2x + 4y - 6z = 16$, then subtract that from row 2 to get $-y + 12z = -12$. The new system is then

$$\begin{aligned} x + 2y - 3z &= 8, \\ -y + 12z &= -12. \end{aligned}$$

We can either retain the free variable z as a parameter, or be more explicit and set $z = t$. Then by backsubstitution, $y = 12(1 + t)$ and $x + 24(1 + t) - 3t = 8$, or $x = -21t - 16$. So the solution set is $\{(x = -21t - 16, y = 12(1 + t), z = t) : t \in \mathbb{R}\}$. In general, barring row dependence, if there are m equations in n variables where $m < n$, the system will reduce to echelon form with $n - m$ free variables.

We might encounter the system

$$\begin{aligned} a_{11}x_1 + a_{12}x_2 + \cdots + a_{1n}x_n &= 0, \\ a_{21}x_1 + a_{22}x_2 + \cdots + a_{2n}x_n &= 0, \\ &\vdots \\ a_{j1}x_1 + a_{j2}x_2 + \cdots + a_{jn}x_n &= 0, \\ &\vdots \\ a_{m1}x_1 + a_{m2}x_2 + \cdots + a_{mn}x_n &= 0, \end{aligned}$$

known as a **homogenous linear system**. Clearly, $x_j = 0$ for all j is a trivial solution. If $m = n$, it could be the *only* solution. For a homogenous system to have a nontrivial solution, there must be a free variable available.

Example 5.6

Step 1 – divide row 1 by (2) to get $\begin{bmatrix} 1 & 5 & 2 & 3 \\ 1 & 3 & 2 & 4 \\ 2 & 0 & 1 & 0 \end{bmatrix}$. Step 2 – subtract row 1 from row 2 to get $\begin{bmatrix} 1 & 5 & 2 & 3 \\ 0 & -2 & 0 & 1 \\ 2 & 0 & 1 & 0 \end{bmatrix}$. Step 3 – subtract twice row 1 from row 3 to get $\begin{bmatrix} 1 & 5 & 2 & 3 \\ 0 & -2 & 0 & 1 \\ 0 & -10 & -3 & -6 \end{bmatrix}$. Step 4 – subtract five times row 2 from row 3 to get $\begin{bmatrix} 1 & 5 & 2 & 3 \\ 0 & -2 & 0 & 1 \\ 0 & 0 & -3 & -11 \end{bmatrix}$. The matrix is now in echelon form. We call the leading nonzero entries **pivots**.

Example 5.7

Start with $\begin{bmatrix} 1 & 5 & 2 & 3 \\ 0 & -2 & 0 & 1 \\ 0 & 0 & -3 & -11 \end{bmatrix}$. Step 1 – divide row 2 by (−2) to get $\begin{bmatrix} 1 & 5 & 2 & 3 \\ 0 & 1 & 0 & -\frac{1}{2} \\ 0 & 0 & -3 & -11 \end{bmatrix}$. Step 2 – divide row 3 by (−3) to get $\begin{bmatrix} 1 & 5 & 2 & 3 \\ 0 & 1 & 0 & -\frac{1}{2} \\ 0 & 0 & 1 & \frac{11}{3} \end{bmatrix}$. Step 3 – subtract (5) times row 2 from row 1 to get $\begin{bmatrix} 1 & 0 & 2 & \frac{11}{2} \\ 0 & 1 & 0 & -\frac{1}{2} \\ 0 & 0 & 1 & \frac{11}{3} \end{bmatrix}$. Step 4 – subtract twice row 3 from row 1 to get $\begin{bmatrix} 1 & 0 & 0 & -\frac{11}{6} \\ 0 & 1 & 0 & -\frac{1}{2} \\ 0 & 0 & 1 & \frac{11}{3} \end{bmatrix}$. The matrix is now in row canonical form.

Example 5.8

Transplant the system to the rows of a 3×4 matrix of scalars preserving the relative positions of the respective coefficients and constants. The result is $\begin{bmatrix} 2 & 10 & 4 & 6 \\ 1 & 3 & 2 & 4 \\ 2 & 0 & 1 & 0 \end{bmatrix}$. It is understood that the coefficients in column 1 apply to x, column 2 to y, column 3 to z, and column 4 to the constants. If we could row reduce this matrix to echelon form, since that would be row equivalent to the original, we would have an easy way to solve the system by restoring the variables and backsubstituting. Even better, if we could row reduce this matrix to its row canonical form, the solution would be immediate with no backsubstitution. Happily, we already have the row canonical form from the preceding example. It is $\begin{bmatrix} 1 & 0 & 0 & -\frac{11}{6} \\ 0 & 1 & 0 & -\frac{1}{2} \\ 0 & 0 & 1 & \frac{11}{3} \end{bmatrix}$ and, after doing the restoration step, the original linear system is equivalent to

$$x = -\frac{11}{6},$$
$$y = -\frac{1}{2},$$
$$z = \frac{11}{3}.$$

Done.

Example 5.9

We pick the low-hanging fruit and note that $z = -2x$ from equation 3. Then equation 1 can be rewritten as $2x + 10y + 4(-2x) = -6x + 10y = 6$. Similarly, equation 2 can be expressed as $x + 3y + 2(-2x) = -3x + 3y = 4$. Multiplying this by 2, we have $-6x + 6y = 8$, then $-6x = 8 - 6y$. Substituting this into the modified equation 1, we get $8 - 6y + 10y = 6$, or $4y = -2$. So $y = -\frac{1}{2}$ as before. Going back to $-6x + 6y = 8$, we substitute $y = -\frac{1}{2}$ and obtain $-6x - 3 = 8$ or $x = -\frac{11}{6}$. Finally, from $z = -2x$, we get $z = -2(-\frac{11}{6}) = \frac{11}{3}$. You can choose which method seems more efficient in a given situation. The fact that the third equation had only two variables and allowed an immediate substitution made it appear simpler than usual. In the case where there are four or more equations with all the variables appearing in every equation, the matrix approach is probably preferable, if for no other reason that it is more systematic and hence less error prone.

Example 5.10

Here

$$A \oplus B = \begin{bmatrix} 5 & 11 & 13 \\ 4 & 17 & 12 \\ 13 & 5 & 8 \end{bmatrix}, \quad A \otimes B = \begin{bmatrix} 39 & 92 & 101 \\ 33 & 110 & 110 \\ 31 & 59 & 61 \end{bmatrix}, \quad \text{and} \quad B \otimes A = \begin{bmatrix} 78 & 63 & 51 \\ 101 & 77 & 51 \\ 86 & 60 & 55 \end{bmatrix}.$$

Notice that $A \otimes B \neq B \otimes A$, so general matrix multiplication is order sensitive or noncommutative.

Example 5.11

$$\begin{bmatrix} 3x + 4y - 8z \\ 9x + 5y + z \\ 6x + 7y - 6z \end{bmatrix}.$$

Example 5.12

$$\begin{bmatrix} 5 & 3 & -2 \\ 8 & -9 & 4 \\ 6 & 1 & -7 \end{bmatrix} \otimes \begin{bmatrix} x \\ y \\ z \end{bmatrix} = \begin{bmatrix} 0 \\ -2 \\ 14 \end{bmatrix}.$$

Example 5.13

Here

$$\begin{bmatrix} 2 & 3 \\ 1 & 4 \end{bmatrix}^2 = \begin{bmatrix} 7 & 18 \\ 6 & 19 \end{bmatrix} \quad \text{and} \quad \begin{bmatrix} 2 & 3 \\ 1 & 4 \end{bmatrix}^3 = \begin{bmatrix} 2 & 3 \\ 1 & 4 \end{bmatrix} \otimes \begin{bmatrix} 7 & 18 \\ 6 & 19 \end{bmatrix} = \begin{bmatrix} 32 & 93 \\ 31 & 94 \end{bmatrix},$$

so

$$P(A) = 2\begin{bmatrix} 32 & 93 \\ 31 & 94 \end{bmatrix} \oplus \begin{bmatrix} 7 & 18 \\ 6 & 19 \end{bmatrix} \ominus \begin{bmatrix} 2 & 3 \\ 1 & 4 \end{bmatrix} + 3\begin{bmatrix} 1 & 0 \\ 0 & 1 \end{bmatrix} = \begin{bmatrix} 72 & 201 \\ 67 & 206 \end{bmatrix}.$$

Example 5.14

Since $e^x = \sum_{k=0}^{\infty} \frac{x^k}{k!} = 1 + x + \frac{x^2}{2!} + \frac{x^3}{3!} + \cdots$, we get

$$\exp\left(\begin{bmatrix} \frac{1}{2} & \frac{1}{4} \\ \frac{1}{3} & \frac{1}{2} \end{bmatrix}\right) \approx \begin{bmatrix} 1 & 0 \\ 0 & 1 \end{bmatrix} \oplus \begin{bmatrix} \frac{1}{2} & \frac{1}{4} \\ \frac{1}{3} & \frac{1}{2} \end{bmatrix} \oplus \frac{1}{2}\begin{bmatrix} \frac{1}{3} & \frac{1}{4} \\ \frac{1}{3} & \frac{1}{3} \end{bmatrix} \oplus \frac{1}{6}\begin{bmatrix} \frac{1}{4} & \frac{5}{24} \\ \frac{5}{18} & \frac{1}{4} \end{bmatrix} = \begin{bmatrix} \frac{41}{24} & \frac{59}{144} \\ \frac{59}{18} & \frac{41}{24} \end{bmatrix}.$$

Example 5.15

Since

$$\begin{bmatrix} a & b \\ c & d \end{bmatrix} \otimes \begin{bmatrix} d & -b \\ -c & a \end{bmatrix} = \begin{bmatrix} ad - bc & -ab + ab \\ cd - cd & -bc + ad \end{bmatrix},$$

we have

$$\frac{1}{ad - bc}\begin{bmatrix} ad - bc & -ab + ab \\ cd - cd & -bc + ad \end{bmatrix} = \begin{bmatrix} 1 & 0 \\ 0 & 1 \end{bmatrix}.$$

Example 5.16

It is enough to show this for two matrices:

$$\begin{aligned} (A_1 \otimes A_2) \otimes (A_2^{-1} \otimes A_1^{-1}) &= A_1 \otimes (A_2 \otimes A_2^{-1}) \otimes A_1^{-1} \\ &= A_1 \otimes I_n \otimes A_1^{-1} = A_1 \otimes A_1^{-1} = I_n. \end{aligned}$$

So the inverse of a product is the product of inverses in the reverse order.

Example 5.17

Starting with $\begin{bmatrix} 1 & 0 & 0 \\ 0 & 1 & 0 \\ 0 & 0 & 1 \end{bmatrix}$, we interchange the first two rows to get $\begin{bmatrix} 0 & 1 & 0 \\ 1 & 0 & 0 \\ 0 & 0 & 1 \end{bmatrix}$. Testing it out,

$$\begin{bmatrix} 0 & 1 & 0 \\ 1 & 0 & 0 \\ 0 & 0 & 1 \end{bmatrix} \otimes \begin{bmatrix} 1 & 4 & 7 \\ 2 & 0 & 9 \\ 3 & 6 & 6 \end{bmatrix} = \begin{bmatrix} 2 & 0 & 9 \\ 1 & 4 & 7 \\ 3 & 6 & 6 \end{bmatrix}. \checkmark$$

Example 5.18

Since P is orthogonal, $P^T = P^{-1}$. Then

$$\begin{aligned} D^k &= (P^{-1} \otimes M \otimes P)^k \\ &= (P^{-1} \otimes M \otimes P) \otimes (P^{-1} \otimes M \otimes P) \otimes \cdots \otimes (P^{-1} \otimes M \otimes P). \end{aligned}$$

By associativity, we can regroup the terms as

$$P^{-1} \otimes M \otimes (P \otimes P^{-1}) \otimes M \otimes (P \otimes P^{-1}) \otimes \cdots \otimes M \otimes (P \otimes P^{-1}) \otimes M \otimes P = P^{-1} \otimes M^k \otimes P.$$

Now $D^k = (d_{ii}^k)$ is easy to compute, and since $D^k = P^{-1} \otimes M^k \otimes P$ we conclude $M^k = P \otimes D^k \otimes P^{-1}$. This saves a lot of matrix multiplication. In limit applications, such as finding $\lim_{k\to\infty} M^k$, this method is a lifesaver.

Example 5.19

There are two permutations of $\{1, 2\}$, namely $1 \to 1$, $2 \to 2$ and $1 \to 2$, $2 \to 1$. The product corresponding to the first permutation is $a_{11}a_{22}$, and the second is $a_{21}a_{12}$. The first permutation already is the identity (all fixed points), so its parity is even by default. The second permutation factors as $(1\,2)$, so one transposition yields negative parity. The result is $\det A = a_{11}a_{22} - a_{21}a_{12}$. An implied determinant is written by switching the brackets for a matrix to single vertical bars. So

$$\det \begin{bmatrix} a & b \\ c & d \end{bmatrix} = \begin{vmatrix} a & b \\ c & d \end{vmatrix} = ad - bc$$

by the preceding calculation. Note that this is the same as the factor that appears in Section 5.4 Example 5.15.

Example 5.20

We list the six permutations of $\{1, 2, 3\}$ here with their associated products.

Parity	π	Product
+	identity	$a_{11}a_{22}a_{33}$
+	(12)(13)	$a_{21}a_{32}a_{13}$
+	(13)(12)	$a_{31}a_{12}a_{23}$
−	(12)	$a_{12}a_{21}a_{33}$
−	(13)	$a_{13}a_{22}a_{31}$
−	(23)	$a_{11}a_{23}a_{32}$

Then $\det A = a_{11}a_{22}a_{33} + a_{21}a_{32}a_{13} + a_{31}a_{12}a_{23} - a_{12}a_{21}a_{33} - a_{13}a_{22}a_{31} - a_{11}a_{23}a_{32}$. So much for the theory. As a practical matter here is a calculation aid beloved by algebra students. To find $\begin{vmatrix} a & b & c \\ d & e & f \\ g & h & i \end{vmatrix}$ rewrite it with the first two columns adjoined like this:

$$\begin{matrix} a & b & c & a & b \\ d & e & f & d & e \\ g & h & i & g & h \end{matrix}.$$

Starting at the upper left with a, move downward to the right diagonally and form the product aei. Now do the same for the diagonals starting with b and c to create products bfg and cdh. Give these three products positive parity. Now return to c and move downward to the left diagonally to form the product ceg. Do the same for the diagonals starting with a and b in the adjoined columns to collect the products afh and bdi. Give these last three products negative parity. Add all six products to get $\det A = aei + bfg + cdh - ceg - afh - bdi$.

Example 5.21

From the formula in Example 5.20 we have

$$\det \begin{bmatrix} 2 & 0 & 4 \\ 1 & 3 & 5 \\ 6 & 3 & 8 \end{bmatrix} = 48 + 0 + 12 - 72 - 0 - 30 = -42.$$

Example 5.22

Row one is an obvious choice to use in the Laplace expansion. The cofactor of 1 is

$$(-1)^{1+1} \begin{vmatrix} 4 & 0 & 1 \\ 1 & 3 & 3 \\ 5 & 0 & 1 \end{vmatrix} = -3.$$

The cofactor of -2 is

$$(-1)^{1+3} \begin{vmatrix} 3 & 4 & 1 \\ 4 & 1 & 3 \\ 2 & 5 & 1 \end{vmatrix} = -16.$$

The other two cofactors are 0, so the determinant is $1 \cdot (-3) + (-2) \cdot (-16) = 29$.

Example 5.23

$\det A = 1 \cdot 3 \cdot (-1) \cdot (-3) = 9$ and $\det B = 2 \cdot 4 \cdot 7 \cdot (-1) = -56$.

Example 5.24

The coefficient matrix is $\begin{bmatrix} 5 & 3 & -2 \\ 8 & -9 & 4 \\ 6 & 1 & -7 \end{bmatrix}$. We calculate

$$D = \det \begin{bmatrix} 5 & 3 & -2 \\ 8 & -9 & 4 \\ 6 & 1 & -7 \end{bmatrix} = 411,$$

so apparently it is nonsingular. We replace the first column (corresponding to the variable x) with the column vector of constants $\begin{bmatrix} 0 \\ -2 \\ 14 \end{bmatrix}$. This gives

$$D_x = \det \begin{bmatrix} 0 & 3 & -2 \\ -2 & -9 & 4 \\ 14 & 1 & -7 \end{bmatrix} = -122.$$

Likewise, we compute

$$D_y = \det \begin{bmatrix} 5 & 0 & -2 \\ 8 & -2 & 4 \\ 6 & 14 & -7 \end{bmatrix} = -458 \quad \text{and} \quad D_z = \det \begin{bmatrix} 5 & 3 & 0 \\ 8 & -9 & -2 \\ 6 & 1 & 14 \end{bmatrix} = -992,$$

where the column vector has replaced the second or third columns, respectively, in the original determinant. We have $x = \frac{D_x}{D} = -\frac{122}{411} = -0.2968$, $y = \frac{D_y}{D} = -\frac{458}{411} = -1.1144$, and $z = \frac{D_z}{D} = -\frac{992}{411} = -2.4136$. Spot checking the first equation, we have $5 \cdot (-0.2968) + 3 \cdot (-1.1144) - 2 \cdot (-2.4136) = -1.484 - 3.3432 + 4.8272 = 0$. You may place your bets on the other two.

Example 5.25

Let $B_1 = \{e_1, e_2\}$ and $B_2 = \{f_1, f_2\}$. We know by the definiteness of dimension property that any two bases for the same space will have the same number of basis vectors. So let $v = \alpha_1 e_1 + \alpha_2 e_2$. Now e_1 and e_2 can themselves be represented in terms of B_2, so let $e_1 = x_1 f_1 + x_2 f_2$ and $e_2 = x_3 f_1 + x_4 f_2$. Rewriting v with respect to B_2 we have $v = \alpha_1(x_1 f_1 + x_2 f_2) + \alpha_2(x_3 f_1 + x_4 f_2)$. Upon regrouping terms, we get $v = (\alpha_1 x_1 + \alpha_2 x_3) f_1 + (\alpha_1 x_2 + \alpha_2 x_4) f_2$. Evidently, $\beta_1 = \alpha_1 x_1 + \alpha_2 x_3$ and $\beta_2 = \alpha_1 x_2 + \alpha_2 x_4$. This result is easily expressed as a matrix equation in column vector (transpose) form,

$$\begin{bmatrix} x_1 & x_3 \\ x_2 & x_4 \end{bmatrix} \otimes \begin{bmatrix} \alpha_1 \\ \alpha_2 \end{bmatrix} = \begin{bmatrix} \beta_1 \\ \beta_2 \end{bmatrix}.$$

Observe that the columns of the **transition** or **change-of-basis matrix**

$$T=\begin{bmatrix}x_1 & x_3\\ x_2 & x_4\end{bmatrix}$$

are precisely the respective components of e_1 and e_2 expressed in terms of B_2.

Example 5.26

Suppose $e_1 = x_1f_1 + x_2f_2 + x_3f_3$, $e_2 = x_4f_1 + x_5f_2 + x_6f_3$, and $e_3 = x_7f_1 + x_8f_2 + x_9f_3$. Then

$$T=\begin{bmatrix}x_1 & x_4 & x_7\\ x_2 & x_5 & x_8\\ x_3 & x_6 & x_9\end{bmatrix}$$

and for $[v]_{B_1} = [\alpha_1, \alpha_2, \alpha_3]$ we have

$$\begin{bmatrix}x_1 & x_4 & x_7\\ x_2 & x_5 & x_8\\ x_3 & x_6 & x_9\end{bmatrix}\otimes\begin{bmatrix}\alpha_1\\ \alpha_2\\ \alpha_3\end{bmatrix}=\begin{bmatrix}\beta_1\\ \beta_2\\ \beta_3\end{bmatrix}=[v]_{B_2}.$$

Example 5.27

Expressing

$$\begin{bmatrix}1\\0\end{bmatrix}=x_1\begin{bmatrix}1\\1\end{bmatrix}+x_2\begin{bmatrix}-1\\1\end{bmatrix}$$

gives us $x_1 - x_2 = 1$ and $x_1 + x_2 = 0$. Solving simultaneously, we have $x_1 = \frac{1}{2}$ and $x_2 = -\frac{1}{2}$. Likewise

$$\begin{bmatrix}0\\1\end{bmatrix}=x_3\begin{bmatrix}1\\1\end{bmatrix}+x_4\begin{bmatrix}-1\\1\end{bmatrix}$$

implies $x_3 = x_4 = \frac{1}{2}$. So

$$\begin{bmatrix}1\\0\end{bmatrix}=\frac{1}{2}\begin{bmatrix}1\\1\end{bmatrix}-\frac{1}{2}\begin{bmatrix}-1\\1\end{bmatrix}\quad\text{and}\quad\begin{bmatrix}0\\1\end{bmatrix}=\frac{1}{2}\begin{bmatrix}1\\1\end{bmatrix}+\frac{1}{2}\begin{bmatrix}-1\\1\end{bmatrix}.$$

Evidently, in this case

$$T=\begin{bmatrix}x_1 & x_3\\ x_2 & x_4\end{bmatrix}=\begin{bmatrix}\frac{1}{2} & \frac{1}{2}\\ -\frac{1}{2} & \frac{1}{2}\end{bmatrix}=\frac{1}{2}\begin{bmatrix}1 & 1\\ -1 & 1\end{bmatrix}.$$

Example 5.28

$$\frac{1}{2}\begin{bmatrix}1 & 1\\ -1 & 1\end{bmatrix}\otimes\begin{bmatrix}2\\4\end{bmatrix}=\begin{bmatrix}6\\2\end{bmatrix}.$$

Example 5.29

Suppose T changes the basis from B_1 to B_2 and U changes the basis from B_2 to B_1. The net result of $U \otimes T$ is to change the components relative to B_1 back to B_1, in other words the identity transformation. So $U \otimes T = I$ and $U = T^{-1}$.

Example 5.30

Here

$$\lambda I_2 - M = \begin{bmatrix} \lambda - 1 & 3 \\ 2 & \lambda - 4 \end{bmatrix},$$

so $\Delta_M(\lambda) = (\lambda - 1)(\lambda - 4) - 6 = \lambda^2 - 5\lambda - 2$. Then

$$\begin{bmatrix} 1 & 3 \\ 2 & 4 \end{bmatrix}^2 - 5\begin{bmatrix} 1 & 3 \\ 2 & 4 \end{bmatrix} - \begin{bmatrix} 2 & 0 \\ 0 & 2 \end{bmatrix} = \begin{bmatrix} 7 & 15 \\ 10 & 22 \end{bmatrix} - \begin{bmatrix} 5 & 15 \\ 10 & 20 \end{bmatrix} - \begin{bmatrix} 2 & 0 \\ 0 & 2 \end{bmatrix} = \begin{bmatrix} 0 & 0 \\ 0 & 0 \end{bmatrix} \checkmark.$$

Example 5.31

We already have a formula for the ready-made inverse of a nonsingular 2×2 matrix in Section 5.4 Example 5.15. So let us consider again the 3×3 matrix in Section 5.6, namely

$$M = \begin{bmatrix} 2 & 0 & -1 \\ 2 & -1 & -2 \\ 0 & 3 & 2 \end{bmatrix}.$$

We have already verified it is nonsingular so it is invertible. Its characteristic equation is $\lambda^3 - 3\lambda^2 + 6\lambda - 2 = 0$. Applying Cayley–Hamilton, we have $M^3 - 3M^2 + 6M - 2I = 0$. We factor this as $M(M^2 - 3M + 6I) = 2I$. Note that $\frac{1}{2}(M^2 - 3M + 6I)$ is apparently M^{-1}. Doing the calculation we have

$$\begin{aligned} M^{-1} &= \frac{1}{2}\left(\begin{bmatrix} 4 & -3 & -4 \\ 2 & -5 & -4 \\ 6 & 3 & -2 \end{bmatrix} - 3\begin{bmatrix} 2 & 0 & -1 \\ 2 & -1 & -2 \\ 0 & 3 & 2 \end{bmatrix} + 6\begin{bmatrix} 1 & 0 & 0 \\ 0 & 1 & 0 \\ 0 & 0 & 1 \end{bmatrix}\right) \\ &= \frac{1}{2}\begin{bmatrix} 4 & -3 & -1 \\ -4 & 4 & 2 \\ 6 & -6 & -2 \end{bmatrix} = \begin{bmatrix} 2 & -\frac{3}{2} & -\frac{1}{2} \\ -2 & 2 & 1 \\ 3 & -3 & -1 \end{bmatrix}, \end{aligned}$$

which agrees with our earlier result.

Suppose we have an $n \times n$ matrix with elements from a given field (usually $\mathbb{R}$ or $\mathbb{C}$). If a nonzero n dimensional column vector over the same field has the property that $M \otimes v = \lambda v$ for some scalar λ in the field, we say that v is an **eigenvector** of M belonging to the **eigenvalue** λ. One way to conceptualize an eigenvector is to think of M representing a rotation in $\mathbb{R}^3$. Then an eigenvector is still going to point in its original (or exactly opposite) direction but it may be lengthened or shortened by the **scale factor** λ. Any multiple by the scalar c of an eigenvector belonging to λ is also an eigenvector belonging to λ. This can easily be seen by string of equalities $M \otimes (cv) = c(M \otimes v) = c\lambda v = \lambda cv$. The **eigenspace** of λ consists of the span of the eigenvectors associated with λ. The three roots of $\lambda^3 - 3\lambda^2 + 6\lambda - 2 = 0$ in the preceding example are $\{1, 1, 2\}$, so these are the eigenvalues of M. The prefix "eigen-" is a German word meaning "own" or "self." In this context a matrix generates its *own* specific eigenvalues via the characteristic equation and these values determine the eigenvectors belonging to the matrix.

It is possible for a matrix with real entries to have a characteristic equation whose roots are complex, so it would not have any eigenvalues over $\mathbb{R}$, hence no eigenvectors either. An example would be $M = \begin{bmatrix} 2 & -1 \\ 5 & -2 \end{bmatrix}$. The characteristic equation for this matrix is $(\lambda - 2)(\lambda + 2) + 5 = \lambda^2 + 1 = 0$. There are no real roots for this equation, so no real eigenvalues for M. But if we allowed complex eigenvalues, they would be $\pm i = \pm\sqrt{-1}$. The fundamental theorem of algebra says that every polynomial with complex coefficients can be factored into linear terms with complex roots, so this problem does not come up when the field is $\mathbb{C}$. In any case, we would like to have a condition for a matrix that guarantees its eigenvalues are real. Given the matrix $M = (m_{ij})$, where $m_{ij} \in \mathbb{R}$ and $m_{ij} = m_{ji}$ (symmetry), we are assured that all of the eigenvalues of M are real.

Example 5.32

Here

$$\begin{vmatrix} \lambda - 5 & 2\sqrt{2} \\ 2\sqrt{2} & \lambda - 3 \end{vmatrix} = (\lambda - 5)(\lambda - 3) - 8 = \lambda^2 - 8\lambda + 15 - 8 = \lambda^2 - 8\lambda + 7 = 0$$

gives us $(\lambda - 1)(\lambda - 7) = 0$, so the eigenvalues are 1 and 7. For $\lambda = 1$, we have

$$\begin{bmatrix} 5 & 2\sqrt{2} \\ 2\sqrt{2} & 3 \end{bmatrix} \begin{bmatrix} x \\ y \end{bmatrix} = 1 \cdot \begin{bmatrix} x \\ y \end{bmatrix},$$

which gives the linear system $\{5x + 2\sqrt{2}y = x, 2\sqrt{2}x + 3y = y\}$. This is equivalent to $\{4x + 2\sqrt{2}y = 0, 2\sqrt{2}x + 2y = 0\}$, which has a nonzero solution $\begin{bmatrix} 1 \\ -\sqrt{2} \end{bmatrix}$. Similarly,

$$\begin{bmatrix} 5 & 2\sqrt{2} \\ 2\sqrt{2} & 3 \end{bmatrix} \begin{bmatrix} x \\ y \end{bmatrix} = 7 \cdot \begin{bmatrix} x \\ y \end{bmatrix}$$

gives us $\begin{bmatrix} \sqrt{2} \\ 1 \end{bmatrix}$. So the eigenvalue/eigenvector pairs for M are

$$\left\{1, \begin{bmatrix} 1 \\ -\sqrt{2} \end{bmatrix}\right\} \quad \text{and} \quad \left\{7, \begin{bmatrix} \sqrt{2} \\ 1 \end{bmatrix}\right\}.$$

We can verify the eigenvectors are orthogonal by noting

$$\begin{bmatrix} 1 \\ -\sqrt{2} \end{bmatrix} \cdot \begin{bmatrix} \sqrt{2} \\ 1 \end{bmatrix} = 0.$$

Further, we can unitize them to create an orthonormal set

$$\left\{ \frac{1}{\sqrt{3}} \begin{bmatrix} 1 \\ -\sqrt{2} \end{bmatrix}, \frac{1}{\sqrt{3}} \begin{bmatrix} \sqrt{2} \\ 1 \end{bmatrix} \right\}.$$

6 Solving equations

There are four common types of precalculus level equations that we would like to discuss: polynomial, exponential, logarithmic, and trigonometric. Polynomial equations have been studied for perhaps four millennia and have driven the development of a great deal of algebra, mostly in a crescendo beginning with the Renaissance. When people talk about "solving an equation for x" it is usually a polynomial equation they have in mind. Exponential and logarithmic equations, which are two sides of the same coin, became important after the invention of logarithms about 1500 CE. Certain physical, economic, and biological processes and phenomena are efficiently modeled by exponential or logarithmic equations. Hyperbolic functions are essentially exponentials, and we will explore the relationships among them. Trigonometric equations grew out of geometric relationships and have been around, at least implicitly, for two millennia. The usual system of trigonometric functions is a more recent development and they have turned out to be very important in the theory of integration, which is a part of calculus beyond the scope of our discussion. But there is still value in studying equations involving trigonometric expressions per se. A great deal of the basic theory is discussed in Section 7. Finally, we discuss inequations (or inequalities) and Newton's numerical approximation method.

6.1 Polynomial equations

We should confess at the outset that quite a bit of what we present in this section can be better implemented by a computer, or even in some cases a lowly calculator. The human brain is at a severe disadvantage in competing with **computer algebra systems** (CAS) like Maple©, Mathematica©, Wolfram Alpha©, and similar programs to solve more complicated polynomial equations. Nevertheless, there are skills worth knowing that are applicable to problems at the simpler end of the spectrum. Our discussion is complete in the sense that we derive formulas that give pencil-and-paper algorithms to solve all general polynomial equations in one variable that are known to have ordinary solutions. Ordinary in this context is a loaded word – it means solutions in terms of the usual arithmetic operations including radicals applied to the coefficients that appear in the polynomial. We even provide a glimpse of what might be done to handle equations that do not quite fit that profile. The story of how mathematicians developed these methods is fascinating, and we have taken the liberty of weaving historical comments into the technical narrative.

A polynomial in the indeterminate or variable x is an expression of the form $a_n x^n + \cdots + a_k x^k + \cdots + a_1 x + a_0$, where the **coefficients** a_k are either real or complex numbers and $n \in \mathbb{N}$. If they are complex, we would substitute z for x as the variable to remind us. By convention this is the **standard form** of writing a polynomial. Often we give polynomials a nickname like $P(x)$ or $Q(x)$ when we do not need to refer to the detailed structure.

https://doi.org/10.1515/9783111506654-007

We assume $a_n \neq 0$ since we could always move down to the first coefficient that was nonzero and relabel. However, any of the other coefficients could be zero. The leftmost term in the standard form is the **leading term** and the coefficient a_n is the **leading coefficient**. The coefficient a_1 is called the **linear coefficient** and the coefficient a_0 is called the **constant coefficient.** The degree of the polynomial is n, the power of the indeterminate in the leading term. The various terms that are added together to make up the polynomial are called **monomials**. If a polynomial has two monomials it is a binomial, three and it is a trinomial, and so forth. If a particular value r is substituted for x in a polynomial $P(x)$ and the result is zero, we say that r is a **zero** of $P(x)$. Equivalently, the equation $P(x) = 0$ has **root** r. Some mathematicians call r a root of the polynomial equation but a zero of the polynomial outside the context of an equation. This is a distinction without a difference. If the degree of $P(x)$ is one, the equation $P(x) = 0$ is called a **linear equation**. If the degree is two it is a **quadratic equation**, if three it is a **cubic equation**, if four it is a **quartic equation**, and so forth. It is a fact established in advanced algebra that a real or complex polynomial equation of degree n has n roots if we properly account for multiplicities, where one number is a root multiple times.

The desire to solve polynomial equations in radicals has been an important driver in the development of mathematics since antiquity. Solving an equation in radicals means finding a formula for the roots in terms of the basic field functions – addition, subtraction, multiplication, division, nth powers, and nth roots (**radicals**) applied to the coefficients that appear. If a formula can be found the polynomial equation is deemed **solvable in radicals**, and if not it is **unsolvable**. Other methods may yield numerical solutions, but the point of the definition is that algebraic manipulations of the coefficients plus radicals are inadequate.

The Babylonians knew how to solve linear and quadratic equations in the Old Babylonian Period about 2000 BC and there is evidence that the Egyptians were able to solve some quadratic equations at about the same time. The general solution to cubic equations proved more elusive. Up until about 1540, it was believed that no such solution was possible. Several Italian mathematicians of that era made substantial progress on the problem and were able to solve certain special cases. Based on these discoveries, Giralamo Cardano found the last piece of the puzzle and finally was in possession of the general solution. The intrigue and betrayals surrounding this development make a fascinating story in the history of mathematics and, coincidentally, a famous early case in intellectual property rights.

Cardano was a noted intellectual during the Italian Renaissance and author of numerous works on mathematics, astrology, medicine, and gambling. He seems to have been a contentious character, and this kept him from securing any permanent position at a university, but he did have students. He was chronically broke, and supported himself largely by gambling, which led to his publication of a basic theory of probability. His contemporary and countryman Niccoló Fontana discovered the formula for an important special case of the cubic about 1540 and disclosed it after some badgering to

Cardano, who promised to keep it a secret. Several years later, Cardano came into possession of an unpublished manuscript by Scipione del Ferro, which predated by at least a decade the disclosure by Fontana. This manuscript indicated that del Ferro had independently discovered the cubic formula prior to Fontana. It was not uncommon for mathematicians in those days to keep their best discoveries secret, since they were always subject to "problem challenges" from their rivals. This was the mathematical equivalent of a duel. Each participant submitted to the other a problem they could solve but would be difficult for the recipient to solve. Being defeated in the challenge had ramifications for one's university position or patron support, so del Ferro had a good defense with his cubic formula. Cardano then felt that his promise to Fontana of nondisclosure was moot, and he published his treatise *Ars Magna* in 1545 which contained the general formula he deduced from Fontana's disclosure. In *Ars Magna* Cardano gave both Fontana and del Ferro ample credit in the text for their independent discoveries. Nevertheless, this touched off a public dispute with Fontana that lasted a decade. The modern formula we derive below is called the **Cardano–Tartaglia formula**, giving both men credit. Tartaglia is not a new character here. Fontana had the misfortune of being an inhabitant of Brescia during a siege by the French, who decided to massacre the inhabitants as payback for rather effective resistance. A French soldier cut Fontana's jaw and palate badly enough that he found it difficult to speak, and thereby gained the nickname "The Stammerer," which in Italian is "Tartaglia." It is not clear why del Ferro does not receive more modern credit for this discovery.

We should also mention that these Renaissance results were not as conceptually seamless as the modern version. Cardano was the first mathematician to work with negative numbers in a serious way, and although he was aware of imaginary numbers, he did not profess to understand their properties. Another contemporary and countryman Rafael Bombelli eventually worked out the algebra of complex numbers and elucidated their appearance in the Cardano formula. The approach at the time was to separate cases based on the coefficients, and to work towards real solutions. Cubic equations are guaranteed to have at least one real solution because their graphs must cross the real axis due to continuity of the graph and lead term dominance. In an arbitrary cubic expression $ax^3 + bx^2 + cx + d$, for large values of $|x|$ the leading term ax^3 dominates the entire expression, and depending on the sign of a can be either negative for negative x and positive for positive x, or vice versa. The other solutions can be either both real or a pair of conjugate complex solutions. For example, Cardano would have assumed that $p, q > 0$ and treated the cases $x^3 = px + q$, $x^3 + px = q$, and $x^3 = px + q = 0$ separately. His contribution was extending and systematizing the basic method of del Ferro and Fontana. The famous American physicist Richard Feynman claimed that solving the cubic was important because it was a watershed event marking the first time Renaissance mathematicians produced a significant result that was unknown to the ancients.

Shortly after the cubic was conquered, Cardano's student, Lodovico Ferrari, solved the general quartic equation. These successes encouraged the search to solve the quintic (degree five) equation, but progress was not forthcoming. In fact, no headway was made for the next 250 years. The suspicion arose that perhaps the quintic was unsolvable. Of course, the cubic had appeared that way years before and then finally yielded, so all hope was not lost. However, a more recent Italian mathematician, Paolo Ruffini, offered a proof in 1799 that the general quintic was definitely unsolvable. Ruffini's argument was incomplete when published, but eventually improved to satisfy the proof standards of the time. The Norwegian mathematician, Niels Abel, published a more watertight demonstration in 1824 that there existed polynomial equations of degree five and higher with arbitrary coefficients that could not be solved in radicals. This result is known as the Abel–Ruffini theorem. The key word in this statement is "arbitrary." Some quintics like $x^5 - 1 = 0$ are certainly solvable, but if we have complete freedom to choose the coefficients, we may encounter a quintic like $x^5 + x + 1 = 0$ that is not. During this same period a young French mathematician, Évariste Galois, building on the group theory approach of Joseph-Louis Lagrange, provided a stunning general confirmation of the Abel–Ruffini result using clever new ideas that subsequently developed into a whole new branch of algebra. Galois' personal history is famously tragic. He had made his discoveries in the early 1830s and then died soon after in a duel at the age of 20. His notes were edited and published posthumously in 1846.

At any rate, the grand project to solve polynomial equations of ever higher degree in radicals was over at last. Knowing how to solve linear and quadratic equations with pencil and paper is a worthwhile skill. For cubic and quartic equations, not so much. They can be handled efficiently by a CAS, and in some cases by a calculator. As we have noted, there is no general method for solving quintics in radicals, but the sense in which this is impossible needs to be clarified. There are reducible (factorable over $\mathbb{Q}$) quintics and the lower degree factors would necessarily be subject to the formulas already available. In the course of discussing such a possibility, we present some criteria which help to decide if a given quintic (actually any polynomial) might be reducible. So a quintic that is reducible, if we can find the factors, is eminently solvable in radicals. A quintic that is not reducible is a tougher nut to crack solely with algebraic methods and we are not going to try. But we do want to pay some tribute to the monumental effort that mathematicians expended over several centuries in pursuit of this goal, so you may regard the remaining discussion of irreducible quintics as purely historical comment, illustrated with some mathematics. Notably, there are some irreducible quintics that are still solvable with ordinary radicals. Abel–Ruffini does not rule this out, but many are unsolvable. We will introduce a method to make this determination and give an example of both types. If we enlarge the definition of radical to include so-called ultraradicals, we can justifiably say all quintics are "solvable in radicals." Finally, we will see how to approximate a real root of a quintic.

6.1.1 Factoring

Factoring a polynomial into the product of smaller expressions is a staple of high school algebra. Its utility in solving polynomial equations rests on the factor theorem, which says that a is a root of the polynomial equation if and only if the linear factor $(x - a)$ divides the polynomial. So given a polynomial equation of arbitrary degree, if we can somehow see how to tease a linear factor out of the expression, we immediately have a root. Then we can explicitly divide out the linear factor and we have at least reduced the degree of the equation by one. That may unlock another root if we can repeat the factorization, or perhaps lower the degree enough that a suitable formula can give us all of the remaining roots at once. Trying to factor a polynomial is a low-level approach to problem solving and is often just a time-waster. Unless the polynomial has degree two or three, and a linear factor just jumps out at you, there is no shame in appealing to a prepackaged formula. That being said, for the sake of completeness, we should review some helpful tips.

Table 6.1: Some standard factorizations.

1) $x^2 - a^2 = (x + a)(x - a)$
2) $x^2 + a^2$ does not factor over $\mathbb{R}$; however, $x^2 + a^2 = (x + ia)(x - ia)$
3) $x^3 - a^3 = (x - a)(x^2 + ax + a^2)$
4) $x^3 + a^3 = (x + a)(x^2 - ax + a^2)$
5) $x^4 - a^4 = (x^2 + a^2)(x^2 - a^2) = (x^2 + a^2)(x + a)(x - a)$
6) $x^4 + a^4$ does not factor over $\mathbb{R}$; however, $x^4 + a^4 = (x^2 + ia^2)(x^2 - ia^2)$
7) $x^n - a^n = (x - a)(\sum_{k=0}^{n-1} x^{n-1-k} a^k)$ for $n \in \mathbb{N}$
8) $x^n + a^n = (x + a)(\sum_{k=0}^{n-1} x^{n-1-k} (-a)^k)$ for odd $n \in \mathbb{N}$ only

The most common application of factoring is to quadratic equations. In Section 6.1.2 we present the quadratic formula, which can always be used to short-circuit any need for factoring quadratics.

Example 6.1. Solve $30x^2 + 49x + 20 = 0$ by factoring.

Example 6.2. Solve $18x^2 - 9x - 20 = 0$ by factoring.

Example 6.3. Solve $x^3 - x^2 + 3x - 3 = 0$ by factoring.

Example 6.4. Solve $x^4 + x^3 - 3x^2 - x + 2 = 0$ by factoring.

6.1.2 Linear and quadratic equations

A linear equation is one of the form $ax+b=c$, where a, b, and c are real constants and x is a variable to be determined. Constant a cannot be zero or the equation becomes $b=c$. However, b or c can be zero. The solution is to simply subtract b and divide by a (always possible since $a \neq 0$). The solution is then $x=\frac{c-b}{a}$.

A quadratic equation has the form $ax^2+bx+c=0$, where again a, b, and c are real constants ($a \neq 0$ or the quadratic disappears) and x is a variable to be determined. We assume any initial nonzero quantity on the right-hand side is immediately folded into c. The classic **quadratic formula** is derived as follows. Rewrite the equation as $ax^2+bx=-c$ and divide by $a \neq 0$. Now we have $x^2+\frac{b}{a}x=-\frac{c}{a}$ Next we "complete the square." This involves adding just the right constant to $x^2+\frac{b}{a}x$ to artificially make it a square. We can see that $(x+\frac{b}{2a})^2=x^2+\frac{b}{a}x+\frac{b^2}{4a^2}$, so apparently what we need to add is $\frac{b^2}{4a^2}$. So we add this to both sides and get $x^2+\frac{b}{a}x+\frac{b^2}{4a^2}=\frac{b^2}{4a^2}-\frac{c}{a}$. Rewriting the left-hand side as a square, we have $(x+\frac{b}{2a})^2=\frac{b^2}{4a^2}-\frac{c}{a}$. Let us clean up the right-hand side by finding a common denominator. Now we have $(x+\frac{b}{2a})^2=\frac{b^2-4ac}{4a^2}$. Take the square root of both sides to get $x+\frac{b}{2a}=\frac{\pm\sqrt{b^2-4ac}}{2a}$. Now isolate $x=-\frac{b}{2a}\frac{\pm\sqrt{b^2-4ac}}{2a}=\frac{-b\pm\sqrt{b^2-4ac}}{2a}$. The quantity $\Delta=b^2-4ac$ is called the (quadratic) **discriminant** and characterizes the solution(s). If $b^2-4ac>0$, there are two real solutions corresponding to $\frac{-b\pm\sqrt{b^2-4ac}}{2a}$. If $b^2-4ac=0$, there is one real solution $-\frac{b}{2a}$ (a double root). And if $b^2-4ac<0$, there are two complex solutions corresponding to $\frac{-b\pm i\sqrt{4ac-b^2}}{2a}$. The complex roots will be conjugates of one another. A **conjugate** pair of complex numbers are of the form $z=a+bi$ and $\bar{z}=a-bi$, where $i=\sqrt{-1}$. Note that $z\bar{z}=(a+bi)(a-bi)=a^2-(bi)^2=a^2+b^2$.

Table 6.2: Linear & quadratic formulas.

1) If $ax+b=c$, then $x=\frac{c-b}{a}$
2) If $ax^2+bx+c=0$, then $x=\frac{-b\pm\sqrt{b^2-4ac}}{2a}$

Another approach to the solution of the quadratic $ax^2+bx+c=0$ is to divide by the leading coefficient a to create a **monic** polynomial (leading coefficient 1), $x^2+\frac{b}{a}x+\frac{c}{a}=0$. Substitute $p=\frac{b}{a}$ and $q=\frac{c}{a}$ and rewrite it as $x^2+px+q=0$. Now suppose we have, in principle, the roots r_1 and r_2. Recall the factor theorem, which says that if r is a zero of the polynomial $P(x)$, then $P(x)=(x-r)Q(x)$ for some polynomial $Q(x)$. Since a quadratic can only have two roots, it follows that $x^2+px+q=(x-r_1)(x-r_2)=0$. But $(x-r_1)(x-r_2)=x^2-(r_1+r_2)x+r_1r_2$. Evidently, $-p=r_1+r_2$ and $q=r_1r_2$. Note that $(r_1-r_2)^2=(r_1+r_2)^2-4r_1r_2$. It is clear that this expression together with r_1+r_2 and r_1r_2 are **symmetric** expressions in the two roots. This means we could interchange the two roots and it would not change the value of the expressions. Also r_1+r_2 and r_1-r_2 are called the Lagrange **resolvents** for the equation. If we have the resolvents, it will be easy to recover the two roots. At this point, we know $r_1+r_2 = -p$ and evidently $r_1-r_2 = \pm\sqrt{p^2-4q}$. We choose the positive square root, which breaks the symmetry and allows us to write

$$2r_1 = (r_1+r_2)+(r_1-r_2) = -p+\sqrt{p^2-4q}, \quad \text{so} \quad r_1 = \frac{-p+\sqrt{p^2-4q}}{2}.$$

Likewise,

$$2r_2 = (r_1+r_2)-(r_1-r_2) = -p-\sqrt{p^2-4q}, \quad \text{so} \quad r_2 = \frac{-p-\sqrt{p^2-4q}}{2}.$$

Reversing our substitutions,

$$r_1 = \frac{-\frac{b}{a}+\sqrt{(\frac{b}{a})^2-\frac{4c}{a}}}{2} = \frac{-b+\sqrt{b^2-4ac}}{2a} \quad \text{and} \quad r_2 = \frac{-b-\sqrt{b^2-4ac}}{2a},$$

agreeing with our derivation via completing the square.

Finally, a method known to the ninth century Indian mathematician Śrīdhara recovers the quadratic formula very efficiently. We start by noting that

$$(2ax+b)^2 = 4a^2x^2+4abx+b^2.$$

Compare this to the original equation $ax^2+bx+c=0$ multiplied by $4a$. We get $4a^2x^2+4abx+4ac = 0$. We can turn this into the former expression by subtracting $4ac$ and adding b^2. Apparently $4a^2x^2+4abx+4ac+(b^2-4ac) = (b^2-4ac)$ implies $(2ax+b)^2 = b^2-4ac$. Taking square roots, we have $2ax+b = \pm\sqrt{b^2-4ac}$. It follows that $x = \frac{-b\pm\sqrt{b^2-4ac}}{2a}$.

Example 6.5. The fixed cost for a business is \$5000 per month. Revenue is \$12,000. If 1000 units of product are produced per month, what is the highest cost per unit that can be sustained so that the business breaks even?

Example 6.6. A rectangular enclosure is to be fenced off so that it contains 2500 square meters. The owner wants the length to be 15 meters longer than the width. What are the appropriate dimensions?

Example 6.7. The ancient Greeks thought that the most pleasing proportions for a rectangular building resulted if the ratio of length to width was the same as the ratio of width to length minus width; see Fig. 6.1. Find the ratio.

Letting the width be 1 unit, we have the quadratic equation $L^2-L-1=0$. Plugging in to the quadratic formula, we find $L = \frac{1\pm\sqrt{5}}{2}$. We want the length to properly be longer than the width, so we choose $L = \frac{1+\sqrt{5}}{2} \approx 1.618$. The ratio 1.618 to 1 is called the Golden

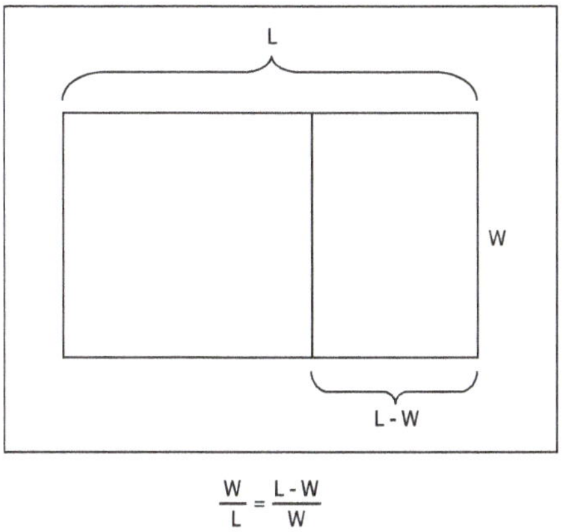

Figure 6.1: A simple rectangular floor plan of an ancient Greek building where the length to width ratio is the "Golden Ratio" 1.618 to 1.0.

Ratio and abbreviated by the Greek letter ϕ. It pops up in completely unrelated contexts, like the limiting ratio of successive Fibonacci numbers.

Example 6.8. There is a phenomenon called catastrophic cancellation that can occur with the quadratic formula when $\sqrt{\Delta}$ is real and close to $|b|$. If $b > 0$, then $-b + \sqrt{\Delta} \approx 0$ and if $b < 0$, then $-b - \sqrt{\Delta} \approx 0$. Although the formula is perfectly correct in theory, using a calculating device or program for a real world approximation can lead to imprecision in computing the root that is smaller in absolute value. The root that is larger in absolute value is not similarly affected. Solve $x^2 - 2200x + 1 = 0$ and compare accuracy of the roots using a pocket calculator and a CAS.

Example 6.9. Suppose we have two sets of numbers

$$A = \{a_1, a_2, \ldots, a_k, \ldots a_n\} \quad \text{and} \quad B = \{b_1, b_2, \ldots, b_k, \ldots b_n\}.$$

Use the quadratic formula to show that $(\sum_{k=1}^n a_k^2) \cdot (\sum_{k=1}^n b_k^2) \geq (\sum_{k=1}^n a_k b_k)^2$.

Example 6.10. Express $144x^2 + 17x - 36 = 0$ in factored form.

6.1.3 Cubic equations

One of the special cases of the cubic equation that was solved prior to Cardano's breakthrough with the general cubic was an equation of the form $x^3 + px + q = 0$, known as a so-called **depressed cubic**, where the quadratic term is missing. The general cubic equation looks like this: $a'x^3 + b'x^2 + c'x + d' = 0$ and it can always be rendered monic by dividing out the leading coefficient. So let us do that and relabel the coefficients without primes as $x^3 + bx^2 + cx + d = 0$. Cardano was led to the solution of the general cubic equation by eventually realizing that the quadratic term could always be eliminated by substituting $x = y - \frac{b}{3}$. The result is

$$\left(y - \frac{b}{3}\right)^3 + b\left(y - \frac{b}{3}\right)^2 + c\left(y - \frac{b}{3}\right) + d = 0.$$

Expanding this expression, we get

$$\left[y^3 - 3y^2\left(\frac{b}{3}\right) + 3y\left(\frac{b}{3}\right)^2 - \left(\frac{b}{3}\right)^3\right] + b\left[y^2 - 2y\left(\frac{b}{3}\right) + \left(\frac{b}{3}\right)^2\right] + c\left[y - \frac{b}{3}\right] + d = 0.$$

Note that $-3y^2(\frac{b}{3})$ in the first bracketed term cancels by^2 in the second, and the quadratic variable is completely eliminated. Cleaning up the remaining terms, we can rewrite the cubic equation as

$$y^3 + \left(\frac{b^2}{3} - \frac{2b^2}{3} + c\right)y + \left(-\frac{b^3}{27} + \frac{b^3}{9} - \frac{bc}{3} + d\right) = 0$$

or

$$y^3 + \left(-\frac{b^2}{3} + c\right)y + \left(\frac{2b^3}{27} - \frac{bc}{3} + d\right) = 0.$$

Every solution y to the depressed cubic implies a solution x to the original cubic. Identifying $(-\frac{b^2}{3} + c)$ as p and $(\frac{2b^3}{27} - \frac{bc}{3} + d)$ as q and reverting to x as the variable, we see the general cubic equation $x^3 + bx^2 + cx + d = 0$ can always be worked with in the depressed form $x^3 + px + q = 0$.

This idea of eliminating some of the intermediate coefficients in a polynomial equation by a suitable transformation was exploited in 1683 by the German mathematician (and coincidentally inventor of porcelain) Ehrenfried von Tschirnhaus. He developed a substitution that simultaneously eliminated the coefficients of the x^{n-1} and x^{n-2} terms in a polynomial equation of degree $n > 2$. The hope was that the depressed equation could be solved in terms of the transformed variable, and then applying the inverse transformation would yield a solution in terms of the original variable. In 1834 the British mathematician George Jerrard took it one step further and discovered a Tschirnhaus-type transformation that got rid of the x^{n-1}, x^{n-2}, and x^{n-3} terms. Jerrard was motivated by his skepticism over Abel's then recent demonstration that the general quintic equation could not be solved in radicals. If he could have extended his method to also eliminate the linear term, he would have solved the general quintic and become a legend. But, alas, that never happened because the Abel–Ruffini theorem was valid. Jerrard's name is honored in the so-called Bring–Jerrard normal form of the reduced quintic $x^5 + \alpha x + \beta = 0$. Erland Bring was a Swedish mathematician who slightly predated Jerrard and worked on the same issues. Incidentally, it was discovered early on that trying to depress an already depressed equation resulted in the x^{n-1} term reappearing even though the x^{n-2}

term was eliminated. The value of the von Tschirnhaus and Bring–Jerrard transformations was that they eliminated multiple powers of x in one step.

One modern strategy even better than a von Tschirnhaus substitution is based on the fact that the graph of an odd degree polynomial $P(x)$ must cross the real axis somewhere due to lead term dominance and continuity. So for cubics and quintics, a graphing calculator can always display to a desired degree of accuracy one guaranteed real root r. Then dividing $P(x)$ by the factor $(x - r)$ creates a complementary factor of lower degree which may be easier to resolve into its roots.

Example 6.11. Convert $2x^3 - 18x^2 + 10x - 12 = 0$ to a depressed equation.

Example 6.12. Find a root of $x^3 + 6x - 20 = 0$. We will introduce Tartaglia's method as we go along.

Example 6.13. Following the special case in Example 6.12 as a model, find a root of $x^3 + px + q = 0$.

Example 6.14. Validate the claim in Example 6.13 that besides $x = u + v$, the remaining solutions to the cubic $x^3 + px + q = 0$ are $\omega u + \omega^2 v$ and $\omega^2 u + \omega v$, where $\omega = \frac{-1+i\sqrt{3}}{2}$.

Example 6.15. Find the other two solutions to the cubic equation in Example 6.12.

Example 6.16. Use the method in Example 6.13 to find all the roots of $x^3 + 24x - 56 = 0$.

Example 6.17. Develop an analog to the quadratic determinant $b^2 - 4ac$ that applies to cubic equations and characterizes cubic roots.

Example 6.18. Solve the cubic $x^3 - \pi x^2 + (e + 1)x - \pi(e + 1) = 0$.

Example 6.19. Solve the cubic $x^3 - 3.95x^2 + 8.345x - 7.5075 = 0$.

6.1.4 Quartic equations

As we mentioned above, Lodovico Ferrari solved the general quartic equation soon after the cubic was "cracked." Cardano published both the cubic and quartic formulas in his Ars Magna in 1545. The quartic solution relies heavily on the cubic solution. The general quartic $a'x^4 + b'x^3 + c'x^2 + d'x + e' = 0$ can immediately be simplified by dividing across-the-board by a'. The resulting monic polynomial $x^4 + bx^3 + cx^2 + dx + e = 0$ can be reduced further by eliminating the cubic term by a suitable transformation $x = y - \frac{b}{4}$. The calculation parallels that which we did for the cubic above. We have

$$\begin{aligned}
&\left(y-\frac{b}{4}\right)^4 + b\left(y-\frac{b}{4}\right)^3 + c\left(y-\frac{b}{4}\right)^2 + d\left(y-\frac{b}{4}\right) + e \\
&= \left[y^4 - 4y^3\left(\frac{b}{4}\right) + 6y^2\left(\frac{b}{4}\right)^2 - 4y\left(\frac{b}{4}\right)^3 + \left(\frac{b}{4}\right)^4\right] \\
&\quad + b\left[y^3 - 3y^2\left(\frac{b}{4}\right) + 3y\left(\frac{b}{4}\right)^2 - \left(\frac{b}{4}\right)^3\right] \\
&\quad + c\left[y^2 - 2y\left(\frac{b}{4}\right) + \left(\frac{b}{4}\right)^2\right] + \left[y - \frac{b}{4}\right] + e.
\end{aligned}$$

Note that $-4y^3(\frac{b}{4})$ is canceled by by^3 and the cubic terms have disappeared. We will skip the straightforward, but tedious, step of rewriting the polynomial in y in standard form. Let us revert to using x as the variable and *redefine* a, b, and c as the coefficients in the depressed quartic $x^4 + ax^2 + bx + c = 0$. The depressed quartic lacks, of course, the cubic term, and this suggests trying to make a perfect square out of the fourth degree and second degree terms.

So given $x^4 + ax^2 + bx + c = 0$, we can immediately write $x^4 + 2ax^2 + a^2 = (x^2 + a)^2 = -bx - c + ax^2 + a^2$. If the right-hand side were a perfect square, we could take square roots and be well on our way to a solution. But we have no basis to assume that it is. Drawing from our experience with the cubic, we introduce a degree of freedom and try to force the right-hand side to be a square. Consider the term $(x^2+z+a)^2 = (x^2+a)^2+2z(x^2+a)+z^2$. Here we have introduced z with the intention of giving it a value later that will allow us to take square roots. Evidently, we have $(x^2 + z + a)^2 = 2z(x^2 + a) + z^2 - bx - c + ax^2 + a^2$. The right-hand side, as a standard quadratic in x, is $(2z + a)x^2 - (b)x + (2az + z^2 - c + a^2)$. This will be a perfect square of the form $(\alpha x + \beta)^2$ provided its discriminant $b^2 - 4 \cdot (2z + a) \cdot (2az + z^2 - c + a^2) = 0$. But this is a cubic equation in z, namely $8z^3 + 20az^2 + z(16a^2 - 8c) + (4a^3 - 4ac - b^2) = 0$. Let z_0 be a real root (recall there is always at least one real root) of this so-called **resolvent cubic**, then we can write $(x^2 + z_0 + a)^2 = (\alpha x + \beta)^2$, where $\alpha = \sqrt{2z_0 + a}$ and $\beta = \pm\sqrt{2az_0 + z_0^2 - c + a^2}$. The sign of β here (denoted by the function $\operatorname{sgn}(\beta)$) must be taken opposite to that of b to ensure $-b = 2\alpha\beta$. Upon taking square roots, we have $x^2 + z_0 + a = \pm(x\sqrt{2z_0 + a} - \operatorname{sgn}(b) \cdot \sqrt{2az + z_0^2 - c + a^2})$. Finally, we seek solutions of $x^2 \pm x\sqrt{2z_0 + a} + (z_0 + a \pm (-\operatorname{sgn}(b))\sqrt{2az_0 + z_0^2 - c + a^2}) = 0$. In general, these two quadratics will spin off two solutions each, and there will be four solutions to the original quartic.

Example 6.20. Solve $x^4 - 10x^2 + 4x + 8 = 0$.

Example 6.21. Here is an original problem from Cardano's Ars Magna Chapter XXXIX (Problem VI): Find a number which is equal to its square root plus twice its cube root.

Some special cases of the quartic equation allow for simpler solutions than the general method of Example 6.20.

Example 6.22. Solve the quartic equation $x^4 + 5x^3 + 8x^2 + 5x + 1 = 0$. This was posed by Leonhard Euler (1770).

Example 6.23. Solve the quartic in Example 6.22 by using the substitution $z = x + \frac{1}{x}$.

Example 6.24. Generalize the method in Example 6.23 to handle the **quasisymmetric** equation $ax^4 + bx^3 + cx^2 + mbx + m^2a = 0$.

Example 6.25. Solve the quartic $x^4 - 13x^2 + 36 = 0$.

6.1.5 Quintic equations

For general polynomial equations of degree five and higher, we no longer have the iron-clad assurance that a solution in radicals is going to be available. Recall this issue was settled by the Abel–Ruffini theorem in 1824. For all polynomials of lower degree, we were able to develop formulas that guaranteed a complete solution, even if some of the formulas tended to be complicated and tedious to apply. For quintics, the situation is more nuanced. If a polynomial is **reducible** over a given field of coefficients, usually the rational numbers $\mathbb{Q}$, it means the polynomial can be factored into two polynomials of lesser positive degree. If a quintic is reducible and the factorization can be found, our methods above would be adequate to separately find the roots of the two factors and hence all the roots of the quintic. If a quintic is **irreducible** over $\mathbb{Q}$, it means no such factorization is possible. In this case there is still a chance that, the Abel–Ruffini theorem notwithstanding, the polynomial roots can be expressed in radicals. Given a specific quintic, there is a test involving functions of its coefficients to determine if it is solvable in radicals. The computation involved is not for the faint of heart. We will discuss this somewhat superficially below. In the list of references for this chapter, we highlight one in particular that explores this subject in proper detail. The first order of business is to decide if a quintic can be factored into two polynomials of lesser degree. Here are two tests for irreducibility which apply to polynomials of any degree.

Example 6.26. Decide if the quintic $3x^5 + 6x^4 - 10x^3 + 4x^2 + 6x + 2$ is reducible.

Example 6.27. Decide if the quintic $9x^5 + 16x^4 - 7x^3 + 6x^2 + 4x + 1$ is reducible.

Example 6.28. Decide if the quintic $x^5 - x^4 - 2x^3 + 2x^2 + x - 1$ is reducible.

Example 6.29. Decide if the quintic $\frac{3}{7}x^5 + \frac{6}{7}x^4 + \frac{1}{5}x^3 - \frac{2}{7}x + \frac{3}{5}$ is irreducible.

The two preceding tests can determine if we have an irreducible quintic. There are two possibilities for such a polynomial. Either it is solvable in the sense that we defined above, where its roots can be written as algebraic functions, including radicals, of its coefficients, or it is not. We emphasize again that the Abel–Ruffini theorem does not say *every* quintic is unsolvable in radicals. The British mathematician Arthur Cayley devised a test, based on the then novel group theory ideas of Galois, which are far beyond the scope of our discussion. Cayley's test decides the solvable/unsolvable issue dispositively since it is an "if and only" if statement. It is not an exaggeration to say that even if we know a quintic is solvable, the methods to actually find the roots are disappointingly complicated. They are a tribute to the ingenuity of their developers and bring a certain degree of closure to the search, but they tend to be so long and involved that they have limited practical value. Eventually, we will explore a more practical way to solve quintics, not with an algebraic approach, but rather with a numerical approximation method. For the moment, we will have to be content with answering the solvable/unsolvable question.

So **Cayley's test** starts with a depressed quintic, where the quartic term is removed by a von Tschirnhaus transformation. Suppose we have $x^5 + px^3 + qx^2 + rx + s$. First, calculate the following cubic function

$$\begin{aligned}P(z) = z^3 - z^2(20r + 3p^2) - z(8p^2r - 16pq^2 - 240r^2 + 400sq - 3p^4)\\ + (p^6 + 28p^4r - 16p^3q^2 - 176p^2r^2 - 80p^2sq + 224prq^2 - 64q^4 + 4000ps^2\\ + 320r^3 - 1600rsq).\end{aligned}$$

Now calculate the discriminant

$$\begin{aligned}\Delta = -128p^2r^4 + 3125s^4 - 72p^4qrs + 560p^2qr^2s + 16p^4r^3 + 256r^5 + 108p^5s^2 - 1600qr^3s\\ + 144pq^2r^3 - 900p^3rs^2 + 2000pr^2s^2 - 3750pqs^3 + 825p^2q^2s^2 + 2250q^2rs^2 + 108q^5s\\ - 27q^4r^2 - 630pq^3rs + 16p^3q^3s - 4p^3q^2r^2.\end{aligned}$$

This is complicated, but apparently not complicated enough. We now form the **Cayley resolvent** $R(z) = P^2(z) - 1024z\Delta$. This is a polynomial that requires about 12 full lines of "code" to express. We will save the reader further eyestrain by not expanding it in detail. Instead, we will use below a vastly simplified version of the resolvent that is available when $p = q = 0$ in the depressed quintic.

To summarize, Cayley's test states that the depressed quintic $x^5 + px^3 + qx^2 + rx + s$ is solvable in radicals if and only if (i) it is reducible (in which case our earlier formulas for quartics and below would give the roots), or (ii) the Cayley resolvent $R(z)$ has some root $z_0 \in \mathbb{Q}$. If such a z_0 can be found, then the roots of the quintic can be expressed in terms of radicals of the coefficients and z_0, although this assurance does not include a method of finding them. Obviously, if the depressed quintic is solvable, so is the original quintic from which it was formed. More obviously, $R(z)$ is a sixth degree polynomial, so we are seemingly faced with a task at least as difficult as solving the quintic. But this turns out to be an illusion. All we really have to do is confirm or deny that the Cayley resolvent $R(z)$ has a rational root. Here is a simple criterion that we can apply to $R(z)$, or in fact a polynomial of any degree, to make this determination.

Example 6.30. Show that if the polynomial $P(x) = a_n x^n + a_{n-1}x^{n-1} + \cdots + a_1 x + a_0$ has a rational root $\frac{p}{q}$ where $a_i, p, q \in \mathbb{Z}$ and $\gcd(p, q) = 1$, then p must divide a_0 and q must divide a_n.

Example 6.31. Find all rational roots, if any, of $P(x) = 3x^4 - 5x^3 + 2x^2 - 4x + 6 = 0$.

Example 6.32. Find all rational roots, if any, of $P(x) = 2x^5 - 4x^4 - 3x^3 + 7x^2 + 2x - 8 + 0$.

Returning to the Cayley's test, there is a clever sequence of two von Tschirnhaus type substitutions, first a quadratic and then a quartic, discovered by Erland Bring and George Jerrard that transform an arbitrary depressed quintic into so-called **Bring–Jerrard normal form** with the x^4, x^3, and x^2 terms removed. Applying Cayley's test to a quintic in this form is much simpler. The resolvent for $x^5 + ax + b$ simplifies to

$$R(z) = z^6 + 8az^5 + 40a^2z^4 + 160a^3z^3 + 400a^4z^2 + (512a^5 - 3125b^4)z - 9375ab^4 + 256a^6.$$

Let us use it on some very simple cases.

Example 6.33. Suppose a quintic equation has been reduced to $x^5 - x - 1 = 0$. Decide if it is solvable in radicals.

Example 6.34. Suppose a quintic equation has been reduced to $x^5 + x \pm 1 = 0$. Decide if it is solvable in radicals.

Example 6.35. Suppose a quintic equation has been reduced to $x^5 - 5x + 12 = 0$. Decide if it is solvable in radicals.

A **Bring radical**, or **ultraradical**, is the unique real root of $x^5 + x + a = 0$, denoted $BR(a)$. There are methods to find this numerically. In Example 6.34 we confirmed that to express a root of $x^5 + x + 1 = 0$, we need to have $BR(1)$. If a quintic is put in Bring–Jerrard normal form and with a further substitution expressed as $x^5 + x + a = 0$, then the overall solution to the quintic can be written in terms of $BR(a)$ and other regular radicals. If we stretch the definition of radical to include Bring radicals, then we can legitimately say that the general quintic can be solved in closed form with “radicals.” We are thankful for CAS.

6.1.6 Recap for polynomials

We have developed methods to do the following:

1) Solve linear equations
2) Solve quadratic equations
3) Solve cubic equations
4) Solve quartic equations
5) Determine if a polynomial is irreducible (factorable) over the rationals using the Eisenstein criterion
6) Determine if a polynomial is irreducible over the rationals using the mod p test
7) Determine if a polynomial has a rational root (and hence a factor) using the rational root test
8) Determine if an irreducible quintic is solvable in radicals using the Cayley’s test

Items 5 through 7 are applicable to polynomials of any degree. Here is an easily applied test for polynomials of any degree that roughly tells us the character of roots with respect to sign.

Example 6.36. Use **Descartes’ rule of signs** to characterize the real roots of $P(x) = x^5 - 6x^4 + 12x^3 - 12x^2 + 11x - 6$.

6.2 Exponential equations

Recall from Section 0.2.2 the basic properties of exponents. A simple exponential equation looks like this: $b^x = b^y$, where $b \in \mathbb{R}$, $b > 0$ and $b \neq 1$ is a base, and x and y are powers or exponents. Since real exponentiation is an injective map, the only way b^x can equal b^y is for x to equal y. Obviously, it is true vice versa. The first step in applying this principle is to make sure the two bases are the same, even if they initially look different. Then equate the exponents and solve for the unknown.

Example 6.37. Solve the following exponential equation for x: $4^x = 8^3$.

Example 6.38. Solve the following exponential equation for x: $2^x = (\sqrt{8})^5$.

Example 6.39. Solve the following exponential equation for x: $4^{3x} = \sqrt[3]{32}$.

Example 6.40. Solve the following exponential equation for x: $9^{2x} = 27^{x+1}$.

Example 6.41. Solve the following exponential equation for x: $(\frac{1}{49})^{3x} = 243$.

Example 6.42. If $A = 3^x$ and $B = \sqrt{A^3}$, express B in terms of x.

If the rate of change of a quantity is directly proportional to the amount of the quantity present, then the growth or decline of the quantity over time or space will exhibit exponential behavior. This is true in several common situations. The amount of money on deposit in an interest bearing account grows according to the interest rate applied to the principal at a given time. The constant of proportionality between the dollars of interest paid and the principal on deposit earning that interest is the interest rate. The amount of a radioactive substance lost due to spontaneous decay at a given instant is proportional to the amount of substance present at that instant. The constant of proportionality is the characteristic decay rate for that isotope.

Example 6.43. The **PERT** formula is often used in finance to relate the present P and future F value of monetary amounts given a time interval t and interest rate r. The formula states that $F = Pe^{rt}$. How much will \$1000 be worth in 8 years at the continuous compounding rate of 6 % per annum?

Example 6.44. How much would have to be invested at $7\frac{3}{4}$ % continuous interest to grow to \$25, 000 in 15 years?

Example 6.45. How much money would have to be put aside today at 6 % continuous interest to support a withdrawal of \$5000 at the beginning of each year in perpetuity?

Example 6.46. In a particular lake divers have noticed that the underwater light intensity decreases 3 % per meter of depth. Find a formula that expresses this fact and compute the light intensity at 25 meters compared to that at the surface.

Example 6.47. One kilogram of a radioactive isotope decays at the rate of 0.5 % per year into a non-radioactive form. How much would be left after 75 years?

In 1949 a method for estimating the age of various carbon-containing artifacts was developed by the chemist Willard Libby at the University of Chicago. It is based on the fact that a tiny fraction of the carbon in the environment is carbon-14 (^{14}C), a weakly radioactive cousin of carbon-12, which accounts for most of the carbon in our world. ^{14}C is actually created in the upper atmosphere from nitrogen that is bombarded by cosmic rays. It eventually combines with oxygen to form carbon dioxide that is incorporated into living things. As long as a living thing remains living, it participates in the carbon cycle and the relative amount of ^{14}C it contains remains static. Once it dies, its ^{14}C is no longer replaced and what is left begins to decay (in fact, it reverts to nonradioactive nitrogen). The rate of decay is constant and by comparing the concentration of ^{14}C in a sample from a dead organism with a sample from a living version of that same organism, an estimate of how long ago the organism died can be obtained. This phenomenon follows the radioactive decay equation presented in Example 6.47 with $\lambda = 0.00012$.

Example 6.48. A sample of oak taken from a Viking ship that sank in 950 CE is analyzed for ^{14}C. What would be the expected amount of ^{14}C as a fraction of that in a modern sample of the same type of oak?

Example 6.49. When the rate of change of a quantity is directly proportional not to the amount present, but rather the difference between the amount present and some maximum possible amount, the growth of the quantity follows the so-called diminishing returns **learning curve**. If $f(t)$ is the amount of such a quantity at time t, and the maximum possible amount of the quantity is M, then $f(t) = M(1 - e^{-kt})$, for some constant k. A typical example is the rate of learning a script for a play. When an actor begins to memorize a script he has $M = 100\,\%$ to learn and his efficiency is high. As he learns more and more of the script he has to keep organized what he already has memorized and connect it with the new material. His rate of acquisition of additional material declines. Suppose it takes the actor one week to memorize 20 % of a play. When can we expect him to have 80 % memorized?

Example 6.50. The **logistic curve** describes a growth pattern where the rate of change of a quantity is directly proportional to both the amount present and the difference between the amount present and some maximum possible amount. In a sense it is a blending of exponential growth and learning curve growth. If $f(t)$ is a function of time t growing logistically, then $f(t) = \frac{M}{1+Be^{-kMt}}$, where M is the maximum amount of $f(t)$, with k and B both constants determining the shape of the growth curve. A typical situation is a fixed group of 50 people at a cocktail party that are circulating freely and talking to one another. Initially two of the people have some juicy gossip between them and as they encounter other guests they spread the gossip by word of mouth. If after 10 minutes 20 people have heard the gossip (including the original two), when might we expect 40 people to know? Intuitively, we would expect the propagation of the gossip to depend jointly on the subpopulation of partygoers "in the know" and on the subpopulation that do not know yet but are available to be told. If nobody except a few people know, then the gossip spreads slowly because the chance that a conversational encounter involves a spreader is low. Conversely, if almost everyone knows, the chance that a conversational encounter involves someone who does not know is also low. The rate of spread is highest when the two subpopulations are equal and there are lots of productive encounters. The logistic equation is sometimes referred to as the ver Hulst equation, after the nineteenth century Belgian demographer who first developed it as a model to predict the population of the Unites States.

Example 6.51. The hyperbolic functions

$$\sinh x = \frac{e^x - e^{-x}}{2} \quad \text{and} \quad \cosh x = \frac{e^x + e^{-x}}{2},$$

together with their reciprocals and ratios, are named by analogy with the six trigonometric functions. Derive a relationship between $\sinh x$ and $\cosh x$ that resembles the Pythagorean relationship between $\sin x$ and $\cos x$.

Example 6.52. Derive a formula for $\sinh(x + y)$ that resembles the addition formula for $\sin(x + y)$.

Example 6.53. Two vertical poles separated by four feet have a rope hanging between them attached at 13.65 feet above the ground to each pole. The rope at the midpoint between the poles is half a foot off the ground. How high is the rope one foot from the midpoint?

Example 6.54. If $y = \cosh x$, find x in terms of y.

6.3 Logarithmic equations

Since logarithms are just special exponents, we would expect logarithmic equations to behave somewhat like exponential equations. If $\log_b A = \log_b B$, then $b^{\log_b A} = b^{\log_b B}$ implies $A = B$, and vice versa. Often it is possible to eliminate logarithms by exponentiating both sides of an equation. Conversely, in some situations we can get rid of exponents by taking logs.

Example 6.55. Given $\log_b(4^{2x}) = \log_b 8$, solve for x.

Example 6.56. Find a formula that relates $\log_b A$ to $\log_c A$, where c is also an acceptable base and convert a common log to a natural log.

Example 6.57. Given that $\ln 2 \approx 0.69315$, estimate $\log 2$.

Example 6.58. Prove the **exchange formula**, $A^{\log_B C} = C^{\log_B A}$.

Example 6.59. Given $8 = 1000^x$, solve for x.

Example 6.60. Find x if $27^{2x} = 16^{x+1}$.

Example 6.61. Given $9^{\log_3 x} = 5x - 6$, solve for x.

Example 6.62. What is logarithm of 36 to the base $6^{\sqrt{2}/3}$?

Example 6.63. If $y = \frac{\log_x 3}{\log_{x^2} 3}$, find y numerically.

Example 6.64. Given $x^{\log 9} + x^{\log 15} = x^{\log 25}$, solve for $x \neq 0$. This is a tricky problem.

Example 6.65. The Weber–Fechner law is a psychological model of how we perceive the intensity of various stimuli. It postulates that $p = \alpha \ln(\frac{S}{S_0})$, where p is the perception of a stimulus level S relative to some base level S_0 and α is a constant. The law is not as exact as that for radioactive decay or light diminution because it subsumes a lot of variable human factors, but it is roughly accurate in the midrange of stimulus ratios. Compare the ratio of perceptions of lifting one pound and then an extra pound versus lifting ten pounds and then an extra pound.

Example 6.66. The Richter scale for measuring the severity of earthquakes is based on the maximum movement A of a seismograph needle compared to a base movement A_0. All measurements are in microns [10^{-6} meter] and $A_0 = 1$ micron. The Richter number is $M_L = \log(\frac{A}{A_0})$. The movement of the seismograph needle is a proxy for the energy released in the earthquake. How far would the needle move during a 5.5 earthquake and how much energy does this represent compared to a 3.2 earthquake?

Example 6.67. Find a simple formula for the doubling time of an investment. This is the time t for \$1 to grow to 2\$ at a continuous rate of $100r$% per annum.

Example 6.68. In the 1960s, Yale University purchased an historical relic called the Vinland Map, which was purportedly created in the fifteenth century and contained information about Viking explorations in North America during the eleventh century. The map raised suspicions about its true provenance, but it was not until 2018 that a scholarly consensus emerged that it was definitely a fake. Radiocarbon dating put the parchment on which the map was drawn around 1430. It was not uncommon for maps to be written on palimpsests in medieval times. These were parchments that had been scraped clean, recycled as it were, so the radiocarbon analysis did not disqualify the map as a fraud outright. Eventually, an analysis of the inks used showed that they were titanium-based and therefore modern. If a radiocarbon analysis of the parchment today showed a 7 % reduction in ^{14}C, is it reasonable to believe the parchment (but not the inks) had its origin about 1430?

Example 6.69. The half-life $t_{1/2}$ of a radioactive substance is the length of time required for a given amount of the substance to decay to half its original amount. Given the decay constant for ^{14}C is 0.00012, calculate its half-life.

6.4 Trigonometric equations

See Section 7 for a discussion of trigonometric functions and various useful formulas involving them.

Example 6.70. Given $f(x) = x + 2\cos x = 0$, show that it has a solution.

Example 6.71. Recall that there is a Pythagorean triple (5, 12, 13) since $5^2 + 12^2 = 13^2$. We could construct this triangle and verify that the angle θ opposite the side of length 5 had tangent $\frac{5}{12}$. Show that θ is twice the angle with tangent $\frac{1}{5}$ and therefore $\arctan\frac{1}{5}$ is a constructible angle (by bisecting θ). Hint: use the formula in Section 7.3.2 Example 7.28.

Example 6.72. Show that $\sin^4\theta - \cos^4\theta + 2\cos^2\theta = 1$.

Example 6.73. For a finite N, the sum $\sum_{n=0}^{N} r^n = \frac{1-r^{N+1}}{1-r}$ for $r \neq 1$. Use the fact that $e^{in\theta} = \cos n\theta + i\sin n\theta$ to find $\sum_{n=0}^{N}\sin n\theta$ and $\sum_{n=0}^{N}\cos n\theta$.

6.5 Inequations

Any equation can be turned into an **inequation**, or **inequality**, by changing the equals sign to an order symbol. Linear inequations are easy to manipulate with the rules outlined in Table 0.5. Let us focus on the more challenging inequations involving rational functions, or ratios of polynomials in a single variable. The general rule is to perform whatever operations are necessary to put a given inequation into the form of a single fraction either greater than or less than zero. If the numerator and denominator are expressible as products of linear factors or factors with a fixed sign, then it is apparent from sign parity considerations what values of the variable cause the inequation to be true or false. Some examples should make this clear.

Example 6.74. For what values of x is $\frac{1}{x} + x > 0$?

Example 6.75. For what values of x is $x^3 - 6x^2 + 11x - 6 < 0$?

Example 6.76. For what values of x is $\frac{x^2-4x+3}{x-2} < 0$?

Example 6.77. For what values of x is $\frac{x+1}{x-2} - 2 \geq 0$?

Example 6.78. For what values of x is $\frac{x^2(x+1)}{x-2} \geq 0$?

Inequations involving absolute values are of two types: (i) if $|x| < a$ then $-a < x$ and $x < a$, and (ii) if $|x| > a$ then $x > a$ or $x < -a$. Clearly, "≤" and "≥" may be substituted appropriately in these expressions.

Example 6.79. For what values of x is $|\frac{x+1}{x-2}| > 1$?

Example 6.80. For what values of x is $|\frac{1}{x-4}| < 1$?

6.6 Approximation

The method is an iterative algorithm that Isaac Newton used to estimate the roots of polynomial equations. The basic idea is to make an informed guess x_1 as to where the graph of a polynomial function $y = P(x)$ crosses the x-axis, then "linearize" the curve by getting the equation of the tangent to the curve at x_1. Usually, the tangent crosses the x-axis at a point x_2 closer to the true root. Then the tangent at x_2 is used to find x_3, and so on. If all goes well, the method generates a sequence of approximations that converge to the true root. The algorithm can be applied to other functions which may have real zeroes, but this invites exposure to various pathologies. These issues include the difficulty of computing successive tangents, the rate of convergence of the iterates, and the possibility that the iterates oscillate or diverge. Newton himself seems to have used the method only for polynomial equations. His original method from about 1670 was cumbersome and involved more computations than the improved version published by Joseph Raphson in 1690. Whether or not we have credence in paleopsychology (personality reconstruction from extant writings), it is known that Newton was unusually capable of sustaining long periods of concentration on mind-numbing calculation. We should be thankful for that, but it has been proposed that he suffered from some degree of autism.

There is a bit of calculus involved in the full-blown presentation of the **Newton–Raphson** method, but as long as we are restricting it to the solution of polynomial equations, we can operate with a formula that has the calculus parts already included. Given the polynomial $P(x) = a_nx^n + a_{n-1}x^{n-1} + \cdots + a_1x + a_0$, we denote the slope of the tangent to the graph of $P(x)$ at any arbitrary point as $P'(x) = na_nx^{n-1} + (n-1)a_{n-1}x^{n-2} + \cdots + 2a_2x + a_1$. If you are familiar with calculus, this is the derivative of $P(x)$. We can evaluate $P(x)$ at various points along the real axis, and if we find two adjacent points where $P(x)$ changes sign, we are assured that a root of $P(x) = 0$ lies in the interval determined by those points. Then we pick an arbitrary x_1 in that interval and construct the equation of the tangent line to the graph of $y = P(x)$ at x_1. The slope of that tangent will be $P'(x_1)$ and it will go through the point $(x_1, P(x_1))$. From the point–slope form (see Section 4.7.1) of the straight line equation, we can write $\frac{y-P(x_1)}{x-x_1} = P'(x_1)$. Here (x, y) is an arbitrary point on the tangent line. We can put this in the form $y = P(x_1) + P'(x_1)(x - x_1)$. Now we ask where does the tangent cross the x-axis? When $y = 0$. Then $\frac{-P(x_1)}{P'(x_1)} = x - x_1$ or $x = x_1 - \frac{P(x_1)}{P'(x_1)}$. Call this value x_2. Repeat the tangent line construction at $(x_2, P(x_2))$, find where it crosses the x-axis and call that x_3. Continuing in this manner, we find $x_{n+1} = x_n - \frac{P(x_n)}{P'(x_n)}$. Of course, we may assume $P'(x_n) \neq 0$, since that would imply a horizontal tangent. The general scheme of the algorithm is depicted in Figure 6.2 below.

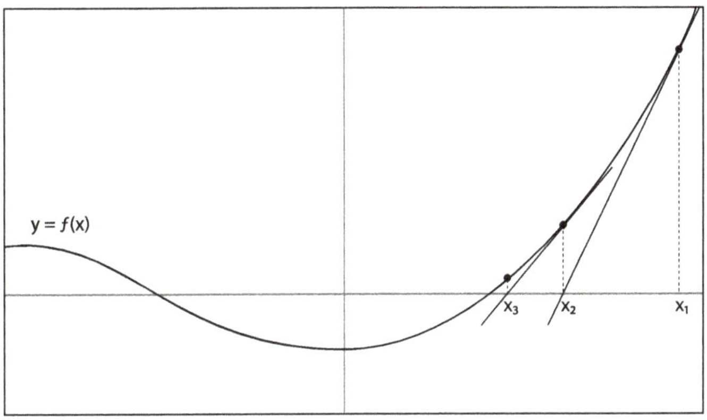

Successive tangents locate points converging to x where $f(x) = 0$

Figure 6.2: A concave piece of a typical polynomial graph crossing the x-axis at a root with three approximants generated by successive tangents given by the algorithm.

Example 6.81. Estimate $\sqrt{5}$ by using the Newton–Raphson method.

Example 6.82. What if we repeat Example 6.81 with a less cooperative initial value, say, $x = 1$.

Example 6.83. Recall from Section 6.1.5 that the Bring radical $BR(1)$ is the real root of $x^5 + x + 1 = 0$. As we know, odd degree polynomial equations always have a real root by lead term dominance. So let us try to find $BR(1)$.

Example 6.84. Recall that the Bring–Jerrard normal form of a quintic $x^5 + \alpha x + \beta = 0$ is the end result of applying several power-elimination transformations to an arbitrary general quintic expression. Estimate a root of the normal form where $\alpha = -5$ and $\beta = 3$.

References & suggested further reading

1) *Contemporary Abstract Algebra* (10th ed.) by Joseph Gallian (2021), Taylor & Francis – *very clear exposition of polynomial properties*
2) *Abstract Algebra* (3rd ed.) by David Dummit and Richard Foote (2004), Wiley – *comprehensive text, indispensable for the material on the quintic, especially Cayley's resolvent*
3) *Abstract Algebra* (4th ed.) by John Beachy and William Blair (2019), Waveland Press

Solutions to Chapter 6 examples

Example 6.1

We can write down trial (guess) factorizations and check them, but the number of factors of the coefficients make this difficult and time-consuming. Here is a reasonably efficient plan (sometimes called "no-fuss factoring"). Start by writing $(30x\quad)(30x\quad)$. Yes, this would cause the quadratic term to be $900x^2$, but do not worry, we will correct for that later. Now make the following list of possible divisors of the quadratic and constant coefficients. Since the linear and constant terms are positive, we need only consider positive divisors.

30	20
15	10
10	5
6	4
5	2
3	1
2	*
1	*

We create pairs of numbers, one from the first column and one from the second, and multiply them together. The object is to find two pairs whose products sum to 49, the linear coefficient. It may seem like there are too many pairs to easily check, but most of the pairs are easily disqualified due to size (or in some problems due to sign). We soon see that $5 \cdot 5 + 6 \cdot 4 = 49$. Now export $5 \cdot 5 = 25$ and $6 \cdot 4 = 24$ to the two incomplete parenthetical expressions: $(30x + 25)(30x + 24)$. The last step is to divide out from each factor the greatest common divisor (gcd) of its terms. For the first factor, the gcd of 30 and 25 is 5. So after dividing that out, we get $(6x + 5)$. Repeating this for the second factor we find the gcd of 30 and 24 is 6, and after dividing we get $(5x + 4)$. Checking $(6x + 5)(5x + 4) = 30x^2 + 24x + 25x + 20 = 30x^2 + 49x + 20$. So the linear factors set individually to zero give the two quadratic solutions $\{-\frac{5}{6}, -\frac{4}{5}\}$. Not too much fuss, really.

Example 6.2

The linear and constant terms are negative, so we have to consider both positive and negative divisors in our list:

18	–20
±9	±10
±6	±5
±3	±4
±2	±2
±1	±1

From a little experience with the method, we first focus on numbers in the middle of the list. It appears that $(5 \cdot 3) + (-6 \cdot 4) = 15 - 24 = -9$. So we write $(18x + 15)(18x - 24)$. Dividing out the highest common factors from each term, we finally get $(6x+5)(3x-4) = 18x^2+15x-24x-20 = 18x^2-9x-20 = 0$. Setting each factor separately equal to zero gives the solution set $\{-\frac{5}{6}, \frac{4}{3}\}$. You can always start with a quadratic where the coefficient of x^2 is positive by multiplying all terms by –1 if necessary.

Example 6.3

This example illustrates the method of factoring by grouping. Note that $x^3 - x^2 + 3x - 3 = x^2(x-1)+3(x-1)$. We have grouped the terms in a way that has produced the factor $(x-1)$ in two places. Factoring it out gives $x^3-x^2+3x-3 = (x-1)(x^2+3) = 0$. We immediately have the real root 1 and two conjugate imaginary roots $\pm i\sqrt{3}$. This cubic equation has been solved with about the least possible expenditure of effort. This situation does not happen often, but it is always worth a quick scan to see if factoring by grouping is available.

Example 6.4

This is a quartic equation, so we would expect a higher degree of difficulty than the previous example. The mechanism of factoring by grouping is fairly well hidden in this example. Since it is a quartic, we may suspect that it is a product of two quadratics. Which two is not obvious. Let us see if a very simple quadratic $x^2 - 2x + 1$ is hiding in there. Tentatively, we write $x^4+x^3-3x^2-x+2 = x^2(x^2-2x+1)+Ax(x^2-2x+1)+B(x^2-2x+1)$ and try to determine A and B. Now the first of these three terms has no parameter, so we are committed to $x^4 - 2x^3 + x^2$ being part of the right-hand side. Deduct it from the right- and left-hand sides to see where we stand. Now we have $3x^3 - 4x^2 - x - 2 = Ax(x^2 - 2x + 1) + B(x^2 - 2x + 1)$. It looks like we are forced to set $A = 3$. Now we have $3x^3 - 4x^2 - x + 2 = 3x^3 - 6x^2 + 3x + B(x^2 - 2x + 1)$. This simplifies to $2x^2 - 4x + 2 = B(x^2 - 2x + 1)$. This last equality is true if $B = 2$. Evidently, $x^4 + x^3 - 3x^2 - x + 2 = x^2(x^2 - 2x + 1) + 3x(x^2 - 2x + 1) + 2(x^2 - 2x + 1)$ and this simplifies to $x^4 + x^3 - 3x^2 - x + 2 = (x^2 + 3x + 2)(x^2 - 2x + 1) = 0$. Factoring the quadratics, we can go further and write $x^4 + x^3 - 3x^2 - x + 2 = (x + 2)(x + 1)(x - 1)^2 = 0$. So the quartic roots are –2, –1, and 1 as a double. This example was reverse-engineered to show the utility of factoring by grouping when it works. It would not be advisable to attack an arbitrary quartic this way unless there were some additional information or clues that validated the approach.

Example 6.5

Let x be the maximum unit cost to break even. We have $1000x + 5000 = 12000$ which reduces to $x = \frac{12000-5000}{1000} = \$7$ per unit.

Example 6.6

Let x be the width in meters. Then the length is $x + 15$. The area is 2500 square meters, so $x(x + 15) = 2500$. This gives $x^2 + 15x - 2500 = 0$. Plugging into the quadratic formula, we get

$$x = \frac{-15 \pm \sqrt{225 - 4 \cdot 1 \cdot (-2500)}}{2} = \frac{-15 \pm \sqrt{10225}}{2} = \frac{-15 \pm 101.1}{2},$$

so the roots are 43.05 and −58.05. We discard the negative root as being **spurious**. It satisfies the equation but has no physical significance. This often happens with quadratics due to the squaring operation. The final dimensions should be 43.05 wide and +58.05 long. Checking: $43.05 \times 58.05 = 2499.1$✓ – close enough.

Example 6.8

From the quadratic formula, we get $x = \frac{2200 \pm 2199.999091}{2}$. The large root is 2199.999546 and the small root is 0.0004545 using a pocket calculator. For this problem, the pocket calculator and Maple® agree on the small root to seven decimals. Not bad, but the potential for the solution to be precise but inaccurate exists. Precision means lots of significant digits, accuracy means the digits reported are true.

Example 6.9

Consider the family of quadratic expressions $\{(a_k x + b_k)^2 : a_k \in A, b_k \in B, 1 \le k \le n\}$. Certainly, $\sum_{k=1}^{n}(a_k x + b_k)^2 \ge 0$ as they are all squares. Expanding each term and summing, we obtain $(\sum_{k=1}^{n} a_k^2)x^2 + 2(\sum_{k=1}^{n} a_k b_k)x + (\sum_{k=1}^{n} b_k^2) \ge 0$. This is a quadratic that is nonnegative and has at most one real root. This means its discriminant cannot be positive. But then its discriminant $\Delta = 4(\sum_{k=1}^{n} a_k b_k)^2 - 4(\sum_{k=1}^{n} a_k^2) \cdot (\sum_{k=1}^{n} b_k^2) \le 0$. It follows immediately that $(\sum_{k=1}^{n} a_k b_k)^2 \le (\sum_{k=1}^{n} a_k^2) \cdot (\sum_{k=1}^{n} b_k^2)$, which establishes the result. This is a famous nineteenth century inequality called the Cauchy–Bunyakovsky–Schwarz inequality. The finite sequence version we have shown here was originally derived by A.-L. Cauchy in 1821.

Example 6.10

The point of this example is to show that the list method for factoring above may be applicable, but due to the profusion of trial divisors is impractical or at least tedious The quadratic formula gives

$$x = \frac{-17 \pm \sqrt{289 - 4 \cdot 144 \cdot (-36)}}{2 \cdot 144} = \frac{-17 \pm 145}{288}.$$

The two roots are then $\frac{4}{9}$ and $-\frac{9}{16}$. Now $(x - \frac{4}{9})(x - (-\frac{9}{16})) = 0$ can be rewritten as $(9x - 4)(16x + 9) = 0$. Checking the expansion, we get $144x^2 + 17x - 36 = 0$. Evidently, the desired factorization is $(9x - 4)(16x + 9) = 0$.

Example 6.11

Dividing by 2 to make it monic gives $x^3 - 9x^2 + 5x - 6 = 0$. Then set $x = y - \frac{(-9)}{3} = y + 3$. Expanding the cubic yields

$$\begin{aligned}(y+3)^3 - 9(y+3)^2 + 5(y+3) + 6 &= (y^3 + \mathbf{9y}^2 + 27y + 27) - \mathbf{9}(\mathbf{y}^2 + 6y + 9) + 5y + 21 \\ &= y^3 - 22y - 33.\end{aligned}$$

The quadratic term is gone.

Example 6.12

We are given the monic depressed equation $x^3 + 6x - 20 = 0$. Here is Tartaglia's big idea – let $x = u + v$. This is counterintuitive in that the introduction of two variables would seem to complicate the problem. But in practice they allow a degree of freedom to eliminate troublesome terms. Expanding $(u+v)^3 + 6(u+v) - 20 = 0$, we obtain $(u^3 + 3u^2v + 3uv^2 + v^3) + 6(u + v) - 20 = 0$. Rearranging we can write $u^3 + 3uv(u + v) + 6(u + v) + v^3 - 20 = 0$. Notice we can reduce this even further to $u^3 + (3uv + 6)(u + v) + v^3 - 20 = 0$. Now is when the big idea bears fruit. Tie u and v together by declaring $3uv + 6$ to be zero and remove it from the equation. This is the step where the degree of freedom is sacrificed. The original equation is now $u^3 + v^3 - 20 = 0$ with the side condition that $3uv + 6$, or $uv = -2$, so $v = \frac{-2}{u}$. Substituting for v yields $u^3 + (\frac{-2}{u})^3 - 20 = 0$ Clearing the fraction results in a quadratic equation in u^3, specifically $(u^3)^2 - 20u^3 - 8 = 0$. Using the quadratic formula, we find $u^3 = \frac{20 \pm \sqrt{400 - 4(1)(-8)}}{2} = 10 \pm \sqrt{108}$. We choose the positive square root for u^3 and the negative one for v^3, which satisfies the exact same quadratic. Then $u = \sqrt[3]{10 + \sqrt{108}}$ and $v = \sqrt[3]{10 - \sqrt{108}}$, so $x = \sqrt[3]{10 + \sqrt{108}} + \sqrt[3]{10 - \sqrt{108}}$. Numerically this works out to $2.7321\ldots - 0.7321\ldots \approx 2$. In fact, the problem was created with 2 in mind as an exact solution. Since cubic equations are supposed to have three roots, where are the other two? We will answer that question as part of the general case in Example 6.13.

Example 6.13

Starting with the general depressed cubic $x^3 + px + q = 0$, we set $x = u + v$ to obtain $(u^3+3u^2v+3uv^2+v^3)+p(u+v)+q = 0$. Rearranging, we have $u^3+v^3+(u+v)(3uv+p)+q = 0$. Now we give up the extra degree of freedom and invoke the constraint $3uv + p = 0$ in order to get rid of the inconvenient term $(u + v)(3uv + p)$. This implies $v = \frac{-p}{3u}$, and the equation can now be written $u^3 + (\frac{-p}{3u})^3 + q = 0$. Multiplying through by u^3, we have a quadratic in u^3 called the **resolvent**, namely $(u^3)^2 + qu^3 - \frac{p^3}{27} = 0$, from which we deduce

$$u^3 = \frac{-q \pm \sqrt{q^2 - 4(-\frac{p^3}{27})}}{2} = \frac{-q \pm \sqrt{\frac{4q^2}{4} + 4(\frac{p^3}{27})}}{2} = -\frac{q}{2} \pm \sqrt{\left(\frac{q}{2}\right)^2 + \left(\frac{p}{3}\right)^3}.$$

The constraint $3uv + p = 0$ is symmetric in u and v, so we take

$$u = \sqrt[3]{-\frac{q}{2} + \sqrt{\left(\frac{q}{2}\right)^2 + \left(\frac{p}{3}\right)^3}} \quad \text{and} \quad v = \sqrt[3]{-\frac{q}{2} - \sqrt{\left(\frac{q}{2}\right)^2 + \left(\frac{p}{3}\right)^3}}.$$

Then

$$x = \sqrt[3]{-\frac{q}{2} + \sqrt{\left(\frac{q}{2}\right)^2 + \left(\frac{p}{3}\right)^3}} + \sqrt[3]{-\frac{q}{2} - \sqrt{\left(\frac{q}{2}\right)^2 + \left(\frac{p}{3}\right)^3}}$$

in terms of the original coefficients. Now if $x = u+v$ is one solution of the cubic equation, then there are in general two others given by $\omega u + \omega^2 v$ and $\omega^2 u + \omega v$, where $\omega = \frac{-1+i\sqrt{3}}{2}$ is a complex cube root of 1. Conveniently enough, $\omega^2 = \frac{-1-i\sqrt{3}}{2}$.

Example 6.14

In Example 6.13, we could have written $u^3 = \omega^3 u^3$ and $v^3 = \omega^3 v^3$, since $\omega^3 = 1$. Upon taking cube roots, we would have obtained ωu and ωv, respectively. Similarly, had we used $\omega^6 = 1$, we would have had $\omega^2 u$ and $\omega^2 v$ as cube roots. Reconstructing x from these cube roots in the manner of Example 6.13 gives four candidates: $\omega u + \omega v$, $\omega^2 u + \omega^2 v$, $\omega u + \omega^2 v$, and $\omega^2 u + \omega v$. How can this be? We cannot have five solutions to a degree three equation. The answer lies in the constraint that we invoked which ties u and v together, namely $3uv + p = 0$. The product uv has to be a real number, so if $(\omega^i u)(\omega^j v) = \omega^{i+j}(uv) \in \mathbb{R}$, $i + j$ must equal 0 or 3. If $i + j = 0$, we have the original solution given in Example 6.13. If $i + j = 3$, then either $i = 1$ and $j = 2$, or vice versa, giving the two additional solutions $\omega u + \omega^2 v$ and $\omega^2 u + \omega v$. These can be written more explicitly as $x_2 = -\frac{u+v}{2} + \frac{u-v}{2}\sqrt{-3}$ and $x_3 = -\frac{u+v}{2} - \frac{u-v}{2}\sqrt{-3}$.

Example 6.15

From Example 6.14, we have

$$x_2 = -\frac{u+v}{2} + \frac{u-v}{2}\sqrt{-3} = -\frac{\sqrt[3]{10+\sqrt{108}} + \sqrt[3]{10-\sqrt{108}}}{2} + \frac{\sqrt[3]{10+\sqrt{108}} - \sqrt[3]{10-\sqrt{108}}}{2}\sqrt{-3}.$$

Note that

$$-\frac{u+v}{2} = -\frac{\sqrt[3]{20.392} + \sqrt[3]{-0.392}}{2} = -\frac{2.732 - 0.732}{2} = -1.$$

Also

$$\frac{u-v}{2} = \frac{2.732 + 0.732}{2} = 1.732 \approx \sqrt{3}.$$

Finally, $x_2 = -1 + \sqrt{3}(\sqrt{-3}) = -1 - 3i$. Similarly, $x_3 = -1 + 3i$. We justify the numerical approximation for $\sqrt{3}$ by testing $-1 \pm 3i$ in the original equation to confirm they are roots.

Example 6.16

Our formula gives

$$u = \sqrt[3]{-\frac{-56}{2} + \sqrt{\left(\frac{-56}{2}\right)^2 + \left(\frac{24}{3}\right)^3}} = \sqrt[3]{28 + \sqrt{1296}} = 4$$

and

$$v = \sqrt[3]{-\frac{-56}{2} - \sqrt{\left(\frac{-56}{2}\right)^2 + \left(\frac{24}{3}\right)^3}} = \sqrt[3]{28 - \sqrt{1296}} = -2,$$

so we recover $x = 4 - 2 = 2$. The other (complex) solutions to the equation are $-1 \pm 3\sqrt{-3}$.

Example 6.17

Let $\Delta = (\frac{q}{2})^2 + (\frac{p}{3})^3$ and consider the term $\sqrt{\Delta}$ which appears in the expression for x in Example 6.13. This term is real if $\Delta \geq 0$ and imaginary if $\Delta < 0$. If $\Delta > 0$, the equation has one real root and two conjugate complex roots. If $\Delta = 0$, the equation has three real roots where at least two are equal. If $\Delta < 0$, the equation has three distinct real roots. So in Example 6.16 we could have predicted the character of the roots by noting $(\frac{q}{2})^2 + (\frac{p}{3})^3 = (\frac{-56}{2})^2 + (\frac{24}{3})^3 > 0$, hence one real and two complex.

Example 6.18

The coefficients here are a little unusual, so let us try the graphing calculator strategy. Plotting this, we see that the graph crosses the real axis three times, once a little above $x = 3$. This raises our suspicion that π might be a root. Magnifying the interval, we see it crosses between 3.1 and 3.2. More evidence. There is no harm in doing a trial division of the cubic by $(x - \pi)$. We find it divides exactly with the complementary factor $x^2 + e + 1$. Applying the quadratic formula to that gives roots $\frac{-e \pm \sqrt{e^2-4}}{2}$. Numerically, the roots are 3.1416, −0.4387, and −2.2796. We did not recognize the two negative roots, but π was enough.

Example 6.19

We could clear fractions and work with $400x^3 - 1580x^2 + 3338x - 3003 = 0$ but that looks like a lot of work. Graphing $x^3 - 3.95x^2 + 8.345x - 7.5075$, we see it crosses the real axis once a little above $x = 1.5$. Magnifying the region around 1.5 we see that $x = 1.65$ looks like an exact hit. So dividing the cubic by $(x - 1.65)$ gives the quotient $x^2 - 2.30x + 4.55$ with no remainder. Applying the quadratic formula to the quotient, we get

$$x = \frac{2.3 \pm \sqrt{5.29 - 4 \cdot 4.55}}{2} = \frac{2.3 \pm \sqrt{5.29 - 4 \cdot 4.55}}{2} = 1.15 \pm i\sqrt{3.23}$$

for the other two roots. If there had been a remainder after the division, we could have adjusted our estimate for the divisor or, depending on our desired accuracy, just ignored a small remainder.

Example 6.20

We identify $a = -10$, $b = 4$, and $c = 8$. Then the resolvent cubic is $8z^3 - 200z^2 + 1536z - 3696 = 0$, which can be simplified to $z^3 - 250z^2 + 192z - 462 = 0$. The solutions to this cubic are $z_0 = 7, 9 \pm \sqrt{15}$. Then using $z_0 = 7$ and noting $\operatorname{sgn}(b) = +$, we have $x^2 \pm x\sqrt{2 \cdot 7 - 10} + 7 - 10 \mp \sqrt{2 \cdot (-10) \cdot 7 + 7^2 - 8 + 1^2} = 0$. More recognizably, these quadratics are $x^2 + 2x - 4 = 0$ and $x^2 - 2x - 2 = 0$, so the solutions to the original quartic are then $1 \pm \sqrt{3}$ and $-1 \pm \sqrt{5}$. For practical calculations, it may be easier once z_0 is determined to immediately substitute everything into $(2z_0 + a)x^2 - (b)x + (2az_0 + z_0^2 - c + a^2)$ and obtain the perfect square that equals $(x^2 + z_0 + a)^2$.

Example 6.21

Let the number be x^6, then its square root is x^3 and its cube root is x^2, hence $x^6 = x^3 + 2x^2$, and we can reduce this to $x^4 = x + 2$, or $x^4 - x - 2 = 0$. Cardano finesses this problem nicely and converts the quartic (he called them biquadratics, which as a modern term refers to something else) to a cubic the following way. Since $x^4 = x + 2$, it must be that $x^4 - 1 = x + 1$, and upon dividing by $x + 1$, which we know is not zero, we are left with

$x^3 - x^2 + x - 1 = 1$. Plugging that into the Cardano–Tartaglia formula, we see that the real root (the only one Cardano would have considered) is

$$x = \frac{1}{3} + \frac{1}{6}\sqrt[3]{188 + 12\sqrt{249}} - \frac{4}{3\sqrt[3]{188 + 12\sqrt{249}}}.$$

Then x^6 solves the problem. Letting Maple© do the heavy lifting, we find $x^6 \approx 6.140321679$. Checking,

$$\sqrt{6.140321679} + 2\sqrt[3]{6.140321679} \approx 6.140321665\,\surd.$$

One wonders how many calves gave their lives to provide enough vellum to write down this calculation in 1545.

Example 6.22

Notice that the coefficients are symmetrically distributed. We employ a trial factorization as $(x^2 + ax + 1)(x^2 + bx + 1) = x^4 + (a + b)x^3 + (ab + 2)x^2 + (a + b)x + 1$ and fit the data to the original expression. We have $a + b = 5$ and $ab + 2 = 8$. These two conditions imply $a(5 - a) = 6$, hence $a^2 - 5a + 6 = 0$, which in turn gives the solutions $a = 2, b = 3$ and $a = 3, b = 2$. Solving the two quadratics will give the four roots of the quartic. By symmetry, we need only consider one (a, b) pair, say $(2, 3)$. Then solving $x^2 + 2x + 1 = 0$ yields by the quadratic formula the double real root $x = \frac{-2\pm\sqrt{4-4\cdot1\cdot1}}{2} = -1$. Next, solving $x^2 + 3x + 1 = 0$ gives two distinct real roots $x = \frac{-3\pm\sqrt{9-4\cdot1\cdot1}}{2} = \frac{-3\pm\sqrt{5}}{2}$. The four roots of this quartic are -1 (double), $\frac{-3+\sqrt{5}}{2}$, and $\frac{-3-\sqrt{5}}{2}$. Checking, we confirm that

$$(x+1)^2\left(x - \frac{-3+\sqrt{5}}{2}\right)\left(x - \frac{-3-\sqrt{5}}{2}\right) = x^4 + 5x^3 + 8x^2 + 5x + 1.$$

Example 6.23

First, we divide $x^4 + 5x^3 + 8x^2 + 5x + 1 = 0$ by x^2 to produce $x^2 + 5x + 8 + \frac{5}{x} + \frac{1}{x^2} = 0$. Now observe that $z^2 = (x + \frac{1}{x})^2 = x^2 + 2 + \frac{1}{x^2}$. Then $x^2 + 5x + 8 + \frac{5}{x} + \frac{1}{x^2} = z^2 + 5z + 6 = 0$. Solving the quadratic in z yields

$$z = \frac{-5 \pm \sqrt{25 - 4\cdot1\cdot6}}{2} = \frac{-5 \pm 1}{2} = -2, -3.$$

Reverting to x, we have two quadratics, $x + \frac{1}{x} = -2$ and $x + \frac{1}{x} = -3$. In standard form, these become $x^2 + 2x + 1 = 0$ and $x^2 + 3x + 1 = 0$. We solved these same quadratics in Example 6.22 to get the four quartic roots.

Example 6.24

Given $ax^4 + bx^3 + cx^2 + mbx + m^2a = 0$, divide by x^2 to get $ax^2 + bx + c + \frac{mb}{x} + \frac{m^2a}{x^2} = 0$. Now substitute $z = x + \frac{m}{x}$ by analogy with Example 6.23. We can regroup the terms in the previous expression as $(ax^2 + \frac{m^2a}{x^2}) + (bx + \frac{mb}{x}) + c = 0$. Note that $z^2 = (x + \frac{m}{x})^2 = x^2 + 2m + \frac{m^2}{x^2}$. Plugging this into our regrouped expression, we get $(az^2 - 2am) + bz + c = 0$. In standard form, the quadratic in z is $az^2 + bz + (c - 2am) = 0$. The roots z_1 and z_2 are $\frac{-b \pm \sqrt{b^2 - 4 \cdot a \cdot (c-2am)}}{2a}$. These roots spawn the quadratics in x given by $z_1 = x + \frac{m}{x}$ or $x^2 - z_1x + m = 0$, and similarly $x^2 - z_2x + m = 0$. The four roots of the two quadratics are the quartic roots.

Example 6.25

A **biquadratic** equation is now defined as a quartic with the cubic and linear terms missing, which is the case in this example. These are the simplest nontrivial quartics. The substitution $z = x^2$ immediately reduces the equation to $z^2 - 13z + 36 = (z - 9)(z - 4) = 0$. So we solve $x^2 = 4$ to get $x = \pm 2$ and $x^2 = 9$ to get $x = \pm 3$. These are the four roots of the quartic.

Example 6.26

There is a test called the **Eisenstein criterion** (1850) that says if you have a polynomial *of any degree* with integer coefficients and you can find a prime that does not divide the leading coefficient but divides all the other coefficients (but not the constant twice), then it is irreducible over the rationals. So looking over the coefficients of the quintic, we see that the prime 2 does not divide the leading coefficient, divides all the others because they are all even, and finally, $2^2 = 4$ does not divide the constant 2. By the Eisenstein criterion, we claim the quintic is irreducible over the rational numbers. So there is no hope of factoring it over the rationals.

We are using an unstated fact here that was proved by the German mathematician Carl Gauss. If you can factor a polynomial using rational numbers, then by suitable adjustments you can express that factorization purely in integers. The contrapositive of this statement is that if you cannot factor it over the integers, then neither can you factor it over the rationals. The Eisenstein criterion initially shows that you cannot factor over the integers, but then the contrapositive allows you to conclude the stronger statement that there is no factorization over the rationals.

Example 6.27

Another irreducibility test is the **mod p test**. Using this effectively requires basic familiarity with modular arithmetic, which we review in Section 10. The idea is to choose a prime modulus p and reduce all coefficients modulo p. Letting $p = 2$, we reduce our quintic mod 2 and get $x^5 + 0x^4 - x^3 + 0x^2 + 0x + 1 = x^5 - x^3 + 1$. We check to see that the reduced polynomial has the *same degree* as the original. If not, we cannot use that particular p. Now the test asserts that if the coefficient-reduced polynomial is irreducible over its field $\mathbb{Z}_p$, in this case $\mathbb{Z}_2 = \{0, 1\}$, then the original polynomial is irreducible over $\mathbb{Q}$. So the problem boils down to checking whether or not $x^5 - x^3 + 1$ can be factored over $\mathbb{Z}_2$. Neither $x = 0$ nor $x = 1$ are zeroes of this polynomial, so apparently it has no linear factors. How about quadratic factors? The only possible quadratic factors would be x^2, $x^2 + 1$, or $x^2 + x + 1$. Remember that in $\mathbb{Z}_2$, plus and minus signs can be interchanged without affecting the value of the expression (this is counterintuitive, but in $\mathbb{Z}_2$ we have $1 + 1 = 0$). The point is that this is the complete collection of possible quadratic factors. Note that x^2 cannot be a factor because then zero would be a root. Likewise, $x^2 + 1$ cannot be a factor because it has 1 as a root. Finally, $x^2 + x + 1$ has no roots itself, but using polynomial long division into $x^5 - x^3 + 1$ shows a nonzero remainder, hence it is not a divisor. There is no need to check for a cubic factor, since the complementary factor would be a quadratic and that possibility has already been eliminated. The conclusion from all of this is that $x^5 - x^3 + 1$ is irreducible over $\mathbb{Z}_2$ and the mod p test then implies $9x^5 + 16x^4 - 7x^3 + 6x^2 + 4x + 1$ is irreducible over $\mathbb{Q}$.

It is important to note that the converse of this test is false, that is, if the polynomial with coefficients reduced mod p is reducible over $\mathbb{Z}_p$, there is no guarantee that the original polynomial is reducible over $\mathbb{Q}$. The classic counterexample is the quartic x^4+1. This polynomial is irreducible over $\mathbb{Q}$ but reducible over every $\mathbb{Z}_p$, so it thoroughly defeats any attempt to apply the mod p test. Interestingly, $x^4+1 = (x^2+\sqrt{2}x+1)(x^2-\sqrt{2}x+1)$, but this factorization does not render it reducible over $\mathbb{Q}$ because $\sqrt{2} \notin \mathbb{Q}$. On the plus side, the mod p test is applicable to a polynomial of any degree with arbitrary integer coefficients. The Eisenstein criterion is easier to use but requires a more stringent condition on the coefficients that may not be available.

Example 6.28

The Eisenstein criterion is not directly applicable. We could let $x = y \pm 1$ or ± 2 and generate a new polynomial that may be testable. Then if that polynomial is irreducible, so is the original one in x. But this looks like it could become a wild goose chase, so let us stick with the mod p test with $p = 2$. Reducing coefficients mod 2 gives us $x^5 - x^4 + x - 1$. Note that 1 is a zero $(1 - 1 + 1 - 1 = 0)$. You can see from factoring by grouping that $x^5 - x^4 + x - 1 = x^4(x - 1) + (x - 1) = (x^4 + 1)(x - 1)$, so it is reducible mod 2. Recalling the caveat in Example 6.27 that this does not mean the original quintic is reducible over $\mathbb{Q}$, we can still take encouragement from the fact that at least the test did not preclude the reducibility of the original. Rerunning the test with other primes will not result in any modular reduction, so we are stuck. Perhaps we should try to factor the original. It will be easy to look for small linear factors using **synthetic division**. Let us see if $(x + 1)$ divides $x^5 - x^4 - 2x^3 + 2x^2 + x - 1$. Here is the template for synthetic division:

Table 6.3: Rapid Test for Divisibility

−1	1	−1	−2	2	1	−1
		−1	2	0	−2	1
	1	−2	0	2	−1	0

The remainder is zero, so $(x+1)$ is a factor of $x^5 - x^4 - 2x^3 + 2x^2 + x - 1$. We can get the other factor without having to do long division simply by taking the middle row of our calculation $(-1\,2\,0\,-2\,1)$ reversing all the signs and attaching them as coefficients to the appropriate descending powers of x. This gives us $x^4 - 2x^3 + 2x - 1$, which can be further reduced to $(x-1)^3(x+1)$. So finally, we discover that not only is $x^5 - x^4 - 2x^3 + 2x^2 + x - 1$ reducible over $\mathbb{Q}$, it splits into linear factors $(x-1)^3(x+1)^2$.

Example 6.29

We need to convert the coefficients to integers to apply one of our tests. Multiplying the quintic by 35 will do the trick. Then we will have $15x^5 + 30x^4 + 7x^3 - 10x + 21$. Note that if $P(x)$ is any polynomial and $c \neq 0$ is any constant, then if cP does not factor nontrivially, neither does P. Let us apply the mod p test with $p = 2$ to the integer-converted quintic. The coefficients reduced modulo 2 give the test quintic $x^5 + x^3 + 1$. Application of the test will be valid since we have not reduced the degree. As mentioned in Example 6.27, it is a quirk of arithmetic modulo 2 that $1 = -1$. So in $\mathbb{Z}_2$ we have $x^5 + x^3 + 1 = x^5 - x^3 + 1$, which we already showed was irreducible. We may now conclude that $15x^5 + 30x^4 - 7x^3 - 10x + 21$ is irreducible over $\mathbb{Q}$ and therefore so is $\frac{3}{7}x^5 + \frac{6}{7}x^4 - \frac{1}{5}x^3 - \frac{2}{7}x + \frac{3}{5}$. If the leading coefficient of a polynomial of any degree is odd, using the mod p test with $p = 2$ is always a good start when trying to show irreducibility.

Example 6.30

Suppose $P(\frac{p}{q}) = 0$. Then

$$a_n\left(\frac{p}{q}\right)^n + a_{n-1}\left(\frac{p}{q}\right)^{n-1} + \cdots + a_1\left(\frac{p}{q}\right) + a_0 = 0.$$

Clear fractions by multiplying by q^n. This leaves $a_np^n + a_{n-1}p^{n-1}q + \cdots + a_1pq^{n-1} + a_0q^n = 0$. Rearrange this and factor out p as follows: $p(a_np^{n-1} + a_{n-1}p^{n-2}q + \cdots + a_1q^{n-1}) = -a_0q^n$. Evidently, p divides a_0q^n. Any factor of p cannot divide q^n because p and q have no common factors. So all the factors of p must divide a_0, but that implies p itself divides a_0. Similarly, rearrange $a_np^n + a_{n-1}p^{n-1}q + \cdots + a_1pq^{n-1} + a_0q^n = 0$ as $q(a_{n-1}p^{n-1} + \cdots + a_1pq^{n-2} + a_0q^{n-1}) = -a_np^n$. Now we see q divides a_np^n. All the factors of q must divide a_n, so q itself must divide a_n. This is called the **rational root test**. The idea is to assemble all the candidate numerators p (factors of a_0) and candidate denominators q (factors of a_n), form all possible fractions and test each one by checking if $P(\frac{p}{q}) = 0$. If $P(x)$ has a rational root, it will be revealed by this process. This will be the key to our application of the Cayley's test.

Example 6.31

Here $p \in \pm\{1, 2, 3, 6\}$ and $q \in \{1, 3\}$. The notation $\pm$ reminds us to test each fraction both positive and negative. The complete set of test fractions is then $\pm\{1, \frac{1}{3}, 2, \frac{2}{3}, 3, 6\}$. All duplicates have been removed. The calculations yield:

$$P(1) = 2, \quad P(-1) = 20, \quad P\left(\frac{1}{3}\right) = \frac{128}{27}, \quad P\left(-\frac{1}{3}\right) = \frac{70}{9},$$
$$P(2) = 14, \quad P(-2) = 110, \quad P\left(\frac{2}{3}\right) = \frac{10}{3}, \quad P\left(-\frac{2}{3}\right) = \frac{314}{27},$$
$$P(3) = 120, \quad P(-3) = 414, \quad P(6) = 2862, \quad \text{and} \quad P(-6) = 5070.$$

No luck, so we conclude $P(x)$ has no rational roots.

Example 6.32

Here $p \in \pm\{1, 2, 4, 8\}$ and $q \in \{1, 2\}$. The complete set of test fractions modulo duplicates is then $\pm\{1, \frac{1}{2}, 2, 4, 8\}$. The calculations yield:

$$P(1) = -4, \quad P(-1) = -6, \quad P\left(\frac{1}{2}\right) = -\frac{93}{16}, \quad P\left(-\frac{1}{2}\right) = -\frac{115}{16},$$
$$P(2) = 0, \quad P(-2) = -88, \quad P(4) = 944, \quad P(-4) = -2784,$$
$$P(8) = 48072, \quad \text{and} \quad P(-8) = -79960.$$

We discover that $x = 2$ is a rational root. If we were interested in getting the other roots, we could divide $P(x)$ by $(x - 2)$ and use our method to solve quartics for the rest of the solutions. In this case the rational root test served as a reducibility criterion yielding the factor $(x - 2)$.

Example 6.33

We have $a = -1$ and $b = -1$, so

$$R(z) = z^6 + 8z^5 + 40z^4 - 160z^3 + 400z^2 - 3637z + 9631.$$

Since 9631 is a prime, the only possible rational roots are ± 9631. Neither one satisfies the equation, so we conclude by Cayley's test that $x^5 - x - 1 = 0$ is not solvable in radicals. If we consider $x^5 - x + 1 = 0$, $R(z)$ is the same, so $x^5 - x + 1 = 0$ is not solvable in radicals either.

Example 6.34

We have $a = 1$ and $b = \pm 1$, so

$$R(z) = z^6 - 8z^5 + 40z^4 + 160z^3 + 400z^2 - 2613z - 9119.$$

Since $9119 = 11 \cdot 829$, the only possible rational roots are ± 11 and ± 829. None of these satisfy $R(z) = 0$, so again we conclude that $x^5 + x + 1$ and $x^5 + x - 1$ are not solvable in radicals.

Example 6.35

We have $a = -5$ and $b = 12$, so

$$R(z) = z^6 - 40z^5 + 1000z^4 + 20000z^3 + 250000z^2 - 66400000z + 976000000.$$

Now $976000000 = 2^{10} \cdot 5^6 \cdot 61$, so there are $11 \cdot 7 \cdot 2 \cdot 2 = 308$ rational numbers that could possibly be a root of $R(z) = 0$. One of our references kindly saves us time and reveals that $z_0 = 40 = 2^3 \cdot 5$ is a root. Then by Cayley's test, $x^5 - 5x + 12$ is solvable in radicals. Once again, since $R(z)$ only has b^4 in its formula, we get a two-for-one confirmation that $x^5 - 5x - 12$ is also solvable.

Example 6.36

Descartes' rule states that the number of positive zeroes, including multiplicities, of a polynomial $P(x) \in \mathbb{Z}[x]$ is either equal to the number of sign changes of its coefficients or a multiple of two less. As a corollary, the number of negative real zeroes is either equal to the number of sign changes of $p(-x)$ or a multiple of two less. Starting at the left, we count five sign changes in $P(x)$, so the number of positive real roots must be 5, 3, or 1. Calculating $P(-x) = -x^5 - x^4 - 12x^3 - 12x^2 - 11x - 6$, and applying the rule, apparently there are no negative real roots. We conclude that there must be one, three, or five positive real roots and one or two complex conjugate pairs with no negative real roots. We can be certain that the graph of $y = P(x)$ does not cross the negative x-axis. In fact, $P(x) = (x^2+1)(x-1)(x-2)(x-3)$, so the positive real roots are 1, 2, and 3, and the single conjugate pair is $\pm i$. We omit the proof of this rule, since it is straightforward but rather tedious. Newton used it and thought it was too obvious to prove. The French geometer Jean Paul de Gua de Malves was the first to offer a satisfactory proof of Descartes Rule in 1740.

Example 6.37

$4^x = (2^2)^x = 2^{2x} = 8^3 = (2^3)^3 = 2^{27}$. Apparently, $2^{2x} = 2^9$, so $2x = 9$ and $x = \frac{9}{2}$.

Example 6.38

We recognize $8 = 2^3$, so the expression on the right can be rewritten with fractional exponents as $2^x = ((2^3)^{1/2})^5 = 2^{15/2}$. Now it is easily seen that $x = \frac{15}{2}$.

Example 6.39

Restated we have $4^{3x} = (2^5)^{1/3} = 2^{5/3}$. Evidently, $2^{6x} = 2^{5/3}$, so $6x = \frac{5}{3}$ and $x = \frac{5}{18}$.

Example 6.40

$9^{2x} = 3^{4x} = (3^3)^{x+1} = 3^{3x+3}$, so $4x = 3x + 3$ or $x = 3$.

Example 6.41

$(\frac{1}{49})^{3x} = (\frac{1}{7^2})^{3x} = (7^{-2})^{3x} = 7^{-6x} = 243 = 7^3$. So $-6x = 3$, or $x = -\frac{1}{2}$.

Example 6.42

$B = \sqrt{A^3} = \sqrt{(3^x)^3} = \sqrt{3^{3x}} = (3^{3x})^{1/2} = 3^{3x/2}$.

Example 6.43

Plugging in values, we have $F = (1000)e^{0.06\cdot 8} = (1000)e^{0.48} = \$1616.07$. Sometimes we may choose to write e^x as $\exp(x)$ to improve readability if necessary. So $e^{0.48} = \exp(0.48)$.

Example 6.44

Here the unknown is P. We have $25000 = P\exp(0.0775 \cdot 15)$. Rearranged this gives

$$P = \frac{25000}{\exp(0.0775 \cdot 15)} = 25000 \cdot \frac{1}{3.197918} = \$7817.59.$$

Example 6.45

We need to calculate the present value of the entire stream of payments assuming that money is worth 6 % *per annum*. The present values of the payments in dollars are $5000 + 5000r + 5000r^2 + \cdots$, where $r = 0.9418$. The sum of an infinite series of the form $A + Ar + Ar^2 + \cdots$ where $0 \le r < 1$ is $\frac{A}{1-r}$. So the total present value of the stream of payments is $5000 \cdot \frac{1}{1-0.9418} = 5000 \cdot 17.1821 = \$85,910.50$. This is the amount on deposit at 6 % continuous interest that would earn enough to support the annuity indefinitely. After the initial \$5000 draw down, it will take a year at 6 % interest to recover for the next and subsequent payments to be made sustainably.

Example 6.46

The rate of decrease of illumination at depth x is proportional to the light intensity $L(x)$ at x. The constant of proportionality is 0.03. We have $L(x) = L(0)e^{-0.03x}$, where $L(0)$ is the light intensity at the surface of the lake. The ratio $\frac{L(x)}{L(0)} = e^{-0.03x}$ is the fraction of surface illumination available at depth x. Plugging in $x = 25$, we get $\frac{L(25)}{L(0)} = e^{-0.75} = 47.2\,\%$. The light level 25 meters down is roughly half of what it is at the surface.

Example 6.47

The governing formula for radioactive decay is $Q(t) = Q(0)e^{-\lambda t}$, where $Q(t)$ is the amount of radioactive substance after t years, $Q(0)$ is the amount at the beginning of the decay period, and $\lambda > 0$ is the decay constant for the isotope. A higher λ means faster decay. Plugging our data into the formula, $Q(75) = Q(0)\exp(-0.005 \cdot 75) = 0.687$ [kg]. So roughly a third of the original mass would be transmuted.

Example 6.48

Let $Q(t) = Q(0)e^{-0.00012t}$, where time t is in years and $Q(t)$ is the quantity of ^{14}C t years after the oak was cut down to make the ship. Then $\frac{Q(1075)}{Q(0)} = e^{-0.1307} = 87.7\,\%$. When we discuss logarithmic equations, we will turn this calculation around and ask how old something is given a certain percentage of ^{14}C. The limiting factor in using radiocarbon dating is the ability to accurately assess tiny amounts of ^{14}C. It is a chemical problem, not mathematical. For every trillion atoms of carbon in the environment, approximately 1 is ^{14}C. The maximum reliable age detectable by radiocarbon dating is about 50, 000 years. There are various potential sources of error in the method due to circumstances beyond anyone's control, such as variations in atmospheric concentrations of ^{14}C due to solar activity. All in all the method is a "silver bullet" for archaeologists, who previously had to rely on tree rings and sedimentation rates to guesstimate the ages of prehistorical artifacts. Now it is possible to accurately date Bronze Age items to within ±100 years or so.

Example 6.49

Here $f(t)$ is the percentage of the play the actor has memorized at time t in weeks. We do not know the applicable k, but we have data that we can plug in to get it. Evidently, $20 = 100(1 - e^{-k(1)})$. Rearranging this, we get $0.2 = (1 - e^{-k})$ or $e^{-k} = 0.8$. Taking natural logs, we find $k = 0.2231$. Now we can put that back into the formula to find $f(t)$ for any t. We want to solve $80 = 100(1 - e^{-0.2231t})$ for t. Following our previous calculation, we have $e^{-0.2231t} = 0.2$. Taking natural logs again, we have $0.2231t = 1.6094$, which yields $t = 7.2$ weeks to learn 80 %. The diminishing returns are quite evident. If he continued to learn 20 % per week, he would have needed only four weeks.

Example 6.50

Let $f(t)$ be the number of people who know the gossip at time t. Here $M = 50$, so $f(t) = \frac{50}{1+Be^{-50kt}}$. We have two undetermined constants, so we need two data points to find them. We know at $t = 0, f(0) = 2$, so

$$2 = f(t) = \frac{50}{1 + Be^0}, \quad \text{or} \quad 2(1 + B) = 50,$$

which implies $B = 24$. We also know

$$f(10) = 20 = \frac{50}{1 + 24e^{-50k(10)}}, \quad \text{or} \quad 20(1 + 24e^{-500k}) = 50.$$

This implies $e^{-500k} = \frac{3}{48} = 0.0625$. Taking logs, $-500k = -2.773$, so $k = 0.005545$. Now reassembling the general formula, we have

$$f(t) = \frac{50}{1 + 24e^{-0.277t}},$$

where we have multiplied M and k together. We set

$$f(t) = 40 = \frac{50}{1 + 24e^{-0.277t}}$$

and solve for t. So $40(1 + 24e^{-0.277t}) = 50$. Then $e^{-0.277t} = 0.0104$ and, after taking logs, we get $-0.277t = -4.564$, which implies $t = 16.5$ minutes. This is time from the beginning, so it took ten minutes for 20 people to know and only 6.5 minutes after that for another 20 to find out.

Example 6.51

1)

$$\begin{aligned}\sinh^2 x + 1 &= \left(\frac{e^x - e^{-x}}{2}\right)^2 + 1 = \frac{1}{4}(e^{2x} - 2 + e^{-2x}) + \frac{4}{4} = \frac{1}{4}(e^{2x} + 2 + e^{-2x}) \\ &= \left(\frac{e^x + e^{-x}}{2}\right)^2 = \cosh^2 x.\end{aligned}$$

So $\cosh^2 x = \sinh^2 x + 1$. Compare this with $\sin^2 x + \cos^2 x = 1$.

Example 6.52

By definition, $\sinh(x + y) = \frac{e^{x+y} - e^{-(x+y)}}{2}$. Look at the expression

$$\left(\frac{e^x - e^{-x}}{2}\right)\left(\frac{e^y + e^{-y}}{2}\right) = \frac{1}{4}(e^x e^y + e^x e^{-y} - e^{-x} e^y - e^{-x} e^{-y}).$$

Now interchange x and y to get

$$\left(\frac{e^y - e^{-y}}{2}\right)\left(\frac{e^x + e^{-x}}{2}\right) = \frac{1}{4}(e^y e^x + e^y e^{-x} - e^{-y} e^x - e^{-y} e^{-x}).$$

Add these two expressions to get

$$\left(\frac{e^x - e^{-x}}{2}\right)\left(\frac{e^y + e^{-y}}{2}\right) + \left(\frac{e^y - e^{-y}}{2}\right)\left(\frac{e^x + e^{-x}}{2}\right) = \frac{1}{4}(2e^x e^y - 2e^{-x}e^{-y}).$$

The intermediate terms cancel in pairs. This leaves

$$\left(\frac{e^x - e^{-x}}{2}\right)\left(\frac{e^y + e^{-y}}{2}\right) + \left(\frac{e^y - e^{-y}}{2}\right)\left(\frac{e^x + e^{-x}}{2}\right) = \frac{e^x e^y - e^{-x}e^{-y}}{2}.$$

We conclude $\sinh x \cosh y + \cosh x \sinh y = \sinh(x+y)$. Had we taken the difference of the two pairs, the same cancellations occur and we get the companion formula $\sinh(x-y) = \sinh x \cosh y - \cosh x \sinh y$. Formulas for $\cosh(x \pm y)$ and $\tanh(x \pm y) = \frac{\sinh(x\pm y)}{\cosh(x\pm y)}$ are verified similarly.

Example 6.53

This problem is a classic and one of a few instances where Galileo's analysis was incorrect. A rope or chain hanging with only its self-weight as the load assumes the shape of a catenary. Galileo initially thought the rope would hang in a parabolic shape. It does not matter if the points of suspension are at unequal heights, but in this example they are. Imagine the poles set at $x = \pm 2$ in the xy-plane. The equation of this catenary is $y(x) = 0.5\cosh(\frac{x}{0.5})$. Checking the problem data, we have $y(0) = 0.5\cosh(0) = 0.5$ at the midpoint, and $y(2) = 0.5\cosh(4) = 13.65$, at the points of attachment. The cosh function is even, so this applies to both poles. Now it is simply a matter of evaluating $y(1) = 0.5\cosh(2) = 1.88$ feet. This is the height of the rope at $x = \pm 1$ foot.

Example 6.54

Since $y = \frac{e^x + e^{-x}}{2}$, we can multiply both sides by e^x to get $ye^x = \frac{e^{2x}+1}{2}$, or $e^{2x} - 2ye^x + 1 = 0$. This is a quadratic in e^x with solution $e^x = \frac{2y \pm \sqrt{4y^2-4}}{2} = y \pm \sqrt{y^2 - 1}$. The convention is to take the positive radical with the understanding that $y \geq 1$. Taking natural logs, we finally get $x = \ln(y + \sqrt{y^2 - 1})$.

Example 6.55

Exponentiate both sides relative to b. Then $4^{2x} = 8$ or $2^{4x} = 2^3$, which gives $4x = 3$, hence $x = \frac{3}{4}$.

Example 6.56

We have $A = b^{\log_b A} = c^{\log_c A}$. Now $c = \log_b c$, so we can write $A = b^{\log_b A} = (b^{\log_b c})^{\log_c A} = b^{\log_b c \cdot \log_c A}$. Evidently, $b^{\log_b A} = b^{\log_b c \cdot \log_c A}$, and we conclude $\log_b A = \log_b c \cdot \log_c A$. Sometimes this is written

$$\log_c A = \frac{\log_b A}{\log_b c}.$$

Now suppose we are given $\log A$. By the formula with $b = 10$ and $c = e$,

$$\ln A = \frac{\log A}{\log e} = \frac{\log A}{0.43429} = 2.30258 \log A.$$

So natural logs are 2.30258 times as big as corresponding common logs, which makes intuitive sense as the base of natural logs is much smaller than the base of common logs. This equation is called the **change of base formula**, or Euler's golden rule.

Example 6.57

By the change of base formula with $b = e$ and $c = 10$, we have $\log 2 \approx \frac{0.69315}{\ln 10} = \frac{0.69315}{2.30258} = 0.30103$.

Example 6.58

Note that $A^{\log_B C} = (B^{\log_B A})^{\log_B C} = B^{(\log_B A)(\log_B C)} = B^{(\log_B C)(\log_B A)} = (B^{\log_B C})^{\log_B A} = C^{\log_B A}$.

Example 6.59

We cannot easily align the bases as we did in Section 6.2 Example 6.37, but we can use the logarithm laws Section 0.2.3 to write $\log(2^3) = \log(1000^x)$. This is equivalent to $3\log 2 = x \log 1000 = 3x$. So $x = \log 2$.

Example 6.60

We can immediately write $(3^3)^{2x} = (2^4)^{x+1}$ or $3^{6x} = 2^{4x+4}$. Taking logs, we get $6x \log 3 = (4x + 4)\log 2$. Substituting values gives $6x(0.47712) = (4x + 4)(0.30103)$. Simplifying we have $(2.86272 - 1.20412)x = 1.20412$, so $x = \frac{1.20412}{1.65860} = 0.72599$.

Example 6.61

Note that $9^{\log_3 x} = (3^2)^{\log_3 x} = 3^{2\log_3 x} = (3^{\log_3 x})^2 = x^2$. So we are really solving $x^2 - 5x + 6 = (x - 3)(x - 2) = 0$. We finally have $x = 2, 3$.

Example 6.62

Let this log be x, then $(6^{\sqrt{2}/3})^x = 6^2$. This means $\frac{\sqrt{2}}{3}x = 2$, which is the same as $x = \frac{6}{\sqrt{2}} = 3\sqrt{2}$.

Example 6.63

Using the change of base formula on both numerator and denominator to 10, we have

$$\log_x 3 = \frac{\log 3}{\log x} \quad \text{and} \quad \log_{x^2} 3 = \frac{\log 3}{\log x^2}.$$

Then

$$\frac{\log_x 3}{\log_{x^2} 3} = \frac{\log 3}{\log x} \cdot \frac{\log x^2}{\log 3} = \frac{2\log x}{\log x} = 2.$$

Example 6.64

Use the exchange formula (Example 6.58) to write this as $9^{\log x} + 15^{\log x} = 25^{\log x}$. Now divide by $9^{\log x}$ across the board to get

$$1 + \frac{15^{\log x}}{9^{\log x}} = \frac{25^{\log x}}{9^{\log x}}.$$

Since $\frac{a^r}{b^r} = (\frac{a}{b})^r$, we can reduce this further to $1 + (\frac{15}{9})^{\log x} = (\frac{25}{9})^{\log x}$. Putting $\frac{15}{9}$ in lowest terms and expressing $\frac{25}{9}$ as $(\frac{5}{3})^2$, we get

$$1 + \left(\frac{5}{3}\right)^{\log x} = \left(\left(\frac{5}{3}\right)^2\right)^{\log x}.$$

Writing

$$\left(\left(\frac{5}{3}\right)^2\right)^{\log x} = \left(\left(\frac{5}{3}\right)^{\log x}\right)^2$$

and expressing the result as a quadratic in $(\frac{5}{3})^{\log x}$, we have

$$\left(\left(\frac{5}{3}\right)^{\log x}\right)^2 - \left(\frac{5}{3}\right)^{\log x} - 1 = 0.$$

Applying the quadratic formula, we obtain

$$\left(\frac{5}{3}\right)^{\log x} = \frac{1 \pm \sqrt{1 - 4(-1)}}{2} = \frac{1 \pm \sqrt{5}}{2}.$$

Now $1 - \sqrt{5} < 0$ is a spurious solution because $(\frac{5}{3})^{\log x} > 0$, so the only valid solution is $(\frac{5}{3})^{\log x} = \frac{1+\sqrt{5}}{2}$. So as awkward as it may seem, we can take logs to the base $\frac{5}{3}$ and get

$$\log_{5/3}\left(\frac{5}{3}\right)^{\log x} = \log x = \log_{5/3}\left(\frac{1+\sqrt{5}}{2}\right).$$

Exponentiating finally gives us

$$10^{\log x} = 10^{\log_{5/3}(\frac{1+\sqrt{5}}{2})},$$

so $x = 10^{0.94203} = 8.7504$. Checking numerically: $9^{8.7504} + 15^{8.7504} - 25^{8.7504} = 6.8391 + 10.6936 - 16.7207 = 0.812$✓. There are some rounding errors with the calculator, but we are pretty close. We have every confidence in the formal algebraic solution. Interestingly, our computer algebra system refuses to attempt to solve $9^{\log x} + 15^{\log x} = 25^{\log x}$ and returns only the trivial solution $x = 0$ for $x^{\log 9} + x^{\log 15} = x^{\log 25}$. Humans won one!

Example 6.65

Here $p_1 = \alpha \ln(\frac{2}{1}) = 0.693\alpha$. But $p_{10} = \alpha \ln(\frac{11}{10}) = 0.0953$. The perceived effort in lifting an extra pound from 1 to 2 is $\frac{0.693}{0.0953} = 7.3$ times the perceived effort in lifting an extra pound from 10 to 11.

Example 6.66

Here $5.5 = \log(A)$, so $A = 10^{5.5} = 316228\,[\mu\text{m}] = 31.6\,[\text{cm}]$. Also $3.2 = \log(A)$ implies $A = 1585\,[\mu\text{m}]$, so the energy ratio is $\frac{316228}{1585} \approx 200$.

Example 6.67

From the Pert formula, we have $2 = 1e^{rt}$. Taking logs, $0.69315 = rt$, so $100rt = 69.3$. We conclude that the interest rate in percent times the time in years for money to double is 69.3. This is exact and an improvement over the accountant's old "rule of 72" which was approximate and based on periodic rather than continuous compounding.

Example 6.68

From Section 6.2 Example 6.48, we have $Q(t) = Q(0)e^{-0.00012t}$. Then $\frac{Q(t)}{Q(0)} = 1 - 0.07 = 0.93 = e^{-0.00012t}$. Taking natural logs we get $-0.0725 = -0.00012t$, so $t \approx 605$ years seems reasonable. Yale paid \$300,000 for the map, which is now valued in excess of \$25 million, perhaps as much or more than if it were legitimate, paradoxically.

Example 6.69

Here $\frac{Q(t_{1/2})}{Q(0)} = 0.5 = \exp(-0.00012t_{1/2})$. Taking natural logs, we get $-0.69315 = -0.00012t_{1/2}$ which implies $t_{1/2} = \frac{0.69315}{0.00012} = 5776$ years. Without doing any detailed calculation, it is possible to estimate the age of a carbonaceous artifact if the percentage of retained ^{14}C is known. If an object had only about 12 % of normal ^{14}C, then it has gone through about three half-lives, so it would be about 17,000 years old.

Example 6.70

Here $f(x)$ is a continuous function because x and $\cos x$ are such separately and the sum of continuous functions is also continuous. We calculate $f(\frac{\pi}{3}) = 2.0472$ and $f(-\frac{\pi}{3}) = -2.0472$. The graph of $y = f(x)$ changes sign from $-\frac{\pi}{3}$ to $+\frac{\pi}{3}$ and therefore the graph must cross the real axis where $f(x) = 0$ at some point in the interval $[-\frac{\pi}{3}, \frac{\pi}{3}]$. This is a transcendental equation and we cannot solve it in closed form, but at least we know it has a solution. Our CAS declines to give us an estimate, but a graphing calculator with sufficient resolution says $x = -1.03$ is close. Checking $-1.03 + 2 \cdot \cos(-1.03) \approx 0.0004$ – good enough✓.

Example 6.71

The necessary formula is $\arctan u + \arctan v = \arctan \frac{u+v}{1-uv}$. Set $u = v$ to get $2 \arctan u = \arctan \frac{2u}{1-u^2}$. Setting $u = \frac{1}{5}$, we get

$$\frac{2u}{1-u^2} = \frac{\frac{2}{5}}{1-(\frac{2}{5})^2} = \frac{\frac{2}{5}}{1-\frac{1}{25}} = \frac{\frac{2}{5}}{\frac{24}{25}} = \frac{5}{12}.$$

So $2 \arctan \frac{1}{5} = \arctan \frac{5}{12}$. This implies $\arctan(\frac{1}{5}) = \frac{1}{2} \arctan \frac{5}{12}$.

Example 6.72

Note that $\sin^4\theta - \cos^4\theta = (\sin^2\theta - \cos^2\theta)(\sin^2\theta + \cos^2\theta) = (\sin^2\theta - \cos^2\theta) = (1-\cos^2\theta) - \cos^2\theta = 1 - 2\cos^2\theta$. The statement follows.

Example 6.73

Here $\sum_{n=0}^{N} r^n$, where $r = e^{i\theta}$, gives

$$\sum_{n=0}^{N} e^{in\theta} = \sum_{n=0}^{N} (\cos n\theta + i \sin n\theta) = \sum_{n=0}^{N} \cos n\theta + i \sum_{n=0}^{N} \sin n\theta = \frac{1-e^{i(N+1)\theta}}{1-e^{i\theta}}.$$

This is the same as $\frac{1-(\cos(N+1)\theta + i\sin(N+1)\theta)}{(1-\cos\theta) - i\sin\theta}$. We can separate the real and imaginary parts by multiplying top and bottom by $(1-\cos\theta) + i\sin\theta$ to get

$$\begin{aligned} &\frac{[1-(\cos(N+1)\theta + i\sin(N+1)\theta)]((1-\cos\theta) + i\sin\theta)}{(1-\cos\theta)^2 + \sin^2\theta} \\ &= \frac{(1-\cos(N+1)\theta)(1-\cos\theta) - (\sin(N+1)\theta)(\sin\theta)}{2(1-\cos\theta)} \\ &\quad + i\frac{(1-\cos\theta)(\sin(N+1)\theta) + \sin\theta(1-\cos(N+1)\theta)}{2(1-\cos\theta)}. \end{aligned}$$

We conclude

$$\sum_{n=0}^{N} \sin n\theta = \frac{(1-\cos\theta)(\sin(N+1)\theta) + \sin\theta(1-\cos(N+1)\theta)}{2(1-\cos\theta)} \quad \text{and}$$

$$\sum_{n=0}^{N} \cos n\theta = \frac{(1-\cos(N+1)\theta)(1-\cos\theta) - (\sin(N+1)\theta)(\sin\theta)}{2(1-\cos\theta)}.$$

Example 6.74

$\frac{1}{x} + x = \frac{x^2+1}{x}$. Since $x^2 + 1$ is always positive, the sign of x in the denominator controls the sign of the entire expression. Certainly, x cannot be zero, but the inequation is true for all $x > 0$.

Example 6.75

After some trial divisions with small linear factors, we see that $f(x) = x^3 - 6x^2 + 11x - 6 = (x-1)(x-2)(x-3)$. So the problem is to trace the sign parity of this product. If $x = 1, 2,$ or $3, f(x) = 0$, so those values are excluded from the solution set. For $x < 1$, all three factors are negative, so $f(x) < 0$. For $1 < x < 2$, the factor $(x-1) > 0$ and this causes $f(x) > 0$. For $2 < x < 3$, the factor $(x-1) > 0$ and $(x-2) > 0$, so $f(x) < 0$. Finally, if $x > 3$ then $f(x) > 0$. We see the solution set for the inequation is $\{x : f(x) < 0\} = (-\infty, 1) \cup (2, 3)$.

Example 6.76

Here $f(x) = \frac{x^2-4x+3}{x-2} = \frac{(x-1)(x-3)}{x-2}$, so the same factors as in Example 6.75 create the same sign parity analysis and again $\{x : f(x) < 0\} = (-\infty, 1) \cup (2, 3)$. Note that $x = 2$ is disqualified for a different reason here than in the prior example.

Example 6.77

Here $f(x) = \frac{x+1}{x-2} - 2 = \frac{(x+1)-2(x-2)}{x-2} = -\frac{x-5}{x-2}$. The parity points are 2 and 5. Note that x can be 5 but not 2. If $x < 2$ then both $x - 5 < 0$ and $x - 2 < 0$, so their ratio is positive, hence $f(x) < 0$. If $2 < x < 5$ then $x - 5 < 0$ but $x - 2 > 0$, so their ratio is negative, hence $f(x) > 0$. If $x = 5$, then $f(5) = 0$. If $x > 5$ then both $x - 5 > 0$ and $x - 2 > 0$, so their ratio is positive, hence again $f(x) < 0$. The solution set for the inequation is $\{x : f(x) \geq 0\} = (2, 5]$.

Example 6.78

Here $f(x) = \frac{x^2(x+1)}{x-2}$. The parity points are -1, 0, and 2, with 0 a double. Passing a double (or quadruple or any even tuple) parity point does not actually change parity of the expression, although it may be prohibited from taking on certain values at the point itself. If $x \leq -1$ then $x^2 > 0$ and both $(x + 1) \leq 0$ and $(x - 2) < 0$, so $f(x) \geq 0$. If $-1 < x < 2$ then $x^2 \geq 0$, $(x + 1) > 0$, and $(x - 2) < 0$. So $f(x) \leq 0$. Note $x \neq 2$. If $x > 2$, all factors are positive and $f(x) \geq 0$. The solution set for the inequation is $\{x : f(x) \geq 0\} = (-\infty, -1] \cup (2, \infty) \cup \{0\}$.

Example 6.79

This is equivalent to $\frac{x+1}{x-2} > 1$ or $\frac{x+1}{x-2} < -1$. For $\frac{x+1}{x-2} > 1$, we convert it to $\frac{x+1}{x-2} - 1 > 0$ or $\frac{3}{x-2} > 0$. Then the solution set is $x \in (2, \infty)$. The other case is handled similarly.

Example 6.80

We need to find all values of x such that $-1 < \frac{1}{x-4} < 1$. The left-hand inequality can be written as $\frac{1}{x-4} + 1 > 0$ or $\frac{x-3}{x-4} > 0$. Note that the parity points are $3 < 4$. If $x < 3$, the numerator and denominator are both negative, so $\frac{x-3}{x-4} > 0$. If $3 < x < 4$, the fraction goes negative; and if $x > 4$, it goes back to positive. So the solution set for $-1 < \frac{1}{x-4}$ is $(-\infty, 3) \cup (4, \infty)$. The right-hand inequality can be written as $\frac{1}{x-4} - 1 < 0$ or $-\frac{x-5}{x-4} < 0$. Flipping the sense of the inequality by multiplying by -1, we get $\frac{x-5}{x-4} > 0$. Then for $x < 4$ the fraction is positive, for $4 < x < 5$ it becomes negative, and for $x > 5$ it goes back to positive. So the solution set for $\frac{1}{x-4} < 1$ is $(-\infty, 4) \cup (5, \infty)$. The solution set for $|\frac{1}{x-4}| < 1$ is the intersection of the two previous solution sets, hence $(-\infty, 3) \cup (5, \infty)$.

Example 6.81

The polynomial $P(x) = x^2 - 5$ has a parabolic graph that crosses the x-axis at the roots $\pm\sqrt{5}$. We know $2 < \sqrt{5} < 2.5$ (note $(2.5)^2 = 6.25$). So we pick $x_1 = 2.3$ as the **seed** or **initialization** value. Note $P'(x) = 2x$, so the slope of the tangent at $x_1 = 2.3$ is $P'(x_1) = 2x_1 = 4.6$. Using the bootstrap formula, we get $x_2 = 2.3 - \frac{(2.3)^2-5}{4.6} = 2.3 - \frac{0.29}{4.6} = 2.237$. Repeating, we have $x_3 = 2.237 - \frac{(2.237)^2-5}{4.474}$. This reduces to $x_3 = 2.237 - \frac{0.00417}{4.474} = 2.236$. If we had saved more decimals we could go further, but $\sqrt{5} = 2.236068$ to six decimals, so with just an initial (reasonable) guess and two iterations of the algorithm we are accurate to three decimals.

Example 6.82

So we redo Example 6.81 with $x_1 = 1$. Here $P'(x) = 2x$ does not change, so $P'(x_1) = 2$. The formula gives us $x_2 = 1 - \frac{(1)^2-5}{2} = 1 - (-2) = 3$. Suddenly we are higher than the x_2 from Example 6.81, so we are hoping the next iteration brings us back down. Repeating, $x_3 = 3 - \frac{(3)^2-5}{6} = 3 - \frac{2}{3} = 2.333$. That is better. One more time, $x_4 = 2.333 - \frac{(2.333)^2-5}{4.666} = 2.333 - 0.0949 = 2.238$. So the approximants are getting closer but with a little more work this time because we disadvantaged the algorithm with a more extreme seed value.

Example 6.83

Here $P(x) = x^5 + x + 1$, so $P'(x) = 5x^4 + 1$. Any positive x will not be close to the root, but a good seed value guess might be negative, a little greater than -1. The fifth power will reduce it considerably in absolute value and then adding the first power might be close to cancelling the 1. All of this is speculative at the moment. Set $x_1 = -0.9$. Then $P'(-0.9) = 4.2805$. Now

$$x_2 = -0.9 - \frac{(-0.9)^5 - 0.9 + 1}{4.2805} = -0.78541.$$

One more iteration gives

$$x_3 = -0.78541 - \frac{(-0.78541)^5 - 0.78541 + 1}{5(-0.78541)^4 + 1} = -0.78541 - \frac{-0.08429}{2.9027}$$
$$= -0.78541 + 0.02904 = -0.7564.$$

How close are we to a root? We plug it into the polynomial and get $(-0.7564)^5 - 0.7564 + 1 = 0.003937$. So the verdict is "pretty close." A few more iterations would be an improvement, but we already have a serviceable estimate of $BR(1)$.

Example 6.84

Here $P(x) = x^5 - 5x + 3$, so $P'(x) = 5x^4 - 5 = 5(x^4 - 1)$. Note that $P(0) = 3 > 0$ and $P(1) = -1 < 0$, so there is a root in the interval $(0, 1)$. Let us split the difference and initialize with $x_1 = 0.5$. Then $P'(0.5) = -4.6875$. So $x_2 = 0.5 - \frac{0.53125}{-4.6875} = 0.61333$. Another iteration gives $x_3 = 0.61333 - \frac{0.020140}{-4.29247} = 0.61802$. One more is $x_4 = 0.61802 - \frac{0.000051183}{-4.27057} = 0.61803$. How close are we to a root? Now $(0.61803)^5 - 5(0.61803) + 3 = 0.000017$. So we are good to about two parts in $100,000$. Recalling the golden ratio calculation from Section 6.2, we are suspicious that the root might be related to the solution we discarded in Example 6.83, namely $x = \frac{1-\sqrt{5}}{2} = -0.61803$. Let us try $x = \frac{\sqrt{5}-1}{2}$ in the equation, $(\frac{\sqrt{5}-1}{2})^5 - 5(\frac{\sqrt{5}-1}{2}) + 3 = 0$. This was just blind luck, but even a blind squirrel finds an acorn occasionally. In principle, we could now divide $x^5 - 5x + 3$ by $(x - \frac{1-\sqrt{5}}{2})$ to obtain a quartic, which could then be solved with Ferrari's method in Section 6.1.4.

7 Trigonometric methods

Trigonometry is based on the pleasant fact that the ratios between the lengths of corresponding sides of similar triangles are independent of the size of the specific triangles. In particular, if the triangles contain a right angle, then it is worthwhile to tabulate once and for all the ratios of the lengths between the six possible pairs of the three sides. These ratios can be expressed as functions of the angles appearing in a triangle. With this information compiled it is possible to solve all manner of surveying problems, terrestrial or astronomical. In fact, early trigonometry grew out of a desire to extend classical Greek geometry to problems of astronomy. The Hellenic mathematicians framed their theory in terms of the length of chords of circles rather than the ratios of lengths of triangle sides, but ultimately the results obtained by Euclid and Archimedes were equivalent to the modern theory.

The first systematic tabulation of trigonometric information was made by Hipparchus of Nicaea in the second century BC. He is usually regarded as the "father of trigonometry." However, the opus magnus of chord-based trigonometry was written by Ptolemy of Alexandria in the second century AD. His work, the Almagest, was an influential treatise on mathematical astronomy. So influential, in fact, that it cemented the notion of a geocentric universe in the heads of astronomers until Copernicus made a successful case for a heliocentric system roughly 1300 years later. Towards the end of the first millennium, several noted Persian mathematicians developed trigonometry in a direction that would be more familiar to us today. They explicitly defined the six modern trigonometric functions, created detailed tables of values, and extended the subject to spherical trigonometry. The mathematician al-Tusi should perhaps then be called the "godfather of trigonometry," as he was the first to treat trigonometry as a branch of mathematics separate from its applications to astronomy and surveying.

Translations of the Greek and Persian works on trigonometry reached Europe by the early Renaissance, and it was fortuitous that the age of exploration was just beginning. The desire for more dependable methods of navigation and improved cartography drove further refinement of trigonometry into a robust applied mathematical discipline. In the 1460s the German mathematician Regiomontanus (aka Johannes Müller of Königsberg) assembled a table of sine values to five decimals for angles at intervals of one minute. This was a tour de force in hand calculation. Over the years "trig" has remained important for its practical use in supporting the surveying profession, the physics and engineering disciplines, and its original focus on astronomy. But it also has had an important behind-the-scenes role in higher mathematics. In the course of studying the physics of heat, the French mathematician Joseph Fourier invented the idea of expressing arbitrary functions in terms of series of trigonometric functions. This grew into the field of harmonic analysis. The early work of the German mathematician Georg Cantor on the convergence properties of trigonometric series led him to develop the basic notions of set theory, which eventually blossomed into another foundational branch of mathematics. Trig is everywhere, even if it is not obvious.

https://doi.org/10.1515/9783111506654-008

7.1 Basic ideas & definitions

7.1.1 Angle measure

We assume the reader is familiar with the very basic definitions from Euclidean geometry. The angle θ between two different radii of a circle is called a **central angle**. It can be anything between 0° and 360°, where $x°$ is read "x degrees." A more natural way to measure central angles for a given circle is to consider the length of the arc that the angle subtends (is across from) and compute the ratio of this arc length to the entire circumference of the circle. We define an **angle measure** of one **radian** if the arc subtended by the angle has the same length as the radius of the circle. Clearly, a circle has total central angle measure 2π radians since the circumference C is 2π times the radius R. If an angle is given as θ radians, the length L of the arc subtending it is given by the **central angle formula** $L = R\theta$. By its definition as a proportion, the radian is a dimensionless quantity. Note that a 360° angle then has measure 2π radians. So 1 radian equals $\frac{360}{2\pi} \approx 57.2958°$ and, conversely, $1° \approx 0.01745$ radians.

Example 7.1. A company wants to put its name on the side of a large cylindrical tank near the highway. The tank is 40 feet in diameter and to be visible from the highway the sign cannot extend more than 30° around the circumference. How long can the sign be?

7.1.2 Trigonometric functions

Figure 7.1 below can be considered the ur-diagram for trigonometry. We have drawn a **right triangle** embedded in a circle of radius 1 in the xy-plane centered at the origin.

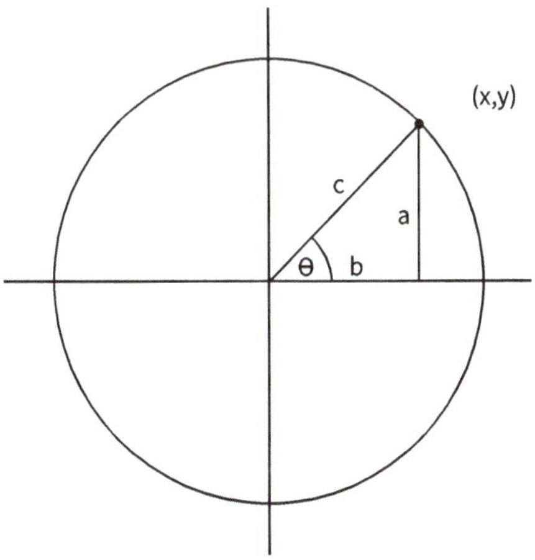

Figure 7.1: A circle with an embedded triangle in the first quadrant defining the central angle θ between two radii.

This is the **unit circle**. The **hypotenuse** of the right triangle will correspond to a radius of the unit circle. This drawing will allow us to visualize various definitions involving angles greater than those available in a drawing of a triangle alone.

The angle θ is the central angle formed by the positive x-axis and the hypotenuse of the right triangle. Of the two legs of the right angle, one coincides with the x-axis and is called the **adjacent side**, and the other is the **opposite side**. Every point on the circle corresponds to a unique right triangle. Note that θ ranges from 0 to 360 in degrees or 0 to 2π in radians. We should try to be flexible in referring to the measure of an angle in both degrees and radians. Granted, degrees are probably more ingrained in our everyday thinking, but radians are more useful mathematically. Until this becomes second nature, you may feel like a person accustomed to degrees Fahrenheit being told it is 20° Celsius, and then not being sure about wearing a sweater.

For the triangle in Figure 7.1, we see the side opposite angle θ is labeled a, the adjacent side b, and the hypotenuse c. We make the following definitions:

Table 7.1: The six trig functions of angle θ.

function	designation	definition	domain	range
sine	$\sin\theta$	$\frac{a}{c}$	$\mathbb{R}$	$[-1, 1]$
cosine	$\cos\theta$	$\frac{b}{c}$	$\mathbb{R}$	$[-1, 1]$
tangent	$\tan\theta$	$\frac{a}{b}$	$\mathbb{R} - \mathbb{O}$	$(-\infty, \infty)$
secant	$\sec\theta$	$\frac{c}{b}$	$\mathbb{R} - \mathbb{O}$	$(-\infty, \infty) - (-1, 1)$
cosecant	$\csc\theta$	$\frac{c}{a}$	$\mathbb{R} - \mathbb{E}$	$(-\infty, \infty) - (-1, 1)$
cotangent	$\cot\theta$	$\frac{b}{a}$	$\mathbb{R} - \mathbb{E}$	$(-\infty, \infty)$

Here $\mathbb{O}$ is the set of odd integer multiples of $\frac{\pi}{2}$; $\mathbb{E}$ is the set of even integer multiples of π. Recall that in the context of sets the minus sign indicates relative complement.

Some key facts are stated below:

1) By the Pythagorean theorem, we have $1 = c = \sqrt{a^2 + b^2}$, so $1 \geq |a|$ and $1 \geq |b|$. Note a or b (but not both together) may be zero.
2) If (i) $0 \leq \theta \leq \frac{\pi}{2}$, then $0 \leq a, b \leq 1$; (ii) if (i) $\frac{\pi}{2} \leq \theta \leq \pi$, then $-1 \leq b \leq 0$ but $0 \leq a \leq 1$; (iii) if $\pi \leq \theta \leq \frac{3\pi}{2}$, then $-1 \leq a, b \leq 0$; (iv) if $\frac{3\pi}{2} \leq \theta \leq 2\pi$, then $0 \leq b \leq 1$ but $-1 \leq a \leq 0$.
3) An important consequence of the preceding inequalities is that if we reflect the angle θ to $-\theta$, b remains positive and then $\cos(-\theta) = \cos\theta$. Unlike b, a changes sign, so $\sin(-\theta) = -\sin\theta$. The upshot is that cosine is an even function and sine is odd.
4) If a and b correspond to a particular θ, then the same a and b correspond to the angle $\theta + 2\pi k$ for any $k \in \mathbb{Z}$. This gives trigonometric functions their characteristic property of periodicity. A function $f(x)$ is periodic with period τ if $f(x + \tau) = f(x)$, where $\tau > 0$ is the minimal number for which this is true. The sine, cosine, and their reciprocals have period 2π. The tangent and cotangent have period π.

5) From Figure 7.1 and Table 7.1, it is easy to see the relationship between the functions sine, tangent, and secant, as well as their respective cofunctions cosine, cotangent, and cosecant. The sine of an angle θ is the cosine of the **complementary angle** $(\frac{\pi}{2} - \theta)$, and likewise for the others.
6) Also from Table 7.1, we may construct reciprocal relationships that may be used to reduce general trig expressions to those featuring only sines, cosine, and tangents. We can even go further and write $\tan\theta = \frac{\sin\theta}{\cos\theta}$. It is often a good strategy to simplify complicated expressions to just sines and cosines to assess what cancellations can be made.

Table 7.2: Reciprocal formulas.

$\sin\theta = \frac{1}{\csc\theta}$
$\cos\theta = \frac{1}{\sec\theta}$
$\tan\theta = \frac{1}{\cot\theta}$

7) It is clear from the definitions in Table 7.1 that the sine and cosine are defined for any $\theta \in \mathbb{R}$, but we have to be careful with tangent and secant when $a = 0$ and with cotangent and cosecant when $b = 0$. Apparently, the angles where tangent and secant are undefined are $\frac{\pi}{2} + 2\pi k$ and $\frac{3\pi}{2} + 2\pi k$. Likewise, cotangent and cosecant are undefined for angles $2\pi k$ and $(2k + 1)\pi$. You can see this also from the reciprocal relationships in Table 7.2. The secant and cosecant can never be zero, so their respective reciprocals cosine and sine are always defined everywhere. Conversely, the sine and cosine can be zero, and at those angles their reciprocals are undefined. Likewise, for the tangent and its reciprocal cotangent, each of which can be zero at selected angles forcing the other one to be undefined there.
8) The following table presents the values of the six trig functions for commonly encountered angles in both degrees and radians.

Table 7.3: Common trig function values (θ in deg/rad).

$\theta \rightarrow$	$0°/0$	$30°/\frac{\pi}{6}$	$36°/\frac{\pi}{5}$	$45°/\frac{\pi}{4}$	$60°/\frac{\pi}{3}$	$90°/\frac{\pi}{2}$	$120°/\frac{2\pi}{3}$	$135°/\frac{3\pi}{4}$	$180°/\pi$
$\sin\theta$	0	$\frac{1}{2}$	$\frac{1}{2}\sqrt{\frac{5-\sqrt{5}}{2}}$	$\frac{\sqrt{2}}{2}$	$\frac{\sqrt{3}}{2}$	1	$\frac{\sqrt{3}}{2}$	$\frac{\sqrt{2}}{2}$	0
$\cos\theta$	1	$\frac{\sqrt{3}}{2}$	$\frac{\sqrt{5}+1}{4}$	$\frac{\sqrt{2}}{2}$	$\frac{1}{2}$	0	$-\frac{1}{2}$	$-\frac{\sqrt{2}}{2}$	-1
$\tan\theta$	0	$\frac{\sqrt{3}}{3}$	$\frac{\sqrt{5-\sqrt{5}}(\sqrt{5}-1)}{2\sqrt{2}}$	1	$\sqrt{3}$	∞	$-\sqrt{3}$	-1	0
$\sec\theta$	1	$\frac{2\sqrt{3}}{3}$	$\sqrt{5}-1$	$\sqrt{2}$	2	∞	-2	$-\sqrt{2}$	-1
$\csc\theta$	∞	2	$\frac{2\sqrt{2}}{\sqrt{5-\sqrt{5}}}$	$\sqrt{2}$	$\frac{2\sqrt{3}}{3}$	1	$\frac{2\sqrt{3}}{3}$	$\sqrt{2}$	∞
$\cot\theta$	∞	$\sqrt{3}$	$\frac{2\sqrt{2}}{\sqrt{5-\sqrt{5}}(\sqrt{5}-1)}$	1	$\frac{\sqrt{3}}{3}$	0	$-\frac{\sqrt{3}}{3}$	-1	∞

Phenomena that vary in a cyclical pattern with respect to time or space can often be modeled by sines and cosines. A simple example is $f(t) = A\sin(\omega t)$, where t is time in seconds. This represents a **sinusoidal waveform** with A as a constant called the **amplitude** that determines the scale of the waveform. Amplitude is defined as half the height of the overall waveform. As the wave progresses $f(t)$ oscillates between **peaks** at $+A$ and **troughs** at $-A$. How often this occurs is governed by the frequency ω of the wave. Usually, ω is given in radians per second, so the product ωt is a pure angle. The time between successive peaks (or troughs) is the **period** T of the wave. The relationship between period and frequency is $\omega T = 2\pi$. The way to see this is to note that the argument of the sine must increase by 2π for $f(t)$ to return to the same value. Also, T is the time interval necessary for this to happen, so $\omega\ [\frac{\text{radians}}{\text{s}}] \times T\ [\text{s}] = 2\pi$, which corresponds to one complete cycle of the wave. In the literature, it is customary to use ω as frequency in radians per second and ν as frequency in cycles per second. So $\omega = 2\pi\nu$. The metric unit hertz (Hz) for oscillatory phenomena is *cycles* per second.

With just the basics considered above, we can solve a number of simple problems. You can provide the easy sketches if needed.

Example 7.2. The sun is 60° above the horizon and a 100-foot tall building casts a shadow on level ground. How long is the shadow?

Example 7.3. Emperor Hirohito was being fitted for a new suit by his tailor. Since the Emperor was considered divine and the tailor was one of the common people, the tailor was not actually allowed to touch the Emperor, or, in fact, get within two meters of him. This presented an acute sartorial problem. The solution was to place a surveyor's theodolite (angle measuring instrument) across the room from the Emperor and take angle measurements based on parts of the Emperor's body as he faced the instrument. If the theodolite was lined up two meters from the Emperor's chin (the centerpoint of the Emperor) and the total angle between the outside of his shoulders was 13.75°, how wide were his shoulders?

Example 7.4. The largest single-stem tree in the world, measured by volume, is the General Sherman tree, a sequoia in a California national park. The base diameter is 36.5 [ft]. If you were able to stand 400 [ft] from the base of the tree and sight the top of it at an angle of 33.3°, what would your estimate of the height be?

Example 7.5. If an object is viewed from two different viewpoints that are not in line with the object, there is an apparent shift in the direction of the object. The angle of the spurious shift is called **parallax**. A celestial object exhibits parallax when being viewed from the Earth as the Earth orbits the Sun. A **parsec** (parallax of one second) is a unit of distance used in astronomy. If the radius of the Earth's orbit (one astronomical unit) subtends an angle of one second of arc at a point in space, then by definition that point is one parsec distant from the Earth. How many kilometers are in a parsec? Recall one degree (°) equals 60 minutes (′), and one minute equals 60 seconds (″). Incidentally, Hipparchus was the first to measure the parallax of the Moon.

Example 7.6. Write an equation for the output of an alternating current generator with maximum voltage 240 [V] at 50 [Hz].

7.2 Trigonometric formulas

There are quite a few formulas that facilitate working with trig functions. We will review them from simplest to most complex. The reciprocal relationships in Table 7.2 have already been discussed.

7.2.1 Pythagorean identities

With the notation and setup in Figure 7.1, the Pythagorean theorem gives us $a^2+b^2=c^2$. Dividing all terms by c^2, we get $\frac{a^2}{c^2}+\frac{b^2}{c^2}=\frac{c^2}{c^2}$. With the definitions in Table 7.1, this is equivalent to $\sin^2\theta+\cos^2\theta=1$. Similarly, if we divide by b^2 and convert to trig functions, we have $\tan^2\theta+1=\sec^2\theta$, and if we divide by a^2, we recover $1+\cot^2\theta=\csc^2\theta$. These are summarized below, written in the form most usually encountered in textbooks. It is hard to overstate the importance of $\sin^2\theta+\cos^2\theta=1$. This identity seems to show up everywhere. The first two identities are hugely important in advanced calculus because they facilitate integration by substitution. That is beyond the scope of our text, but suffice it to say they are frequently encountered and often indispensable.

Table 7.4: Pythagorean identities.

$\sin^2\theta+\cos^2\theta=1$
$1+\tan^2\theta=\sec^2\theta$
$1+\cot^2\theta=\csc^2\theta$

7.2.2 Addition/subtraction of angles

It will be useful to have formulas expressing the sine, cosine, and tangent of the sum or difference of two angles. These will allow us to get the corresponding formulas for cosecant, secant, and cotangent by using the reciprocal relationships. Let α and β be two arbitrary angles. For strategic reasons, we start with the formula for $\cos(\alpha-\beta)$. Justifying these types of formulas from scratch with complicated drawings and geometrical arguments is doable but long-winded and messy. Here is a more efficient way to get them.

1) Consider $\cos(\alpha-\beta)$. Referring to the usual setup in Figure 7.1, let $x_1=\cos\alpha$ and $y_1=\sin\alpha$. Also let $x_2=\cos\beta$ and $y_2=\sin\beta$. Now $P_1=(\cos\alpha,\sin\alpha)$ is the point on the unit circle defined by α and $P_2=(\cos\beta,\sin\beta)$ is the corresponding point defined by β. For the sake of definiteness, let us assume $\alpha\geq\beta$. By the two-dimensional distance

formula, the distance between P_1 and P_2 is then $d(P_1P_2) = \sqrt{(x_1 - x_2)^2 + (y_1 - y_2)^2} = \sqrt{(\cos\alpha - \cos\beta)^2 + (\sin\alpha - \sin\beta)^2}$. Hold that thought. Now rotate P_1 and P_2 together along the unit circle through the angle $-\beta$ keeping them the same distance apart until P_2 coincides with the point $(1, 0)$. The original angular (not linear) separation between P_1 and P_2 was $\alpha - \beta$. It still is after the rotation, but now identifying them as P_1' and P_2' we have $P_1' = (\cos(\alpha - \beta), \sin(\alpha - \beta))$ and $P_2' = (1, 0)$. What is the distance between P_1' and P_2'? Now $d(P_1'P_2') = \sqrt{(\cos(\alpha - \beta) - 1)^2 + (\sin(\alpha - \beta) - 0)^2}$. Recalling that $d(P_1P_2) = d(P_1'P_2')$, we can square both distances and finally write $(\cos\alpha - \cos\beta)^2 + (\sin\alpha - \sin\beta)^2 = (\cos(\alpha - \beta) - 1)^2 + (\sin(\alpha - \beta) - 0)^2$. Multiplying everything out on the left, we get $\cos^2\alpha - 2\cos\alpha\cos\beta + \cos^2\beta + \sin^2\alpha - 2\sin\alpha\sin\beta + \sin^2\beta$. Doing likewise on the right, we get $\cos^2(\alpha - \beta) - 2\cos(\alpha - \beta) + 1 + \sin^2(\alpha - \beta)$. Already we can use the $\sin^2\theta + \cos^2\theta = 1$ identity to rewrite the equation as $2 - 2\cos\alpha\cos\beta - 2\sin\alpha\sin\beta = 2 - 2\cos(\alpha - \beta)$. This reduces immediately to $\cos\alpha\cos\beta + \sin\alpha\sin\beta = \cos(\alpha - \beta)$ as required.

2) Now we can do $\cos(\alpha + \beta) = \cos(\alpha - (-\beta))$ simply by invoking the odd/even nature of sine and cosine as explained in the comments after Table 7.1 above. So $\cos(\alpha - (-\beta)) = \cos\alpha\cos(-\beta) + \sin\alpha\sin(-\beta)$. Since the cosine is an even function and the sine is odd, this simplifies to $\cos(\alpha + \beta) = \cos\alpha\cos\beta - \sin\alpha\sin\beta$.
3) For $\sin(\alpha + \beta)$, we note that

$$\sin(\alpha + \beta) = \cos\left(\frac{\pi}{2} - (\alpha + \beta)\right) = \cos\left(\left(\frac{\pi}{2} - \alpha\right) - \beta\right)$$

by the cofunction property. But

$$\cos\left(\left(\frac{\pi}{2} - \alpha\right) - \beta\right) = \cos\left(\frac{\pi}{2} - \alpha\right)\cos\beta + \sin\left(\frac{\pi}{2} - \alpha\right)\sin\beta.$$

Using the cofunction property again, we finally have

$$\sin(\alpha + \beta) = \cos\left(\frac{\pi}{2} - \alpha\right)\cos\beta + \sin\left(\frac{\pi}{2} - \alpha\right)\sin\beta = \sin\alpha\cos\beta + \cos\alpha\sin\beta.$$

4) Last but not least, we write $\sin(\alpha - \beta) = \sin(\alpha + (-\beta)) = \sin\alpha\cos(-\beta) + \cos\alpha\sin(-\beta)$. Since the cosine is an even function and the sine is odd, this reduces to $\sin(\alpha - \beta) = \sin\alpha\cos\beta - \cos\alpha\sin\beta$.
5) The tangent is a little trickier. We can write

$$\tan(\alpha + \beta) = \frac{\sin(\alpha + \beta)}{\cos(\alpha + \beta)} = \frac{\sin\alpha\cos\beta + \cos\alpha\sin\beta}{\cos\alpha\cos\beta - \sin\alpha\sin\beta}$$

from 2) and 3) above. Dividing numerator and denominator both by $\cos\alpha\cos\beta$, we get

$$\tan(\alpha+\beta) = \frac{\left(\frac{\sin\alpha\cos\beta+\cos\alpha\sin\beta}{\cos\alpha\cos\beta}\right)}{\left(\frac{\cos\alpha\cos\beta-\sin\alpha\sin\beta}{\cos\alpha\cos\beta}\right)}.$$

Now simplifying by using $\frac{\sin\alpha}{\cos\alpha} = \tan\alpha$ and $\frac{\sin\beta}{\cos\beta} = \tan\beta$, we have $\tan(\alpha+\beta) = \frac{\tan\alpha+\tan\beta}{1-\tan\alpha\tan\beta}$.

6) Finally, $\tan(\alpha-\beta) = \tan(\alpha+(-\beta)) = \frac{\tan\alpha+\tan(-\beta)}{1-\tan\alpha\tan(-\beta)}$. Since the tangent is the ratio of an odd function to an even function, it behaves as an odd function. So $\tan(\alpha-\beta) = \frac{\tan\alpha-\tan\beta}{1+\tan\alpha\tan\beta}$. These formulas are summarized in the following table.

Table 7.5: Angle addition/subtraction formulas.

$\sin(\alpha+\beta) = \sin\alpha\cos\beta + \cos\alpha\sin\beta$
$\sin(\alpha-\beta) = \sin\alpha\cos\beta - \cos\alpha\sin\beta$
$\cos(\alpha+\beta) = \cos\alpha\cos\beta - \sin\alpha\sin\beta$
$\cos(\alpha-\beta) = \cos\alpha\cos\beta + \sin\alpha\sin\beta$
$\tan(\alpha+\beta) = \frac{\tan\alpha+\tan\beta}{1-\tan\alpha\tan\beta}$
$\tan(\alpha-\beta) = \frac{\tan\alpha-\tan\beta}{1+\tan\alpha\tan\beta}$

Example 7.7. What are the sine, cosine, and tangent of 75°? Do not use a calculator.

7.2.3 Double & half angle formulas

The addition formulas give us the double angle formulas immediately:

1) $\sin 2\theta = \sin(\theta+\theta) = \sin\theta\cos\theta + \cos\theta\sin\theta = 2\sin\theta\cos\theta$.
2) $\cos 2\theta = \cos(\theta+\theta) = \cos^2\theta - \sin^2\theta$. Using the identity $\sin^2\theta+\cos^2\theta = 1$, we also have $\cos 2\theta = \cos^2\theta-(1-\cos^2\theta) = 2\cos^2\theta-1$ and $\cos 2\theta = (1-\sin^2\theta)-\sin^2\theta = 1-2\sin^2\theta$.
3) $\tan 2\theta = \tan(\theta+\theta) = \frac{2\tan\theta}{1-\tan^2\theta}$.
4) Using $\cos 2\theta = 1-2\sin^2\theta$ from above, we can solve for $\sin\theta = \pm\sqrt{\frac{1-\cos 2\theta}{2}}$. Letting $\alpha = 2\theta$, we get the half-angle sine formula $\sin\frac{\alpha}{2} = \pm\sqrt{\frac{1-\cos\alpha}{2}}$.
5) Likewise, we can solve $\cos 2\theta = 2\cos^2\theta - 1$ for $\cos\theta = \pm\sqrt{\frac{1+\cos 2\theta}{2}}$. Letting $\alpha = 2\theta$, we get the half-angle cosine formula $\cos\frac{\alpha}{2} = \pm\sqrt{\frac{1+\cos\alpha}{2}}$.
6) The half-angle formula for the tangent is an easy consequence of the two preceding half-angle formulas,

$$\tan\frac{\alpha}{2} = \frac{\sin\frac{\alpha}{2}}{\cos\frac{\alpha}{2}} = \pm\sqrt{\frac{1-\cos\alpha}{1+\cos\alpha}}.$$

Some discussion is warranted concerning the action of doubling or halving an angle on the sign of the output of one of our formulas. For example, $\cos 120^\circ = \cos^2 60^\circ - \sin^2 60^\circ$. We are expecting $\cos 120^\circ$ to be negative because for angles between 90° and 180° the hypotenuse in Figure 7.1 is in the second quadrant of the xy-plane where b is negative. The doubling formula automatically makes the adjustment. Once an angle between 0° and 90° exceeds 45°, its sine exceeds its cosine. So if we double an angle θ between 0° and 45°, the result 2θ will be between 0° and 90°, and since $\cos^2\theta - \sin^2\theta \geq 0$ for $0^\circ \leq \theta \leq 45^\circ$, $\cos 2\theta \geq 0$, as it should be. If we double an angle θ between 45° and 90°, the result 2θ will be between 90° and 180°, and since $\cos^2\theta - \sin^2\theta \leq 0$ for $45^\circ \leq \theta \leq 90^\circ$, $\cos 2\theta \leq 0$, also as it should be. This observation extends to all $0^\circ \leq \theta \leq 360^\circ$ and the sine and tangent exhibit similar behavior.

The half-angle formulas are not so self-correcting. Depending on the quadrant of the xy-plane in which the terminal side (hypotenuse in Figure 7.1) of the angle $\frac{\alpha}{2}$ falls, we may have to manually adjust the sign in front of the square root, which, of course, is always considered positive. For illustration, suppose the terminal side of the angle $\frac{\alpha}{2}$ is in the third or fourth quadrant. Then its sine has to be negative and we would use the formula $\sin\frac{\alpha}{2} = -\sqrt{\frac{1-\cos\alpha}{2}}$. Otherwise, $\sin\frac{\alpha}{2} = +\sqrt{\frac{1-\cos\alpha}{2}}$. Here is a chart indicating which sign to use for sine, cosine, and tangent in all four quadrants.

Table 7.6: Half-angle sign parities.

Quadrant ↓	$\sin\frac{\alpha}{2}$	$\cos\frac{\alpha}{2}$	$\tan\frac{\alpha}{2}$
1st	+	+	+
2nd	+	–	–
3rd	–	–	+
4th	–	+	–

Example 7.8. Find the sine, cosine, and tangent of 15°.

Example 7.9. Find the cosine of 795°.

Example 7.10. Find a recursion formula for sine and cosine.

7.2.4 Product & sum formulas

We turn our attention to formulas for products like $\sin\alpha\sin\beta$. These are called generically "product-to-sum" formulas:

1) Consider

$$\cos(\alpha - \beta) - \cos(\alpha + \beta) = (\cos\alpha\cos\beta + \sin\alpha\sin\beta) - (\cos\alpha\cos\beta - \sin\alpha\sin\beta)$$
$$= 2\sin\alpha\sin\beta.$$

We immediately have the formula $\sin\alpha\sin\beta = \frac{1}{2}(\cos(\alpha-\beta) - \cos(\alpha+\beta))$.

2) Now consider

$$\cos(\alpha - \beta) + \cos(\alpha + \beta) = (\cos\alpha\cos\beta + \sin\alpha\sin\beta) + (\cos\alpha\cos\beta - \sin\alpha\sin\beta)$$
$$= 2\cos\alpha\cos\beta.$$

This gives us $\cos\alpha\cos\beta = \frac{1}{2}(\cos(\alpha-\beta) + \cos(\alpha+\beta))$.

3) Finally, compute

$$\sin(\alpha + \beta) + \sin(\alpha - \beta) = (\sin\alpha\cos\beta + \cos\alpha\sin\beta) + (\sin\alpha\cos\beta - \cos\alpha\sin\beta)$$
$$= 2\sin\alpha\cos\beta.$$

So $\sin\alpha\cos\beta = \frac{1}{2}(\sin(\alpha+\beta) + \sin(\alpha-\beta))$.
Conversely, we can derive expressions for sums like $\sin\alpha + \sin\beta$. Unsurprisingly, these are called "sum-to-product" formulas.

4) Let $u = \frac{\alpha+\beta}{2}$ and $v = \frac{\alpha-\beta}{2}$. Now use the formula from 1) above to write $\sin u\cos v = \frac{1}{2}(\sin(u+v) + \sin(u-v))$. Note that $u+v = \alpha$ and $u-v = \beta$. Backsubstituting we have

$$\sin\left(\frac{\alpha+\beta}{2}\right)\cos\left(\frac{\alpha-\beta}{2}\right) = \frac{1}{2}(\sin\alpha + \sin\beta).$$

Then we see that

$$\sin\alpha + \sin\beta = 2\sin\left(\frac{\alpha+\beta}{2}\right)\cos\left(\frac{\alpha-\beta}{2}\right).$$

5) Starting with 4), change β to $-\beta$. Then

$$\sin\alpha + \sin(-\beta) = 2\sin\left(\frac{\alpha+(-\beta)}{2}\right)\cos\left(\frac{\alpha-(-\beta)}{2}\right).$$

Since the sine is an odd function, this gives

$$\sin\alpha - \sin\beta = 2\sin\left(\frac{\alpha-\beta}{2}\right)\cos\left(\frac{\alpha+\beta}{2}\right).$$

6) With u and v as in 4), we have $\cos u\cos v = \frac{1}{2}(\cos(u-v) + \cos(u+v))$ from 2). Backsubstituting gives

$$\cos\left(\frac{\alpha+\beta}{2}\right)\cos\left(\frac{\alpha-\beta}{2}\right) = \frac{1}{2}(\cos\beta + \cos\alpha),$$

or

$$\cos\alpha + \cos\beta = 2\cos\left(\frac{\alpha+\beta}{2}\right)\cos\left(\frac{\alpha-\beta}{2}\right).$$

7) With u and v as in 4), we have $\sin u \sin v = \frac{1}{2}(\cos(u-v) - \cos(u+v))$ from 1). Back-substituting gives

$$\sin\left(\frac{\alpha+\beta}{2}\right)\sin\left(\frac{\alpha-\beta}{2}\right) = \frac{1}{2}(\cos\beta - \cos\alpha).$$

To be consistent and write this as a formula for $\cos\alpha - \cos\beta$, we have

$$\cos\alpha - \cos\beta = -2\sin\left(\frac{\alpha+\beta}{2}\right)\sin\left(\frac{\alpha-\beta}{2}\right).$$

We summarize the sum/product identities in Table 7.7A/7.7B below.

Table 7.7A: Product-to-sum.

$\sin\alpha\sin\beta = \frac{1}{2}(\cos(\alpha-\beta) - \cos(\alpha+\beta))$
$\cos\alpha\cos\beta = \frac{1}{2}(\cos(\alpha-\beta) + \cos(\alpha+\beta))$
$\sin\alpha\cos\beta = \frac{1}{2}(\sin(\alpha+\beta) + \sin(\alpha-\beta))$

Table 7.7B: Sum-to-product.

$\sin\alpha + \sin\beta = 2\sin(\frac{\alpha+\beta}{2})\cos(\frac{\alpha-\beta}{2})$
$\sin\alpha - \sin\beta = 2\sin(\frac{\alpha-\beta}{2})\cos(\frac{\alpha+\beta}{2})$
$\cos\alpha + \cos\beta = 2\cos(\frac{\alpha+\beta}{2})\cos(\frac{\alpha-\beta}{2})$
$\cos\alpha - \cos\beta = -2\sin(\frac{\alpha+\beta}{2})\sin(\frac{\alpha-\beta}{2})$

Example 7.11. Find $\sin(\frac{\pi}{3})\sin(\frac{\pi}{4})$ and $\sin(\frac{\pi}{3}) + \sin(\frac{\pi}{4})$.

Example 7.12. Show that

$$\frac{\sin 6x + \sin 2x}{\cos 6x - \cos 2x} = -\cot 2x.$$

Example 7.13. Show that

$$\cot 2x = \cot x - \frac{1}{\sin 2x}.$$

Example 7.14 (See Section 7.1.2 Example 7.6). A waveform is created by the superposition of two others, specifically, $f(t) = A\sin\omega t + B\cos\omega t$. Express this as a single trig function.

7.2.5 De Moivre's theorem

Abraham DeMoivre was a French mathematician who upon religious exile became an English mathematician. He is famous for his eponymous theorem or identity that facilitates the calculation of sines and cosines of higher integer multiples of angles. DeMoivre has another slightly morbid claim to fame in correctly predicting the day of his death. Late in life he noticed he was sleeping a quarter hour more each night. He projected when he would sleep for 24 hours, and surely on that day he passed away. He left us an amazing formula. To fully appreciate it, we need to review some very basic facts about complex numbers, which we discuss briefly below.

The number $\sqrt{-1} \in \mathbb{C} - \mathbb{R}$ is the imaginary unit denoted by i. A typical complex number $z = a+bi$ has a real part a and an imaginary part b multiplied by i. If two complex numbers $z_1 = a_1+b_1i$ and $z_2 = a_2+b_2i$ are equal, it must be the case that the two real parts are equal and the two imaginary parts are equal. Symbolically, $a_1 + b_1i = a_2 + b_2i$ forces $a_1 = a_2$ and $b_1 = b_2$. DeMoivre's theorem states that $(\cos\theta + i\sin\theta)^n = \cos n\theta + i\sin n\theta$. It can be proved trivially using Euler's identity (the proof is trivial, Euler's identity is not), $e^{i\theta} = \cos\theta + i\sin\theta$. Simply note that $\cos n\theta + i\sin n\theta = e^{i(n\theta)} = (e^{i\theta})^n = (\cos\theta + i\sin\theta)^n$. Of course, DeMoivre developed his identity long before Euler discovered his, so priority properly goes to DeMoivre. Be careful, n must be an integer. Fractions do not work. The notation $\operatorname{cis}\theta = \cos\theta + i\sin\theta$ is common in the literature. DeMoivre's theorem can then be compressed to $\operatorname{cis}^n\theta = \operatorname{cis}(n\theta)$.

Example 7.15. Compute $\cos 3\theta$ solely in terms of $\cos\theta$ by expanding $\cos(2\theta + \theta)$.

Example 7.16. Find $\cos 3\theta$ using DeMoivre's formula.

7.2.6 Viète's formulas

An even earlier method for calculating sines and cosines of integer multiples of angles is due to another French mathematician François Viète. These formulas are somewhat less convenient than DeMoivre's because they are more calculation intensive. The symbol $\binom{n}{k} = \frac{n!}{k!(n-k)!}$ is the familiar binomial coefficient. Viète's formulas are:

1) $\sin n\theta = \sum_{k=0}^{n} \binom{n}{k}(\cos\theta)^k(\sin\theta)^{n-k}\sin((n-k)\frac{\pi}{2})$
2) $\cos n\theta = \sum_{k=0}^{n} \binom{n}{k}(\cos\theta)^k(\sin\theta)^{n-k}\cos((n-k)\frac{\pi}{2})$

Example 7.17. Find $\cos 3\theta$ using Viète's cosine formula.

7.2.7 Law of cosines & law of sines

Triangles are classified as **acute** (all angles less than 90°), **right** (one angle equal to 90°), or **obtuse** (one angle greater than 90°). Acute and obtuse triangles together are called **oblique** triangles. In any case, the Pythagorean theorem does not apply to a general oblique triangle as shown in Figure 7.2 below.

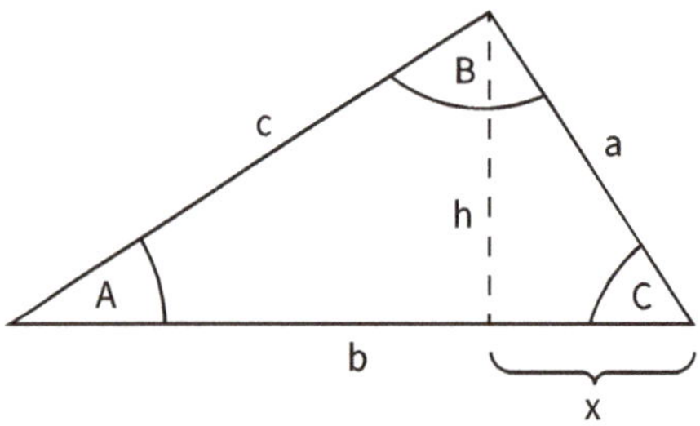

Figure 7.2: An arbitrary scalene triangle used to derive the law of cosines and the law of sines.

Law of cosines

We are interested in restoring as much of the convenience of the Pythagorean theorem to oblique triangles as we can. This will be possible by introducing a correction factor that allows us to write a Pythagorean-like expression for the length of any side of the triangle. In the drawing the sides are a, b, and c, while the angles opposite the respective sides are A, B, and C. Let us develop a formula for c. We construct an altitude h perpendicular to side b as shown that cuts b into two pieces, x and $b - x$. By the Pythagorean theorem, we have $c^2 = (b - x)^2 + h^2$. Also, $a^2 = x^2 + h^2$. Expanding the first expression, we get $c^2 = b^2 - 2bx + x^2 + h^2$, which simplifies to $c^2 = b^2 - 2bx + a^2$ by using the second expression for a^2. Now observe that $x = a \cos C$. Putting it all together, we have $c^2 = a^2 + b^2 - 2ab \cos C$, which is the law of cosines for c. The correction factor that needs to be applied to the raw Pythagorean expression is "minus twice the product of the other two sides times the cosine of their included angle." Table 7.8 shows the versions for all three sides.

Table 7.8: Law of cosines.

$a^2 = b^2 + c^2 - 2bc \cos A$
$b^2 = a^2 + c^2 - 2ac \cos B$
$c^2 = a^2 + b^2 - 2ab \cos C$

Law of sines

Referring again to Figure 7.2, we see immediately that $\frac{h}{c} = \sin A$ and $\frac{h}{a} = \sin C$. Evidently $c \sin A = a \sin C$, hence $\frac{a}{\sin A} = \frac{c}{\sin C}$. By changing the altitude to the one perpendicular to side c, we can verify in the same fashion that $\frac{a}{\sin A} = \frac{b}{\sin B}$, then the law of sines states that $\frac{a}{\sin A} = \frac{b}{\sin B} = \frac{c}{\sin C}$. The angles in a proper (nondegenerate) triangle cannot be 0 or π radians, so dividing by zero is not an issue here.

A typical use of the law of cosines or the law of sines is to "solve a triangle." What is meant by that is to determine all three sides and three angles of a triangle based on partial information. Any three of these data elements is sufficient to determine the other three with one exception noted below. The nature of the partial information is traditionally abbreviated as follows. If three sides are known (and no angles) it is an SSS case. If two sides are known and the angle between them (the included angle), it is an SAS case, or if two sides are known and a nonincluded angle, it is an SSA case. If one side is known and two angles, it is an SAA or ASA case. Note that the third angle in either of these cases can be gotten right away since every triangle has a total of 180°, so either case can be converted to the other immediately if necessary. And finally, if three angles only are known it is an AAA case. The first four cases above represent the minimum amount of partial information to actually solve triangles. The fifth case at first seems like it should be enough, but recall the definition of similar triangles. Two triangles are **similar** if their angles are the same. There is no way to infer the scale of similar triangles without knowing at least the length of one side. We will solve an example for each feasible case below and summarize the general strategy for each case in a table.

Example 7.18. A triangular garden is laid out with sides measuring 20, 30, and 40 meters. What are the angles at the corners of the garden?

Example 7.19. A triangular piece of land is being surveyed. The following data is available. From one corner of the property, a boundary line runs 450 [ft] due north and another from the same corner 600 [ft] due northeast. What is the length of the third side, what are the other two angles, and how many acres are in the parcel?

Example 7.20. Show that there are four subcases for the SSA case, where the angle is acute ($\theta < 90°$), one resulting in no triangle, one resulting in a right triangle, one resulting in a scalene or isoceles triangle, and one resulting in two possible triangles.

Example 7.21. Show that there are two subcases for the SSA case, where the angle is obtuse ($\theta > 90°$), one resulting in no triangle, and one resulting in a scalene or isoceles triangle.

Example 7.22. Suppose we are given sides $a = 2.0$ [m] and $b = \sqrt{3} = 1.732$ [m], and assume the angle a makes with the third side c is 60°. Solve the triangle.

Example 7.23. A triangular frame to support a winch is to be made from two steel tubes, one 20 feet long and one 15 feet long. The 20-foot tube will be welded to the steel floor at an angle of 45° and the 15-foot tube will be welded to it at the top and also to the floor in a triangular configuration that minimizes the spacing of the two floor welds. How far apart should the floor welds be?

Example 7.24. The method of triangulation involves two stations (receivers or viewpoints) separated by a known distance along a baseline drawn between them. Each station detects some signal or activity between them at a certain angle from the baseline. This gives two angles and a side of a triangle which can be solved to give the distances of the activity from each station. So suppose two fire towers are separated by five miles and Tower #1 detects smoke at an angle of 27° from the baseline and Tower #2 sees the smoke at 38°. How far is the smoke from each station?

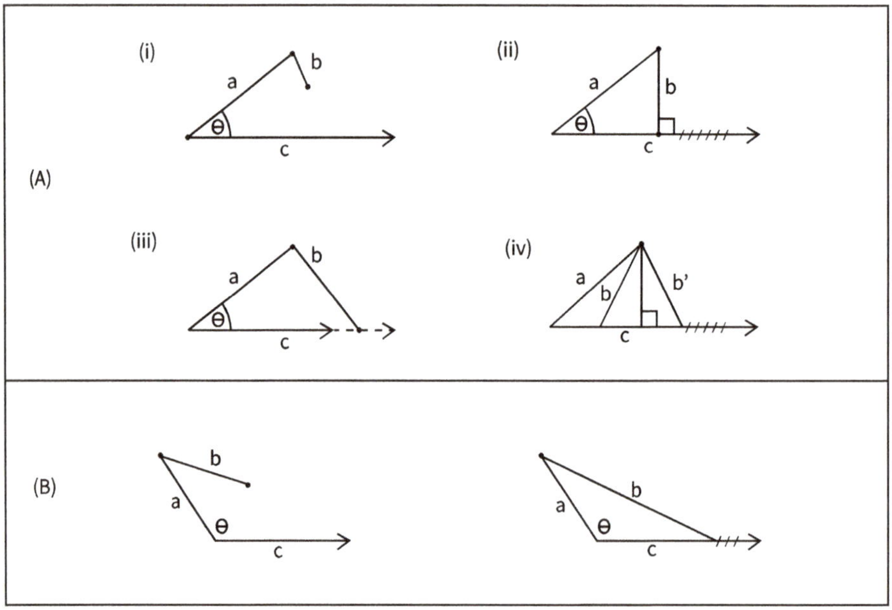

Figure 7.3: The triangular configurations described in Examples 7.20 and 7.21.

Here is a summary of the strategies and formulas for solving triangles. The sides are a, b, and c understood to be opposite angles α, β, and γ, respectively.

Notes for Table 7.9:

1) For the SSS case, always check that the sum of any two sides exceeds the third. Start with the angle opposite the longest side in the law of cosines.
2) The SAS case is straightforward.
3) For the SSA case, check the conditions given in Example 7.20. If a solution is possible, but only one triangle can be constructed, use the sine and cosine laws to recover the two unknown angles and one unknown side. If a solution is possible but two

triangles can be constructed, solve the quadratic in the table. If there is only one root, the triangle is a right triangle with $\alpha = 90^\circ$. If there are two roots, denote the larger by c_1 and the smaller by c_2. Use the law of sines to determine α_1, then γ_1. This solves the acute triangle. Note that $\alpha_2 = 180^\circ - \alpha_1$. Use c_2 in the law of sines to determine γ_2. This completes the solution of the obtuse triangle. See Example 7.23 for a worked problem.

4) The ASA or AAS case is straightforward.

Table 7.9: Solving triangles.

Case	Knowns	Unknown 1	Unknown 2	Unknown 3
SSS	a, b, c	$\alpha = \frac{b^2+c^2-a^2}{2bc}$	$\beta = \frac{a^2+c^2-b^2}{2ac}$	$\gamma = \frac{a^2+b^2-c^2}{2ab}$
SAS	a, β, c	$b = a^2 + c^2 - 2ac\cos\beta$	$\sin\alpha = \frac{a\sin\beta}{b}$	$\sin\gamma = \frac{c\sin\beta}{b}$
SSA	a, b, β	$c^2 - 2ac\cos\beta + (b^2 - a^2) = 0$	$\sin\alpha_1 = \frac{a\sin\beta}{b}$	$\sin\gamma_1 = \frac{c_1\sin\beta}{b}$
ASA	α, b, γ	$\beta = 180 - \alpha - \gamma$	$a = \frac{b\sin\alpha}{\sin\beta}$	$c = \frac{b\sin\gamma}{\sin\beta}$
AAA	α, β, γ	no information	*	*

7.2.8 Mollweide's formulas

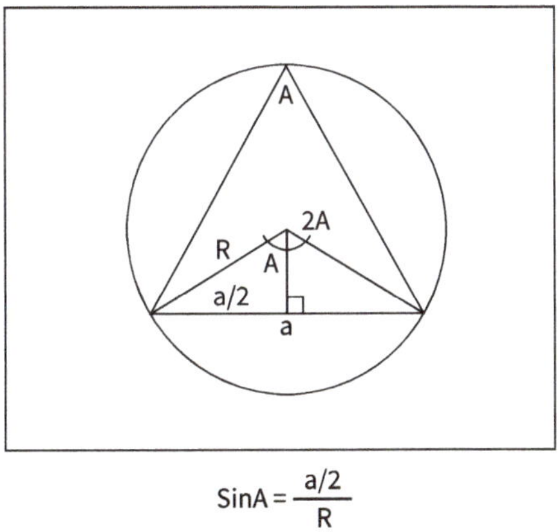

Figure 7.4: The triangles used to prove the extended law of sines and Mollweide's formulas.

Two historical trig formulas were developed by the German mathematician Karl Mollweide, although they seem to have been known earlier by mathematicians including Newton. They are not as useful as the laws of sines and cosines, but they do have the virtue of providing an automatic check on triangle solutions. Rreferring to the notation in Figure 7.4 above, we first need to show that $\frac{a}{\sin A} = 2R$ where R is the radius of the circumcircle of the triangle. The circumcircle is the unique circle that goes through all three vertices of the triangle. From the drawing, we see that angle A is inscribed in the circle and therefore half the corresponding central angle $2A$ formed by the two radii. Bisecting $2A$ with a perpendicular drawn from side a to the center of the circle, we have created a right triangle with hypotenuse R and side of length $\frac{a}{2}$ opposite the angle A. Evidently, $\sin A = \frac{a}{2R}$ which establishes that $\frac{a}{\sin A} = 2R$. By the regular law of sines, this must hold for the other two angles B and C in the triangle, so we have the **extended law of sines**, which says that $\frac{a}{\sin A} = \frac{b}{\sin B} = \frac{c}{\sin C} = 2R$.

Now starting with the facts that $\frac{a}{\sin A} = \frac{b}{\sin B} = 2R$, we can write $a = 2R\sin A$, $b = 2R\sin B$, and $c = 2R\sin C$. So

$$\frac{a+b}{c} = \frac{2R(\sin A + \sin B)}{2R\sin C} = \frac{\sin A + \sin B}{\sin C}.$$

Now use our sum-to-product formula $\sin A + \sin B = 2\sin\frac{A+B}{2}\cos\frac{A-B}{2}$ and the double angle formula $\sin C = 2\sin\frac{C}{2}\cos\frac{C}{2}$ to write

$$\frac{a+b}{c} = \frac{2\sin\frac{A+B}{2}\cos\frac{A-B}{2}}{2\sin\frac{C}{2}\cos\frac{C}{2}} = \frac{\sin\frac{A+B}{2}\cos\frac{A-B}{2}}{\sin\frac{C}{2}\cos\frac{C}{2}}.$$

We know another fact that will allow additional simplification of this expression: $A+B+C = 180^\circ$. So $\frac{A+B}{2} = 90^\circ - \frac{C}{2}$. Then by the cofunction identities, $\sin\frac{A+B}{2} = \sin 90^\circ - \frac{C}{2} = \cos\frac{C}{2}$. Cancelling $\cos\frac{C}{2}$ against $\sin\frac{A+B}{2}$ in our expression yields

$$\frac{a+b}{c} = \frac{\cos\frac{A-B}{2}}{\sin\frac{C}{2}}.$$

This is **Mollweide's first formula**. **Mollweide's second formula**,

$$\frac{a-b}{c} = \frac{\sin\frac{A-B}{2}}{\cos\frac{C}{2}},$$

follows with nearly identical reasoning.

Example 7.25. Use Mollweide's first formula to check the solved triangle in Section 7.2.7 Example 7.24.

7.3 Inverse trigonometric functions

7.3.1 Notational conventions

In view of the periodic nature of the trigonometric functions, inverting them is not a well-defined operation. We will fix this by restricting the domain of the functions. There is also a potential source of confusion with the notation for inverse trig functions. Ordinarily, if $f(x)$ is an injective function and therefore invertible, we denote the inverse of $f(x)$ as $f^{-1}(x)$. Shall we interpret $\sin^{-1}\theta$ as the inverse of $\sin\theta$ or as $\frac{1}{\sin\theta} = \csc\theta$? Textbooks do not handle this issue consistently. Our recommendation is to use the prefix "arc-" and call the inverse of sine, for example, the *arcsine* and similarly with the five other trig functions. Then $\sin^{-1}\theta = \frac{1}{\sin\theta}$ is consistent with our usual understanding that negative powers are reciprocals. We will write expressions like $\arcsin\frac{1}{2} = \frac{\pi}{6}$ and $\arccos\frac{\sqrt{2}}{2} = \frac{\pi}{4}$.

7.3.2 Principal trigonometric functions

A principal trigonometric function is one whose domain has been restricted from $\mathbb{R}$ to an interval upon which all the range values of the function are taken only once. We write $\operatorname{Sin}\theta$ for the principal sine, $\operatorname{Cos}\theta$ for the principal cosine, and so forth. We define $\operatorname{Sin}\theta$ as $\sin\theta$ restricted to the interval $[-\frac{\pi}{2}, \frac{\pi}{2}]$. As θ goes from $-\frac{\pi}{2}$ to $\frac{\pi}{2}$, $\operatorname{Sin}\theta$ takes on every value from -1 to $+1$ *but only once* on this interval. So it is invertible on this interval and $\arcsin x$ makes sense to be defined on $[-1, 1]$. On the other hand, $\operatorname{Cos}\theta$ requires a different restriction. Here $\operatorname{Cos}\theta$ is $\cos\theta$ restricted to the interval $[0, \pi]$. Again, as θ goes from 0 to π, $\operatorname{Cos}\theta$ takes on every value from $+1$ to -1 *but only once*. This validates the definition of $\arccos x$ also on $[-1, 1]$. There is no magic in the following selection of domain restrictions, since any domain that allows the principal function to be invertible will do. However, we have selected those which conform to common practice. Here is a table of the six principal trigonometric functions and their domains as well as the corresponding inverse functions and their domains. We have capitalized the inverse functions just for Table 7.10 to emphasize that they are inverses only for the principal trig functions. For example, although $\tan\frac{4\pi}{3} = \tan\frac{\pi}{3}$, it is certainly not true that $\arctan(\tan\frac{4\pi}{3}) = \frac{4\pi}{3}$. By convention, the arctangent function must return a value in the domain of the principal tangent, which is between $-\frac{\pi}{2}$ and $\frac{\pi}{2}$. In fact, $\arctan(\tan\frac{4\pi}{3}) = \frac{\pi}{3}$. There is no angle other than $\frac{\pi}{3}$ which resides in the domain of the principal tangent and also has the same tangent as $\frac{4\pi}{3}$.

Table 7.10: Principal & inverse trigonometric functions.

Principal	Domain (θ) & Range (x)	Inverse	Domain (x) & Range (θ)
$\operatorname{Sin}\theta = x$	$[-\frac{\pi}{2}, \frac{\pi}{2}]$	$\operatorname{Arcsin}x = \theta$	$[-1, 1]$
$\operatorname{Cos}\theta = x$	$[0, \pi]$	$\operatorname{Arccos}x = \theta$	$[-1, 1]$
$\operatorname{Tan}\theta = x$	$(-\frac{\pi}{2}, \frac{\pi}{2})$	$\operatorname{Arctan}x = \theta$	$\mathbb{R}$
$\operatorname{Csc}\theta = x$	$[-\frac{\pi}{2}, 0) \cup (0, \frac{\pi}{2}]$	$\operatorname{Arccsc}x = \theta$	$\lvert x\rvert \geq 1$
$\operatorname{Sec}\theta = x$	$[0, \frac{\pi}{2}) \cup (\frac{\pi}{2}, \pi]$	$\operatorname{Arcsec}x = \theta$	$\lvert x\rvert \geq 1$
$\operatorname{Cot}\theta = x$	$(0, \pi)$	$\operatorname{Arccot}x = \theta$	$\mathbb{R}$

Inverse trig functions are unexpectedly useful in calculus where they allow handling of some otherwise difficult functions.

Example 7.26. Find $\arcsin x$, $\arccos x$, $\arctan x$, and $\operatorname{arccot} x$ for $x = \frac{1}{2}$.

Example 7.27. Find (i) $\arcsin(\operatorname{Sin}\frac{1}{2})$, (ii) $\arctan(\operatorname{Tan}\frac{4\pi}{3})$, (iii) $\operatorname{arccsc}(\operatorname{Sin} 89.9^\circ)$.

Example 7.28. Show that $\arctan x + \arctan y = \arctan(\frac{x+y}{1-xy})$.

Example 7.29. Apply the formula in Example 7.28 to obtain an expression for π.

Example 7.30. Find $\operatorname{Sin}(\arcsin\frac{1}{2} + \arcsin\frac{1}{3})$.

Example 7.31. Find $\operatorname{Tan}(\arccos\frac{1}{2} + \arcsin\frac{1}{3})$.

References & suggested further reading

1) *Analysis by Its History*, by E. Hairer and G. Wanner (1996), Springer, Berlin – *good section on trig functions*
2) *College Trigonometry* (5th ed.) by Richard Aufmann, Richard Nation, and Vernon Barker (2004), Brooks-Cole – *standard college intro text*

Solutions to Chapter 7 examples

Example 7.1

The radius of the tank is 20 feet and $30^\circ = 0.524$ radians. Using the central angle formula, we get maximum $L = 20 \cdot 0.524 = 10.48$ feet.

Example 7.2

From the end of the shadow to the base of the building is the adjacent side of a right angle. The building height is the opposite side. The angle is 60°. Letting the length of the shadow be A, we can write $\frac{100}{A} = \tan 60^\circ$. So $A = \frac{100}{\tan 60^\circ} = \frac{100}{1.732} = 57.74$ [ft]. This makes sense because the higher the sun in the sky, the shorter the shadows on the ground. If the sun is 45° above the horizon, the shadow of an object has the same length as its height.

Example 7.3

Let S be the Emperor's shoulder width in meters. Then $\frac{S/2}{2} = \tan(\frac{13.75}{2})^\circ$. The right triangle here goes from the theodolite to the Emperor's chin, then 90° to his shoulder, then back to the theodolite. Then $\frac{S}{4} = \tan 6.875^\circ = 0.1205$. So $S = 0.482$ [m], or about 19 [in]. Supposedly, this actually happened!

Example 7.4

The adjacent side of the right angle is $400 + \frac{36.5}{2} = 418.25$ [ft]. If H is the height of the tree, then $\frac{H}{418.25} = \tan 33.3^\circ = 0.6569$. Solving for H, we estimate the height of the tree to be $(418.25)(0.6569) = 274\frac{3}{4}$ [ft].

Example 7.5

We create a right triangle with the adjacent side L being the distance from the Earth to the distant point and the opposite side being the mean distance d of the Earth from the Sun, namely 1.4960×10^8 [km]. We have $\frac{d}{L} = \tan 1''$, so $L = \frac{1.4960\times10^8}{\tan 1''}$. What is the tangent of a second? There are 60 seconds in a minute (of arc) and 60 minutes in a degree. So $\tan 1'' = \tan(\frac{1}{3600})^\circ = 4.848 \times 10^{-6}$. Evidently, $L = \frac{1.4960\times10^8}{4.8481\times10^{-6}} = 3.085 \times 10^{13}$ [km]. So the conversion factor is 1 parsec = 3.085×10^{13} [km]. The distance light travels in one year is 9.461×10^{12} [km], so 1 parsec = 3.262 light-years.

Example 7.6

Let $v(t)$ be the voltage function and assume $v(0) = 0$. Then $v(t) = 240\sin(2\pi)(50)t = 240\sin 100\pi t$ [V].

Example 7.7

Since $75^\circ = 30^\circ + 45^\circ$, the data we need from Table 7.3 are: $\sin 30^\circ = \frac{1}{2}$, $\cos 30^\circ = \frac{\sqrt{3}}{2}$, and $\sin 45^\circ = \cos 45^\circ = \frac{\sqrt{2}}{2}$. Assembling $\sin(30^\circ + 45^\circ) = \sin 30^\circ \cos 45^\circ + \cos 30^\circ \sin 45^\circ$, we can write

$$\sin 75^\circ = \left(\frac{1}{2}\right)\left(\frac{\sqrt{2}}{2}\right) + \left(\frac{\sqrt{3}}{2}\right)\left(\frac{\sqrt{2}}{2}\right) = \frac{\sqrt{2}}{4}(1 + \sqrt{3}).$$

We could mimic this calculation for $\cos 75^\circ$, but let us use one of the Pythagorean identities. Rearranging $\sin^2\theta + \cos^2\theta = 1$ into $\cos\theta = \sqrt{1 - \sin^2\theta}$ and applying it to this example, we have

$$\cos 75^\circ = \sqrt{1 - \left(\frac{\sqrt{2}}{4}(1 + \sqrt{3})\right)^2} = \sqrt{1 - \frac{1}{4}(2 + \sqrt{3})} = \frac{1}{2}\sqrt{2 - \sqrt{3}}.$$

Now for the tangent. Note

$$\tan 75^\circ = \frac{\sin 75^\circ}{\cos 75^\circ} = \frac{\frac{\sqrt{2}}{4}(1+\sqrt{3})}{\frac{1}{2}\sqrt{2-\sqrt{3}}} = \frac{\sqrt{3}+1}{\sqrt{3}-1}.$$

Example 7.8

Let $\alpha = 30^\circ$, then $\frac{\alpha}{2} = 15^\circ$. So

$$\sin 15^\circ = +\sqrt{\frac{1-\cos 30^\circ}{2}} = \sqrt{\frac{1-\sqrt{3}/2}{2}} = \frac{1}{2}\sqrt{2-\sqrt{3}} \approx 0.2588.$$

Similarly,

$$\cos 15^\circ = +\sqrt{\frac{1+\cos 30^\circ}{2}} = \sqrt{\frac{1+\sqrt{3}/2}{2}} = \frac{1}{2}\sqrt{2+\sqrt{3}} \approx 0.9659.$$

And also

$$\tan 15^\circ = +\sqrt{\frac{1-\cos 30^\circ}{1+\cos 30^\circ}} = \sqrt{\frac{1-\sqrt{3}/2}{1+\sqrt{3}/2}} = \frac{\sqrt{2-\sqrt{3}}}{\sqrt{2+\sqrt{3}}} \approx 0.2679.$$

Example 7.9

The first order of business is to find the angle between 0° and 360° that 795° corresponds to. We note that $795 = 2 \cdot 360 + 75$, so apparently $\cos 795^\circ = \cos 75^\circ$. We could use the cosine addition formula for angles 30° and 45°, but let us do this a different way. The angle 150° has the same opposite side and the negative of the same side as the angle 30°. So $\sin 150^\circ = \sin 30^\circ$ and $\cos 150^\circ = -\cos 30^\circ$. From Table 7.3 (or hopefully memory), we know $\cos 30^\circ = \frac{\sqrt{3}}{2}$, hence $\cos 150^\circ = -\frac{\sqrt{3}}{2}$. By the half-angle formula,

$$\cos 75^\circ = +\sqrt{\frac{1+\cos 150^\circ}{2}} = \sqrt{\frac{1-\sqrt{3}/2}{2}} = \frac{1}{2}\sqrt{2-\sqrt{3}} \approx 0.2588.$$

Note that this is the same as $\sin 15^\circ$ in Example 7.8. We should not be surprised because a trig function of one angle is the cofunction of its complementary angle, and $15^\circ + 75^\circ = 90^\circ$. Bottom line, $\cos 795^\circ \approx 0.2588$.

Example 7.10

Write $\sin(n+1)\theta = \sin(n\theta + \theta)$ and use the addition formula to get $\sin(n+1)\theta = \sin n\theta \cos\theta + \cos n\theta \sin\theta$. Likewise, $\cos(n+1)\theta = \cos n\theta \cos\theta - \sin n\theta \sin\theta$.

Example 7.11

Here

$$\sin\left(\frac{\pi}{3}\right)\sin\left(\frac{\pi}{4}\right) = \frac{1}{2}\left(\cos\left(\frac{\pi}{3}-\frac{\pi}{4}\right)-\cos\left(\frac{\pi}{3}+\frac{\pi}{4}\right)\right)$$
$$= \frac{1}{2}\left(\cos\left(\frac{\pi}{12}\right)-\cos\left(\frac{7\pi}{12}\right)\right), \quad \text{and}$$
$$\sin\left(\frac{\pi}{3}\right)+\sin\left(\frac{\pi}{4}\right) = 2\sin\left(\frac{7\pi}{24}\right)\cos\left(\frac{\pi}{12}\right).$$

Example 7.12

We have

$$\frac{\sin 6x+\sin 2x}{\cos 6x-\cos 2x} = \frac{2\sin\frac{6x+2x}{2}\cos\frac{6x-2x}{2}}{-2\sin\frac{6x+2x}{2}\sin\frac{6x-2x}{2}} = -\frac{\cos 2x}{\sin 2x} = -\cot 2x.$$

Example 7.13

We can write

$$\cot 2x = \frac{1}{\tan 2x} = \frac{\cos 2x}{\sin 2x} = \frac{2\cos^2 x-1}{2\sin x\cos x}$$
$$= \frac{2\cos^2 x}{2\sin x\cos x} - \frac{1}{2\sin x\cos x} = \cot x - \frac{1}{\sin 2x}.$$

Example 7.14

We may assume A, B, and θ are real numbers with $A > 0$. Let $C = \sqrt{A^2+B^2}$. Then

$$f(t) = \sqrt{A^2+B^2}\left[\frac{A\sin\omega t}{\sqrt{A^2+B^2}} + \frac{B\cos\omega t}{\sqrt{A^2+B^2}}\right].$$

Now pick θ so that $\tan\theta = \frac{B}{A}$. Since

$$\sec^2\theta = 1+\tan^2\theta = 1+\left(\frac{B}{A}\right)^2 = \frac{A^2+B^2}{A^2},$$

upon inversion we have $\cos^2\theta = \frac{A^2}{A^2+B^2}$. It follows that $\cos\theta = \frac{A}{\sqrt{A^2+B^2}}$. Similarly, since $\cot\theta = \frac{A}{B}$ and $\csc^2\theta = 1+\cot^2\theta$, we have

$$\csc^2\theta = 1+\left(\frac{A}{B}\right)^2 = \frac{A^2+B^2}{B^2}.$$

After inverting, $\sin^2\theta = \frac{B^2}{A^2+B^2}$, whereupon $\sin\theta = \frac{B}{\sqrt{A^2+B^2}}$. Taking these expressions for $\sin\theta$ and $\cos\theta$ back to the square-bracketed term, we see that $f(t) = C[\cos\theta\sin\omega t +$

$\sin\theta\cos\omega t]$. But this is the same as $C\sin(\omega t+\theta)$ by the angle sum formula for sine. This general result is sometimes referred to as the **phase shift formula**. If we graph $f(t)$ versus t, the presence of positive θ shifts the graph to the left relative to $C\sin\omega t$, and conversely to the right for negative θ; θ itself is called the **phase shift**.

Example 7.15

From the angle addition formula for cosine, we have

$$\cos(2\theta+\theta) = \cos 2\theta\cos\theta - \sin 2\theta\sin\theta.$$

Now plug in the double angle formulas for cosine and sine. We get

$$\begin{aligned}\cos 2\theta\cos\theta - \sin 2\theta\sin\theta &= (2\cos^2\theta - 1)(\cos\theta) - (2\sin\theta\cos\theta)(\sin\theta)\\ &= (2\cos^3\theta - \cos\theta) - (2\sin^2\theta\cos\theta).\end{aligned}$$

Since $\sin^2\theta = 1-\cos^2\theta$, we can write

$$\cos 3\theta = 2\cos^3\theta - \cos\theta - 2\cos\theta(1-\cos^2\theta) = 4\cos^3\theta - 3\cos\theta.$$

Example 7.16

Here $\cos 3\theta + i\sin 3\theta = (\cos\theta + i\sin\theta)^3$. Notice we will get the formula for $\sin 3\theta$ as a bonus. With a little algebra, we have $(\cos\theta+i\sin\theta)^3 = (\cos^3\theta - 3\cos\theta\sin^2\theta) + i(3\cos^2\theta\sin\theta - \sin^3\theta)$. Equating real and imaginary parts between $\cos 3\theta + i\sin 3\theta$ and the previous expansion, we get (i) $\cos 3\theta = \cos^3\theta - 3\cos\theta\sin^2\theta$ and (ii) $\sin 3\theta = 3\cos^2\theta\sin\theta - \sin^3\theta$. Using $\sin^2\theta + \cos^2\theta = 1$, we convert these respectively to (i) $\cos 3\theta = \cos^3\theta - 3\cos\theta\sin^2\theta = \cos^3\theta - 3\cos\theta(1-\cos^2\theta) = 4\cos^3\theta - 3\cos\theta$ and (ii) $\sin 3\theta = 3\cos^2\theta\sin\theta - \sin^3\theta = \sin 3\theta = 3(1-\sin^2\theta)\sin\theta - \sin^3\theta = 3\sin\theta - 4\sin^3\theta$.

Example 7.17

We compute

$$\begin{aligned}\cos 3\theta = {} & \binom{3}{0}(\cos\theta)^0(\sin\theta)^3\cos\left((3)\frac{\pi}{2}\right) + \binom{3}{1}(\cos\theta)^1(\sin\theta)^2\cos\left((2)\frac{\pi}{2}\right)\\ & + \binom{3}{2}(\cos\theta)^2(\sin\theta)^1\cos\left((1)\frac{\pi}{2}\right) + \binom{3}{3}(\cos\theta)^3(\sin\theta)^0\cos\left((0)\frac{\pi}{2}\right).\end{aligned}$$

Note that for odd multiples of $\frac{\pi}{2}$, the cosine is zero. So using the surviving terms

$$\cos 3\theta = \binom{3}{1}(\cos\theta)^1(\sin\theta)^2\cos(\pi) + \binom{3}{3}(\cos\theta)^3(\sin\theta)^0\cos(0) = 3\cos\theta\sin^2\theta + \cos^3\theta.$$

Using $\sin^2\theta = 1-\cos^2\theta$, this reduces, as we expect, to $4\cos^3\theta - 3\cos\theta$.

Example 7.18

This is an SSS case. The very first order of business is to check if a triangle with these sides is even possible. The sum of the lengths of any two sides has to exceed the third side. If that sum does not exceed the third side, the sides cannot be assembled into a triangle. If the sum should happen to equal the third side, then the triangle is "degenerate" and collapses into a straight line. In the given problem, the data pass the test and we expect a proper triangle. So let us fit the law of cosines to the three sides. We should have $40^2 = 30^2 + 20^2 - 2(20)(30)\cos A$, where A is the angle between the 20- and 30-meter sides (opposite the 40-meter side). Solving for A, we get $\cos A = \frac{1600-900-400}{-1200} = -\frac{1}{4}$. Borrowing from Section 7.3, we determine $A = 104.5^\circ$. Now we could set up the law of cosines to determine the other two angles, or perhaps more easily use the law of sines with the new information. Letting B and C be the angles opposite the 20- and 30-meter sides, respectively, we have $\frac{40}{\sin 104.5} = \frac{20}{\sin B} = \frac{30}{\sin C}$. Then $\frac{40}{0.968} = 41.32 = \frac{20}{\sin B} = \frac{30}{\sin C}$. This gives $\sin B = \frac{20}{41.32} = 0.484$ so $B = 28.9^\circ$. Also $\sin C = \frac{30}{41.32} = 0.726$ so $C = 46.6^\circ$. Checking, $104.5^\circ + 28.9^\circ + 46.6^\circ = 180.0^\circ$. Obviously, after two angles were computed we could have subtracted their sum from 180° as a shortcut to the third angle.

Example 7.19

This is an SAS case. The law of sines will not help us immediately because we are lacking the length of the side opposite the 45° angle at the corner formed by the two known sides. So we set up the law of cosines to get that third side, call it c. We can write $c^2 = 450^2 + 600^2 - 2(450)(600)\cos 45^\circ = 180{,}662$, so $c = 425$ [ft]. Using the law of sines, $\frac{425}{\sin 45^\circ} = 601.1 = \frac{450}{\sin A} = \frac{600}{\sin B}$. Solving for the angle A opposite the 450 [ft] side, we get 48.5°, and likewise for angle B opposite the long side we have 86.5°. Checking, $45^\circ + 48.5^\circ + 86.5^\circ = 180.0^\circ$. There are several options at this point to calculate the area: (i) an altitude perpendicular to the 450 [ft] side is given by $600 \sin 45^\circ = 424.2$ [ft]. Note that this is very close to our calculated value of c, which means the shape of the parcel is almost a right triangle. Using Area $= \frac{1}{2}$ Base $\times$ Height, we get $\frac{1}{2}(450)(424.2) = 95445$ [sf]. Since there are 43,560 square feet in an acre, the parcel is 2.19 acres; (ii) going the other way, an altitude perpendicular to the 600 [ft] side is given by $450 \sin 45^\circ = 318.2$ [ft]. The area formula then returns $\frac{1}{2}(600)(318.2) = 95{,}445$ [sf] as before; (iii) **Heron's formula** dates from the Golden Age of Greek geometry. If a triangle has sides a, b, and c, then the semiperimeter $s = \frac{a+b+c}{2}$, and the area $A = \sqrt{s(s-a)(s-b)(s-c)}$. Plugging in our data $A = \sqrt{(737.5)(287.5)(137.5)(312.5)} = 95{,}445$ [sf] again; (iv) note that options (i) and (ii) are subsumed under the vector cross product formula for area. To use this, we would locate the origin at the corner, express both sides $a = 450$ and $b = 600$ as vectors and calculate $A = \frac{1}{2}|(\mathbf{a} \times \mathbf{b})| = \frac{1}{2}|\mathbf{a}||\mathbf{b}| \sin\theta$ where θ is the angle between vectors $\mathbf{a}$ and $\mathbf{b}$.

Example 7.20

See Figure 7.3(A).

Assume we are given two sides a and b, as well as the angle θ that a makes with the unspecified third side c. Without loss of generality, we can immediately draw side a meeting side c at the angle θ. Side c can be extended as much as we want from this intersection to the right, since we are going to "fit" side b running from the free end of side a to somewhere on side c. Then we will terminate c where it meets b to complete the triangle. We have to be careful with this case as there are four outcomes possible:

(i) $a \sin\theta > b$. So b is not long enough to reach c and there is no triangle, hence no solution.

(ii) $a \sin\theta = b$. This means b is just long enough to extend perpendicularly to c. There is no ambiguity regarding its placement.

(iii) $a < b$. There is no way b can meet c to the left of the perpendicular drawn in case (ii). It would then necessarily be to the left of the intersection of a with c, contrary to our model because c would vanish. So it meets c to the right of the perpendicular forming a single triangle.

(iv) $a \sin\theta < b < a$. This is the interesting ambiguous case. Indeed, b can meet c to the left or right of the perpendicular, resulting in two possible triangles.

Example 7.21

See Figure 7.3(B). Two outcomes are possible:

(i) $a > b$. Again b cannot meet c so there is no triangle.

(ii) $a \leq b$. Now b meets c to the right of the perpendicular. If $a = b$, the triangle is isoceles.

Example 7.22

This is an SSA case with θ acute. We calculate $a \sin\theta = 2 \sin 60^\circ = 1.732$. That looks like case (ii) in Example 7.20, so we conclude this is a right triangle with third side $\sqrt{2^2 - (1.732)^2} = 1$ and third angle $180^\circ - 60^\circ - 90^\circ = 30^\circ$.

Example 7.23

This is another SSA case with θ acute. Let $a = 20$ [ft] and $b = 15$ [ft]. Evidently, $a \sin\theta = 20 \sin 45^\circ = 14.14$ [ft]. This is the ambiguous case (iv) in Example 7.20 because $14.14 < 15 < 20$. The law of cosines will give us the possible lengths of side c that conform to the data we have. So $b^2 = a^2 + c^2 - 2ac \cos\theta$ becomes $15^2 = 20^2 + c^2 - 2(20)(c) \cos 45^\circ$. Rearranging we get $c^2 - 20\sqrt{2}c + 175 = 0$. There are two solutions for c, $c_1 = 19.14$ [ft] and $c_2 = 9.14$ [ft]. Since a condition of the problem is to minimize the third side (the weld spacing), we choose 9.14 [ft].

Note that $\frac{\sin\beta}{b} = \frac{\sin 45^\circ}{15} = 0.0471$, regardless of which value of c we choose, hence which triangle we ultimately obtain. If we form the triangle using c_1, we will have α_1 opposite the 20-foot side, and if we use c_2 we will have α_2 instead. It looks like there is going to be a contradiction in applying the law of sines because $\frac{\sin 45^\circ}{15} = \frac{\sin\alpha_1}{20} = \frac{\sin\alpha_2}{20}$

but $\alpha_1 \neq \alpha_2$. But we are saved by the fact that $\sin\alpha = \sin(180^\circ - \alpha)$. To see this note that $\sin(180^\circ - \alpha) = \sin 180^\circ \cos\alpha - \cos 180^\circ \sin\alpha = 0 \cdot \cos\alpha - (-1) \cdot \sin\alpha = \sin\alpha$. What happens in these cases is that $\alpha_2 = 180^\circ - \alpha_1$. In this example, using the "long" value c_1, we find $\alpha_1 = 70.5^\circ$, and using the "short" value c_2, we get $\alpha_2 = 109.5^\circ$. Certainly, $\alpha_2 = 180^\circ - \alpha_1$, so there is no contradiction. If we needed γ_1 corresponding to c_1, we would use the law of sines again and set $\frac{\sin\beta}{b} = \frac{\sin\gamma_1}{c_1}$, and similarly for γ_2.

Example 7.24

This is an ASA case. Let the baseline be c, the distance of the smoke from Tower #1 be a, and from Tower #2 be b. The third angle is $180^\circ - 27^\circ - 38^\circ = 115^\circ$. From the law of sines, we have $\frac{5.00}{\sin 115^\circ} = 5.52 = \frac{a}{\sin 38^\circ} = \frac{b}{\sin 27^\circ}$. Note the angle opposite side a is the one reported by Tower #2, and vice versa. Solving we have $a = 3.40$ [mi] and $b = 2.51$ [mi].

Example 7.25

Collecting data, we have $a = 3.40$, $b = 2.51$, and $c = 5.00$ for the sides. Also $A = 38^\circ$, $B = 27^\circ$, and $C = 115^\circ$. Plugging into the formula and simplifying, we find

$$1.18 = \frac{3.40 + 2.51}{5.00} = \frac{\cos\frac{38-27}{2}}{\sin\frac{115}{2}} = \frac{\cos 5.5}{\sin 57.5} = \frac{0.995}{0.843} = 1.18.$$

So we have additional confidence in our earlier result.

Example 7.26

$\arcsin(\frac{1}{2}) = 30^\circ$, $\arccos(\frac{1}{2}) = 60^\circ$, $\arctan(\frac{1}{2}) = 26.57^\circ$, and $\operatorname{arccot}(\frac{1}{2}) = 63.43^\circ$. Note that $\frac{1}{2}$ is out-of-range for cosecant and secant.

Example 7.27

(i) $\frac{1}{2}$ is in the domain of Sine, so $\arcsin(\text{Sin}\,\frac{1}{2})$ returns $\frac{1}{2}$.

(ii) See the discussion just before Table 7.10.

(iii) Any value of sine that is not either plus or minus 1 will not be in the domain of arccosecant. We can see that $\text{Sin}\,89.9^\circ$ is very close to 1 because 89.9° is very close to 90°. But close is not good enough. We conclude $\text{arccsc}(\text{Sin}\,89.9^\circ)$ is undefined.

Example 7.28

This looks a little like the formula for the tangent of a sum of angles, $\text{Tan}\,(\alpha + \beta) = \frac{\text{Tan}\,\alpha + \text{Tan}\,\beta}{1 - \text{Tan}\,\alpha \text{Tan}\,\beta}$. Let us try the substitutions $x = \text{Tan}\,\alpha$ and $y = \text{Tan}\,\beta$. Then we have $\text{Tan}\,(\arctan x + \arctan y) = \frac{x+y}{1-xy}$. Now take the arctangent of both sides to get $\arctan x + \arctan y = \arctan\frac{x+y}{1-xy}$.

Example 7.29
We know $\operatorname{Tan}\frac{\pi}{4} = 1$, so if we can find x and y such that $\frac{x+y}{1-xy} = 1$, we will have $4\arctan\frac{x+y}{1-xy} = \pi$. So we test some fractions with small integer denominators and eventually stumble upon $\frac{1}{2}$ and $\frac{1}{3}$, since

$$\frac{\frac{1}{2}+\frac{1}{3}}{1-\frac{1}{2}\cdot\frac{1}{3}} = \frac{\frac{5}{6}}{1-\frac{1}{6}} = 1.$$

Evidently, $4(\arctan\frac{1}{2}+\arctan\frac{1}{3}) = \pi$. In conjunction with a series expression for arctangent, formulas like this were exploited in the pre-calculator era to find ever more digits of π.

Example 7.30
Let $u = \arcsin\frac{1}{2}$ and $v = \arcsin\frac{1}{3}$. Then $\operatorname{Sin} u = \frac{1}{2}$ and $\operatorname{Sin} v = \frac{1}{3}$. Note that u and v are in the principal domain for sine. We have $\operatorname{Sin}(u+v) = \operatorname{Sin} u \operatorname{Cos} v + \operatorname{Cos} u \operatorname{Sin} v$. To evaluate this we need $\operatorname{Cos} u$ and $\operatorname{Cos} v$. From the Pythagorean identities,

$$\operatorname{Cos} u = \sqrt{1-\operatorname{Sin}^2 u} = \sqrt{1-\frac{1}{4}} = \frac{\sqrt{3}}{2} \quad \text{and} \quad \operatorname{Cos} v = \sqrt{1-\operatorname{Sin}^2 v} = \sqrt{1-\frac{1}{9}} = \frac{2\sqrt{2}}{3}.$$

Assembling everything, we have

$$\operatorname{Sin}\left(\arcsin\frac{1}{2}+\arcsin\frac{1}{3}\right) = \left(\frac{1}{2}\right)\left(\frac{2\sqrt{2}}{3}\right)+\left(\frac{\sqrt{3}}{2}\right)\left(\frac{1}{3}\right) = \frac{2\sqrt{2}+\sqrt{3}}{6}.$$

Example 7.31
Let $u = \arccos\frac{1}{2}$ and $v = \arcsin\frac{1}{3}$. Then $\operatorname{Cos} u = \frac{1}{2}$ and $\operatorname{Sin} v = \frac{1}{3}$. Again note that $\frac{1}{2}$ and $\operatorname{Sin} v = \frac{1}{3}$. Now $\operatorname{Tan}(u+v) = \frac{\operatorname{Tan} u+\operatorname{Tan} v}{1-\operatorname{Tan} u \operatorname{Tan} v}$. If we recover $\operatorname{Sin} u$ and $\operatorname{Cos} v$ we can construct the tangents that we need. So

$$\operatorname{Sin} u = \sqrt{1-\operatorname{Cos}^2 u} = \sqrt{1-\frac{1}{4}} = \frac{\sqrt{3}}{2} \quad \text{and} \quad \operatorname{Cos} v = \sqrt{1-\operatorname{Sin}^2 v} = \sqrt{1-\frac{1}{9}} = \frac{2\sqrt{2}}{3}.$$

Finding the tangents,

$$\operatorname{Tan} u = \frac{\operatorname{Sin} u}{\operatorname{Cos} u} = \left(\frac{\sqrt{3}}{2}\right)(2) = \sqrt{3} \quad \text{and} \quad \operatorname{Tan} v = \frac{\operatorname{Sin} v}{\operatorname{Cos} v} = \left(\frac{1}{3}\right)\left(\frac{3}{2\sqrt{2}}\right) = \frac{1}{2\sqrt{2}} = \frac{\sqrt{2}}{4}.$$

Plugging into the tangent of sum formula, we finally get

$$\operatorname{Tan}\left(\arccos\frac{1}{2}+\arcsin\frac{1}{3}\right) = \frac{\operatorname{Tan} u+\operatorname{Tan} v}{1-\operatorname{Tan} u \operatorname{Tan} v} = \frac{\sqrt{3}+\frac{\sqrt{2}}{4}}{1-\sqrt{3}(\frac{\sqrt{2}}{4})} = \frac{4\sqrt{3}+2}{4-\sqrt{6}}.$$

8 Generating function methods

A sequence of numbers $\{a_0, a_1, a_2, \ldots, a_n, \ldots\}$ may be converted into a formal power series in the indeterminate x by forming $\sum_{n=0}^{\infty} a_n x^n$. The idea is to use the structure of the series to organize and subsequently manipulate or transform the coefficients, which will eventually turn out to be solutions to various types of problems. The series is called a **generating function** denoted by $G(a_n; x)$. A power series is *formal* if its convergence properties are irrelevant to the manner in which it is used. Professional opinion on this point in the literature is mixed. We adhere to the purely formalist approach; however, some authors feel the need to explicitly consider convergence issues. A convenient occurrence, which by no means is guaranteed, is that $G(a_n; x)$ corresponds to some closed form expression $f(x)$. In that case we say both $G(a_n; x)$ and $f(x)$ are the generating function for the sequence. The Anglo–French mathematician Abraham deMoivre seems to have been the first to use generating functions in connection with solving general linear recurrences. Leonhard Euler thoroughly exploited the method, as well, but it was finally the French mathematician Pierre Laplace who christened the formal power series as a "generating function." We will use them to solve some classic enumeration and finite difference problems that we have already addressed with other methods in this book. The generating function concept is also important in applications such as differential equations, but that is beyond the scope of our discussion. The American mathematician Herbert Wilf wrote a seminal text on the subject and informally defined a generating function to be "a clothesline on which we hang up a sequence of numbers for display." That seems to capture the spirit.

8.1 Basic definitions

Let us back up and be both more general and more precise than the introductory definition. Even though we call them functions, it is important to note that generating functions are not what we technically identify as functions. There is no domain or range. Any variable in the series expansion is to be regarded as simply a formal indeterminate or place holder. So suppose we are given a sequence of numbers $\{a_0, a_1, a_2, \ldots, a_n, \ldots\}$, finite or infinite. We create a **formal series** with these numbers as coefficients by multiplying them respectively with another sequence of so-called **indicator functions** $\{\phi_0(x), \phi_1(x), \phi_2(x), \ldots, \phi_n(x), \ldots\}$. Define the **ordinary generating function (OGF)** associated with the given sequence of numbers as $G(a_n; \phi_n(x)) = \sum_{n=0}^{\infty} a_n \phi_n(x)$. If the sequence of numbers is finite and stops at $n = m$, then set $a_n = 0$ for $n > m$. It would be confusing if this construction were not injective, that is, if it were the case that $\sum_{n=0}^{\infty} a_n \phi_n(x) = \sum_{n=0}^{\infty} b_n \phi_n(x)$ but $a_n \neq b_n$ for some n. By judicious choice of the indicator functions, we can make sure this does not happen, so if two formal series with the same indicator functions are equal, we may safely conclude that the respective sequences represented by the two series are the same. The assignment $\phi_n(x) = x^n$ is a common

https://doi.org/10.1515/9783111506654-009

choice for an indicator function because it ensures that a generating function represents a unique ordered set of coefficients. Other serviceable choices include $\phi_n(x) = \cos nx$ or $\phi_n(x) = \sin nx$.

In addition to ordinary generating functions, for some purposes it may be advantageous to modify the indicator functions. Given the sequence of numbers $\{a_0, a_1, a_2, \ldots, a_n, \ldots\}$ and the set of indicator functions

$$\{\phi_0(x), \phi_1(x), \phi_2(x), \ldots, \phi_n(x), \ldots\},$$

we define the **exponential generating function (EGF)** associated with the sequence as $E(a_n; \phi_n(x)) = \sum_{n=0}^{\infty} a_n \frac{\phi_n(x)}{n!}$. Unless otherwise noted, we assume in the following that any OGF is based on $\phi_n(x) = x^n$ and any EGF on $\frac{\phi_n(x)}{n!} = \frac{x^n}{n!}$.

There are other more exotic choices of modified indicator functions that have been developed to allow generating functions in special situations to yield closed form expressions. As usual, the only requirement is that equal generating functions represent the exact same sequence of coefficients. If we are able to perform algebraic or other types of operations on an OGF or EGF and express the result in closed form as $f(x)$, we often need to extract the nth coefficient of the result to answer some question. We write the extraction operator $[[f(x)]]_{x^n}$ to indicate we want the nth coefficient of $f(x) = G(a_n; x)$ or $[[f(x)]]_{\frac{x^n}{n!}}$ for the nth coefficient of $f(x) = E(a_n; x)$. We will present several shortcuts below to make this process easier.

Example 8.1. Show that the set of indicator functions $\{1, x+1, x-1, x^2+2, x^2-2, \ldots, x^n+n, x^n-n, \ldots\}$ is not acceptable for an OGF.

Example 8.2. Find the OGF for the sequence of combination statistics ${}_nC_k$ for k-subsets of an n-set.

Example 8.3. Find the OGF for the sequence of permutation statistics ${}_nP_k$ for k-lists of an n-set.

Example 8.4. Find the EGF for the sequence $\{1, 1, 1, \ldots\}$ where $\phi_n(x) = x^n$.

There are several rules for combining OGF's.

Table 8.1a: OGF algebra rules. Let $f(x) = \sum_{n=0}^{\infty} a_n x^n$, $g(x) = \sum_{n=0}^{\infty} b_n x^n$, and $h(x) = \sum_{n=0}^{\infty} c_n x^n$.

(1) $f(x) + g(x) = h(x)$ provided $c_n = a_n + b_n$ for all n
(2) $f(x) \cdot g(x) = h(x)$ provided $c_n = \sum_{k=0}^{n} a_{n-k} b_k$ for all n

Rule 2 is known as the **Cauchy product** or **convolution product** of the two power series.

Table 8.1b: EGF algebra rules. Let $f(x) = \sum_{n=0}^{\infty} a_n \frac{x^n}{n!}$, $g(x) = \sum_{n=0}^{\infty} b_n \frac{x^n}{n!}$, and $h(x) = \sum_{n=0}^{\infty} c_n \frac{x^n}{n!}$

(3) $f(x) + g(x) = h(x)$ provided $c_n = a_n + b_n$ for all n
(4) $f(x) \cdot g(x) = h(x)$ provided $c_n = \sum_{k=0}^{n} \binom{n}{k} a_{n-k} b_k$ for all n

Example 8.5. Given $G(n; x)$, find $G(n + 2; x)$.

Example 8.6 (∫). Given $G(n; x)$, find $G(n^2; x)$.

8.2 Summation

Example 8.7. Show that $\frac{1}{1-x}$ is a **summation operator** in the sense that if

$$g(x) = \sum_{n=0}^{\infty} b_n x^n$$

is the OGF for $b_0, b_1, \ldots, b_n, \ldots$, then $\frac{1}{1-x} \cdot g(x)$ is the OGF for $\{b_0, b_0 + b_1, b_0 + \cdots + b_n, \ldots\}$.

Example 8.8. Find the closed form of the OGF for $\{0, 1, 2, 3, \ldots, n, \ldots\}$ with and without differentiation.

Example 8.9. Find a general formula for the sum of the first n integers.

It is difficult to overstate the usefulness and importance of the generalized binomial theorem. We will appeal to it often, since it is the basis for frequently needed extraction operators. Unlike many of our generating functions,

$$(1 + x)^r = 1 + \sum_{n=1}^{\infty} \frac{r(r-1)\cdots(r-n+1)}{n!} x^n$$

actually converges for arbitrary r as long as $|x| < 1$. We pause our obsession with formal power series to show the power of this theorem in an analytical context where convergence is available.

Example 8.10. The famous Einstein equation $E = mc^2$ leads to the equivalence of mass and energy the following way. Suppose m_0 is the mass of an object measured in a reference frame in which the object is at rest. If the object has velocity v as measured in some laboratory frame the apparent mass of the object will be $\frac{m_0}{\sqrt{1-\frac{v^2}{c^2}}}$, where c is the speed of light. The factor $\frac{1}{\sqrt{1-\frac{v^2}{c^2}}}$ is close to 1 if $v \ll c$, but becomes a significant amplification as v becomes an appreciable fraction of c. Letting $\beta = \frac{v}{c}$, we can express the lab frame mass as a function of velocity, $m(v) = m_0(1-\beta^2)^{-1/2}$. Now $0 \le v < c$, so $0 \le \beta^2 < 1$. The *generalized binomial theorem now applies* and we have

$$m(v) = m_0\left[1 + \sum_{n=1}^{\infty} \frac{(-\frac{1}{2})(-\frac{3}{2})(-\frac{5}{2})\cdots(-\frac{1}{2}-n+1)}{n!}(-\beta^2)^n\right].$$

Writing out the first few terms, it is clear that

$$m(v) = m_0\left[1 + \frac{-\frac{1}{2}}{1!}(-\beta^2) + \frac{(-\frac{1}{2})(-\frac{3}{2})}{2!}(-\beta^2)^2 + \frac{(-\frac{1}{2})(-\frac{3}{2})(-\frac{5}{2})}{3!}(-\beta^2)^3 + \cdots\right].$$

This simplifies to $m(v) = m_0[1 + \frac{\beta^2}{2} + \frac{3\beta^4}{8} + \frac{15\beta^6}{48} + \cdots]$. Now multiplying by c^2 and reverting to $\beta = \frac{v}{c}$, we have

$$E = m(v)c^2 = m_0c^2\left[1 + \frac{v^2}{2c^2} + \frac{3v^4}{8c^4} + \frac{15v^6}{48c^6} + \cdots\right].$$

Simplifying again, we get $E = m_0c^2 + \frac{m_0v^2}{2} + \frac{3m_0v^4}{8c^2} + \frac{15m_0v^6}{48c^4} + \cdots$. The term m_0c^2 is the energy available by converting mass to energy (this makes nuclear power possible). The term $\frac{m_0v^2}{2} + \frac{3m_0v^4}{8c^2} + \frac{15m_0v^6}{48c^4} + \cdots$ is the (relativistic) kinetic energy (energy of motion). Whenever $v \ll c$, the only appreciable term is $\frac{m_0v^2}{2}$, which we identify as the classical kinetic energy. Pause concluded.

Example 8.11 ($\int$). Repeat the method of Example 8.9 to obtain the sum of the squares of the first n integers.

Table 8.2: Extraction operator rules.

(1) $[[c \cdot f(x)]]_{x^n} = c \cdot [[f(x)]]_{x^n}$
(2) $[[f(x) + g(x)]]_{x^n} = [[f(x)]]_{x^n} + [[g(x)]]_{x^n}$
(3) $[[f(x)]]_{x^{n-j}} = [[x^j f(x)]]_{x^n}$
(4) $[[\frac{1}{1-cx}]]_{x^n} = c^k$
(5) $[[(1+x)^n]]_{x^n} = \binom{n}{k}$
(6) $[[(1-x)^{-n}]]_{x^n} = \left(\!\binom{n}{k}\!\right)$ (n multichoose k)
(7) $[[f(x) \cdot g(x)]]_{x^n} = \sum_{k=0}^{n} a_k b_{n-k}\, f(x) = \sum_{n=0}^{\infty} a_n x^n,\ g(x) = \sum_{n=0}^{\infty} b_n x^n$

8.3 Enumerators – general

In Chapter 1 we introduced the Twelve-Fold Way, which organized the most common formulas applicable to enumeration problems. We derived these formulas with elementary arguments based on counting functions with finite domains and ranges. The formulas and useful variants can also be derived by means of generating functions. Generating functions that count things are called **enumerators**. For example, we discovered in Section 8.1 Example 8.2 above that the function $f(x) = (1+x)^n = \sum_{k=0}^{n} \binom{n}{k} x^k$ is an enumerator that gives the ways of choosing a k-set from an n-set. Since it is an OGF, we call it an **ordinary enumerator**. In Section 8.1 Example 8.3, we rearranged the terms in this power series as $f(x) = (1+x)^n = \sum_{k=0}^{n} (\frac{n!}{(n-k)!}) \frac{x^k}{k!}$ to form an EGF that expresses the sequence of numbers $\frac{n!}{(n-k)!}$, which we recognize as the permutation statistics $_nP_k$, the number of permutations of n objects taken k at a time. In this context, the same $f(x)$ is an **exponential enumerator**.

One obvious benefit of a generating function that happens to be a finite sum is that the indeterminate x can be treated as a variable if it suits our purpose.

Example 8.12. Find the sum of the binomial coefficients for k-subsets of an n-set.

Example 8.13. Show that the number of ways of selecting an even number of objects from a pool of n overall is the same as the number of ways of selecting an odd number of objects.

Example 8.14 ($\int$). Find the weighted sum of the binomial coefficients where $\binom{n}{k}$ is weighted by the size of the subset k.

Example 8.15. Find the sum of the squares of the binomial coefficients.

Example 8.16. Given two each of two kinds of objects, a and b, find how many selections of two or three objects are possible if a can be selected any number of times but b must be selected twice or not at all.

The previous example was very simplified, but it suggests a powerful general method. We did not really need the detailed composition of a specific selection, just the number of possible selections, so we could have simply set $a = 1$ and written the enumerator as $1 + x + x^2$. Likewise, the enumerator for b could have been $1+x^2$. Then the overall enumerator $(1+x+x^2)(1+x^2) = 1+x+2x^2+x^3+x^4$ would have been adequate to answer the question posed in Example 8.16. The ordinary enumerator of possible outcomes of a compound selection can be constructed by multiplying together the ordinary enumerators of its constituent selections as long as they are independent.

Example 8.17. What are the numbers of distinct outcomes obtainable by rolling three fair standard dice? We always assume coins and dice are fair unless otherwise noted. The enumerator does not depend at all on fairness, only feasibility. However, relative likelihood of a feasible outcome does depend on fairness.

Example 8.18. Given two fair dice, one a dodecahedron (12-sided regular polyhedron) with faces numbered 1 through 12 and the other a standard cubical die with 1 through 6, find the most likely up face sum when both dice are rolled.

Example 8.19. I have five pennies, three nickels, four dimes, and two quarters totaling $1.10. Can I make change for every amount less than $1.10?

Example 8.20. Continuing Example 8.19, what if the pennies were reduced to three, the nickels to one, the dimes to two, and the quarters remained the same? Now the total is $0.78. For how many amounts less than 25 ¢ can I not make change?

Example 8.21. In American football a team compiles a score in various increments: an extra point can be either 1 or 2 points, but they only happen after a touchdown worth 6 points. So really the scoring possibilities for a touchdown are 6, 7, or 8 points. A field goal is worth 3 points. There is also a fairly rare scoring opportunity called a safety, worth 2 points which can be ignored. Assuming a team does not score more than 40 points total, and that no more than six field goals will be made, how many ways can each point total be achieved and which are the most feasible?

Example 8.22. Suppose we have two each of p different colored marbles and also that we have an additional q marbles, each with a color different from all the rest, including the original p pairs. So we have $2p + q$ marbles in all, representing $p + q$ different colors. Find the ordinary enumerator for selections from this collection of marbles.

Example 8.23. Derive the multiset numbers $(\binom{n}{k})$ using ordinary enumerators. See Table 8.2 (6) above.

Example 8.24. Specialize the enumerator in Example 8.23 where we want to select k objects out of a pool of $n \leq k$ distinct objects with unlimited repetitions permitted but with each of the n objects guaranteed to be selected at least once. Find how many ways six objects can be selected from four objects with replacement and every object in the pool represented at least once.

We have constructed ordinary enumerators for making selections from a pool of several objects by multiplying together the ordinary enumerators for the individual objects. We can emulate this construction for exponential enumerators, but the result will be an overall exponential enumerator for permutations of a pool of objects. In Section 8.1 Example 8.3, we established that the exponential enumerator for the number of permutations of n distinct objects taken k at a time was $(1 + x)^n = \sum_{k=0}^{n} \frac{{}_nP_k x^k}{k!}$. This can also be viewed the following way. The exponential enumerator for one object is $(1 + x)$. Either the object appears in a possible permutation or not. Multiplying together one copy of this enumerator for each of n distinct objects gives us $(1 + x)^n$.

We can extend this to the case where not all the objects being permuted are distinct. This is equivalent to allowing permutations with repetition or replacement. The number of permutations of k identical objects taken k at a time is obviously 1, so the exponential enumerator for exactly that situation is $\frac{x^k}{k!}$. Then the exponential enumerator for taking any number from 0 to k is $\frac{x^0}{0!} + \frac{x^1}{1!} + \frac{x^2}{2!} + \cdots + \frac{x^k}{k!}$. So if we had m identical objects of type 1 and n identical objects of type 2, the exponential enumerator for the number of permutations of all $m+n$ objects would be $(1+x+\frac{x^2}{2!}+\cdots+\frac{x^m}{m!})(1+x+\frac{x^2}{2!}+\cdots+\frac{x^n}{n!})$. Recall the multinomial formula gives the number of permutations of $m+n$ objects of which m are of one type and n of another type as $\frac{(m+n)!}{m!\,n!}$. Assembling an exponential enumerator based on this formula, we would find the $(m+n)$th term to be $\frac{(m+n)!}{m!\,n!}\frac{x^{m+n}}{(m+n)!} = \frac{x^{m+n}}{m!\,n!}$, which agrees perfectly with the $(m+n)$th term of our product of enumerators above.

Example 8.25. If we are given three red (R) flags and four blue (B) flags, find the exponential enumerator for permutations of the flags and check the number if all flags are flown.

Example 8.26. If we have an unlimited supply of gold, red, black, white, and blue flags, find how many distinct permutations of ten flags are possible if at least one red flag is used.

Example 8.27. We derived the Stirling numbers of the second kind using the inclusion/exclusion formula in Section 1.4 Example 1.38. Recall these numbers give how many ways a set of k distinct objects can be distributed into n distinct cells with no cell empty. Show how the formula for the Stirling number $S(k,n)$ can be derived by considering exponential enumerators for permutations.

8.4 Enumerators – partitions

In the Twelve-Fold Way, several formulas require use of the integer partition numbers $p_m(k)$ and it will be convenient to have a way of calculating them. Recall a partition of the integer $k \geq 1$ into m parts is a set of positive integers $k_1, k_2, \ldots, k_m$ that sum to k and which are customarily listed in weakly decreasing order. Technically, the order of listing the parts of the partition is immaterial, but adopting our scheme for organizing them makes it easier to identify important features. The following few examples show how generating functions can answer questions about partitions.

Example 8.28. How many ways can the number 5 be partitioned? Here we are being asked for all possible partitions.

In the preceding example, we actually determined the number of partitions of the integers 1 through 4 as well, since the factor enumerators represent all possible parts that might appear in those partitions. For the numbers greater than 5, we have determined how many partitions of those numbers there are if the partition elements are limited to five 1's, two 2's, and one each of 3, 4, and 5. We have considerable flexibility in adjusting the factor enumerators to answer more specific questions.

Example 8.29. Find the number of partitions of 7 that have no part bigger than 4.

There is an interesting bijective correspondence between the number of ways to partition n into parts, the largest of which is k, and the number of ways to partition n into k or fewer parts. They are the same number. Let us test this claim on the result of Example 8.23. We have determined that there are 11 partitions of 7 with no part bigger than 4. They are $(4,3)$, $(4,2,1)$, $(4,1,1,1)$, $(3,3,1)$, $(3,2,2)$, $(3,2,1,1)$, $(3,1,1,1,1)$, $(2,2,2,1)$, $(2,2,1,1,1)$, $(2,1,1,1,1,1)$, and $(1,1,1,1,1,1,1)$. And we can list the partitions of 7 with 4 or fewer parts. They are (7), $(6,1)$, $(5,2)$, $(5,1,1)$, $(4,3)$, $(4,2,1)$, $(4,1,1,1)$, $(3,3,1)$, $(3,2,2)$, $(3,2,1,1)$, and $(2,2,2,1)$. So there are 11 of these and our claim checks out. Why does this work? We can represent the partition $(4,2,1)$, for example, by the following drawing of dots: $\begin{array}{llll} \bullet & \bullet & \bullet & \bullet \\ \bullet & \bullet & & \\ \bullet & & & \end{array}$. Each part of the partition from largest to smallest is converted to that number of dots in successive rows from top to bottom. Now construct the transpose of the dot diagram (interchange rows and columns) We get $\begin{array}{lll} \bullet & \bullet & \bullet \\ \bullet & \bullet & \\ \bullet & & \\ \bullet & & \end{array}$.

This diagram corresponds to the partition $(3,2,1,1)$, the so-called **conjugate partition** to $(4,2,1)$. Evidently, any partition with maximum part k has as its conjugate a partition with k parts. Conversely, if we start with a partition with k parts, then its conjugate will have no part greater than k. As an illustration of this consider $(5,2)$ which corresponds to $\begin{array}{lllll} \bullet & \bullet & \bullet & \bullet & \bullet \\ \bullet & \bullet & & & \end{array}$ and its transpose is $\begin{array}{ll} \bullet & \bullet \\ \bullet & \bullet \\ \bullet & \\ \bullet & \\ \bullet & \end{array}$ which represents $(2,2,1,1,1)$. Conjugate partitions come in pairs, so if a number has an odd number of partitions, then some partition is its own conjugate. Since there are 15 partitions of 7, we would expect this to happen, and in fact $(4,1,1,1)$ depicted by $\begin{array}{llll} \bullet & \bullet & \bullet & \bullet \\ \bullet & & & \\ \bullet & & & \\ \bullet & & & \end{array}$ is obviously self-conjugate. The diagram of dots is called a **Ferrers graph.**

Example 8.30. Find the number of partitions of 10 that contain no more than four parts.

Example 8.31. Find the number of partitions of 7 that do not contain any 2's.

Example 8.32. Write an enumerator for partitions of an integer n into (i) distinct parts, and (ii) find the number of partitions of 8 into distinct parts

Enumerators can be infinite power series and this allows them to be expressed in compact form. The enumerator for selecting any number of integer k's is $\frac{1}{1-x^k} = 1 + x^k + x^{2k} + \cdots + x^{mk} + \cdots$. So the total number of partitions of any number n would be the coefficient of x^n in

$$\prod_{k=1}^{n} \frac{1}{1-x^k} = \frac{1}{1-x} \cdot \frac{1}{1-x^2} \cdot \frac{1}{1-x^3} \cdots \frac{1}{1-x^m} \cdots \frac{1}{1-x^n}.$$

Formulas such as this have perhaps more theoretical interest than practical. But we can use them to illustrate some interesting facts about partitions originally discovered using generating functions by letting the index of the preceding product approach infinity. The number of partitions of an integer into distinct parts has enumerator

$$\prod_{k=1}^{\infty}(1+x^k) = (1+x)(1+x^2)(1+x^3)\cdots(1+x^m)\cdots.$$

Note that $\frac{1-x^2}{1-x} = 1 + x$, $\frac{1-x^4}{1-x^2} = 1 + x^2$, $\frac{1-x^6}{1-x^3} = 1 + x^3$, and so forth. So

$$\prod_{k=1}^{\infty}(1+x^k) = \prod_{k=1}^{\infty} \frac{1-x^{2k}}{1-x^k}.$$

Can this product of fractions be simplified? Yes, as k ranges from 1 to ∞, every numerator $1 - x^{2k}$ will eventually be cancelled as k takes on even integer values. This will leave only factors of the form $\frac{1}{1-x^m}$, where m ranges only over the odd integers. The conclusion from these observations is that the number of permutations of an integer into distinct parts is equal to the number of partitions into just odd parts. Who knew? Euler, apparently.

Another identity courtesy of the magic of formal power series is

$$\begin{aligned}(1-x)&(1+x)(1+x^2)(1+x^4)(1+x^8)\cdots(1+x^{2^m})\cdots\\&= (1-x^2)(1+x^2)(1+x^4)(1+x^8)\cdots(1+x^{2^m})\cdots\\&= (1-x^4)(1+x^4)(1+x^8)\cdots(1+x^{2^m})\cdots.\end{aligned}$$

Eventually, this process absorbs every power of x and we conclude

$$(1-x)(1+x)(1+x^2)(1+x^4)(1+x^8)\cdots(1+x^{2^m})\cdots = 1,$$

hence

$$\frac{1}{1-x} = \sum_{k=0}^{\infty} x^k = \prod_{k=1}^{n} (1 + x^{2^k}).$$

The generating function on the left states that any number n can be written, of course, as the sum of n 1's. The generating function on the right states that any number n can be written as the sum of unrepeated powers of 2. The upshot is that each positive integer base 10 can be written uniquely base 2.

Example 8.33. Find the number of partitions of an integer into exactly m parts.

Example 8.34. Use the formula in Example 8.33 to find the number of partitions of 7 into exactly 3 parts.

8.5 Rook polynomials

There is an interesting class of ordinary generating functions that solve a special type of permutation problem, specifically enumerating all possible permutations where certain ones are disallowed. These problems can have a very practical flavor. For example, suppose we are managing an auto repair shop that services Ford, Chevy, VW, and Honda. Al works only on VWs and Hondas, Bob on Chevys and VWs, Charlie on Fords and Chevys, and Dave on Chevys and Hondas. How many ways can we have all the service technicians working on a backlog of cars of all four types if we make sure the work assignments are compatible with the technicians' specialties? We might make a chart as in Figure 8.1 showing all the incompatible assignments blacked out. Our task is to turn this into a generating function that tells us how many assignments are feasible.

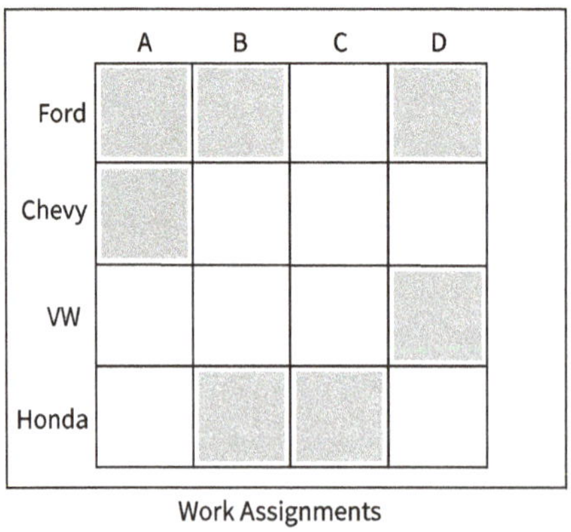

Figure 8.1: A square array showing possible work assignments for four technicians covering four tasks.

To attack this problem, we borrow an idea from chess. The rook (or castle) in chess is a piece that moves any number of squares along a row or file on the chessboard. We are going to be creative in defining our boards. We will retain the rectilinear grid, but the size and shape of the board is going to be arbitrary. All rooks will be considered equivalent and they will move along the rows and files of these generalized boards as if they were operating on part of a regular chessboard. What we will be interested in knowing is how many rooks can be placed on a board so they would not be in a position to capture or "take" each other. Think about how many rooks could be placed on a regular chessboard so that no rook can capture any other rook. It appears to be eight. Each rook disqualifies one row and one file from being occupied by any other rook. There are enough rows and files to accommodate eight rooks, but the ninth would cause a conflict. Now how many configurations could these eight rooks occupy. We are committed to one rook per row, but there are $8! = 40{,}320$ permutations of the files they are sitting on. On a smaller scale, how many ways could three interchangeable rooks be placed on a regular chessboard so that they cannot take each other? They would certainly have to be in three different rows, so there are $\binom{8}{3}$ ways of arranging that. Then they cannot be in the same file, so we have to permute their positions along the files. There are ${}_8P_3$ ways of arranging this. We conclude there are $\binom{8}{3} \cdot {}_8P_3 = 56 \cdot 336 = 18{,}816$ placements of three nontaking rooks on a standard chessboard.

Let us imitate this calculation on a small generalized board as shown in Figure 8.2.

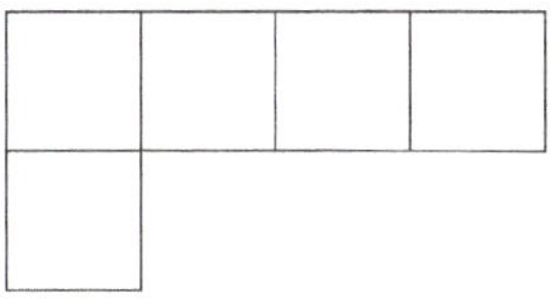

Figure 8.2: A generalized board consisting of a long "L" formed from five equal squares.

There is one way to place no rooks on the board, five ways to place one rook, four ways to place two, and obviously no ways to place three or more. Given an arbitrary board B, let $r_k(B)$ be the number of ways k nontaking rooks can be put on the board. Define the **rook polynomial** as the OGF $R(x, B) = \sum_{k=0}^{n} r_k(B)x^k$. Obviously, the degree of $R(x, B)$ cannot exceed the number of squares on B. If there is only one board under discussion, we usually suppress B and simply write $R(x) = \sum_{k=0}^{n} r_k x^k$. For the board in Figure 8.2, we have $R(x) = 1 + 5x + 4x^2$.

It will be very helpful to have a formula that allows us to calculate the rook polynomial for a given board by breaking it into smaller boards where we already know or can easily calculate the rook polynomials for the constituent boards. The way to do this is to pick a special square on the given board and mark it as the pivot. Now we create two smaller boards by (i) deleting the pivot square, and (ii) deleting the row and file of the pivot square. Denote the board with the pivot square only deleted as $\overline{B}$. Denote the board with the row and file of the pivot square deleted as $\widetilde{B}$. Note that the number of ways of placing rooks on B can be sorted into two mutually exclusive types. The first type are those configurations which have no rook on the pivot and the second type are those configurations which do. Every placement of r_k rooks on B with no rook on the pivot square corresponds to a placement of r_k rooks on $\overline{B}$ which has the pivot square removed. Likewise, every allowed placement of r_k rooks on B with a rook on the pivot square corresponds to a placement of r_{k-1} rooks on $\widetilde{B}$ which has the row and file of the pivot square removed. We can add the results of these disjoint cases and get $r_k(B) = r_k(\overline{B}) + r_{k-1}(\widetilde{B})$. Then the relation of the three corresponding rook polynomials is $R(x,B) = R(x,\overline{B}) + xR(x,\widetilde{B})$. This formula is called the **expansion formula for rook polynomials**. It leads to a graphic algebra of deconstructed boards resembling puzzle pieces.

For example, consider the three-square board B in Figure 8.3.

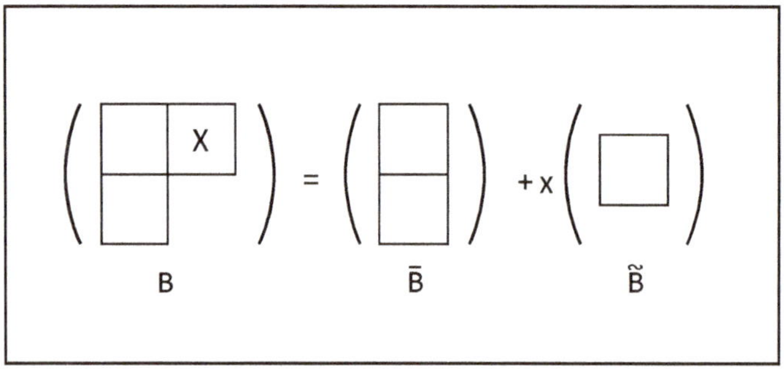

$R(x,B) = R(x,\bar{B}) + x\,R(x,\tilde{B})$

(a) $1+3x+x^2 = (1+2x) + x(1+x)$

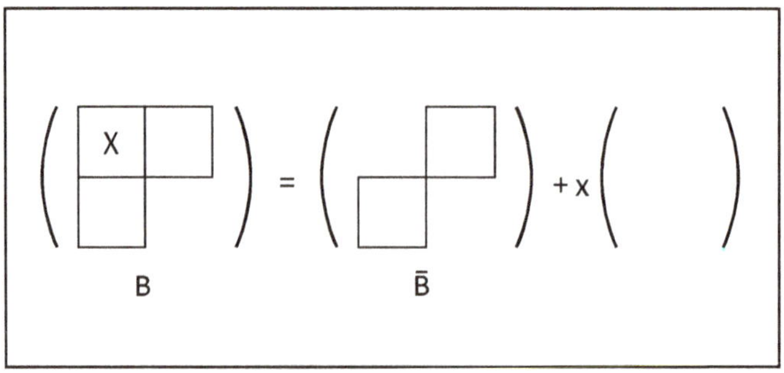

(b) $1+3x+x^2 = (1+2x+x^2) + x(1)$

Figure 8.3: A graphic depiction of the two possible paths to the generating function for the three-square board.

We can let the upper right square be the pivot and decompose B using the expansion formula as shown in Figure 8.3 (a). The board $\overline{B}$ is the two-square column resulting from deleting the pivot square from B and the board $\widetilde{B}$ is the single square resulting from deleting the row and column of the pivot square from B. We can easily find the rook polynomials for $\overline{B}$ and $\widetilde{B}$. For $\overline{B}$, there is one way of placing no rooks, two ways of placing one rook, and no ways of placing two or more rooks. So $R(x, \overline{B}) = 1 + 2x$. For $\widetilde{B}$, there is one way of placing no rooks, one way of placing one rook, and no ways of placing two or more rooks. So $R(x, \widetilde{B}) = 1 + x$. The expansion formula then tells us that $R(x, B) = (1 + 2x) + x(1 + x) = 1 + 3x + x^2$.

Alternatively, if we let the pivot be the upper left square we obtain the decomposition in Figure 8.3 (b). This time the board $\overline{B}$ is the two-square diagonal board that has one way of placing no rooks, two ways of placing one rook, one way of placing two rooks, and no ways of placing three or more rooks, so $R(x, \overline{B}) = 1 + 2x + x^2$. The empty board $\widetilde{B}$ has exactly one way of placing no rooks on it, so $R(x, \widetilde{B}) = 1$. Again the expansion formula gives us $R(x, B) = (1 + 2x + x^2) + x(1) = 1 + 3x + x^2$, as we expected.

Example 8.35. Find the rook polynomial for the board B in Figure 8.4.

Example 8.36. Show that if a board can be decomposed into two sub-boards B_1 and B_2 with the property that no square of B_1 is in the same row or file as any square of B_2, then $R(x, B) = R(x, B_1)R(x, B_2)$. In other words, if the subboards are disjoint in terms of alignment, then the rook polynomials are multiplicative.

Example 8.37. Find the rook polynomial for the board B in Figure 8.5.

Now that we have the basic mechanics of rook polynomials, let us develop a way to connect them to permutations of distinct objects. If we had a four horse race and the results were: first prize – Horse B, second prize – Horse D, third prize – Horse A, and fourth prize – Horse C, we could encode this permutation in the following array:

∘	A	B	C	D
1		x		
2				x
3	x			
4			x	

Each x in a certain row denotes the prize won by the horse at the top of the column where the x appears. Since the horses and the prizes are distinct, no two x's can be in the same row or column. This corresponds exactly to a valid placement of four nontaking rooks on a 4×4 board. Treating this array as board B, we can see that $r_4(B) = \binom{4}{4} \cdot 4! = 24$. This follows the calculation in the introduction where we first chose four out of four rows and then permuted the four columns. In general, we see that each possible per-

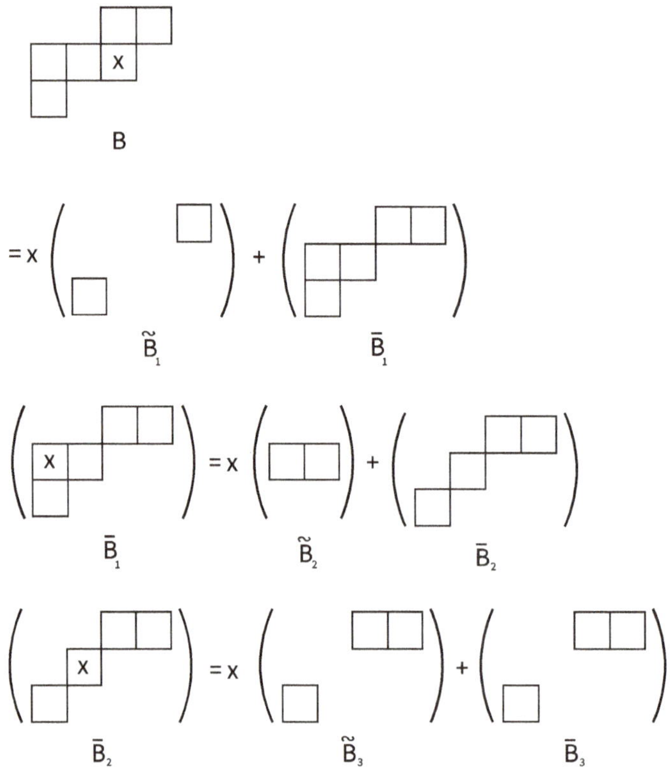

(a)

(i) $R(x,B) = xR(x,\tilde{B}_1) + R(x,\bar{B}_1)$

(ii) $R(x,\bar{B}_1) = xR(x,\tilde{B}_2) + R(x,\bar{B}_2)$

(iii) $R(x,\bar{B}_2) = xR(x,\tilde{B}_3) + R(x,\bar{B}_3)$

So, $R(x,B) = xR(x,\tilde{B}_1) + xR(x,\tilde{B}_2) + xR(x,\tilde{B}_3) + R(x,\bar{B}_3)$

$= x(1+2x+x^2) + x(1+2x) +$

$x(1+3x+2x^2) + (1+3x+2x^2)$

$= (x+2x^2+x^3) + (x+2x^2) +$

$(x+3x^2+2x^3) + (1+3x+2x^2)$

(b) $= 1+6x+9x^2+3x^3$

Figure 8.4: The graphic and algebraic steps in finding the generating function for the given board.

mutation of n distinct objects can be represented as an array and therefore associated with a unique configuration of n nontaking rooks on an $n \times n$ board. Studying these permutations, especially when we may wish to disqualify certain ones, will be equivalent to studying rook placements on boards that are designed to reflect those disqualifying

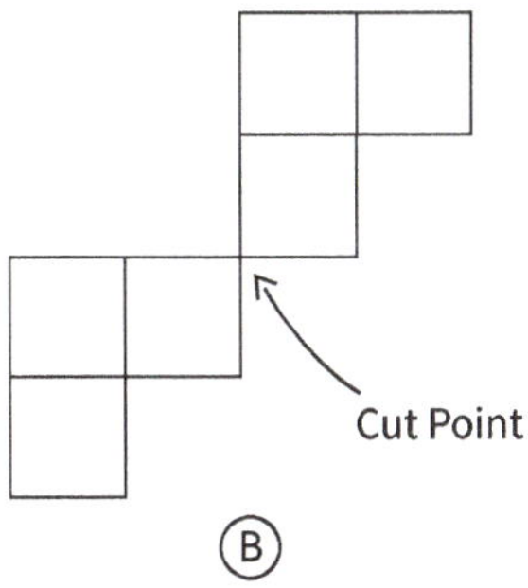

Figure 8.5: A board that can be cut to simplify the calculation of the rook polynomial.

conditions. These types of problems fall under the category of **permutations with forbidden positions**. A practical example will make this clear.

Example 8.38. Returning to the auto service technician chart in Figure 8.1, answer the question posed regarding possible work assignments for the group of technicians.

We find $R(x,B) = 1+8x+19x^2+14x^3+2x^4$, so $r_4(B) = 2$ and there apparently are only two work assignments where all four technicians can be matched with the cars they can fix. Note that $r_3(B)$ gives the number of three-person crews possible if all four technicians

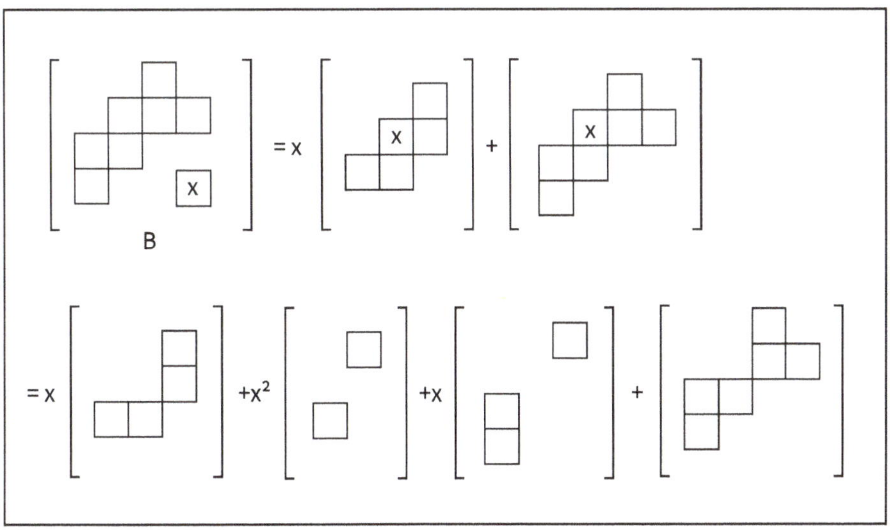

$$= x(1+2x)^2 + x^2(1+2x+x^2) + x(1+2x)(1+x) + (1+3x+x^2)^2$$

$$= 1 + 8x + 19x^2 + 14x^3 + 2x^4$$

Figure 8.6: The car technicians' problem solved graphically.

are available. If one technician is unavailable, not all of the three person crews can be formed. So if we asked how many compatible work assignments would there be if Bob took the day off, the answer is not 14, but depends on a recalculation with all four squares under Bob blacked out.

Example 8.39. Four piano students are giving a recital where they each play one of four pieces. The music teacher would like the four pieces to be different. Student A prefers to play concerto #1 or #2. So does Student B. Student C is a little more versatile and likes #1, #3, or #4. Finally, Student D prefers to play #3 or #4. How many programs can the music teacher arrange that match the students' preferences?

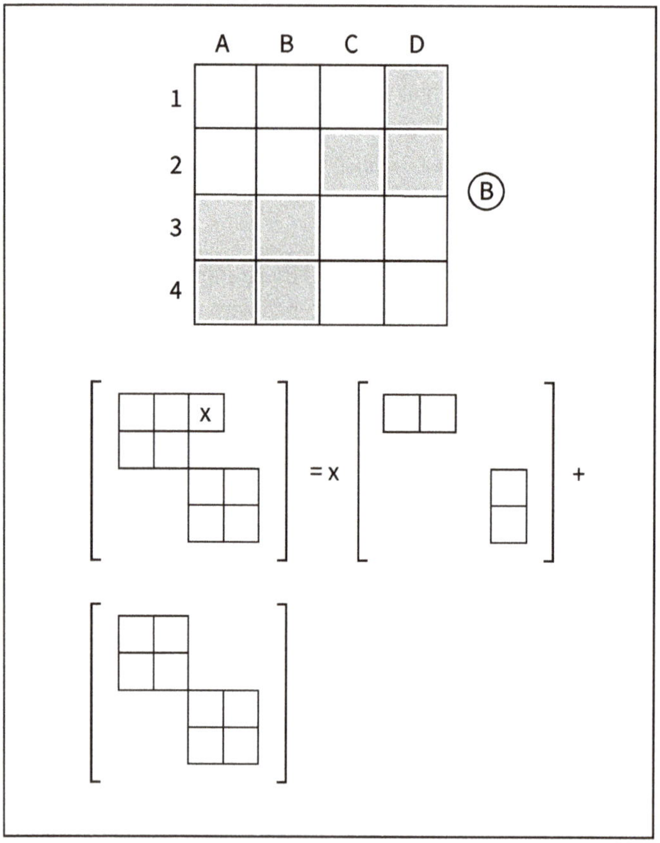

$$R(x,B) = x(1+2x)^2 + (1+4x+2x^2)^2$$
$$= 1 + 9x + 24x^2 + 20x^3 + 4x^4$$

Figure 8.7: Board for Example 8.39.

Thus far we have focused on the boards composed of the white squares – the allowed positions. We can turn this around and look at the boards composed of black squares – the forbidden positions. Recall the inclusion/exclusion formula from Section 1.4, $N_{=}(\varnothing) = \sum_{M\subset P}(-1)^{|M|}N_{\geq}(M)$. If we have a collection of objects with various properties chosen from the set P, we can find the number of objects that have no properties $N_{=}(\varnothing)$ by effectively deducting the number of objects in the various sets that do have some properties. The parity factor $(-1)^{|M|}$ ensures that objects with one or more properties are deducted only once overall. In the context of rook polynomials, we can consider permutations that send objects to one or more forbidden positions. A permutation has property m_i if it sends an object to the ith forbidden position. So $M = \{m_i, m_j, \ldots\}$ is a set of properties of this type and $N_{\geq}(M)$ is the number of permutations having at least all the properties in M.

Now given an $n \times n$ board representing the possible permutations of n distinct objects, we can identify the forbidden positions and construct the white square subboard B and the complementary black square subboard B^c. Let $R(x, B^c) = \sum_{k=0}^{n} r_k^c(B^c)x^k$ be the rook polynomial of the board B^c of black squares. Then $N_{=}(\varnothing) = \sum_{k=0}^{n}(-1)^k r_k^c(n-k)!$. To see this, note that if M has k properties it means that k out of n objects have been sent to forbidden positions r_k^c ways and the additional $(n-k)$ objects have been permuted among the $(n-k)$ positions remaining in $(n-k)!$ ways. Some of these remaining positions may also be forbidden positions, but the term $N_{\geq}(M)$ in the inclusion/exclusion formula represents the number of objects with *at least* the properties in M. So we have $N_{\geq}(M) = r_k^c(n-k)!$. Plugging into the inclusion/exclusion formula gives the result.

Example 8.40. Redo Example 8.39 using the forbidden square rook polynomial.

We can create an OGF for the terms $r_k^c(n-k)!$ that appear in Example 8.40. Define $E(x) = \sum_{k=0}^{n} r_k^c(n-k)!(x-1)^k$. The coefficient of x^k in the expansion of this sum gives the number of permutations of n objects where exactly k objects are sent to forbidden positions. Since these forbidden positions are "hit" in a sense, $E(x)$ is called a **hit polynomial**.

Example 8.41. Find the hit polynomial for Example 8.40 and determine how many ways the recital could be given with various numbers of denied preferences (forbidden positions).

References & suggested further reading

1) *Enumerative Combinatorics, Vol. 1* by Richard P. Stanley (1986), Wadsworth & Brooks
2) *Introduction to Combinatorial Mathematics* by C. L. Liu (1968), McGraw-Hill – *good discussion of generating functions*
3) *Combinatorics, A Guided Tour* by David R. Mazur (2010), Mathematical Assoc. of America

Solutions to Chapter 8 examples

Example 8.1

Consider the two sequences $\{1, 2, 1, 6, 6\}$ and $\{1, 0, 3, 7, 5\}$. Forming the OGF for the first sequence, we have $1{\cdot}1+2{\cdot}(x+1)+1{\cdot}(x-1)+6{\cdot}(x^2+2)+6{\cdot}(x^2-2)$. Simplified this yields $f(x) = 2+3x+12x^2$. Now for the second sequence, we have $1{\cdot}1+0{\cdot}(x+1)+3{\cdot}(x-1)+7(x^2+2)+5{\cdot}(x^2-2)$, which yields $f(x) = 2 + 3x + 12x^2$. Evidently, two different sequences correspond to the same generating function. So we have to disqualify $\{1, x+1, x-1, x^2+2, x^2-2, \ldots, x^n + n, x^n - n, \ldots\}$ as a viable set of indicator functions.

Example 8.2

Of course, ${}_nC_k$ is the binomial coefficient $\binom{n}{k}$, so $\binom{n}{0} + \binom{n}{1}x + \binom{n}{2}x^2 + \binom{n}{3}x^3 + \cdots + \binom{n}{n}x^n = (1+x)^n$ by the binomial theorem. So the OGF has a convenient closed form expression $G({}_nC_k; x) = (1+x)^n$.

Example 8.3

Mimicking Example 8.2, we know ${}_nP_k = \frac{n!}{(n-k)!}$, so

$$\frac{n!}{(n-0)!} + \frac{n!}{(n-1)!}x + \frac{n!}{(n-2)!}x^2 + \frac{n!}{(n-3)!}x^3 + \cdots + \frac{n!}{(n-n)!}x^n + \cdots.$$

This is a fine OGF as far as it goes, but it is not working out as neatly as it did for the combination statistics. There is no simple closed form expression. If we want an OGF, we have to settle for a clumsy polynomial that we cannot shrink. So we try an EGF and write

$$\frac{n!}{(n-0)!}\frac{1}{0!} + \frac{n!}{(n-1)!}\frac{1}{1!}x + \frac{n!}{(n-2)!}\frac{1}{2!}x^2 + \frac{n!}{(n-3)!}\frac{1}{3!}x^3 + \cdots + \frac{n!}{(n-n)!}\frac{1}{n!}x^n = (1+x)^n$$

again. So $E({}_nP_k; x) = G({}_nC_k; x) = (1+x)^n$. The rationale for having EGF's as an option is clear.

Example 8.4

$E(1; x) = \sum_{n=0}^{\infty}(1)\frac{x^n}{n!}$. We recognize $E(1; x) = e^x$.

Example 8.5

$G(n+2; x) = G(n; x) + G(2; x)$, where $a_n = 2$ for all n in the second OGF.

Example 8.6

$G(n;x) = \sum_{n=0}^{\infty} nx^n$, so simple differentiation of each term gives

$$\frac{d}{dx}\sum_{n=0}^{\infty} nx^n = \sum_{n=0}^{\infty} n^2x^{n-1}.$$

Then

$$x\sum_{n=0}^{\infty} n^2x^{n-1} = \sum_{n=0}^{\infty} n^2x^n = G(n^2;x).$$

So $x\frac{d}{dx}G(n;x) = G(n^2;x)$. Note that we could have started the summation index at 1, but to conform to our definition we have kept it at 0.

Example 8.7

We have

$$\frac{1}{1-x} = 1 + x + x^2 + \cdots = \sum_{n=0}^{\infty} x^n.$$

By Rule 2 of the preceding table,

$$\frac{1}{1-x}\cdot g(x) = \sum_{n=0}^{\infty}\left(\sum_{k=0}^{n} a_{n-k}b_k\right)x^n.$$

Now

$$\sum_{k=0}^{n} a_{n-k}b_k = \sum_{k=0}^{n} b_k$$

since $a_{n-k} = 1$ for all k. We conclude that

$$\frac{1}{1-x}\cdot g(x) = \sum_{n=0}^{\infty}\left(\sum_{k=0}^{n} b_k\right)x^n$$

is the OGF for the sequence of sums $\{\sum_{k=0}^{n} b_k\}_{n\in\mathbb{N}}$.

Example 8.8

1: ($\int$) We know

$$\frac{1}{1-x} = 1 + x + x^2 + \cdots = \sum_{n=0}^{\infty} x^n$$

Then

$$\frac{d}{dx}\sum_{n=0}^{\infty} x^n = \sum_{n=0}^{\infty} nx^{n-1}.$$

Adjusting the powers of x allows us to write

$$x\frac{d}{dx}\sum_{n=0}^{\infty} x^n = \sum_{n=0}^{\infty} nx^n.$$

We conclude $x\frac{d}{dx}(\frac{1}{1-x}) = \frac{x}{(1-x)^2}$ is the closed form representation of the OGF $\sum_{n=0}^{\infty} nx^n$.

2: We can apply the summation operator from Example 8.7 to the generating function for the sequence $\{1, 1, \ldots, 1, \ldots\}$, which is coincidentally the same as the summation operator $\frac{1}{1-x} = 1 + x + x^2 + \cdots$. There is a minor technical issue here. We want the integer sequence to start with zero, but the coefficient of x^0 in the summation series will obviously be 1 and this will persist so that the coefficient of x^n will be $n+1$. This is obvious from the expansion $\frac{1}{(1-x)^2} = 1 + 2x + 3x^2 + \cdots$. To correct this, we will need to premultiply $\frac{1}{1-x}$ by another factor of x to get the correct matchup of integer coefficients with exponents. Applying the summation operator, we have

$$\frac{1}{1-x} \cdot \frac{x}{1-x} = \frac{x}{(1-x)^2} = 0x^0 + 1x^1 + 2x^2 + 3x^3 + \cdots = \sum_{n=0}^{\infty} nx^n.$$

Example 8.9

From Example 8.8, the OGF for $\{0, 1, 2, 3, \ldots, n, \ldots\}$ is $\frac{x}{(1-x)^2}$. Our strategy is to use the summation operator yet again, so the OGF for the sum of the first n integers should be $\frac{1}{1-x} \cdot \frac{x}{(1-x)^2} = \frac{x}{(1-x)^3}$. If we can expand $\frac{x}{(1-x)^3}$ in a power series and extract the nth coefficient, we will have the formula for $\sum_{k=0}^{n} k$. To carry out this plan, we will need the industrial strength version of the binomial theorem. The regular version was introduced in Section 1.1.2 (Example 1.13). The **generalized binomial theorem** allows r to be negative or fractional in the following expansion:

$$(1+x)^r = 1 + \sum_{n=1}^{\infty} \frac{r(r-1)\cdots(r-n+1)}{n!}x^n.$$

Note that this sum is finite if r is a nonnegative integer, so the falling factorial terminates. Applying this to $(1-x)^{-3}$, we find $[[\frac{1}{(1-x)^3}]]_{x^n}$ to be $\frac{(-3)(-3-1)\cdots(-3-n+1)}{n!}(-1)^n$. Note that the factor $(-1)^n$ compensates for the fact that we are expanding $(1-x)^{-3}$, not $(1+x)^{-3}$. Distributing this factor over the numerator gives us

$$\frac{(3)(4)\cdots(n+2)}{n!} = \frac{(1)(2)(3)(4)\cdots(n)(n+1)(n+2)}{(1)(2)n!} = \frac{(n+1)(n+2)}{2}.$$

Now we need to account for the additional factor x that we suppressed to do the extraction. We conclude that $\frac{(n+1)(n+2)}{2}$ would be the coefficient of x^{n+1} in a series expansion of $\frac{x}{(1-x)^3}$. To get the coefficient of x^n, we deduct 1 from each n. Finally, we have

$$\frac{x}{(1-x)^3} = \sum_{n=0}^{\infty} \frac{n(n+1)}{2} x^n.$$

It follows that

$$\sum_{n=0}^{\infty} \left(\sum_{k=0}^{n} k \right) x^n = \sum_{n=0}^{\infty} \frac{n(n+1)}{2} x^n,$$

so $\sum_{k=0}^{n} k = \frac{n(n+1)}{2}$.

Example 8.11

Our strategy will be to apply the summation operator to $\sum_{n=0}^{\infty} n^2 x^n$. From Section 8.1 (Example 8.6), we know

$$\sum_{n=0}^{\infty} n^2 x^n = G(n^2; x) = x \frac{d}{dx} G(n; x).$$

From Section 8.2 (Example 8.8), we have $G(n; x) = \frac{x}{(1-x)^2}$. Putting these facts together, we get

$$\sum_{n=0}^{\infty} n^2 x^n = x \frac{d}{dx} \left(\frac{x}{(1-x)^2} \right) = \frac{x(1+x)}{(1-x)^3}.$$

We see that the OGF of the sequence $0^2, 1^2, 2^2, \ldots$ is $\frac{x(1+x)}{(1-x)^3}$. Then multiplying by the summation operator, we have $\frac{1}{1-x} \cdot \frac{x(1+x)}{(1-x)^3} = \frac{x(1+x)}{(1-x)^4}$ as the OGF for the corresponding sequence of sums $0^2, 0^2 + 1^2, 0^2 + 1^2 + 2^2$, and so forth. Now we turn to the generalized binomial theorem to expand $\frac{x(1+x)}{(1-x)^4}$ and then extract the coefficient of x^n. Really, all we need is the expansion of $\frac{1}{(1-x)^4}$. It will be a simple matter to adjust for the $x(1+x)$ in the numerator. Proceeding with $(1-x)^{-4}$, we find $[[\frac{1}{(1-x)^4}]]_{x^n}$ to be $\frac{(-4)(-4-1)\cdots(-4-n+1)}{n!}(-1)^n$. As in Example 8.9, the term $(-1)^n$ accounts for the fact that we are expanding $(1-x)^n$ instead of $(1+x)^n$. Simplifying as before, the nth coefficient is

$$\begin{aligned} \frac{(4)(5)\cdots(n+3)}{n!} &= \frac{(1)(2)(3)(4)(5)\cdots(n)(n+1)(n+2)(n+3)}{(1)(2)(3)n!} \\ &= \frac{(n+1)(n+2)(n+3)}{6}. \end{aligned}$$

We are almost there. This is still the nth coefficient of $\frac{1}{(1-x)^4}$, and we need the nth coefficient of $\frac{x(1+x)}{(1-x)^4} = \frac{x}{(1-x)^4} + \frac{x^2}{(1-x)^4}$. Note that $\frac{(n+1)(n+2)(n+3)}{6}$ would be the $(n+1)$st coefficient of $\frac{x}{(1-x)^4}$ and the $(n+2)$nd coefficient of $\frac{x^2}{(1-x)^4}$. So the nth coefficient of $\frac{x}{(1-x)^4}$ would instead be $\frac{(n)(n+1)(n+2)}{6}$ and then the nth coefficient of $\frac{x^2}{(1-x)^4}$ would be $\frac{(n-1)(n-)(n+1)}{6}$. Therefore, the overall nth coefficient of $\frac{x(1+x)}{(1-x)^4}$ is

$$\frac{(n)(n+1)(n+2)}{6} + \frac{(n-1)(n)(n+1)}{6}.$$

Finally, this simplifies to $\frac{2n^3+3n^2+n}{6} = \frac{n(n+1)(2n+1)}{6}$. Summarizing the argument,

$$\frac{x(1+x)}{(1-x)^4} = \sum_{n=0}^{\infty}\left(\frac{n(n+1)(2n+1)}{6}\right)x^n = \sum_{n=0}^{\infty}\left(\sum_{k=0}^{n} k^2\right)x^n,$$

so $\sum_{k=0}^{n} k^2 = \frac{n(n+1)(2n+1)}{6}$. We solve this problem more elegantly in Chapter 9 using Newton's forward difference theorem.

Example 8.12

Since $f(x) = (1+x)^n = \sum_{k=0}^{n}\binom{n}{k}x^k$, we have $f(1) = (1+1)^n = \sum_{k=0}^{n}\binom{n}{k} = 2^n$. This counts all possible subsets of an n-set, null set included.

Example 8.13

We use the same approach as in Example 8.12, except we set $x = -1$. Then $(1-1)^n = 0 = \sum_{k=0}^{n}\binom{n}{k}(-1)^k$, so $\binom{n}{0} - \binom{n}{1} + \binom{n}{2} - \cdots + \binom{n}{k=2j} - \binom{n}{k=2j+1}\cdots = 0$. Evidently, we can move all the negative odd selections to the right-hand side of the equation and it is then obvious why the claim is true.

Example 8.14

We are seeking $\sum_{k=0}^{n} k\binom{n}{k}$. Note that

$$\frac{d}{dx}\sum_{k=0}^{n}\binom{n}{k}x^k = \sum_{k=0}^{n} k\binom{n}{k}x^{k-1} = \frac{d}{dx}(1+x)^n = n(1+x)^{n-1}.$$

Setting $x = 1$, we get $\sum_{k=0}^{n} k\binom{n}{k} = n2^{n-1}$.

Example 8.15

As $(1+x)^n = \sum_{k=0}^{n}\binom{n}{k}x^k$ and $(1+\frac{1}{x})^n = \sum_{k=0}^{n}\binom{n}{k}(\frac{1}{x})^k$, multiplying $(1+x)^n$ by $(1+\frac{1}{x})^n$ results in an expression of the form $\sum_{k=-n}^{n} A_k x^k$ where $A_k = A_{-k}$ and, in particular, $A_0 = \sum_{k=0}^{n}\binom{n}{k}\binom{n}{n-k} = \sum_{k=0}^{n}\binom{n}{k}^2$. Note that $(1+\frac{1}{x})^n = x^{-n}(1+x)^n$, so $(1+\frac{1}{x})^n(1+x)^n = x^{-n}(1+x)^{2n}$.

Expanding $(1+x)^{2n}$ in a power series gives $(1+x)^{2n} = \sum_{k=0}^{2n} \binom{2n}{k} x^k$, so the coefficient of x^0 in $x^{-n}(1+x)^{2n}$ occurs when $k = n$ and is therefore $\binom{2n}{n}$. Equating this to A_0 finally gives $\sum_{k=0}^{n} \binom{n}{k}^2 = \binom{2n}{n}$.

Example 8.16

We construct the enumerator for each type of object and then multiply them together to get an overall enumerator. Symbolize the possible selections of a as follows: $1 + ax + a^2x^2$. Here 1 means a is not selected, ax means a is selected once, and $aax^2 = a^2x^2$ means a is selected twice. The exponent of x indicates how many selections are being made and the coefficient of a power of x tells what is being selected. Likewise, the enumerator for b is $1 + b^2x^2$, since the term bx will be missing by the statement of the problem. We interpret the overall enumerator $1 + (a)x + (a^2 + b^2)x^2 + (ab^2)x^3 + (a^2b^2)x^4$ using the convention that "and" is symbolized by a product and "or" by a sum. So we have the following possibilities: $1x^0$ means no objects were selected, $(a)x^1$ means a only was selected once, $(a^2 + b^2)x^2$ means either a was selected twice or b twice, $(ab^2)x^3$ means a was selected once and b twice, and $(a^2b^2)x^4$ means both objects were selected twice. We conclude that there were two ways two objects could be selected and one way three objects could be selected.

Example 8.17

We construct the ordinary enumerator for this situation by finding the enumerator for the outcomes of one die and then cubing the result. The exponent of x keeps track of the number of dots on the up face of the die and the coefficient of x^n gives the number of ways this can be attained with one die (one each, obviously). So the enumerator for one die is $x + x^2 + x^3 + x^4 + x^5 + x^6$. Since rolling any one die is independent of any other one, the enumerator for the sum of the dots on the up faces of three dice is $(x + x^2 + x^3 + x^4 + x^5 + x^6)^3$. This works out to be $x^{18} + 3x^{17} + 6x^{16} + 10x^{15} + 15x^{14} + 21x^{13} + 25x^{12} + 27x^{11} + 27x^{10} + 25x^9 + 21x^8 + 15x^7 + 10x^6 + 6x^5 + 3x^4 + x^3$. This is a symmetric polynomial, as we might have expected. It gives the complete picture of the possible outcomes, of which there are 16 distinct numerical results which are determined by 216 physical results. The various coefficients give the number of ways particular numerical results encoded by the exponents could be obtained. The probability of rolling either 10 or 11 is 12.5 %, slightly less than the probability of rolling a seven with two dice.

Example 8.18

The enumerator for the dodecahedral die is $\sum_{k=1}^{12} x^k$ and the regular die $\sum_{k=1}^{6} x^k$. The enumerator for rolling both is $x^{18} + 2x^{17} + 3x^{16} + 4x^{15} + 5x^{14} + 6x^{13} + 6x^{12} + 6x^{11} + 6x^{10} + 6x^9 + 6x^8 + 6x^7 + 5x^6 + 4x^5 + 3x^4 + 2x^3 + x^2$. Strangely enough, the sum of the up faces from 7 to 13 are both most likely and equally likely.

Example 8.19

The enumerator for the pennies is $\sum_{k=0}^{5} x^k$. For the nickels, $\sum_{k=0}^{3} x^{5k}$. For the dimes, $\sum_{k=0}^{4} x^{10k}$. And for the quarters, $\sum_{k=0}^{2} x^{25k}$. The overall enumerator for making change is $\sum_{k=0}^{5} x^k \cdot \sum_{k=0}^{3} x^{5k} \cdot \sum_{k=0}^{4} x^{10k} \cdot \sum_{k=0}^{2} x^{25k}$. Using a computer algebra system to expand this cumbersome expression, we find every amount up to \$1.10 can be made in at least one way.

Example 8.20

Scanning the revised enumerator, there are five amounts under 25 ¢ that cannot be made, 4 ¢, obviously the smallest. This type of problem is called a Frobenius problem after the German mathematician Georg Frobenius, who first studied it formally.

Example 8.21

The enumerator for touchdowns is $\sum_{k=0}^{6} x^{6k}$, touchdowns and one extra point $\sum_{k=0}^{5} x^{7k}$, touchdowns and two extra points $\sum_{k=0}^{5} x^{8k}$, and field goals $\sum_{k=0}^{6} x^{3k}$. We multiply these together to get the overall enumerator for scoring configurations: $1 + x^3 + 2x^6 + x^7 + x^8 + 2x^9 + x^{10} + x^{11} + 3x^{12} + 2x^{13} + 3x^{14} + 4x^{15} + 3x^{16} + 3x^{17} + 5x^{18} + 4x^{19} + 5x^{20} + 6x^{21} + 6x^{22} + 6x^{23} + 8x^{24} + 7x^{25} + 8x^{26} + 9x^{27} + 9x^{28} + 9x^{29} + 12x^{30} + 11x^{31} + 12x^{32} + 13x^{33} + 13x^{34} + 13x^{35} + 16x^{36} + 15x^{37} + 17x^{38} + 18x^{39} + 18x^{40}$. It appears that the only impossible scores are 1, 2, 4, and 5. Of course, if we allowed the two point play all of these except 1 would become feasible scores. We note that there are more ways to achieve the higher scores, so the trend is generally up, but not consistently so. Apparently 39 and 40 are the most numerically feasible scores, and 11 is the highest score achievable in only one way.

Example 8.22

The enumerator for selections from the pairs of marbles is $(1 + x + x^2)^p$. The enumerator for selections from the singletons' group is $(1 + x)^q$. The overall enumerator for the number of selections from the collection of $2p + q$ marbles is then $(1 + x + x^2)^p(1 + x)^q$. This can be written $\sum_{r=0}^{2p+q}(\sum_{k=0}^{\lfloor r/2 \rfloor} \binom{p}{k}\binom{p+q-k}{r-2k})x^r$, where $\lfloor r/2 \rfloor$ is the floor function of $r/2$. Note $\lfloor r/2 \rfloor = r/2$ if r is even and $\lfloor r/2 \rfloor = (r-1)/2$ if r is odd. The rationale for the product of binomial coefficients giving the overall coefficient of x^r is that we first select the x^2 in the term $(1 + x + x^2)$ exactly k times. This consumes k colored pairs or $2k$ marbles. We have $r - 2k$ selections of single marbles left. Next we select an x term from either the $(p - k)$ factors of $(1 + x + x^2)$ remaining or the q factors of $(1 + x)$ which have thus far not been used. Clearly, there are $r - 2k$ such selections remaining to be made from this pool of $p + q - k$ single-color marbles. By the multiplicative principle, the product of binomials gives the overall number of ways to make r total selections.

Example 8.23

The enumerator for selecting a particular object any number of times without restriction ("with replacement" perhaps gives a more satisfying mental image) is $1+x+x^2+x^3+\cdots = \frac{1}{1-x}$. If we have n objects, the overall enumerator would be $(\frac{1}{1-x})^n = (1-x)^{-n}$. Expanding as usual with the generalized binomial theorem, we get

$$(1-x)^{-n} = 1 + \sum_{k=1}^{\infty}(-1)^k \frac{(-n)(-n-1)\cdots(-n-k+1)}{k!}x^k.$$

Distributing the minus signs gives us $1 + \sum_{k=1}^{\infty} \frac{(n)(n+1)\cdots(n+k-1)}{k!}x^k$, and finally, we rewrite this in more recognizable form

$$\begin{aligned} 1 + \sum_{k=1}^{\infty} \frac{(n-1)!}{(n-1)!} \cdot \frac{(n)(n+1)\cdots(n+k-1)}{k!}x^k &= \sum_{k=0}^{\infty} \frac{(n+k-1)!}{(n-1)!k!}x^k \\ &= \sum_{k=0}^{\infty} \binom{n+k-1}{k}x^k. \end{aligned}$$

Evidently,

$$\sum_{k=0}^{\infty} \binom{n+k-1}{k}x^k = \sum_{k=0}^{\infty} \left(\!\binom{n}{k}\!\right)x^k$$

and we see that

$$\left(\!\binom{n}{k}\!\right) = \binom{n+k-1}{k}.$$

Example 8.24

The enumerator for one object is $x + x^2 + x^3 + \cdots = \frac{x}{1-x}$, so for n objects it becomes $\frac{x^n}{(1-x)^n} = x^n(1-x)^{-n}$. We utilize the result in Example 8.23 to write this enumerator as

$$x^n(1-x)^{-n} = x^n \sum_{k=0}^{\infty} \binom{n+k-1}{k}x^k = \sum_{k=0}^{\infty} \binom{n+k-1}{k}x^{k+n}.$$

Reindexing with $r = k+n$, we rewrite this as $\sum_{r=n}^{\infty} \binom{r-1}{r-n}x^r$. Note that

$$\binom{r-1}{r-n} = \frac{(r-1)!}{(r-n)!(n-1)!} = \binom{r-1}{n-1}.$$

The final enumerator is then $\sum_{r=n}^{\infty} \binom{r-1}{n-1}x^r$. For the specific example, $k = 6$ and $n = 4$, so $r = 10$. Then there are $\binom{9}{3} = 84$ ways.

Example 8.25

From the preceding discussion, the exponential enumerator would be

$$\left(1+x+\frac{x^2}{2!}+\frac{x^3}{3!}\right)\left(1+x+\frac{x^2}{2!}+\frac{x^3}{3!}+\frac{x^4}{4!}\right)$$
$$=1+2x+4\frac{x^2}{2!}+8\frac{x^3}{3!}+15\frac{x^4}{4!}+25\frac{x^5}{5!}+35\frac{x^6}{6!}+35\frac{x^7}{7!}.$$

The number of distinct permutations of all the flags is $\frac{7!}{3!4!}=35$. Note that this enumerator automatically handles the cases where not all the flags are deployed. For example, if we wanted to know how many distinct pairs of flags could be presented, and we decided to use the multinomial formula, we would first have to separate the cases according to what subset of two flags was to be used. The enumerator says four (RR, RB, BR, BB) right away.

Example 8.26

The exponential enumerator for each color except red is

$$\left(1+x+\frac{x^2}{2!}+\frac{x^3}{3!}+\frac{x^4}{4!}+\cdots\right).$$

For red, it is $(x+\frac{x^2}{2!}+\frac{x^3}{3!}+\frac{x^4}{4!}+\cdots)$. Recall that $e^x=\sum_{k=0}^{\infty}\frac{x^k}{k!}$, so the exponential enumerator for every color except red is e^x and for red it is e^x-1. Evidently, the overall exponential enumerator for all permutations of k flags is $e^{5x}(e^x-1)=e^{6x}-e^{5x}$. Since $e^{rx}=\sum_{k=0}^{\infty}\frac{(rx)^k}{k!}=\sum_{k=0}^{\infty}r^k\frac{x^k}{k!}$, we see that $e^{6x}-e^{5x}=\sum_{k=0}^{\infty}(6^k-5^k)\frac{x^k}{k!}$. So for $k=10$, the number of permutations is $6^{10}-5^{10}=50,700,551$. Checking the case $k=2$, we can be more explicit. Letting R stand for red and X for any one of the other five colors, the possible permutations are XR, RX, and RR. Since X can take on five values, this gives 11 permutations in all, agreeing with $6^2-5^2=11$.

Example 8.27

View the problem as finding the number of k-permutations with unlimited repetition of the n distinct cells with each cell included at least once in each permutation. So, for example, if we had three objects $\{a,b,c\}$ and two cells C_1 and C_2, a representative distribution could be the 3-permutation with repetition (C_1,C_1,C_2), signifying a and b are in cell C_1 and c is in cell C_2. The exponential enumerator for one (nonempty) cell is $(x+\frac{x^2}{2!}+\frac{x^3}{3!}+\cdots)=e^x-1$. So for n cells the enumerator is $(e^x-1)^n$. Developing this in a series expansion, we get

$$\sum_{i=0}^{\infty}\binom{n}{i}(-1)^i e^{(n-i)x}.$$

Now substitute the series expansion of $e^{(n-i)x}$ to get

$$\sum_{i=0}^{\infty}\binom{n}{i}(-1)^i\left[\sum_{k=0}^{\infty}\frac{1}{k!}(n-i)^k x^k\right].$$

This is the point where we apply the magic potion and reverse the order of summation (they are being treated as formal sums, after all). Now we arrive at the following exponential enumerator:

$$\sum_{k=0}^{\infty}\frac{x^k}{k!}\left[\sum_{i=0}^{n}\binom{n}{i}(-1)^i(n-i)^k\right].$$

The term in square brackets is the extracted coefficient of $\frac{x^k}{k!}$. It follows that the number of k-permutations of n distinct cells with unlimited reselection permitted and each cell selected at least once is $\sum_{i=0}^{n}\binom{n}{i}(-1)^i(n-i)^k$. This is $n!S(k,n)$. We see that the exponential enumerator of the Stirling numbers of the second kind is $\frac{(e^x-1)^n}{n!}$.

Example 8.28

Here is a list of the enumerators for the number of times n can be chosen and contributed to a sum not exceeding 5:

$n = 1, 1 + x + x^2 + x^3 + x^4 + x^5$, 1 can be chosen zero through five times
$n = 2, 1 + x^2 + x^4$, 2 can be chosen once, twice, or not at all
$n = 3, 1 + x^3$, 3 can be chosen once or not at all
$n = 4, 1 + x^4$, 4 can be chosen once or not at all
$n = 5, 1 + x^5$, 5 can be chosen once or not at all

The product of these enumerators is $1 + x + 2x^2 + 3x^3 + 5x^4 + 7x^5 + \cdots$, where we have truncated the 21st-degree polynomial at a level that answers the question posed. The coefficient 7 of x^5 tells us that there are 7 partitions of 5. They are (5), (4, 1), (3, 2), (3, 1, 1), (2, 2, 1), (2, 1, 1, 1), and (1, 1, 1, 1, 1). Consider how this coefficient 7 is created. We pick a specific power of x from each of the five factor enumerators so that the total of the exponents is 5. Here is an example where the selection made is underlined: $(1+x+x^2+\underline{x^3}+x^4+x^5)(1+\underline{x^2}+x^4)(\underline{1}+x^3)(\underline{1}+x^4)(\underline{1}+x^5)$. Note that this selection adds 1 to the coefficient of x^5 in the product enumerator. It also represents the partition (2, 1, 1, 1) of 5 because the x^3 in the first factor represents three 1's and the x^2 in the second factor represents one 2. Of course, the remaining three factors can only contribute an $x^0 = 1$ to the product to keep the total of the exponents at 5. It should be clear that each distinct selection made from the factor enumerators in this manner corresponds to a distinct partition of 5. It is a welcome convenience that the multiplication of factor enumerators automatically accounts for all the relevant selections. So to find the total number of partitions of 5, it is simply a matter of finding the coefficient of x^5 in the product enumerator.

Example 8.29

Guided by Example 8.28, we can write the overall enumerator as $(1+x+x^2+x^3+x^4+x^5+x^6+x^7)(1+x^2+x^4+x^6)(1+x^3+x^6)(1+x^4)$. We purposely omit factor enumerators for any integer larger than 4 so they will not appear in any partition.. Expanding this and truncating at x^7, we get $1+x+2x^2+3x^3+5x^4+6x^5+9x^6+11x^7+\cdots$. Evidently, there are 11 partitions of 7 with no part bigger than 4. The overall enumerator for <u>all</u> partitions of 7 works out to be $1+x+2x^2+3x^3+5x^4+7x^5+11x^6+15x^7+\cdots$, so there are $15-11=4$ partitions of 7 which feature a part greater than 4. They happen to be (7), (6, 1), (5, 2), and (5, 1, 1).

Example 8.30

We answer the question by calculating the number of partitions of 10 that have largest part 4. The enumerator is $(\sum_{k=0}^{10} x^k)\cdot(\sum_{k=0}^{5} x^{2k})\cdot(\sum_{k=0}^{3} x^{3k})\cdot(\sum_{k=0}^{2} x^{4k})$. Using a computer algebra system to do the expansion, we find the coefficient of x^{10} is 23. We conclude by the Ferrers graph conjugation result that there are 23 partitions of 10 that contain no more than four parts. Partition problems for small integers can often by solved by brute force. This problem showcases the usefulness of the generating function method where a brute force enumeration would be impractical, or at the very least, annoyingly time consuming.

Example 8.31

Omit the factor enumerator for 2's in the overall enumerator

$$(1+x+x^2+x^3+x^4+x^5+x^6+x^7)(1+x^3+x^6)(1+x^4)(1+x^5)(1+x^6)(1+x^7).$$

Expanding and truncating this at x^7, we have $1+x+x^2+2x^3+3x^4+4x^5+6x^6+8x^7+\cdots$. So there are 8 partitions of 7 that do not involve 2 as a part.

Example 8.32

An enumerator for a single copy of the integer k is just $1+x^k$. So $\prod_{k=1}^{n}(1+x^k)$ would be the enumerator for partitions of n (and any number less than n) into distinct parts. So for 8 we need to expand

$$\prod_{k=1}^{8}(1+x^k)=1+x+x^2+2x^3+2x^4+3x^5+4x^6+5x^7+6x^8+\cdots.$$

Apparently there are 6 partitions of 8 where all the parts are different. They are (8), (7, 1), (6, 2), (5, 3), (5, 2, 1), and (4, 3, 1).

Example 8.33

It will suffice to find the number of partitions into at most m parts and then subtract the same statistic for $m-1$ parts. Using the Ferrers bijection, we find the enumerator for number of partitions into at most m parts is the same as the enumerator for the number of partitions into parts no greater than m, which is $\frac{1}{(1-x)(1-x^2)(1-x^3)\cdots(1-x^m)}$. Repeating the calculation for $m-1$, we find the enumerator for the number of partitions with no more than $m-1$ parts is $\frac{1}{(1-x)(1-x^2)(1-x^3)\cdots(1-x^{m-1})}$. The difference after determining a common denominator is $\frac{x^m}{(1-x)(1-x^2)(1-x^3)\cdots(1-x^m)}$, which is the enumerator for exactly m parts.

Example 8.34

Expanding the enumerator with a computer algebra system we have $x^3+x^4+2x^5+3x^6+4x^7+\cdots$. Evidently, there are exactly four partitions of 7 into 3 parts. They are $(5,1,1)$, $(4,2,1)$, $(3,3,1)$, and $(3,2,2)$.

Example 8.35

See the application of the expansion formula and the resulting calculations in Figure 8.4.

Example 8.36

Since each subboard places no restriction on the allowed rook placements for the other board, $r_k(B_1)$ is independent of $r_j(B_2)$ for all pairs of indices i, j. If

$$R(x, B_1) = \sum_{k=0}^{N} r_k(B_1)x^k \quad \text{and} \quad R(x, B_2) = \sum_{k=0}^{M} r_k(B_2)x^k,$$

then we can take the convolution product according to Table 8.1a and write

$$R(x, B) = \sum_{k=0}^{N+M}\left[\sum_{j=0}^{k} r_j(B_1)r_{k-j}(B_2)\right]x^k = R(x, B_1)R(x, B_2).$$

Example 8.37

We split B at the cutpoint into two identical but disjoint subboards. We already know that the three-square subboard has rook polynomial $1+3x+x^2$. Evidently, $R(x, B) = (1+3x+x^2)^2 = 1+6x+11x^2+6x^3+x^4$.

Example 8.38

The forbidden positions are blacked out in the square array. The permitted assignments correspond to the white squares, so call the collection of white squares board B. Our task is to put rooks on B and find $r(x, B)$. Then $r_4(B)$ answers the question about all four technicians. So we use the tools available for our graphic algebra to find $r(x, B)$.

Example 8.39

Figure 8.7 gives the possibilities.

The rook polynomial for this board is $R(x,B) = 1+9x+24x^2+20x^3+4x^4$, so $r_4(B) = 4$ and four different programs could be scheduled.

Example 8.40

Referring to the black square board B^c in Figure 8.7, and noting that they can be viewed as two disjoint subboards, we can easily calculate the overall rook polynomial using the multiplicative principle from Example 8.36. The three-square subboard has rook polynomial $1+3x+x^2$, and by inspection the 2×2 subboard has rook polynomial $1+4x+2x^2$. Evidently, $R(x,B^c) = (1+3x+x^2)(1+4x+2x^2)$ which simplifies to $1+7x+15x^2+10x^3+2x^4$. Exporting these r_k^c coefficients to the formula $N_=(\varnothing) = \sum_{k=0}^{n}(-1)^k r_k^c(n-k)!$, we have $N_=(\varnothing) = n!-7(n-1)!+15(n-2)!-10(n-3)!+2(n-4)!$. In this case, $n = 4$ so the expression reduces to $24 - 7\cdot 6 + 15\cdot 2 - 10\cdot 1 + 2 = 4$. This is the number of permutations of four objects which do not send any object to a black square. This agrees with Example 8.39. Depending on a given situation, it may be easier to compute $R(x,B^c)$ than $R(x,B)$, so this approach allows that option.

Example 8.41

Using the calculations in Example 8.40, $E(x) = 1\cdot 4!(x-1)^0\cdot 7\cdot 3!(x-1) + 15\cdot 2!(x-1)^2 + 10\cdot 1!(x-1)^3 + 2\cdot 0!(x-1)^4$. This simplifies to $E(x) = 4 + 4x + 12x^2 + 2x^3 + 2x^4$. The interpretation is that there are four programs where every student gets their preference, four programs where one student is denied their preference, 12 programs where two students are denied, two programs where three students are denied, and finally two programs where nobody gets their preference. Spot checking this result, we can see the two programs where everyone is denied: {A4, B3, C2, D1} and {A3, B4, C2, D1}.

9 Finite difference methods

Some branches of mathematics germinate from a seed in the form of a particular foundational problem. One such problem was proposed more than 800 years ago. Leonardo Pisano, whose nickname was Fibonacci (figlio di Bonacci) published a question in his 1202 book *Liber Abaci* (Book of Calculation) that featured the analysis of an idealized growth model for a population of rabbits. We will explore this problem and its ramifications, but its importance rests on the fact that it has inspired the study of recurrence formulas. This subject has expanded over the years into a modern discipline that has found application to a wide range of problems in higher mathematics. Our foray into this area will be very basic – just a little beyond counting rabbits.

Incidentally, a case can be made that Fibonacci was the greatest mathematician of pre-Renaissance Europe. We base this claim on a contribution far more influential than recurrence theory. Fibonacci was educated in what is now Algeria and was exposed to the workings of the Hindu–Arabic decimal number system. He brought this back to Italy and it spread to the rest of Europe. We can thank him for the Arabic numerals $0, 1, 2, \ldots, 9$ supplanting the Roman I, II, III, …, IX. The Roman system had a concept of zero (nulla), but no symbol for it. As if that were not enough, he also appears to have introduced the sign $\sqrt{n}$ for the square root of n, as well as the fraction bar $\frac{a}{b}$ separating numerator from denominator. Electronic calculators certainly would be interesting to operate today without Fibonacci's contributions.

9.1 Recurrences – general

Suppose we have a sequence of numbers, often integers but not necessarily, $\{a_0, a_1, \ldots, a_n, \ldots\}$ It may be possible to express a_n as a function of some of the terms preceding it. For example, given the sequence $\{1, 2, 4, 8, 16, \ldots\}$ we recognize that each term is twice the one immediately before it. We can capture this observation in the formula $a_n = 2a_{n-1}$. Some information about the first term is necessary as well, or we might get off on the wrong foot and regenerate the sequence $\{3, 6, 12, 24, \ldots\}$. So we define $a_0 = 1$. Now with the first term specified and the term-to-term relationship given, we have provided an algorithm for constructing exactly the sequence we were given. We define the **recurrence relation** (or **difference equation**) for the sequence to be the formula $a_n = 2a_{n-1}$ and the **initial value** (**seed value** or **boundary condition**) of the recurrence to be $a_0 = 1$.

More complicated recurrences may use more than one preceding value to construct the next term. In these cases, enough initial values have to be given to work the algorithm. Here is the famous Fibonacci sequence, $\{1, 1, 2, 3, 5, 8, 13, \ldots\}$. Can you guess the recurrence? It is $F_n = F_{n-1} + F_{n-2}$. The initial values are $F_1 = F_2 = 1$. In words, the next term is the sum of the two preceding terms. How this sequence is connected to the rabbits will be addressed shortly.

https://doi.org/10.1515/9783111506654-010

Completely solving a recurrence problem involves two things. First, we need to deduce the recurrence relation from the given sequence. This may not be obvious. Second, we need to find a general expression for the nth term of the sequence. The nth term can always be found by brute force calculation – grinding out the terms by successive application of the recurrence algorithm until we finally get to the one we want. A more genteel approach is to use the form of the recurrence relation to obtain a simple (in the eye of the beholder) formula for the nth term. To illustrate, the formula corresponding to the first recurrence presented above would be $a_n = 2^n$. (Remember we start with $n = 0$.) The formula for the nth Fibonacci number is

$$F_n = \frac{\sqrt{5}}{5}\left(\frac{1+\sqrt{5}}{2}\right)^n - \frac{\sqrt{5}}{5}\left(\frac{1-\sqrt{5}}{2}\right)^n.$$

Here F_n must be an integer and how the presence of $\sqrt{5}$ permits this is an interesting phenomenon of algebra. We now turn to developing a method for solving these problems.

9.2 Linear recurrences – substitution method

A general **linear recurrence** relation with constant coefficients can be written as $c_0 a_n + c_1 a_{n-1} + \cdots + c_k a_{n-k} = f(n)$. Here $c_0, c_1, \ldots, c_k$ are constants and $a_{n-k}, a_{n-k+1}, \ldots, a_n$ are terms of a sequence. The number of successive prior terms necessary to calculate the term a_n is k, which is called the **order of the recurrence.** Finally, $f(n)$ is an arbitrary integer function. The Fibonacci recurrence has order two. A typical example of order three would be $a_n - 2a_{n-1} + 5a_{n-2} - 3a_{n-3} = n^3$. We could write this as $a_n = 2a_{n-1} - 5a_{n-2} + 3a_{n-3} - n^3$, and given initial values a_0, a_1, and a_2, we could find a_3. Actually, any three successive values of a_i would allow us to calculate the very next one in the sequence. In fact, if we had a_3, a_2, and a_1, we could turn the equation around to $3a_{n-3} = a_n - 2a_{n-1} + 5a_{n-2} - n^3$ and solve for a_0. So as long as we have k successive values of a_i, we can do the same operations on the general recurrence of order k. Initial conditions would be convenient to have, but we can always go backwards from the arbitrary k successive values to recover earlier terms in the sequence.

The most general solution to $c_0 a_n + c_1 a_{n-1} + \cdots + c_k a_{n-k} = f(n)$ consists of two parts, the **homogenous solution** and the **particular solution**, which are added together. The terminology is the same as that encountered in the theory of differential equations. The homogenous solution solves $c_0 a_n + c_1 a_{n-1} + \cdots + c_k a_{n-k} = 0$ and the particular solution solves the entire original equation $c_0 a_n + c_1 a_{n-1} + \cdots + c_k a_{n-k} = f(n)$. To see why this is the case, let a_n^h denote a solution to the homogenous equation and a_n^p denote a solution to the particular equation. We would have: (i) $c_0 a_n^h + c_1 a_{n-1}^h + \cdots + c_k a_{n-k}^h = 0$ and (ii) $c_0 a_n^p + c_1 a_{n-1}^p + \cdots + c_k a_{n-k}^p = f(n)$. Adding these two equations and regrouping, we obtain $c_0(a_n^p + a_n^h) + c_1(a_{n-1}^p + a_{n-1}^h) + \cdots + c_k(a_{n-k}^p + a_{n-k}^h) = f(n)$. Evidently, $a_n^p + a_n^h$ is the **total**

solution of the original equation. The ease of finding a particular solution hinges on the properties of $f(n)$. The homogenous solution, however, is subject to a more deterministic method, so we will discuss that first.

Given the homogenous equation $c_0a_n + c_1a_{n-1} + \cdots + c_ka_{n-k} = 0$, we substitute $a_n = Cr^n$, where C is a constant to be determined by the initial conditions, and r is a variable. Plugging in our formal solution, we get $c_0Cr^n + c_1Cr^{n-1} + \cdots + c_kCr^{n-k} = 0$. We can immediately divide out C because the right-hand side is zero. Now we have $c_0r^n + c_1r^{n-1} + \cdots + c_kr^{n-k} = 0$. Finally, dividing by r^{n-k} we arrive at a polynomial of degree k in r called the **characteristic equation** of the given homogenous difference equation, $c_0r^k + c_1r^{k-1} + \cdots + c_k = 0$. In general, a polynomial of degree k will have k roots counting multiplicity and in this context they are called **characteristic roots**. If the k roots are distinct, say $\{r_i : 1 \le i \le k\}$, then each $a_n = Cr_i^n$ is individually a solution of the homogenous equation. We can add them all together to form $C_1r_1^n + C_2r_2^n + \cdots + C_kr_k^n$, which is the general solution to the homogenous difference equation. The k coefficients C_i will be determined by the initial conditions of the recurrence. If $f(n) = 0$, as it often is in these problems, we are done. If $f(n) \ne 0$, we need to do a little extra work to get the particular solution.

Example 9.1. Even though it is trivial, apply our method to the recurrence $a_n = 2a_{n-1}$, where $a_0 = 1$.

Example 9.2. Fibonacci's rabbits. The translation of the original problem from Liber Abaci is: "A certain man put a pair of rabbits in a place surrounded on all sides by a wall. How many pairs of rabbits can be produced from that pair in a year if it is supposed that every month each pair begets a new pair, which from the second month on becomes productive?" Some clarification: (i) we may assume Fibonacci intended a pair of rabbits to always consist of one male and one female, (ii) the initial pair of rabbits was not mature, (iii) a pair of rabbits becomes mature after one month of being immature, (iv) a mature pair produces one immature pair by the end of every month starting with the second month of being mature, (v) the rabbits are immortal, (vi) Fibonacci's question is how many rabbit pairs overall will there be at the end of month n, not how many will be born during month n.

Example 9.3. Estimate F_{100}.

Example 9.4. Find the limiting value as $n \to \infty$ of $\frac{F_n}{F_{n-1}}$.

Example 9.5. Solve the difference equation $a_n + 3a_{n-1} = n + 5$, where $a_1 = 1$

Thus far the characteristic equations we have encountered featured distinct roots. This need not happen and we need a modification of our method to handle the cases when it does not. Let r_1 be a root of multiplicity k in the characteristic equation $\phi(r) = C_0r^n + C_1r^{n-1} + \cdots + C_mr^{n-m} = 0$. This means that the characteristic equation has a factor of

$(r - r_1)^k$ where, of course, $n \geq k$. Our previous construction of the total homogenous solution from distinct roots is now unavailable. The contribution of the multiple root r_1 to the overall homogenous solution turns out to be $(B_1 n^{k-1} + B_2 n^{k-2} + \cdots + B_{k-1} n + B_k) r_1^n$, instead of simply Br_1^n as it was previously in the distinct root case. The constants B_i will be determined by the initial conditions. Certainly, $a_n^h = B_k r_1^n$ is going to be a solution as it would be in the distinct case. We claim the terms of form $B_i n^{k-i} r_1^n$ are also individual solutions of the difference equation. To see this, consider the **formal derivative** $\phi'(r) = nC_0 r^{n-1} + (n-1)C_1 r^{n-2} + \cdots + (n-m)C_m r^{n-m-1} = 0$. There is a fact from the theory of equations that if a polynomial has a multiple root, then that root is also a root of the corresponding formal derivative. ($\int$) Here is a thumbnail sketch of why that is true. Suppose the polynomial $f(x)$ factors as $g(x)(x-r)^k$, where $g(x)$ is another polynomial and r is a root of multiplicity $k \geq 2$. The formal derivative of $f(x)$ using the product rule would be

$$f'(x) = g'(x)(x-r)^k + kg(x)(x-r)^{k-1}.$$

Note $(x - r)$ is still a factor of $f'(x)$, hence r is a root of $f'(x) = 0$. Returning to the formal derivative of the characteristic equation $\phi(r)$, let $r = r_1$ and multiply each term of $\phi'(r)$ by $B_{k-1} r_1$ to get $C_0(nB_{k-1} r_1^n) + C_1((n-1)B_{k-1} r_1^{n-1}) + \cdots + C_m((n-m)B_{k-1} r_1^{n-m}) = 0$. Evidently, $nB_{k-1} r_1^n$ is also a solution to $\phi(r) = 0$. By similar arguments, we can show $n^2 B_{k-2} r_1^n$, $n^3 B_{k-3} r_1^n, \ldots, B_2 n^{k-2} r_1^n$ are also individually solutions, hence the combined expression

$$(B_1 n^{k-1} + B_2 n^{k-2} + \cdots + B_{k-1} n + B_k) r_1^n$$

is as well. An example will make this clear.

Example 9.6. Solve the difference equation $a_n + 4a_{n-1} + 4a_{n-2} = 0$. Initial conditions are $a_0 = 1$ and $a_1 = 1$.

Example 9.7. Solve the difference equation $a_n - 17a_{n-1} + 105a_{n-2} - 275a_{n-3} + 250a_{n-4} = 0$. Initial conditions are $a_0 = 1$, $a_1 = 2$, $a_2 = 1$, and $a_3 = 2$.

Example 9.8. The Tower of Hanoi puzzle was introduced by the French mathematician François Edouard Lucas in 1883. Lucas was also the person who named the rabbit population sequence in Example 9.2 the "Fibonacci sequence." The puzzle consists of three pegs and a number of disks that can be stacked on any of the pegs. The set of disks has one each of decreasing diameter and the starting position for the puzzle is that all the disks are stacked on one peg, largest at the bottom decreasing uniformly to smallest at the top. The problem consists in moving all the disks from the starting peg onto a different peg and ending with the same bottom to top configuration as the starting position. The rules for moving the disks are: (i) only one disk at a time can be moved to another peg, and (ii) no larger disk can be placed on a smaller one. Occasionally, the background for the puzzle is romanticized. One version has Tibetan monks in a remote monastery moving 64 emerald disks onto golden pegs. When they finish the puzzle the world will end. Assuming the monks have just started and make the most efficient move every time at the rate of one move per second, how long do we have?

Example 9.9. Find a formula for the nth Lucas number. The Lucas numbers satisfy the recurrence $L_n = L_{n-1} + L_{n-2}$ with $L_0 = 2$ and $L_1 = 1$.

Example 9.10. This example illustrates how to handle a characteristic equation where the coefficients are real but the roots are complex numbers. Solve the recurrence $a_n - a_{n-1} + a_{n-2} = 0$ subject to the initial conditions $a_0 = 1$ and $a_1 = 0$.

9.3 Recurrences – generating function method

We can borrow the idea of generating function from Chapter 8 and apply it to solving recurrences. This section gives another method to solve the linear Fibonacci recurrence and also features a nonlinear recurrence, which cannot be handled with the method of Section 9.2.

Example 9.11. Solve the Fibonacci recurrence using generating functions. If F_n is the nth Fibonacci number, the recurrence is $F_{n+2} = F_{n+1} + F_n$ with initial conditions $F_1 = F_2 = 1$.

Example 9.12. Consider the expression $x_1 + x_2 + x_3 + \cdots + x_k + \cdots + x_n$. If only two terms can be added at a time, how many ways can this be done?

9.4 Difference tables

An old problem that perplexed mathematicians in the seventeenth century involved finding the sum of the mth powers of the first n natural numbers. We can try our method from the previous section to derive a formula for the sum of the first n numbers ($m = 1$) $S_n = \sum_{k=1}^{n} k$.

Example 9.13. Find

$$S_n = \sum_{k=1}^{n} k.$$

There must be a way that does not require pulling the particular solution out of a hat. Actually, there is a very intuitive demonstration of this formula.

Figure 9.1 shows a grid of points representing all the numbers from 1 to n laid out in a triangle. We take a copy of the grid and turn it upside down, then mesh it with the original grid so that the height (in points) of the resulting rectangle is n and the width

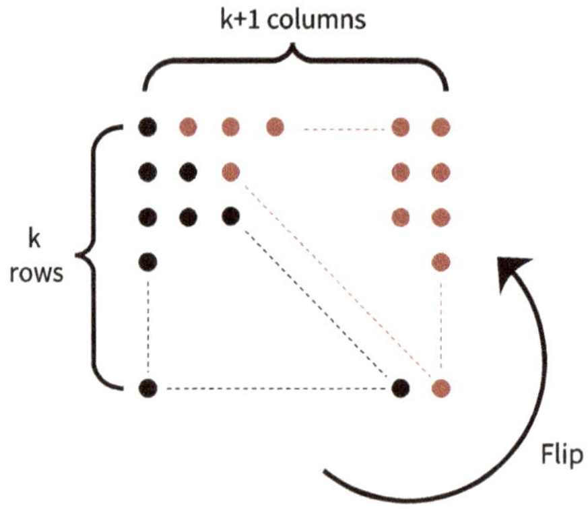

Figure 9.1: Grid for Sum of Integers.

is $n+1$. The number of points in the new rectangle is $2S_n$. We can count these points another way by multiplying the rectangle width (in grid points) $n+1$ by its height n to get $n(n+1)$. Equating these two numbers, we find $S_n = \frac{n(n+1)}{2}$ as before. Let us develop a general method that does not hinge on the quirks of geometry.

The calculus of finite differences (often referred to as calculus without limits) has grown from early work by Isaac Newton and Leonhard Euler into an especially useful and significant branch of mathematics. Suppose we have a function $f : \mathbb{R} \to \mathbb{R}$. We define the operator $\Delta f(x) := f(x+h) - f(x)$ to be the **forward difference** of f at x with increment h understood to be constant. We also have a **backward difference** $\Delta f(x) := f(x) - f(x-h)$ and even a **central difference** $\Delta f(x) := f(x+\frac{h}{2}) - f(x-\frac{h}{2})$. Our interest will be in forward differences considered as operators with their own rules of combination. In particular, these operators are linear, so $\Delta(\alpha f + \beta g)(x) = \alpha\Delta f(x) + \beta\Delta g(x)$. We also define $\Delta^n f(x) := \Delta(\Delta^{n-1} f(x))$ and refer to that as the nth **forward difference of f at x**. Note $\Delta^0 f(x) = f(x)$. From this point on, we set the increment $h = 1$. **Newton's theorem** for forward differences says that if $f(x)$ is a polynomial of degree d, it can be written in the form $f(x) = \sum_{k=0}^{d} \Delta^k f(0) \frac{x^{\underline{k}}}{k!}$. The expression $x^{\underline{k}}$ is called a **falling factorial** and means $x(x-1)\cdots(x-(k-1)) = x(x-1)\cdots(x-k+1)$. There is little consistency in the general literature regarding the notation for falling (or rising) factorials. Here is a simple application of Newton's formula.

Example 9.14. Find a polynomial that has the following values: $f(0) = 1, f(1) = 4, f(2) = 10, f(3) = 20$, $f(4) = 35$, and $f(5) = 56$.

Example 9.15. A sequence generated by a polynomial is

$$\{3, 14, 40, 86, 157, 258, \dots\}.$$

What is the next term?

Let us now slightly narrow our focus to polynomials with integer arguments, so instead of $f(x)$ we will have $f(n)$. There is another operator E called the forward **shift operator** that does the following: $Ef(n) = f(n+1)$. Successive applications of the shift operator just keep moving the argument of f along one unit at a time, so $E^2 f(n) = f(n+2)$ and so forth. Note that $E^n f(0) = f(n)$. We have the operator equation $\Delta f(n) = Ef(n) - If(n)$, where I is the **identity operator**. It follows that $Ef(n) = If(n) + \Delta f(n)$, or $Ef(n) = (I + \Delta)f(n)$. Now we can write $f(n) = E^n f(0) = (I + \Delta)^n f(0)$. By the binomial theorem,

$$(I + \Delta)^n f(0) = \sum_{k=0}^{n} \binom{n}{k} \Delta^k I^{n-k} f(0) = \sum_{k=0}^{n} \binom{n}{k} \Delta^k f(0) = f(n).$$

We have derived a version of Newton's theorem where the polynomial domain is limited to the natural numbers, so the general term $\frac{n^{\underline{k}}}{k!}$ from Newton's theorem can now be expressed as $\frac{n(n-1)\cdots(n-k+1)}{k!} \cdot \frac{(n-k)!}{(n-k)!} = \binom{n}{k}$.

Example 9.16. Find

$$S_n = \sum_{k=1}^{n} k \quad \text{and} \quad S_{n^2} = \sum_{k=1}^{n} k^2.$$

The Swiss mathematician Jakob Bernoulli wrote *Ars Conjectandi*, the first modern treatise on probability and combinatorics in 1705. It was published posthumously by his nephew in 1713. Bernoulli's book contained formulas for S_n all the way to $S_{n^{10}}$. We can only marvel at this result considering his resources at the time. Let us check his result for S_{n^5}.

Example 9.17. Use the difference table method to find S_{n^5}.

Example 9.18. Show why Newton's method cannot be applied to exponential functions.

Example 9.19. Suppose we draw an oval on a sheet of paper. It divides the sheet into two regions, inside and out. Now draw another oval that intersects the first oval in two points. So in addition to the two regions defined by the first oval, the arc created on the second oval by the two points splits the original inside region of the first oval into two parts. Also, the arc on the second oval complementary to the one inside the first oval creates another region. Now we have four regions, as shown in Figure 9.2.

Continuing this construction, find a formula for the number of regions after the nth oval is drawn. We insist that no three ovals intersect at a single point.

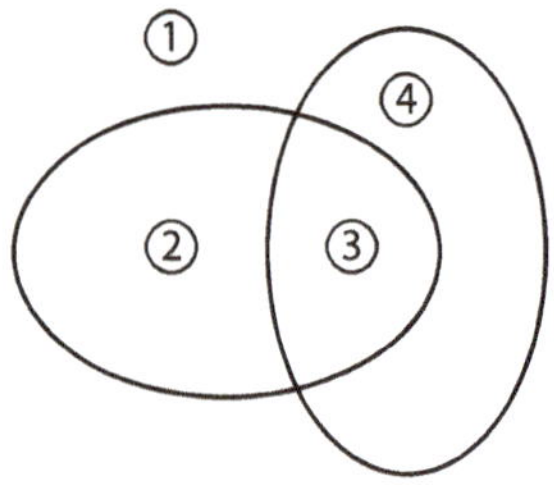

Figure 9.2: Regions for Oval Problem.

9.5 Transform methods

There is a powerful calculus-based method in applied mathematics called **Laplace transforms**. Its primary use is to solve problems in the field of differential equations, which is beyond the scope of this text. There is, however, a self-contained portion of Laplace transform theory that is accessible to us and can assist us in solving difference equations without any calculus involvement. The basic idea of Laplace transforms is to convert a differential calculus problem (or in our case a finite difference problem) into an algebra problem. If we are able to solve the resulting algebra problem, the next step is to use an **inverse transformation** to recover the solution to the original differential or difference equation. We can exploit this method without doing any calculus if we accept on faith the validity of a short list of transforms that apply specifically to finite difference equation problems. Bear in mind that mathematicians have expended considerable effort over the years to plant and nourish this intellectual tree. We are just going to pick the lowest-hanging fruit.

Given a function $f(t)$, we write $F(s) = \mathcal{L}(f(t))$ to indicate that it is the Laplace transform of $f(t)$. The operator $\mathcal{L}(\cdot)$ takes functions of t and converts them into functions of s. Here is a table of recurrence-friendly Laplace transforms, already calculated, that we will want to know. Note that the transform of a sum or difference is the respective sum or difference of transforms, and the transform of a constant times a function is the constant times the transform of the function. Put more concisely, the Laplace transform operation is linear. One feature (bug?) of this method is that the formulas seem peculiar at first and the calculations tend to ramble, but usually things work out smoothly.

Table 9.1 doubles as a table of Laplace transforms and also inverse transforms. Reading left to right gives the transform, while right to left gives the inverse transform. The general procedure is to transform all the terms of the recurrence using the table. Next we solve for the transform function $F(s)$. Finally (the hard part), we try to simplify the expression for $F(s)$ by driving the calculation towards creating sums and differences of individual terms of the type that appear on the right-hand side of Table 9.1. If we are successful in doing this, we inverse transform each term simply by finding its twin on the left-hand side of the table. An example will hopefully make this clear.

Table 9.1: Given the sequence $a_0, a_1, a_2, \ldots$ and $f(t) = a_n$ for $n \le t < n+1$, $n \in \mathbb{N}$, r a constant.

1) If $f(t) = r^n$ for $n \le t < n+1$	$\mathcal{L}(f(t)) = F(s) = \frac{1-e^{-s}}{s(1-re^{-s})}$
2) If $f(t) = r^{n-1}$ for $n \le t < n+1$	$\mathcal{L}(f(t)) = F(s) = \frac{(1-e^{-s})e^{-s}}{s(1-re^{-s})}$
3) If $f(t) = n$ for $n \le t < n+1$	$\mathcal{L}(f(t)) = F(s) = \frac{e^{-s}}{s(1-e^{-s})}$
4) If $f(t) = \sum_{k=1}^{n} r^k = \frac{r(r^n-1)}{r-1}$ for $r \neq 1$	$\mathcal{L}(f(t)) = F(s) = \frac{e^{-s}}{s(1-re^{-s})}$
5) If $f(t) = a_n$	$\mathcal{L}(f(t)) = F(s)$
6) If $f(t+1) = a_{n+1}$	$\mathcal{L}(f(t+1)) = e^s F(s) - \frac{(a_0 e^s)(1-e^{-s})}{s}$
7) If $f(t+2) = a_{n+2}$	$\mathcal{L}(f(t+2)) = e^{2s} F(s) - \frac{e^s(1-e^{-s})(a_0 e^s + a_1)}{s}$

Example 9.20. Back to the rabbits. Solve the Fibonacci recurrence using transforms. Recall the difference equation is $F_n = F_{n-1} + F_{n-2}$, where $F_0 = F_1 = 1$.

Example 9.21. Solve the recurrence $a_{n+2} = 8a_{n+1} - 15a_n$, where $a_0 = 0$ and $a_1 = 1$.

Example 9.22. Solve the recurrence $a_{n+1} - 5a_n = 3^n$, where $a_0 = 0$.

Example 9.23. Solve the recurrence $a_{n+2} + 2a_{n+1} - 3a_n = 0$, where $a_0 = 0$ and $a_1 = 1$.

Example 9.24. Solve the recurrence $3a_{n+2} - 5a_{n+1} + 2a_n = 0$, where $a_0 = 1$ and $a_1 = 0$.

References & suggested further reading

1) *Enumerative Combinatorics, Vol. 1* by Richard P. Stanley (1986), Wadsworth & Brooks
2) *Introduction to Combinatorial Mathematics* by C. L. Liu (1968), McGraw-Hill
3) *Combinatorics, A Guided Tour* by David R. Mazur (2010), Mathematical Assoc. of America.
4) *Laplace Transforms* by Murray Spiegel (1965), Schaum Publishing

Solutions to Chapter 9 examples

Example 9.1

Here $a_n - 2a_{n-1} = 0$ is a homogenous equation, so we let $a_n = Cr^n$ and form the characteristic equation $Cr^n - 2Cr^{n-1} = r^n - 2r^{n-1} = 0$. Upon dividing by r^{n-1}, this reduces to $r - 2 = 0$, so $r = 2$ as we knew all along. There is only one characteristic root, so the total solution is $a_n = C2^n$. Applying the initial condition we have $a_0 = C2^0 = 1$, so $C = 1$ and the general term of the recurrence is $a_n = 2^n$.

Example 9.2

First, the recurrence. Let F_n be the number of rabbit pairs at the end of month n. This equals the number of pairs at the end of the prior month F_{n-1} plus the number of new pairs produced during month n. But the number of new pairs in month n must equal the number of pairs, mature or not, from two months prior F_{n-2}, all of which are by month n mature and producing new pairs during the month. Symbolically, $F_n = F_{n-1} + F_{n-2}$, which is the recurrence we would like to solve. The initial conditions are $F_0 = 0$ at the end of month #0 just before the initial immature pair are put in the enclosure. Then $F_1 = 1$ for the initial immature pair growing during month #1. Note that $F_2 = 1$ because the initial pair are now mature in month #2. During month #3 they are in their second month of maturity and therefore produce one immature pair by the end of that month, giving $F_3 = 2$.

We put $F_n = Cr^n$ into the homogenous recurrence $F_n - F_{n-1} - F_{n-2} = 0$ and generate the characteristic equation $Cr^n - Cr^{n-1} - Cr^{n-2} = 0$ which gives $r^n - r^{n-1} - r^{n-2} = 0$, and upon dividing by r^{n-2}, we finally have $r^2 - r - 1 = 0$. Solving this quadratic equation, we see that the characteristic roots are $\frac{1 \pm \sqrt{5}}{2}$. The total solution is then

$$F_n = C_1\left(\frac{1+\sqrt{5}}{2}\right)^n + C_2\left(\frac{1-\sqrt{5}}{2}\right)^n.$$

We need to evaluate C_1 and C_2. Set $n = 0$ and write $F_0 = C_1 + C_2 = 0$. Also

$$F_1 = C_1\left(\frac{1+\sqrt{5}}{2}\right) + C_2\left(\frac{1-\sqrt{5}}{2}\right) = 1.$$

Solving these two equations simultaneously yields $C_1 = \frac{\sqrt{5}}{5}$ and $C_2 = -\frac{\sqrt{5}}{5}$. Reassembling the solution, we get

$$\frac{\sqrt{5}}{5}\left(\frac{1+\sqrt{5}}{2}\right)^n - \frac{\sqrt{5}}{5}\left(\frac{1-\sqrt{5}}{2}\right)^n = F_n = \frac{\sqrt{5}}{5}\left[\left(\frac{1+\sqrt{5}}{2}\right)^n - \left(\frac{1-\sqrt{5}}{2}\right)^n\right].$$

We can spot check this by calculating

$$F_1 = \frac{\sqrt{5}}{5}\left[\left(\frac{1+\sqrt{5}}{2}\right) - \left(\frac{1-\sqrt{5}}{2}\right)\right] = \frac{\sqrt{5}}{5}\cdot\sqrt{5} = 1\checkmark$$

and

$$\begin{aligned} F_2 &= \frac{\sqrt{5}}{5}\left[\left(\frac{1+\sqrt{5}}{2}\right)^2 - \left(\frac{1-\sqrt{5}}{2}\right)^2\right] = \frac{\sqrt{5}}{5}\left[\left(\frac{3+\sqrt{5}}{2}\right) - \left(\frac{3-\sqrt{5}}{2}\right)\right] \\ &= \frac{\sqrt{5}}{5}\cdot\sqrt{5} = 1\checkmark. \end{aligned}$$

Since the rabbits started in month #1, Fibonacci's original question is answered by how many rabbit pairs there are after month #12, which is $F_{12} = 144$. Our expression for F_n is called Binet's formula after the French mathematician Jacques Binet, although it seems to have been known prior, but not by Fibonacci.

Example 9.3

Note $\frac{1-\sqrt{5}}{2} \approx 0.61803$, and $(0.61803)^{100} \approx 1.26 \times 10^{-21}$, so the term $(\frac{1-\sqrt{5}}{2})^{100}$ is negligible. Apparently, we need only consider $\frac{\sqrt{5}}{5}(\frac{1+\sqrt{5}}{2})^{100}$ to estimate F_{100}. But

$$\frac{\sqrt{5}}{5}\left(\frac{1+\sqrt{5}}{2}\right)^{100} \approx (0.447213595)(1.618033989)^{100} = 3.542248478 \times 10^{20}.$$

The exact value (by recursive addition) is $F_{100} = 354224848179261915075$. At any rate, not a bad estimate for a calculator.

Example 9.4

$$\frac{F_n}{F_{n-1}} = \frac{\frac{\sqrt{5}}{5}[(\frac{1+\sqrt{5}}{2})^n - (\frac{1-\sqrt{5}}{2})^n]}{\frac{\sqrt{5}}{5}[(\frac{1+\sqrt{5}}{2})^{n-1} - (\frac{1-\sqrt{5}}{2})^{n-1}]}.$$

By the observation in Example 9.3 that $\lim_{n\to\infty}(\frac{1-\sqrt{5}}{2})^n = 0$, we can write

$$\lim_{n\to\infty}\frac{F_n}{F_{n-1}} = \lim_{n\to\infty}\frac{(\frac{1+\sqrt{5}}{2})^n}{(\frac{1+\sqrt{5}}{2})^{n-1}} = \frac{1+\sqrt{5}}{2} \approx 1.618.$$

This number was important to the ancient Greeks, who thought the proportions of a building, length to width, were most pleasing if they were in the ratio of $\frac{1+\sqrt{5}}{2}$ to 1. They called this "golden ratio" φ. So the ratio of successive Fibonacci numbers approaches the golden ratio as the numbers increase.

Example 9.5

This is an inhomogenous equation since $f(n) \neq 0$. So we are expecting to get a homogenous solution and a particular solution. For the homogenous equation, we have $a_n + 3a_{n-1} = 0$. Plugging in $a_n = Cr^n$, we get $Cr^n + 3Cr^{n-1} = 0$ which yields the characteristic equation $r + 3 = 0$, hence $r = -3$. Then $a_n^h = C(-3)^n$. We must fit C to the initial condition *after* we have the total solution. There is no magic formula for the particular solution, but a good guess is often to set a_n^p equal to a general expression resembling $f(n)$. Our trial solution is then $a_n^p = An + B$. Plugging into the recurrence, we get $An + b + 3(A(n-1) + b) = n + 5$. We see that $4An = n$, so $A = \frac{1}{4}$. Then and $4b - 3(\frac{1}{4}) = 5$, so $b = \frac{23}{16}$. Now we have $a_n^p = \frac{n}{4} + \frac{23}{16}$, and adding the homogenous and particular solutions together, the total solution to the difference equation is $a_n = C(-3)^n + \frac{n}{4} + \frac{23}{16}$. Setting $a_1 = 1$, we find $C = \frac{11}{48}$, so the final solution is $a_n = \frac{11}{48}(-3)^n + \frac{n}{4} + \frac{23}{16}$. The sequence this generates is $1, 4, -4, 21, -53, \ldots$ Note that the sequence quickly begins to oscillate between positive and negative entries as the term $(-3)^n$ becomes dominant.

Example 9.6

Letting $a_n = Cr^n$, we get the characteristic equation $Cr^n + 4Cr^{n-1} + 4Cr^{n-2} = 0$, or $r^2 + 4r + 4 = 0$. But this is $(r+2)^2 = 0$, so -2 is a root of multiplicity two. From the previous discussion, we form the total homogenous solution $a_n = B_1(-2)^n + nB_2(-2)^n = (B_1 + nB_2)(-2)^n$. To find B_1 and B_2 from the initial conditions, we have $(B_1 + 0B_2)(-2)^0 = a_0 = 1$, so $B_1 = 1$. Also, $(1 + 1B_2)(-2)^1 = 1$ implies $1 + B_2 = -\frac{1}{2}$, or $B_2 = -\frac{3}{2}$. The final result is $a_n = (1 - \frac{3n}{2})(-2)^n$.

Example 9.7

Letting $a_n = Cr^n$, the characteristic equation simplifies to $r^4 - 17r^3 + 105r^2 - 275r + 250$. Do not be discouraged if the factorization of this polynomial eludes you. We reverse-engineered the problem from the known factorization $(r-2)(r-5)^3$. So we have two roots, $r_1 = 2$ with no multiplicity and $r_2 = 5$ with multiplicity three. The total solution will have four pieces, $a_n = B_0(2^n) + (B_1n^2 + B_2n + B_3)(5^n)$. Using the initial conditions, we generate the following system of equations: (i) $a_0 = B_0 + B_3 = 1$, (ii) $a_1 = 2 = 2B_0 + 5B_1 + 5B_2 + 5B_3$, (iii) $a_2 = 1 = 4B_0 + 100B_1 + 50B_2 + 25B_3$, and (iv) $a_3 = 2 = 8B_0 + 1125B_1 + 375B_2 + 125B_3$. After a little algebra, we find $B_0 = -\frac{4}{9}$, $B_1 = \frac{1}{5}$, $B_2 = -\frac{16}{15}$, and $B_3 = \frac{13}{9}$. So the overall solution is $a_n = -\frac{4}{9}(2^n) + (\frac{n^2}{5} - \frac{16n}{15} + \frac{13}{9})(5^n)$.

Example 9.8

Let a_n be the number of moves required to transfer n disks from one peg to another according to the rules. At some point in the disk transfer, we must reach the configuration where the largest disk is uncovered and sits on its original peg and all of the $n-1$ smaller disks have been moved to one of the other pegs. This is necessary to have the third peg empty in order to receive the transferred largest disk. That will require a_{n-1} moves. The largest disk can now be moved to the empty peg to allow reconstruction of the entire starting configuration on that peg. That is one additional move. Finally, the $n-1$ smaller disks have to be moved from their current peg to the peg which now has the largest disk. That will need another a_{n-1} moves. This completes the puzzle. We have used $a_{n-1} + 1 + a_{n-1}$ moves to transfer n disks conforming to the rules. So the recurrence equation is $a_n - 2a_{n-1} = 1$. The initial condition involves moving just one disk, so $a_1 = 1$.

This recurrence is not homogenous, so we are looking for both homogenous and particular solutions. The homogenous equation is the same as Example 9.1, so $a_n^h = C2^n$. For the particular solution, we try $a_n^p = B$. Then $B - 2B = 1$ implies $B = -1$. Hence $a_n = a_n^h + a_n^p = C2^n - 1$. Fitting C to the initial condition gives $2C - 1 = 1$, so $C = 1$. Finally, $a_n = 2^n - 1$.

As far as the end of the world is concerned,

$$a_{64} = 2^{64} - 1 = 18446744073709551615.$$

Using the Gregorian year, this works out to 5.845×10^8 millennia. We are safe for a while.

Example 9.9

This looks very familiar. It is the Fibonacci recurrence with a slight change to the initial conditions. We can import our intermediate solution to the Fibonacci recurrence from Example 9.2,

$$L_n = C_1\left(\frac{1+\sqrt{5}}{2}\right)^n + C_2\left(\frac{1-\sqrt{5}}{2}\right)^n,$$

and evaluate new constants C_1 and C_2. For $n = 0$,

$$C_1\left(\frac{1+\sqrt{5}}{2}\right)^0 + C_2\left(\frac{1-\sqrt{5}}{2}\right)^0 = 2$$

and for $n = 1$,

$$C_1\left(\frac{1+\sqrt{5}}{2}\right)^2 + C_2\left(\frac{1-\sqrt{5}}{2}\right)^2 = 1.$$

The second expression reduces to

$$C_1\left(\frac{6+2\sqrt{5}}{4}\right) + C_2\left(\frac{6-2\sqrt{5}}{4}\right) = 1.$$

Solving these equations simultaneously yields $C_1 = C_2 = 1$. The final solution is then

$$L_n = \left(\frac{1+\sqrt{5}}{2}\right)^n + \left(\frac{1-\sqrt{5}}{2}\right)^n.$$

This is a little cleaner than the Binet formula in Example 9.2.

Example 9.10

The characteristic equation is $r^2-r+1 = 0$ with conjugate roots by the quadratic formula $r_1 = \frac{1+\sqrt{-3}}{2}$ and $r_2 = \frac{1-\sqrt{-3}}{2}$, or $r_1 = \frac{1}{2} + \frac{i\sqrt{3}}{2}$ and $r_2 = \frac{1}{2} - \frac{i\sqrt{3}}{2}$. As usual, the partial solution can be expressed as

$$a_n = C_1\left(\frac{1}{2} + \frac{i\sqrt{3}}{2}\right)^n + C_2\left(\frac{1}{2} - \frac{i\sqrt{3}}{2}\right)^n.$$

To determine C_1 and C_2, we have the system

$$a_0 = 1 = C_1 + C_2 \quad \text{and} \quad a_1 = 0 = C_1\left(\frac{1}{2} + \frac{i\sqrt{3}}{2}\right) + C_2\left(\frac{1}{2} - \frac{i\sqrt{3}}{2}\right).$$

We find $C_1 = \frac{1}{2} + \frac{i\sqrt{3}}{6}$ and $C_2 = \frac{1}{2} - \frac{i\sqrt{3}}{6}$. So the final solution is

$$a_n = \left(\frac{1}{2} + \frac{i\sqrt{3}}{6}\right)\left(\frac{1}{2} + \frac{i\sqrt{3}}{2}\right)^n + \left(\frac{1}{2} - \frac{i\sqrt{3}}{6}\right)\left(\frac{1}{2} - \frac{i\sqrt{3}}{2}\right)^n.$$

Much like the irrational $\sqrt{5}$ in Examples 9.2 and 9.9, the imaginary unit disappears in the evaluation of this formula for any n. To illustrate, using the formula (recursive addition would be much easier) $a_2 = -1$, $a_3 = -1$, and $a_4 = 0$. This is not a very exciting sequence.

We can expunge i from our formula by appealing to DeMoivre's theorem. Recall $(\cos\theta + i\sin\theta)^n = \cos n\theta + i\sin n\theta$. A typical complex number $x + iy$ can be written in polar form as $\rho(\cos\theta + i\sin\theta)$, where the modulus $\rho = \sqrt{x^2+y^2}$ and $\theta = \arctan(\frac{y}{x})$. For r_1, we have $\rho = \sqrt{(\frac{1}{2})^2 + (\frac{\sqrt{3}}{2})^2} = 1$, and

$$\theta = \arctan\left(\frac{\frac{\sqrt{3}}{2}}{\frac{1}{2}}\right) = \arctan\sqrt{3} = \frac{\pi}{3}.$$

So $r_1 = (\cos\frac{\pi}{3} + i\sin\frac{\pi}{3})$. Similarly, $r_2 = (\cos\frac{\pi}{3} - i\sin\frac{\pi}{3})$. Now we can write the formula as

$$\begin{aligned} a_n &= C_1\left(\cos\frac{\pi}{3} + i\sin\frac{\pi}{3}\right)^n + C_2\left(\cos\frac{\pi}{3} - i\sin\frac{\pi}{3}\right)^n \\ &= C_1\left(\cos\frac{n\pi}{3} + i\sin\frac{n\pi}{3}\right) + C_2\left(\cos\frac{n\pi}{3} - i\sin\frac{n\pi}{3}\right). \end{aligned}$$

Separating real and imaginary parts,

$$a_n = (C_1 + C_2)\cos\frac{n\pi}{3} + i(C_1 - C_2)\sin\frac{n\pi}{3}.$$

The new coefficients are $C_1 + C_2 = 1$ and

$$i(C_1 - C_2) = \left(\frac{i}{2} - \frac{\sqrt{3}}{6}\right) - \left(\frac{i}{2} + \frac{\sqrt{3}}{6}\right) = -\frac{\sqrt{3}}{3}.$$

Finally, the solution can be written as a real function $a_n = \cos\frac{n\pi}{3} - \frac{\sqrt{3}}{3}\sin\frac{n\pi}{3}$. We were able to get rid of i explicitly because complex roots always come in conjugate pairs.

Example 9.11

We want to set up a generating function where the summation starts at 0, so we can define $F_0 = 0$ arbitrarily. This is consistent with the stated conditions and recurrence. So consider the OGF $\phi(x) = \sum_{n=0}^{\infty} F_n x^n$. We start by forming power series directly from the recurrence,

$$\sum_{n=0}^{\infty} F_{n+2}x^n = \sum_{n=0}^{\infty} F_{n+1}x^n + \sum_{n=0}^{\infty} F_n x^n.$$

Now we adjust the terms to have $\sum_{n=0}^{\infty} F_n x^n$ appear across-the-board. Note that

$$\sum_{n=0}^{\infty} F_{n+2}x^n = \frac{1}{x^2}\sum_{n=0}^{\infty} F_{n+2}x^{n+2}.$$

Now re-index with $k = n + 2$, so

$$\frac{1}{x^2}\sum_{n=0}^{\infty} F_{n+2}x^{n+2} = \frac{1}{x^2}\sum_{k=2}^{\infty} F_k x^k = \frac{1}{x^2}\left(\sum_{k=0}^{\infty} F_k x^k - F_1 x - F_0\right),$$

where we have moved the new index back to zero and compensated by deducting the two terms that this action added. Finally, we have $\frac{1}{x^2}(\sum_{k=0}^{\infty} F_k x^k - x)$ after plugging in the initial conditions. Next we have

$$\sum_{n=0}^{\infty} F_{n+1}x^n = \frac{1}{x}\sum_{n=0}^{\infty} F_{n+1}x^{n+1}.$$

Again we reindex with $k = n + 1$, so

$$\frac{1}{x}\sum_{n=0}^{\infty} F_{n+1}x^{n+1} = \frac{1}{x}\sum_{k=1}^{\infty} F_k x^k = \frac{1}{x}\left(\sum_{k=0}^{\infty} F_k x^k - F_0\right) = \frac{1}{x}\left(\sum_{k=0}^{\infty} F_k x^k\right)$$

after setting $F_0 = 0$ as before. The third term $\sum_{n=0}^{\infty} F_n x^n$ needs no adjustment, but we will change the dummy summation index to k to agree with what we already have. Now

$$\frac{1}{x^2}\left(\sum_{k=0}^{\infty} F_k x^k - x\right) = \frac{1}{x}\left(\sum_{k=0}^{\infty} F_k x^k\right) + \sum_{k=0}^{\infty} F_k x^k.$$

In terms of $\phi(x)$ this is $\frac{1}{x^2}(\phi(x) - x) = \frac{1}{x}\phi(x) + \phi(x)$. Clearing fractions we have $\phi(x) - x = x\phi(x) + x^2\phi(x)$ which reduces to $\phi(x)(1 - x - x^2) = x$, or $\phi(x) = \frac{x}{1-x-x^2}$. By the quadratic formula, we can factor $1 - x - x^2$ as $(1 - \alpha x)(1 - \beta x)$ where $\alpha = \frac{1+\sqrt{5}}{2}$ and $\beta = \frac{1-\sqrt{5}}{2}$. So $\phi(x) = \frac{x}{(1-\alpha x)(1-\beta x)}$. We split the denominator using a partial fraction decomposition to get $\phi(x) = \frac{1}{\alpha-\beta}[\frac{\alpha}{(1-\alpha x)} - \frac{\beta}{(1-\beta x)}]$. Now the task is to apply extraction operators to the separated fractions. From Table 8.2(4) above, $[[\frac{\alpha}{(1-\alpha x)}]]_{x^k} = \alpha^{k+1}$ and likewise $[[\frac{\beta}{(1-\beta x)}]]_{x^k} = \beta^{k+1}$. So

$$\phi(x) = \sum_{k=0}^{\infty} F_k x^k = \frac{x}{\alpha - \beta}\sum_{k=0}^{\infty}(\alpha^{k+1} - \beta^{k+1})x^n = \frac{1}{\sqrt{5}}\sum_{k=0}^{\infty}(\alpha^{k+1} - \beta^{k+1})x^{k+1}.$$

Reindexing yet again with $n = k + 1$, this becomes $\frac{1}{\sqrt{5}}\sum_{n=1}^{\infty}(\alpha^n - \beta^n)x^n$. This sum can be rewritten as $\frac{\sqrt{5}}{5}\sum_{n=0}^{\infty}(\alpha^n - \beta^n)x^n$ because the $n = 0$ term is 0 and we can also change the dummy index k to n in $\sum_{k=0}^{\infty} F_k x^k$. Now we have

$$\sum_{n=0}^{\infty} F_n x^n = \frac{\sqrt{5}}{5}\sum_{n=0}^{\infty}(\alpha^n - \beta^n)x^n,$$

and matching coefficients of x^n we see that

$$F_n = \frac{\sqrt{5}}{5}\left[\left(\frac{1+\sqrt{5}}{2}\right)^n - \left(\frac{1-\sqrt{5}}{2}^n\right)\right].$$

This agrees with our derivation in Section 9.2 Example 9.2.

Example 9.12

We interpret this as how many ways we can form parenthetical expressions signifying the addition of two adjacent expressions at a time where an adjacent expression can be a single x_k or another already parenthesized group. Here are some examples:

$$((x_1 + x_2) + ((x_3 + x_4) + x_5)),\ (x_1 + (((x_2 + x_3) + x_4) + x_5)), \quad \text{and}$$
$$(x_1 + (x_2 + (x_3 + (x_4 + x_5)))).$$

Let m_k be the number of ways to do this for an expression with k singleton terms. We are seeking m_n. This will turn out to be a nonlinear recurrence.

Any parenthesized group of n singletons can be decomposed into two subexpressions joined by a final addition. How many ways can this happen? A parenthesized group with k terms can be added to one with $n-k$ terms with a final pair of parentheses that embrace both expressions. So there would be $m_k \cdot m_{n-k}$ ways to do this. Apparently, the total number of ways as k ranges from 1 to $n-1$ (no empty expressions) is $\sum_{k=1}^{n-1} m_k m_{n-k}$. This sum is valid for $n > 1$. Note that $m_1 = 1$ and we can arbitrarily define $m_0 = 0$. This allows us to write $m_n = m_0 m_n + m_1 m_{n-1} + m_2 m_{n-2} + \cdots + m_{n-1} m_1 + m_n m_0$, where we have artificially extended the range of k from 0 to n by adding the zero terms $m_0 m_n$ and $m_n m_0$. Now let the OGF for m_n be $M(x) = \sum_{n=0}^{\infty} m_n x^n. = \sum_{n=2}^{\infty} (m_0 m_n + m_1 m_{n-1} + \cdots + m_{n-1} m_1 + m_n m_0) x^n$. This is equivalent to $M(x) - m_1 x - m_0 = [M(x)]^2 - (m_0 m_1 + m_1 m_0)x - m_0^2$. Simplifying, we get $[M(x)]^2 - M(x) + (1)x = 0$, since all the terms containing m_0 vanish. Solving using the quadratic formula, we find $M(x) = \frac{1 \pm \sqrt{1-4x}}{2}$. We know the m_k's cannot be negative, so we choose the root that will ensure $\frac{1 \pm \sqrt{1-4x}}{2}$ is positive. Let us take a moment to expand $\sqrt{1-4x} = (1-4x)^{1/2}$ using the generalized binomial theorem. We get

$$(1-4x)^{1/2} = 1 + \sum_{n=1}^{\infty} \frac{\frac{1}{2}(-\frac{3}{2}) \cdots (\frac{1}{2} - n + 1)}{n!} (-4x)^n.$$

Rewrite the nth term as

$$\frac{-\frac{1}{2}(\frac{3}{2}) \cdots (n - \frac{3}{2})}{n!} (2^n)(2^n) x^n = \frac{(-1)(3) \cdots (2n-3)}{n!} (2^n) x^n.$$

Finally, we factor out $-\frac{2}{n}$ and multiply top and bottom by $(n-1)!$ to reduce everything to

$$-\frac{2}{n}\cdot\frac{(3)\cdots(2n-3)[(2^{n-1})(n-1)!]}{(n-1)!(n-1)!}x^n = -\frac{2}{n}\binom{2n-2}{n-1}x^n.$$

Since we want $\frac{1\pm\sqrt{1-4x}}{2}$ to yield positive terms, we select $M(x) = \frac{1-\sqrt{1-4x}}{2}$. Then $m_n = \frac{1}{n}\binom{2n-2}{n-1}$ for $n \geq 1$. So, for example, eight terms can be grouped for addition $m_8 = \frac{1}{8}\binom{14}{7} =$ 429 different ways.

Example 9.13

The recurrence that describes this situation is $S_n - S_{n-1} = n$, so we are looking for both a homogenous solution and particular solution. We know $S_0 = 0$. The characteristic equation for the homogenous equation is simply $r - 1 = 0$. This leads to $S_n^h = C(1^n) = C$. This homogenous solution does not look promising. Apparently, the particular solution will have to provide all the growth in S_n. We soon find that our previous trick to find a particular solution is not much help. It turns out that the particular solution is $S_n^p = \frac{(n+2)(n-1)}{2}$. This would have been time-consuming to discover. We should check it:

$$S_n^p - S_{n-1}^p = \frac{(n+2)(n-1)}{2} - \frac{(n+1)(n-2)}{2} = \frac{n^2+n-2}{2} - \frac{n^2-n-2}{2} = \frac{2n}{2} = n.$$

So the particular solution behaves correctly and the total solution must be $S_n = \frac{(n+2)(n-1)}{2} + C$. Now $S_1 = 1$ implies $C = 1$, so finally we have the solution $S_n = \frac{(n+2)(n-1)}{2} + 1 = \frac{n^2+n}{2} = \frac{n(n+1)}{2}$.

Example 9.14

We organize our data in a (Newton) **difference table**.

Table 9.2: Difference Table for Example 9.14.

x	$f(x)$	$\Delta f(x)$	$\Delta^2 f(x)$	$\Delta^3 f(x)$	$\Delta^4 f(x)$	
0	1	3	3	1	0	⟸ differences @0
1	4	6	4	1	0	
2	10	10	5	1		
3	20	15	6			
4	35	21				
5	56					

Assembling the result, we have

$$\begin{aligned} f(x) &= (1)\left(\frac{x^0}{0!}\right) + (3)\left(\frac{x}{1!}\right) + (3)\left(\frac{x(x-1)}{2!}\right) + (1)\left(\frac{x(x-1)(x-2)}{3!}\right) \\ &= \frac{1}{6}x^3 + x^2 + \frac{11}{6}x + 1. \end{aligned}$$

Newton's method works for all polynomials and many analytic functions. When a polynomial generated by a Newton table exists it is unique.

Example 9.15

We set up the difference table and plug in $x = 7$ (for the next term) directly into Newton's formula.

Table 9.3: Difference Table for Example 9.15.

x	$f(x)$	$\Delta f(x)$	$\Delta^2 f(x)$	$\Delta^3 f(x)$	$\Delta^4 f(x)$	
0	3	11	15	5	0	⟸ differences @0
1	14	26	20	5	0	
2	40	46	25	5		
3	86	71	30			
4	157	101				
5	258					

Assembling the result we have

$$f(x) = (3)\left(\frac{x^0}{0!}\right) + (11)\left(\frac{x}{1!}\right) + (15)\left(\frac{x(x-1)}{2!}\right) + (5)\left(\frac{x(x-1)(x-2)}{3!}\right).$$

Then

$$f(7) = 3 + (11)(7) + (15)\left(\frac{7\cdot 6}{2}\right) + (5)\frac{7\cdot 6\cdot 5}{6} = 570.$$

Example 9.16

For S_n, we construct the appropriate **difference table.**

Table 9.4: Difference Table for Example 9.16.

n	$f(n)$	$\Delta f(n)$	$\Delta^2 f(n)$	$\Delta^3 f(n)$	
0	0	1	1	0	⟸ differences @0
1	1	2	1	0	
2	3	3	1	0	
3	6	4	1		
4	10	5			
5	15				

We see that $\Delta^0 f(0) = 0, \Delta^1 f(0) = 1, \Delta^2 f(0) = 1, \Delta^n f(0) = 0$ if $n \geq 3$. Now

$$f(n) = \sum_{k=0}^{n} \binom{n}{k} \Delta^k f(0) = \binom{n}{1}\Delta^1 f(0) + \binom{n}{2}\Delta^2 f(0) = (1)\binom{n}{1} + (1)\binom{n}{2}.$$

But

$$n + \frac{n(n-1)}{2} = \frac{n^2 - n + 2n}{2} = \frac{n(n+1)}{2} = f(n) = S_n$$

as we saw earlier.

Now for the sum of the squares S_{n^2}. Mimicking the above, we have

Table 9.5: Difference Table for Example 9.16.

n	$f(n)$	$\Delta f(n)$	$\Delta^2 f(n)$	$\Delta^3 f(n)$	$\Delta^4 f(n)$	
0	0	1	3	2	0	$\Longleftarrow$ differences @0
1	1	4	5	2	0	
2	5	9	7	2		
3	14	16	9			
4	30	25				
5	55					

This time we have $\Delta^0 f(0) = 0$, $\Delta^1 f(0) = 1$, $\Delta^2 f(0) = 3$, $\Delta^3 f(0) = 2$, $\Delta^n f(0) = 0$ if $n \geq 4$. Assembling yields

$$\begin{aligned} f(n) &= \binom{n}{0}\Delta^0 f(0) + \binom{n}{1}\Delta^1 f(0) + \binom{n}{2}\Delta^2 f(0) + \binom{n}{3}\Delta^3 f(0) \\ &= n + 3 \cdot \frac{n(n-1)}{2} + 2 \cdot \frac{n(n-1)(n-2)}{6}. \end{aligned}$$

This works out to

$$n + \frac{3n^2}{2} - \frac{3n}{2} + \frac{n^3}{3} - n^2 + \frac{2n}{3} = \frac{n^3}{3} - \frac{n^2}{2} + \frac{n}{6} = \frac{n(2n-1)(n-1)}{6} = f(n) = S_{n^2}.$$

Example 9.17

Setting up the difference table, we have

Table 9.6: Difference Table for Example 9.17.

n	$f(n)$	$\Delta f(n)$	$\Delta^2 f(n)$	$\Delta^3 f(n)$	$\Delta^4 f(n)$	$\Delta^5 f(n)$	$\Delta^6 f(n)$	$\Delta^7 f(n)$	
0	0	1	31	180	390	360	120	0	$\Longleftarrow$ differences @0
1	1	32	211	570	750	480	120	0	
2	33	243	781	1320	1230	600	120		
3	276	1024	2101	2550	1830	720			
4	1300	3125	4651	4380	2550				
5	4425	7776	9031						
6	12201	16807							
7	29008								

Assembling, we have

$$f(n) = 0\binom{n}{0} + 1\binom{n}{1} + 31\binom{n}{2} + 180\binom{n}{3} + 390\binom{n}{4} + 360\binom{n}{5} + 120\binom{n}{6}.$$

After simplification this gives $S_{n^5} = \frac{n^6}{6} + \frac{n^5}{2} + \frac{5n^4}{12} - \frac{n^2}{12}$, which agrees with Bernoulli's table of *Summæ Potestatum (sum of powers)*. Bernoulli seems to have been aware of Newton's theorem.

Example 9.18

Let $f(n) = 2^n$ and construct the difference table.

Table 9.7: Difference Table for Example 9.18.

n	$f(n)$	$\Delta f(n)$	$\Delta^2 f(n)$	$\Delta^3 f(n)$	
0	1	1	1	1	⋯
1	2	2	2	2	⋯
2	4	4	4	4	⋯
3	8	8	8	8	⋯
4	16	16	16	16	⋯
5	32	32	32	32	⋯
⋮	⋮	⋮	⋮	⋮	⋱

The table never reaches a zero column. Our formula would have an unlimited number of terms, so there is no polynomial. We might have suspected this on the basis of knowing that exponential functions grow faster than any polynomial function.

Example 9.19

Let R_n be the number of regions after drawing the nth oval. If we have drawn $n - 1$ ovals and have R_{n-1} regions, consider what drawing the nth oval accomplishes. The nth oval intersects every previous oval in two points, forming $2(n - 1)$ arcs on the nth oval that create $2(n - 1)$ new regions in the existing ovals in addition to the previous R_{n-1} regions. The recurrence is $R_n = R_{n-1} + 2(n - 1)$. This looks very much like Example 9.13 in Section 9.4, where we determined that all the growth in the result from the substitution method would have to come from the inhomogenous term. We can attack this problem with a difference table. First, we need enough values in the sequence to make the table work. Iterating the recurrence starting with $R_1 = 2$, we get successive values $4, 8, 14, 22, 32, 44, 56, \ldots$ We extend the sequence backwards to $R_0 = 2$ to be consistent with the recurrence. Creating the difference table, we obtain

Table 9.8: Difference Table for Example 9.19.

n	$f(n)$	$\Delta f(n)$	$\Delta^2 f(n)$	$\Delta^3 f(n)$	
0	2	0	2	0	⟸ differences @0
1	2	2	2	0	
2	4	4	2	0	
3	8	6	2	0	
4	14	8	2	0	
5	22	10	2	0	
⋮	⋮	⋮	⋮	⋮	

The differences at zero are $\Delta^0 f(0) = 2$, $\Delta^1 f(0) = 0$, $\Delta^2 f(0) = 2$, and $\Delta^k f(0) = 0$ for all $k \geq 3$. Assembling the polynomial,

$$f(n) = 2\binom{n}{0} + 0\binom{n}{1} + 2\binom{n}{2} = 2 + 2\frac{n(n-1)}{2} = 2 + n(n-1).$$

So $R_n = n(n-1) + 2$.

Example 9.20

Let us recast the equation as $a_{n+2} = a_{n+1} + a_n$ with $a_0 = 1$ and $a_1 = 1$. Then viewing $f(t) = a_n$ we have from Table 9.1(5)

$$\mathcal{L}(a_{n+2}) = \mathcal{L}(f(t+2)) = e^{2s}F(s) - \frac{e^s(1-e^{-s})(e^s+1)}{s}.$$

From Table 9.1, we have

$$\mathcal{L}(a_{n+1}) = \mathcal{L}(f(t+1)) = e^{s}F(s) - \frac{e^s(1-e^{-s})}{s}.$$

Putting everything together, we get

$$e^{2s}F(s) - \frac{e^s(1-e^{-s})(e^s+1)}{s} = e^sF(s) - \frac{e^s(1-e^{-s})}{s} + F(s).$$

Solving for $F(s)$, we first group the terms as

$$F(s)(e^{2s} - e^s - 1) = \frac{e^s(1-e^{-s})(e^s+1)}{s} - \frac{e^s(1-e^{-s})}{s}.$$

Then

$$F(s) = \frac{e^s(1-e^{-s})(e^s+1) - e^s(1-e^{-s})}{s(e^{2s}-e^s-1)} = \frac{e^s(1-e^{-s})(e^s+1-1)}{s(e^{2s}-e^s-1)}$$

$$= \frac{e^{2s}(1-e^{-s})}{s(e^{2s}-e^{s}-1)} = \frac{(1-e^{-s})}{s(1-e^{-s}-e^{-2s})}.$$

Now what? We may recognize that the quadratic in the denominator is somewhat famous since it gives rise to the golden ratio $\phi = \frac{1+\sqrt{5}}{2}$ which we have encountered before. At any rate

$$1-e^{-s}-e^{-2s} = \left(1-\frac{1+\sqrt{5}}{2}e^{-s}\right)\left(1-\frac{1-\sqrt{5}}{2}e^{-s}\right).$$

With a little algebra, we can show that

$$\frac{1}{(1-e^{-s}-e^{-2s})} = \frac{1}{1-(\frac{1+\sqrt{5}}{2})e^{-s}} - \frac{1}{1-(\frac{1-\sqrt{5}}{2})e^{-s}}.$$

Finally,

$$\frac{(1-e^{-s})}{s(1-e^{-s}-e^{-2s})} = \frac{(1-e^{-s})}{s(1-(\frac{1+\sqrt{5}}{2})e^{-s})} - \frac{(1-e^{-s})}{s(1-(\frac{1-\sqrt{5}}{2})e^{-s})}.$$

These two fractions fit the exact mold of the right-hand side of Table 9.1, so to find the inverse transform, we simply read off the corresponding left-hand side. It appears that $r_1 = \frac{1+\sqrt{5}}{2}$ and $r_2 = \frac{1-\sqrt{5}}{2}$. So $a_n = A(\frac{1+\sqrt{5}}{2})^n - B(\frac{1-\sqrt{5}}{2})^n$ where we use the initial values to find A and B. When $n = 0$, we have $A - B = 0$, so $A = B$. Then checking $n = 1$, we have

$$A\left[\left(\frac{1+\sqrt{5}}{2}\right) - \left(\frac{1-\sqrt{5}}{2}\right)\right] = 1,$$

or $A\sqrt{5} = 1$, so $A = B = \frac{1}{\sqrt{5}} = \frac{\sqrt{5}}{5}$. Finally, we conclude that

$$F_n = a_n = \frac{\sqrt{5}}{5}\left(\frac{1+\sqrt{5}}{2}\right)^n - \frac{\sqrt{5}}{5}\left(\frac{1-\sqrt{5}}{2}\right)^n,$$

as we had in Section 9.2 Example 9.2.

Example 9.21

Transforming everything according to Table 9.1, we get

$$e^{2s}F(s) - \frac{e^{s}(1-e^{-s})(a_0e^{s}+a_1)}{s} = 8\left[e^{s}F(s) - \frac{(a_0e^{s})(1-e^{-s})}{s}\right] - 15F(s).$$

Rewriting so that the terms involving $F(s)$ appear on the left, we have

$$e^{2s}F(s) - 8e^{s}F(s) + 15F(s) = \frac{e^{s}(1-e^{-s})(a_0e^{s}+a_1)}{s} - 8\frac{(a_0e^{s})(1-e^{-s})}{s}.$$

Then after substituting $a_0 = 0$ and $a_1 = 1$, this reduces to $F(s)(e^{2s} - 8e^s + 15) = \frac{e^s(1-e^{-s})(1)}{s}$. Finally, solving for $F(s)$ gives $F(s) = \frac{e^s(1-e^{-s})}{s(e^{2s}-8e^s+15)}$. We see that $(e^{2s} - 8e^s + 15)$ factors as $(e^s - 3)(e^s - 5)$, so the immediate task is to split $\frac{1}{(e^{2s}-8e^s+15)}$ into two fractions. We keep $\frac{e^s(1-e^{-s})}{s}$ in reserve to help us drive the calculation towards something we will be able to inverse transform. We write $\frac{1}{(e^{2s}-8e^s+15)} = \frac{A}{(e^s-3)} + \frac{B}{(e^s-5)}$ and solve for A and B. Clearing fractions, we have $1 = A(e^s - 5) + B(e^s - 3)$, so evidently $Ae^s + Be^s = 0$ and $-5A - 3B = 1$. The first equation implies $A = -B$, and then the second implies $2B = 1$, or $B = \frac{1}{2}$, and then $A = -\frac{1}{2}$. Restating $F(s)$ with $\frac{e^s(1-e^{-s})}{s}$ distributed over the fraction decomposition, we have

$$F(s) = -\frac{1}{2}\frac{e^s(1-e^{-s})}{s(e^s-3)} + \frac{1}{2}\frac{e^s(1-e^{-s})}{s(e^s-5)} = \frac{1}{2}\left[\frac{(1-e^{-s})}{s(1-5e^{-s})} - \frac{(1-e^{-s})}{s(1-3e^{-s})}\right].$$

Inverting using Table 9.1, it appears that $a_n = \frac{1}{2}(5^n - 3^n)$. Checking a few values, we have $a_0 = 0$, $a_1 = 1$, and $a_2 = 8$, which happily agree with the recurrence.

Example 9.22

Transforming all terms using Table 9.1, we have

$$e^sF(s) - \frac{(a_0e^s)(1-e^{-s})}{s} - 5F(s) = \frac{1-e^{-s}}{s(1-3e^{-s})}.$$

The fact that $a_0 = 0$ is convenient and allows us to simplify this to $e^sF(s) - 5F(s) = \frac{1-e^{-s}}{s(1-3e^{-s})}$. Solving for $F(s)$ gives

$$F(s) = \frac{1}{(e^s-5)} \cdot \frac{1-e^{-s}}{s(1-3e^{-s})} = \frac{(1-e^{-s})}{s} \cdot \frac{1}{(e^s-5)(1-3e^{-s})}.$$

Note that

$$\frac{1}{(e^s-5)(1-3e^{-s})} = \frac{e^{-s}}{(1-5e^{-s})(1-3e^{-s})}.$$

Splitting $\frac{e^{-s}}{(1-5e^{-s})(1-3e^{-s})}$ into two fractions we have

$$\frac{e^{-s}}{(1-5e^{-s})(1-3e^{-s})} = \frac{1}{2}\left[\frac{1}{(1-5e^{-s})} - \frac{1}{(1-3e^{-s})}\right].$$

Restoring the full expression we get

$$F(s) = \frac{1}{2}\left[\frac{1-e^{-s}}{s(1-5e^{-s})} - \frac{1-e^{-s}}{s(1-3e^{-s})}\right].$$

Taking the inverse transforms using Table 9.1 yields $a_n = \frac{1}{2}(5^n - 3^n)$. This recurrence is inhomogenous, but it has exactly the same solution as the second order homogenous recurrence in Example 9.21. How can this happen? The inhomogenous part of Example 9.22 was already part of the homogenous solution in Example 9.21. We basically solved the homogenous recurrence $a_{n+1} - 5a_n = 0$ and then subtracted $\frac{3^n}{2}$ from the homogenous solution.

Example 9.23

Transforming all terms using Table 9.1 and plugging in initial values, we have

$$\left[e^{2s}F(s) - \frac{e^s(1-e^{-s})(1)}{s}\right] + 2e^sF(s) - 3F(s) = 0.$$

Isolating $F(s)$ yields $F(s)(e^{2s} + 2e^s - 3) = \frac{e^s(1-e^{-s})}{s}$, so

$$F(s) = \frac{e^s(1-e^{-s})}{s(e^{2s} + 2e^s - 3)} = \frac{e^s(1-e^{-s})}{s(e^s - 1)(e^s + 3)} = \frac{e^s(1-e^{-s})}{s} \cdot \frac{1}{(e^s - 1)(e^s + 3)}.$$

Finding the fraction decomposition, we have

$$\frac{e^s(1-e^{-s})}{s} \cdot \left[\frac{1}{4(e^s - 1)} - \frac{1}{4(e^s + 3)}\right] = \frac{1}{4}\frac{(1-e^{-s})}{s(1-e^{-s})} - \frac{1}{4}\frac{(1-e^{-s})}{s(1+3e^{-s})}.$$

Finally, by Table 9.1, we conclude that $a_n = \frac{1}{4}[1 - (-3)^n]$. It may seem odd that we did not simplify $\frac{(1-e^{-s})}{s(1-e^{-s})}$ to $\frac{1}{s}$. It is clear that the r in Table 9.1 for this term is 1, so our little table covers this case. Actually, the inverse transform of $\frac{1}{s}$ in a more comprehensive table of Laplace transforms is 1, but we have tried to keep our table as economical in terms of entries as possible.

Example 9.24

Transforming all terms using Table 9.1 and plugging in initial values, we have

$$3\left[e^{2s}F(s) - \frac{e^s(1-e^{-s})(e^s)}{s}\right] - 5\left[e^sF(s) - \frac{(e^s)(1-e^{-s})}{s}\right] + 2F(s) = 0.$$

Isolating $F(s)$ yields

$$F(s)(3e^{2s} - 5e^s + 2) = 3\left[\frac{e^{2s}(1-e^{-s})}{s}\right] - 5\left[\frac{(e^s)(1-e^{-s})}{s}\right],$$

so

$$F(s) = \frac{3e^{2s}(1-e^{-s}) - 5e^s(1-e^{-s})}{s(3e^{2s} - 5e^s + 2)} = \frac{e^s(1-e^{-s})}{s}\left[\frac{3e^s - 5}{(3e^{2s} - 5e^s + 2)}\right].$$

Note that $(3e^{2s} - 5e^{s} + 2) = (3e^{s} - 2)(e^{s} - 1)$, so decomposing $\frac{3e^{s}-5}{(3e^{2s}-5e^{s}+2)}$ into two fractions yields $\frac{3e^{s}-5}{(3e^{2s}-5e^{s}+2)} = \frac{9}{(3e^{s}-2)} - \frac{2}{(e^{s}-1)}$. Reassembling the full expression,

$$F(s) = \frac{e^{s}(1-e^{-s})}{s}\left[\frac{9}{(3e^{s}-2)} - \frac{2}{(e^{s}-1)}\right] = 3\frac{e^{s}(1-e^{-s})}{s(e^{s}-\frac{2}{3})} - 2\frac{e^{s}(1-e^{-s})}{s(e^{s}-1)}.$$

Dividing by e^{s} top and bottom yields

$$F(s) = 3\frac{(1-e^{-s})}{s(1-\frac{2}{3}e^{-s})} - 2\frac{e^{s}(1-e^{-s})}{s(1-e^{-s})}.$$

Inverting using Table 9.1 gives us $a_n = 3(\frac{2}{3})^n - 2$.

10 Modular arithmetic methods

Modular or "clock" arithmetic was invented by the German mathematician C. F. Gauss. It was first used in his number theory treatise, *Disquisitiones Arithmeticæ*, published in 1801. Number theory is a branch of mathematics concerned with the properties and relationships among numbers per se, and for most of its history has been considered the purest of pure mathematics, unsullied by any practical application. However, recently this reputation has changed, as mathematicians have devised several important applications of number theory, in particular, to cryptography – the science of secret communication. The nickname "clock" arithmetic refers to the fundamental idea of letting the integers start over or wrap around a particular integer called the modulus, just like the hours of an ordinary clock start over at 12. All sorts of interesting and useful number theory related problems can be solved with the aid of this idea. We will develop the basics and attempt even some of the more intimidating problems.

10.1 Modular operations

We can declare an integer $m \geq 1$ to be a **modulus** for a special kind of arithmetic. Two numbers $a, b \in \mathbb{N}$ are defined to be **congruent modulo** m whenever their difference $a - b$ is divisible by m. You can see immediately that $m = 1$ is a valid but unproductive choice, so assume $m \geq 2$. The condition that a and b are congruent modulo m is denoted as $a \equiv b (\bmod m)$. We write $\bmod m$ in parentheses to indicate it applies to the entire congruence expression. If we write it without parentheses after a single number, such as $a \bmod m$, it will mean the unique number, called a **residue**, from 0 to $m-1$ to which a is congruent. Integers in this range constitute the set of **least residues** for modulus m. They are least in the sense that they are the smallest nonnegative residues and by construction they are a **complete set of residues**. In this context completeness means the residue of any integer modulo m is in the set. There can be other complete sets of residues, where any integer modulo m is congruent to *some* member of the set, but those members need not be least. Be careful to distinguish equality $a = b$ from congruence signified by the triple bar $a \equiv b$. The least residues are the "hours" in our clock arithmetic where zero is noon.

Congruence of numbers relative to a modulus is similar to the geometrical notion of congruence of triangles. Both are expressed as relations between pairs of objects. In both cases, an object (number or triangle) is congruent to itself. This is the property of reflexivity. If object A is congruent to object B, then object B is congruent to object A. So congruence has the property of symmetry. Finally, if object A is congruent to object B, and B is congruent to C, then A is congruent to C. This is transitivity. Any relation that has these three properties is called an equivalence relation (see Section 0.3.1).

https://doi.org/10.1515/9783111506654-011

Example 10.1. Express

$$13 \bmod 6,\ 343 \bmod 7,\ -8 \bmod 12,\ 15 \bmod 4,\ 93 \bmod 16,\ \text{and } 1357 \bmod 2$$

as least residues.

Here is a table of the basic modular arithmetic rules.

Table 10.1: Modular arithmetic operations. $x, y, u, v, k, r \in \mathbb{Z}, k \neq 0$.

1) Addition	if $x \equiv y (\bmod m)$ and $u \equiv v (\bmod m)$ then $x + u \equiv y + v (\bmod m)$
2) Subtraction	if $x \equiv y (\bmod m)$ and $u \equiv v (\bmod m)$ then $x - u \equiv y - v (\bmod m)$
3) Multiplication	if $x \equiv y (\bmod m)$ and $u \equiv v (\bmod m)$ then $xu \equiv yv (\bmod m)$
4) Division	if $x \equiv y (\bmod m)$ and $x = ku, y = kv$ then $\frac{x}{k} \equiv \frac{y}{k} (\bmod m)$ if $\gcd(k, m) = 1$
5) Powers	if $x \equiv y (\bmod m)$ then $x^r \equiv y^r (\bmod m)$

Example 10.2. Justify the entries in Table 10.1.

Example 10.3. Demonstrate a case where division works and one where it does not.

There are several useful secondary properties as well.

Table 10.2: Additional modular arithmetic properties. $x, y, a \in \mathbb{Z}$.

1) Translation	if $x \equiv y (\bmod m)$ then $x + a \equiv y + a (\bmod m)$
2) Additive cancellation	if $x + a \equiv y + a (\bmod m)$ then $x \equiv y (\bmod m)$
3) Change of scale	if $x \equiv y (\bmod m)$ then $ax \equiv ay\ (\bmod m)$
4) Constant multiple	if $x \equiv y (\bmod m)$ then $ax \equiv ay (\bmod am)$
5) Modulus reduction	if $ax \equiv ay (\bmod am)$ then $x \equiv y (\bmod m)$

Example 10.4. Justify the entries in Table 10.2.

Since the coefficients that appear in polynomials over the integers are subject to modular arithmetic, we can reduce polynomials modulo m. Be careful not to interpret this as reducing the exponents of the indeterminate modulo m.

Example 10.5. Reduce the coefficients of $21x^5 - 37x^4 + 234x^3 + 57x^2 + 161x - 5$ with respect to modulus 5.

Example 10.6. It is true that if $x \equiv y (\bmod m)$ then $x^r \equiv y^r (\bmod m)$ for any $r \in \mathbb{Z}$ (see Table 10.1 (5)). For example, $3 \equiv 8 (\bmod 5)$ implies $3^2 \equiv 8^2 (\bmod 5)$ because $9 \equiv 64 (\bmod 5)$, and so forth. However, it is not true that $x \equiv y (\bmod m)$ implies $a^x \equiv a^y (\bmod m)$. Find a counterexample.

Example 10.7. What are the last two digits of 7^{503}?

Example 10.8. There is a convenient test that reveals at a glance whether a numerical sum/difference or multiplication is incorrect. It is called "casting out nines." For multiplication, the procedure is to add the digits of the multiplicand and reduce mod 9, then add the digits of the multiplier and reduce mod 9. Now take those two reduced digit sums and multiply them together and reduce that answer mod 9. That is your check digit. Next, add the digits of the original product and reduce that sum mod 9. If the last sum does not equal the check digit, the original multiplication is definitely incorrect. If it agrees, then it is correct, maybe. Regrettably, the test only definitively detects error and does not guarantee correctness. But you could interpret passing the test at least as somewhat confirmatory. Checking addition and subtraction works the same way. Use the test to see if $15{,}986 \cdot 25{,}632 = 419{,}753{,}152$.

Example 10.9. Show why casting out nines works.

10.2 Linear congruences

In Section 6.1.2 we solved linear equations in one variable. We would now like to write a linear equation and solve it relative to a modulus. This introduces a few quirks into the process. A typical such equation is called a **linear congruence** and has the general form $ax \equiv b(\operatorname{mod} m)$. Now there is no guarantee that this congruence even has a solution. A congruence like $4x \equiv 23(\operatorname{mod} 32)$ is not guaranteed to have a solution. In fact, it is guaranteed not to have one. Why? $(4x-23)$ is always an odd number for any integer x, but then 32 cannot divide it evenly. In general, if there is to be a solution to $ax \equiv b(\operatorname{mod} m)$, we must have $ax - b = cm$ for some $c \in \mathbb{Z}$. Rearranging this, we have $ax - cm = b$. If we can find a c that makes this true, we can find a solution to the congruence, so let us focus on trying to do that.

There is an old (1624) theorem in algebra, called Méziriac's theorem, named after the Renaissance French mathematician Bachet de Méziriac. It states that if a and m are integers with $\gcd(a, m) = d$, then there exist integers x and y such that $ax + my = d$. This result is often attributed in modern texts to another French mathematician Étienne Bézout, who proved an analogous theorem for polynomials nearly 150 years after Méziriac. But let us give credit where credit is due. Méziriac's result looks like just what we need.

Given $ax \equiv b(\operatorname{mod} m)$, we first need to find $\gcd(a, m) = d$. For small numbers, this is an easy matter of finding common factors. If a and m are large, we can appeal to the Euclidean algorithm discussed below. Next we observe that the linear combination $ax - cm = b$ is governed by the theorem, which asserts b must be a multiple of d. So if d divides b the congruence has a solution and if it does not, there is no solution. Here is an example of a linear congruence with a solution.

Example 10.10. Solve $10x \equiv 2(\operatorname{mod} 14)$.

Example 10.11. Verify Méziriac's theorem.

Since we are concerned with greatest common divisors in the rest of this chapter, here are some useful facts:

Table 10.3: GCD properties.

1) $\gcd(m,n) = \gcd(n,m)$
2) $\gcd(m, n \pm m) = \gcd(m,n)$
3) $\gcd(m, n \pm km) = \gcd(m,n)$ – this follows immediately by applying (2) repetitively
4) $\gcd(rm, rn) = \lvert r \rvert \gcd(m,n)$
5) $\frac{mn}{\gcd(m,n)} = \operatorname{lcm}(m,n)$, where the $\operatorname{lcm}(m,n)$ is the *least common multiple* of both m and n

Example 10.12. If $ax \equiv b(\bmod m)$ is a solvable linear congruence, find all distinct (incongruent) solutions modulo m.

Example 10.13. Solve $35x \equiv 21 \pmod{56}$.

Example 10.14. Solve $3184x \equiv 5573 \pmod{14083}$.

Example 10.15. Given $ax \equiv b(\bmod m)$, with $\gcd(a,m) = 1$, it is always possible find an integer a^{-1} with the property that $a \cdot a^{-1} = 1$. This is the inverse of a relative to the modulus m. If m were not too large, it would be feasible to create a table for various commonly used m that gave pairs of mutual inverses. At the moment, our solution method with the extended Euclidean algorithm does this for us, if rather tediously. The algorithm works for all m, but for small m, with a little mental arithmetic, we should be able to make some basic tables. Suppose we know that a^{-1} is the inverse of a relative to modulus m. Then we can solve $ax \equiv b(\bmod m)$ almost instantly. We have $a^{-1} \cdot ax \equiv 1 \cdot x \equiv a^{-1} \cdot b(\bmod m)$, or $x \equiv a^{-1} \cdot b(\bmod m)$. Create tables for $m = 3, 5, 7, 11$, and 13 and use them to solve (i) $2x \equiv 1(\bmod 3)$, (ii) $4x \equiv 3(\bmod 5)$, (iii) $5x \equiv 2(\bmod 7)$, (iv) $8x \equiv 6(\bmod 11)$, and (v) $10x \equiv 8(\bmod 13)$.

10.3 Systems of linear congruences

Just as we solve systems of linear equations, we also can solve systems of linear congruences. So instead of the linear system $\{2x + 3y = 8,\ 3x - 5y = 7\}$, we might encounter the similar looking system of congruences $\{2x + 3y \equiv 8(\bmod 11),\ 3x - 5y \equiv 7(\bmod 11)\}$. However, congruences offer the possibility for a type of system that is not available with ordinary equations. We can have a system like $\{x \equiv 5 \pmod{13},\ x \equiv 2(\bmod 17)\}$ where we have only one variable but two or more modular equations. This second type of problem was inspired a millenium ago by Chinese mathematicians. The key to solving them is the Chinese remainder theorem, which we will now discuss.

10.3.1 Chinese remainder theorem

This result is the key to solving a system of congruences where a single variable is congruent to various residues with respect to different moduli and a simultaneous solution is desired. The rule for deciding whether a system of two or more congruences of this type has a simultaneous solution is contained in the **Chinese remainder theorem**. It states that the system of congruences $\{x \equiv a_i (\text{mod}\, m_i) : 1 \le i \le k\}$, where the moduli are pairwise coprime, has a unique solution modulo $M = \prod_{i=1}^{k} m_i$. Pairwise coprime means that $\gcd(m_i, m_j) = 1$ if $i \neq j$. Here is an example that illustrates the method.

Under certain conditions, there are solutions even if the moduli are not pairwise prime, but we leave that explanation to the references.

Example 10.16. Solve $x \equiv 2 (\text{mod}\, 3)$ and $x \equiv 3 (\text{mod}\, 5)$ simultaneously.

Example 10.17. When the men in the Ancient Order of Hibernia got together to march in the annual St. Patrick's Day parade, they found that if they lined up three across there was one fellow left over, five across there were two left over, and seven across there were three odd men out. What is the minimum number of men in the club?

Example 10.18. Find a multiple of 5 that leaves a remainder of 1 when it is divided by either 2, 3, or 7.

Example 10.19. Explain why the Chinese remainder theorem works.

10.3.2 Multivariable congruences

Suppose we have a pair of linear congruences in two variables with respect to the same modulus. We should be able to eliminate one variable, solve for the remaining variable, and then come back and use one congruence to evaluate the eliminated variable.

Example 10.20. Solve the system

$$\{x + 4y \equiv 2 \,(\text{mod}\, 7),\ 3x + 2y \equiv 3 \,(\text{mod}\, 7)\}.$$

Example 10.21. Extend the above method to solve the system

$$\{x + y - z \equiv 3 \,(\text{mod}\, 11),\ 8x + 5y + 2z \equiv 3 \,(\text{mod}\, 11),\ 4x + y - 6z \equiv 10 \,(\text{mod}\, 11)\}.$$

10.4 Quadratic congruences

An expression of the general form $x^2 \equiv a \pmod{m}$, where $m > 1$ and $\gcd(a, m) = 1$ is a **quadratic congruence**. Whether such a congruence can be solved or not hinges on a. There is no guarantee that if we square the successive numbers from 1 to $m - 1$ that a will be among them. If a is the square of such a number, we call it a **quadratic residue (QR)** and the congruence has a solution. However, if a is not the square of such a number, then it is a **quadratic nonresidue (QNR)** and there is no solution. From that standpoint, solving a quadratic congruence is pretty cut-and-dried. Conveniently, for our basic solution method we need only consider congruences where the modulus is an odd prime. We will dispose of the case $m = 2$ in the examples, and we will also show how the Chinese remainder theorem will allow us to solve congruences where the modulus is composite. Keep in mind that quadratic residues and nonresidues are nonzero. The congruence $x^2 \equiv 0 \pmod{m}$ is trivially solved by zero, so we recognize that 0 is a QR and assume in the following that all QR's are strictly positive.

Example 10.22. Find all the quadratic residues modulo 30.

Example 10.23. Solve $x^2 \equiv a \pmod{2}$.

Example 10.24. Show that if $x = b$ solves $x^2 \equiv a \pmod{p}$, where p is an odd prime, then $(p - b)$ solves it as well.

Example 10.25. Show that modulo p, an odd prime, there are $\frac{(p-1)}{2}$ quadratic residues and $\frac{(p-1)}{2}$ quadratic nonresidues.

Example 10.26. We are going to need some tables of QR's for small primes to solve problems like $x^2 \equiv a \pmod{p}$. Make tables for $p = 5, 7, 11, 13$ and 17.

Example 10.27. Solve (i) $2x^2 \equiv 4 \pmod{5}$ and (ii) $5x^2 \equiv 6 \pmod{17}$.

Example 10.28. Solve (i) $x^2 - 5x + 1 \equiv 0 \pmod{13}$ and (ii) $3x^2 - 5x + 2 \equiv 0 \pmod{11}$.

Example 10.29. Solve $x^2 \equiv 71 \pmod{77}$.

A good first step in solving quadratic residue problems is to decide if there is going to be any chance for a solution. As we have seen above, there is a 50 % chance that a given number from 1 to $m - 1$ will turn out to be a QR modulo m. For a small modulus, we can just check our tables, perhaps expanded a bit to be more comprehensive. For a large modulus, it would be a waste of time to generate all QR's and then find out that the one we need is not even on the list. To be honest, we are pretending to ignore a CAS like Maple®, which could immediately decide if a number was a QR and solve the congruence to boot. But we are trying here to preserve the spirit of wrangling with these problems unassisted. Luckily, there is a method to answer the question "Is this QR problem solvable?" without spinning our computational wheels. It will not give us the exact QR we need, but it will tell us whether to bother trying to find it.

The Legendre symbol, named after the French mathematician Adrien-Marie Legendre, looks like a fraction in brackets, $(\frac{a}{p})$. Here p is assumed to be a prime modulus and a a number between 1 and $p-1$ that may or may not be a QR. If a is a QR mod p then $(\frac{a}{p}) = 1$ and if a is a QNR mod p then $(\frac{a}{p}) = -1$. Effectively, we are using parity to keep track of QR's and QNR's. This will be important as we will have the occasion to compute products of Legendre symbols in the course of reducing one with large a and/or p to one with smaller values where the diagnosis of QR or QNR is obvious. From Example 10.26, we can write down some Legendre symbols: $(\frac{5}{11}) = 1$, $(\frac{2}{13}) = -1$, $(\frac{8}{17}) = 1$, and $(\frac{3}{7}) = -1$. The product of two QR's is a QR, and the product of two QNR's is a QR, surprisingly enough. The product of a QR and a QNR is still a QNR. These facts can be compressed into the equation $(\frac{a}{p})(\frac{b}{p}) = (\frac{ab}{p})$. For example, $(\frac{6}{13}) = -1$, and $(\frac{2}{13})(\frac{6}{13}) = (\frac{12}{13}) = 1$. Suppose we want to find $(\frac{1001}{p})$. We factor it as $(\frac{7\cdot 11\cdot 13}{p}) = (\frac{7}{p})(\frac{11}{p})(\frac{13}{p})$. Knowing the three terms on the right-hand side would give us the answer. This multiplication rule will allow us to make a useful simplification. We do not have to deal with any composite a's in $(\frac{a}{p})$. It suffices to factor a into primes, say $a = q_1^{e_1} \cdots q_n^{e_n}$, where the prime factors are q_i raised to the power e_i. Then

$$\left(\frac{a}{p}\right) = \left(\frac{q_1^{e_1} \cdots q_n^{e_n}}{p}\right) = \left(\frac{q_1}{p}\right)^{e_1} \cdots \left(\frac{q_n}{p}\right)^{e_n}.$$

It is all right if the primes are repeated since they still conform to the multiplication rule. Now we can proceed knowing our Legendre symbols can ultimately be reduced to primes modulo other primes.

At the core of this approach is the **quadratic reciprocity theorem**, one of the famous results of C. F. Gauss, which established as true a phenomenon that had been noticed empirically in the eighteenth century. Given two odd primes p and q, the Legendre symbols $(\frac{p}{q})$ and $(\frac{q}{p})$ are related by the equation

$$\left(\frac{p}{q}\right) \cdot \left(\frac{q}{p}\right) = (-1)^{\frac{(p-1)(q-1)}{4}}.$$

In words, the status of p as a QR mod q and q as a QR mod p depends on whether $\frac{(p-1)(q-1)}{4}$ is even or odd. If it is even, then p and q are either both QR's or both QNR's of the other as a modulus. If odd, then if one is a QR, the other is a QNR. This reciprocity aspect is very useful in reducing the size of the numbers in a Legendre symbol. Parsing this equation gives us a number of distinct cases:

Table 10.4: Quadratic reciprocity properties. p, q primes, $a \in \{1, 2, \ldots, p-1\}$.

1) $(\frac{-1}{p}) = \begin{cases} +1 & \text{if } p \equiv 1 \pmod 4 \\ -1 & \text{if } p \equiv 3 \pmod 4 \end{cases}$
2) $(\frac{2}{p}) = \begin{cases} +1 & \text{if } p \equiv 1, 7 \pmod 8 \\ -1 & \text{if } p \equiv 3, 5 \pmod 8 \end{cases}$
3) $(\frac{q}{p}) = \begin{cases} (\frac{p}{q}) & \text{if } p \equiv 1 \pmod 4 \text{ or } q \equiv 1 \pmod 4 \\ -(\frac{p}{q}) & \text{if } p \equiv q \equiv 3 \pmod 4 \end{cases}$
4) $(\frac{a^2}{p}) = 1$

Let us use our method to decide if several quadratic congruences have solutions.

Example 10.30. Decide if $x^2 \equiv 26 \pmod{73}$ is solvable.

Example 10.31. Decide if $x^2 \equiv 482 \pmod{1357}$ is solvable.

10.5 Diophantine equations

Diophantus of Alexandria was a Greek mathematician who lived in the third century CE. He wrote a treatise entitled *Arithmetica*, much of which has survived to this day. Lagrange called him the "father of algebra," but others disagree. Perhaps the father of number theory is more appropriate. At any rate, he was ahead of his time in using algebraic symbolization. One of the topics that he appears to have investigated in a serious way was the theory of equations that require integer solutions. Examples would be $5x + 13y = 31$, $x^2 + y^2 = z^2$, $x^2 - 5y^2 = 1$, and so forth. Today these types of equations are called **Diophantine equations**. The most famous of them is $x^n + y^n = z^n$ where $n > 2$ and x, y, and z are nontrivial integers. Of course, there is no challenge if $n = 1$ and the Pythagorean formula leads to infinitely many solutions for $n = 2$ (think $3^2 + 4^2 = 5^2$), but no one has ever found a solution for $n \geq 3$. It is a famous story that the French mathematician Pierre de Fermat was reading a section of Diophantus' *Arithmetica* when he wrote in the margin of a page that he believed he had a proof that $x^n + y^n = z^n$ could not be solved in integers for $n > 2$, but the margin was too small to contain his argument. This statement became known as Fermat's last theorem, not because he died or quit, but because it was the last of Fermat's conjectures from the seventeenth century that remained unconquered by the mathematical community in the years after him. In fact, it was not proven until 1995 using very advanced methods and taking hundreds of pages. Because of the simplicity of its statement and its legendary resistance to proof, Fermat's last theorem escaped academia and became known to the educated public. Despite years of trying unsuccessfully to find a general proof, mathematicians developed valuable new ideas in the attempt, so perhaps that is a consolation.

10.5.1 Linear diophantine equations

A typical linear Diophantine equation has the form $ax + by = c$, where a, b, and c are given integers. The problem is to find integers x and y that satisfy the equation. If the given integers are small, it is sometimes possible to solve the equation "by inspection." This works except when it does not, so we would like to find a fail safe method. The extended Euclidean algorithm that helped us with linear congruences is just what we need.

Example 10.32. Derive a test for solvability of a linear Diophantine equation $ax + by = c$.

Example 10.33. Solve the equation $5x + 7y = 19$.

Example 10.34. Find an integer solution to the system of Diophantine equations $\{x+y+z = 6,\ 2x-y+3z = 9\}$ such that $x \geq 10$.

Example 10.35. Solve the Diophantine equation $ax + by = c$ using modular arithmetic.

Example 10.36. Use the modular method to solve $3x + 5y = 7$.

Example 10.37. Alice went to the fruit market and bought some peaches and pears. A pear cost 20 ¢ more than a peach. She paid $10.00 for nine pieces of fruit. How many of each type did she buy and at what prices?

10.5.2 Nonlinear diophantine equations

Pell's equation $x^2 - Ny^2 = 1$

According to Eric Temple Bell, a noted American mathematician and historian of mathematics, John Pell probably never even saw Pell's equation, much less solved it. This type of nonlinear Diophantine equation had been around since the time of the Greeks, who had solutions in special cases. The Indian mathematician Brahmagupta developed an algorithm that handled more special cases in the seventh century CE. A thousand years later, European mathematicians took up the challenge, and the British mathematician Lord Brounckner made some headway on the general equation, which Euler obtained in the course of his attack on the problem. Unfortunately, Euler confused Brounckner with Pell, and the name Pell's equation stuck due to the enormous influence of Euler. There is some sentiment for rechristening it Fermat's equation, since he worked on it as well. Lagrange was the first to solve it in full generality using continued fractions.

Here are some basic facts concerning Pell's equation. If you are not familiar with continued fractions, now would be a good time to review Section 10.6 which discusses them at the end of this chapter.

1) For every $N \in \mathbb{N}$, there is a solution to Pell's equation. We regard a solution as trivial if either $x = 0$ or $y = 0$. If $N < -1$, clearly, $x = \pm 1$ and $y = 0$ is the only possible solution, or the left-hand side would be too big. If $N = -1$, either $x = \pm 1$ and $y = 0$, or vice versa, would do. If $N = 0$, $x = \pm 1$, and $y =$ anything will work. If $N = 1$, you are demanding the difference between two squares be 1, and that can only happen if again, $x = \pm 1$ and $y = 0$. All of these are considered trivial solutions. The game gets interesting for $N > 1$.
2) Lagrange showed that for every nonsquare $N > 1$, there is a solution to Pell's equation. His proof, which we adapt in an algorithm below to obtain a solution, ultimately establishes that every Pell's solution (x, y) pair is a convergent $\frac{x}{y}$ in the continued fraction expansion of $\sqrt{N}$. You can see that this is intuitively plausible by solving for $N = \frac{x^2-1}{y^2}$, from which it follows that $\sqrt{N} = \sqrt{\frac{x^2}{y^2} - \frac{1}{y^2}}$. As x and y get bigger and bigger, $\sqrt{N}$ starts to look like $\frac{x}{y}$. If N is not a perfect square, the continued fraction will be infinite, and the convergents are the simplified rational expressions one gets by truncating the infinite fraction at various steps along the way.
3) Since $x^2 - Ny^2 = 1$, if N were a perfect square, say m^2, we could factor this as $(x - my)(x + my) = 1$, and apparently the factors on the left have to be both +1 or −1. This leads to two linear equations which return a trivial solution.
4) If N is not a perfect square, then $(x - \sqrt{N}y)(x + \sqrt{N}y) = 1$. The convention is that $a + \sqrt{N}b$ "solves" the Pell's equation if $a^2 - Nb^2 = 1$. We can show that if $\alpha = a + \sqrt{N}b$ is a solution, so is $\frac{1}{\alpha}$, and if β is another solution, then so is $\alpha\beta$. Here $a = 1$ and $b = 0$ gives the trivial solution 1, and we see that the solutions to a Pell's equation form a multiplicative subgroup of $\mathbb{R}$. Not only that, but there is a minimum positive a that appears in a solution, and then the entire group is generated by $\theta = a + \sqrt{N}b$. So if β is any solution, there must be an exponent k such that $\beta = (a + \sqrt{N}b)^k$.

Example 10.38. Solve $x^2 - 7y^2 = 1$ with the continued fraction method.

Example 10.39. Solve Brahmagupta's equation $x^2 - 92y^2 = 1$.

Example 10.40. Fermat's last theorem $x^n + y^n = z^n$

As we mentioned in Section 10.5, the $n = 1$ case is trivial and the $n = 2$ case is shown to have solutions by virtue of **Pythagorean triples**. These are right triangles with integer length sides a, b, and c, such as the 3–4–5 triangle. In fact, all the Pythagorean triples (a, b, c) can be recovered by setting $a = u^2 - v^2$, $b = 2uv$, and $c = u^2 + v^2$, for $u, v \in \mathbb{N}$ with the condition $u > v > 0$ to eliminate degenerate triangles. Then expanding these expressions gives $a^2 + b^2 = u^4 - 2u^2v^2 + v^4 + 4u^2v^2 = u^4 + 2u^2v^2 + v^4 = c^2$.

And that is where it stops. There are no nontrivial solutions for $n > 2$. Various n were eliminated piecemeal over the years by many famous mathematicians, and circumstantial support for Fermat's conjecture kept growing, but there was no coup de grâce. Finally, in 1995 the British/American mathematician Andrew Wiles published a definitive confirmation using very advanced methods. The challenge was not laid completely to rest, however, but morphed into a search for an elementary proof. This would be perhaps the sort of proof that Fermat himself had in mind, assuming he was correct.

Interestingly, if we use modular arithmetic, the Fermat equation has infinitely many solutions for $n > 2$. Issai Schur showed (1916) that for any n there is some threshold prime p_0 that for all primes greater than p_0 the equation $x^n + y^n = z^n \pmod p$ has an integer solution. Schur's argument was nonconstructive, so he did not provide any examples, but we can appeal to a more modern result that generates solutions for small n. Given $n \geq 3$ and p an odd prime, the function $\phi(x, n, p) = x^n \pmod p$ permutes the set of residues $\{1, 2, \ldots, p-1\}$ provided $\gcd(n, p-1) = 1$. In particular, if x_1 and x_2 are not congruent modulo p, then neither are x_1^n and x_2^n. This means that subject to the stated condition, if we calculate an $x^n \bmod p$ and a $y^n \bmod p$, we can always be assured that there is some $z^n \bmod p$ such that $x^n + y^n = z^n \pmod p$. Examples abound: $2^3 + 3^3 \equiv 7^3 \pmod{11}$, $7^5 + 12^5 \equiv 34^5 \pmod{37}$, $12^9 + 13^9 \equiv 17^9 \pmod{23}$, and even $3^{101} + 4^{101} \equiv 27^{101} \pmod{29}$. This is radically different behavior than without a modulus.

10.6 Continued fractions

Mathematicians have been playing around with **continued fractions** for a long time. The idea is to write a number as a fraction with a denominator that itself may be a fraction, whose denominator may be yet again a fraction, and so on. Rational numbers result in this procedure terminating after a finite number of such steps. Irrational numbers necessarily correspond to infinite continued fractions. One might ask in what sense does an infinite continued fraction "represent" an irrational number λ. If the infinite continued fraction is truncated after every step, the result is a sequence λ_n of finite continued fractions that approach the irrational number being represented in the usual way. For some arbitrary $\epsilon > 0$, we would run the process out far enough so that truncating anywhere beyond step $N(\epsilon)$ guarantees that $|\lambda_n - \lambda| < \epsilon$. Continued fractions pop up in interesting places. The German mathematician Lambert used them to prove the irrationality of π about 1760. Lagrange used them to show that the nonlinear Diophantine equation $x^2 - Ny^2 = 1$ has a solution for any N. This is known as Pell's equation, which we discussed above.

One may write a simple (canonical) continued fraction as

$$a_0 + \cfrac{1}{a_1 + \cfrac{1}{a_2 + \cfrac{1}{a_3 + \cdots}}}$$

but it gets pretty space consuming very shortly. One convention is to compress this to $[a_0; a_1, a_2, a_3, \ldots]$. The **integer part** is a_0 and this is split off from the fractional part by a semicolon. The a_i's are called **partial denominators**, the 1's are called **partial numerators**, and the result of truncating a given continued fraction after the ith partial denominator is called the ith **partial quotient**. A partial quotient is then a **convergent** that approaches an irrational number in the case of an infinite continued fraction. A generalized continued fraction allows numbers besides 1 as partial numerators, and the idea can be generalized even further to allow functions to take the place of numbers.

Here are some salient facts about continued fractions:

1) A finite continued fraction can be converted to a simple fraction by successive operations of finding gcd's and inverting. Conversely, a simple fraction can be put in canonical continued fraction form by successively separating out the integral part and inverting. For example, to convert $\frac{9}{7}$, we would write it as

$$1 + \frac{2}{7} = 1 + \frac{1}{\left(\frac{7}{2}\right)} = 1 + \frac{1}{3 + \frac{1}{2}}.$$

2) A continued fraction that is infinite but repeating represents a root of a quadratic, and vice versa. For example, if

$$\lambda = a_0 + \cfrac{1}{a_1 + \cfrac{1}{a_0 + \cfrac{1}{a_1 + \cdots}}},$$

then

$$\lambda = a_0 + \frac{1}{a_1 + \frac{1}{\lambda}}, \quad \text{or} \quad \lambda = a_0 + \frac{\lambda}{\lambda a_1 + 1}$$

which implies $\lambda^2 a_1 - a_0 a_1 \lambda - a_0 = 0$. Longer repeating periods clearly do not produce any higher powers of λ when we exploit this "fractal" nature of the continued fraction.

3) If instead of a simple fraction we have a decimal, we can use the algorithm in (1) and organize our results in a table. For example, suppose we want to get the first seven partial denominators of $\pi = 3.14159265\ldots$ We would set up the following table:

denom	**int part**	**frac part**	**recip frac**
a_0	3	0.14159265	7.06251349
a_1	7	0.06251349	
a_2			
a_3			
a_4			
a_5			
a_6			

This corresponds to writing $\pi = 3 + \frac{1}{7+0.06251349}$. We have split out the integer part, namely 3, then take the reciprocal of the fractional part, then split out its integer part, namely 7, and we are ready to take the reciprocal of its fractional part. Continuing in this manner we get the following table:

denom	**int part**	**frac part**	**recip frac**
a_0	3	0.14159265	7.06251330
a_1	7	0.06251330	15.99659440
a_2	15	0.99659440	1.00341723
a_3	1	0.00341723	292.63456745
a_4	292	0.63456745	1.57582693
a_5	1	0.57582693	1.73663288
a_6	1	0.73663288	...

So $\pi = [3; 7, 15, 1, 292, 1, 1, \ldots]$, or in the canonical (standard) form,

$$\pi = 3 + \cfrac{1}{7 + \cfrac{1}{15 + \cfrac{1}{1 + \cfrac{1}{292 + \cfrac{1}{1 + \cfrac{1}{1 + \cdots}}}}}}.$$

4) The partial quotients obtained by the procedure in (3) represent the "best" rational approximation to π in the sense that no fraction with a smaller denominator is closer to π than the partial quotient at any level. This is true for any infinite continued fraction.

5) If we want to calculate the successive partial quotients of a continued fraction, we may develop a formula for every level of truncation. Thus if

$$\lambda = a_0 + \cfrac{1}{a_1 + \cfrac{1}{a_2 + \cfrac{1}{a_3 + \cdots}}},$$

we could write

$$\lambda_0 = a_0, \quad \lambda_1 = a_0 + \frac{1}{a_1} = \frac{a_0 a_1 + 1}{a_1},$$

$$\lambda_2 = a_0 + \frac{1}{a_1 + \frac{1}{a_2}} = a_0 + \frac{1}{\frac{a_1 a_2 + 1}{a_2}} = a_0 + \frac{a_2}{a_1 a_2 + 1} = \frac{a_0(a_1 a_2 + 1) + a_2}{a_1 a_2 + 1},$$

and so forth. This is a little laborious. If we write $\lambda_n = \frac{h_n}{k_n}$, we can develop the simple recursion formula,

$$\frac{h_n}{k_n} = \frac{a_n h_{n-1} + h_{n-2}}{a_n k_{n-1} + k_{n-2}}.$$

By arbitrarily defining $h_{-1} = 1$, $h_{-2} = 0$, $k_{-1} = 0$, $k_{-2} = 1$, we would have the base case

$$\lambda_0 = \frac{h_0}{k_0} = \frac{a_0 \cdot 1 + 0}{a_0 \cdot 0 + 1} = a_0.$$

Now assume

$$\frac{h_{n-1}}{k_{n-1}} = \frac{a_{n-1} h_{n-2} + h_{n-3}}{a_{n-1} k_{n-2} + k_{n-3}}.$$

Note that

$$\frac{h_n}{k_n} = a_0 + \cfrac{1}{a_1 + \cfrac{1}{a_2 + \cfrac{1}{\ddots\, a_{n-1} + \cfrac{1}{a_n}}}}$$

but this would be equal to the formula for $\frac{h_{n-1}}{k_{n-1}}$ with

$$a_{n-1} + \frac{1}{a_n} = \frac{a_{n-1} a_n + 1}{a_n}$$

plugged in for a_{n-1}. In other words,

$$\frac{h_n}{k_n} = \frac{(\frac{a_{n-1}a_n+1}{a_n})h_{n-2} + h_{n-3}}{(\frac{a_{n-1}a_n+1}{a_n})k_{n-2} + k_{n-3}} = \frac{(a_{n-1}a_n + 1)h_{n-2} + a_n h_{n-3}}{(a_{n-1}a_n + 1)k_{n-2} + a_n k_{n-3}}$$
$$= \frac{a_n(a_{n-1}h_{n-2} + h_{n-3}) + h_{n-2}}{a_n(a_{n-1}k_{n-2} + k_{n-3}) + k_{n-2}} = \frac{a_n h_{n-1} + h_{n-2}}{a_n k_{n-1} + k_{n-2}}$$

by the induction hypothesis. This is a recipe for eyestrain, so let us organize our calculations better.

6) Using the formula in (5) in the following table, we can easily crank out the successive partial quotients. Let us do that for π and get the classic rational estimates.

n	−2	−1	0	1	2	3	4	5	6
a_n	X	X	3	7	15	1	292	1	1
h_n	0	1	3	22					
k_n	1	0	1	7					

We have seeded the table with the partial denominators of the canonical continued fraction for π from our earlier calculation and the initial values that get us to $a_0 = 3$. Entering the table at a_1 we find 7, and multiply it by h_0, namely 3, and add h_{-1}, namely 1, to get 22. Mimicking this for the k's, we get 7. Now moving over one column at a time, we generate the following table:

n	−2	−1	0	1	2	3	4	5	6
a_n	X	X	3	7	15	1	292	1	1
h_n	0	1	3	22	333	355	103993	104348	208341
k_n	1	0	1	7	106	113	33102	33215	66317

Note that $\frac{208341}{66317} = 3.141592653$ agrees with the first nine decimals of π. Think about what it means for a large partial denominator to appear in a continued fraction of an irrational. It means that if we truncate just before it, the rational approximation is going to be pretty good. That is why $\frac{355}{113} = 3.141592$ agrees with π to six places. This is amazingly accurate – off by about 0.000008 % – considering the denominator is three digits.

7) The representation of a real number by a continued fraction in canonical form is unique. If we allow generalized continued fractions, that is no longer true. The flip side is that desirable patterns can be made to appear in the partial numerators and denominators. For example,

$$\pi = 3 + \cfrac{1^2}{6 + \cfrac{3^2}{6 + \cfrac{5^2}{6 + \cfrac{7^2}{6 + \cdots}}}} = \cfrac{4}{1 + \cfrac{1^2}{3 + \cfrac{2^2}{5 + \cfrac{3^2}{7 + \cdots}}}}.$$

Here is a worksheet if you would like to try your hand at creating a partial fraction table and solving a Pell's equation:

$x =$ ______________________

$a_n \downarrow$	integer part	fractional part	(fractional part)$^{-1}$
a_0			
a_1			
a_2			
a_3			
a_4			
a_5			
a_6			
a_7			
a_8			
a_9			

$$x = [a_0; a_1, a_2, a_3, a_4, a_5, a_6, a_7, a_8, a_9, \dots]$$

n	−2	−1	0	1	2	3	4	5	6	7	8	9
a_n	*	*										
h_n	0	1										
k_n	1	0										

$$h_n = a_n \cdot h_{n-1} + h_{n-2}$$
$$k_n = a_n \cdot k_{n-1} + k_{n-2}$$

References & suggested further reading

1) *A Friendly Introduction to Number Theory* (2nd edition) by Joseph Silverman (2001), Prentice Hall
2) *Elementary Number Theory* (2nd edition) by Underwood Dudley (1978), W. H. Freeman.

Solutions to Chapter 10 examples

Example 10.1

$13 \bmod 6 = 2$, $343 \bmod 7 = 0$, $-8 \bmod 12 = 4$, $15 \bmod 4 = 3$, $93 = 13 \bmod 16 = 13$, and $1357 \bmod 2 = 1$.

Example 10.2

1) Addition – given that $x \equiv y \pmod m$ and $u \equiv v \pmod m$, we know that m divides both $x - y$ and $u - v$. But then it also divides $(x - y) + (u - v) = (x + u) - (y + v)$. By definition, this implies $x + u \equiv y + v \pmod m$.
2) Subtraction – similar to addition, but now we know m divides $(x - y) - (u - v) = (x - u) - (y - v)$. The definition implies $x - u \equiv y - v \pmod m$.
3) Multiplication – given that $x \equiv y \pmod m$ and $u \equiv v \pmod m$, we know that m divides both $x - y$ and $u - v$. Then we can write $x - y = k_1 m$ and $u - v = k_2 m$. Accordingly, $x = y + k_1 m$ and $u = v + k_2 m$. Then $xu = (y + k_1 m)(v + k_2 m) = yv + yk_2 m + vk_1 m + k_1 k_2 m^2$. It follows that $xu - yv = m(yk_2 + vk_1 + k_1 k_2 m)$. Apparently, m divides $xu - yv$, so again $xu \equiv yv \pmod m$.
4) Division – this rule is equivalent to multiplicative cancellation of a common factor. We have $x - y = ku - kv = k(u - v)$ divisible by m. So for some t, $mt = k(u - v)$. If k and m are coprime (no shared factors except 1), then all of the factors of m must be present in $(u - v)$, so m divides $(u - v)$ and we see $\frac{x}{k} \equiv \frac{y}{k} \pmod m$.
5) Powers – this is just iterated multiplication.

Example 10.3

(i) Valid: Consider $12 \equiv 26 \pmod 7$. The divisor 2 is coprime to 7 ($\gcd(2, 7) = 1$), so we should have $6 \equiv 13 \pmod 7$, which is correct.

(ii) Not valid: Certainly $16 \equiv 6 \pmod{10}$. Dividing both sides by 2, we would have $8 \equiv 3 \pmod{10}$, which is false. The problem is that $\gcd(2, 10) = 2$.

Example 10.4

1) Translation – let $a = u = v$ in the addition formula Table 10.1(1).
2) Additive cancellation – with $x + a \equiv y + a \pmod m$, let $a = u = v$ in the subtraction formula Table 10.1(2).
3) Change of scale – let $a = u = v$ in the multiplication formula Table 10.1(3)
4) Constant multiple – if $x \equiv y \pmod m$ then $(x - y) = mt$ for some t. But then $ax - ay = (am)t$ implying $ax \equiv ay \pmod{am}$.
5) Modulus reduction – if $ax \equiv ay \pmod{am}$ then $ax - ay = (am)t$ for some t, so after dividing by a, $x - y = mt$, or $x \equiv y \pmod m$.

Example 10.5

We find the least residue corresponding to each coefficient. The reduced polynomial is $x^5 - 2x^4 + 4x^3 + 2x^2 + x$. There is an important test to see if a polynomial can be factored (so perhaps its roots can be found) that is based on reducing its coefficients relative to some suitable modulus (see Section 6.1.5).

Example 10.6

Starting with $3 \equiv 8 (\bmod 5)$, consider $2^3 \equiv 2^8 (\bmod 5)$. This is the same as $8 \equiv 256 (\bmod 5)$. But this is false because 5 does not divide $(256 - 8) = 248$.

Example 10.7

Note that $7^4 = 2401 \equiv 1 (\bmod 100)$. Then $7^{503} = 7^{500} \cdot 7^3 = (7^4)^{125} \cdot 7^3$. Finally, $7^{503} \equiv (1)^{125} \cdot 7^3 \equiv 43 (\bmod 100)$.

Example 10.8

Adding digits, 15,986 gives $1 + 5 + 9 + 8 + 6 = 29$, then $29 \bmod 9 = 2$. Likewise 25,632 gives 18, then $18 \bmod 9 = 0$. The check digit is $2 \cdot 0 = 0$. Adding digits for 419,753,152 gives 37, and $37 \bmod 9 = 1$. So there is a mismatch and something is wrong with the original multiplication. In fact, the correct answer is 409,753,152, which after adding the digits and reducing mod 9 gives 0, as we would expect. Note that if we managed to have exactly compensating errors (bolded) in the product, say reporting it as 4**1**9, **6**53,152, the test would not detect the overall error.

Example 10.9

We can rewrite any number A expressed normally as a string of digits $d_n d_{n-1} \ldots d_1 d_0$ as the sum $d_n \cdot 10^n + d_{n-1} \cdot 10^{n-1} + \cdots + d_1 \cdot 10^1 + d_0 \cdot 10^0$. So finding $A \bmod 9$ amounts to calculating the individual terms of this sum mod 9, which immediately becomes $(d_n + d_{n-1} + \cdots + d_1 + d_0) \bmod 9$, the sum of the digits of A reduced mod 9. For this step, we have relied on the fact that $10^k \bmod 9 = 1$ for all k. Now suppose we have a multiplication problem $M \cdot N = MN$ that we would like to check. From Table 10.1(3), we know $MN \bmod 9 \equiv (M \bmod 9 \cdot N \bmod 9) \bmod 9$. The extra mod 9 on the right takes care of the possibility that the product of the two least residues may exceed 9. From the preceding, this is equivalent to saying that the sum of digits of MN equals the product of the sum of digits of M times the sum of digits of N, all reduced mod 9. This is the principle of casting out nines. If the modular multiplication is not true, then the original multiplication cannot be true. Almost identical arguments establish casting out nines as a check on addition and subtraction.

Example 10.10

Here $a = 10$, $b = 2$, and $m = 14$. We can see immediately that $d = \gcd(10, 14) = 2$, so there will be a solution, but let us run forward through the Euclidean algorithm to verify the gcd and then backward through its so-called "extension" to find x and c. The general scheme of the Euclidean algorithm applied to two integers $m > n$ follows on the left (q_i is a quotient, r_i is a remainder) and the specific application to our problem on the right:

$$\begin{aligned} m &= q_1 n + r_1 \\ n &= q_2 r_1 + r_2 \\ r_1 &= q_3 r_2 + r_3 \\ &\vdots \\ r_{n-2} &= q_n r_{n-1} + r_n \\ r_{n-1} &= q_{n+1} r_n + 0 \end{aligned} \qquad \begin{aligned} 14 &= 1 \cdot 10 + 4 \\ 10 &= 2 \cdot 4 + 2 \\ 4 &= 2 \cdot 2 + 0 \end{aligned}$$

The remainder r_n is the gcd of m and n. This is the remainder in the step just before the step where the zero remainder appears, when r_n evenly divides $r_{n-1.}$ The coefficients x and c so that $ax - cm = \gcd(a, m)$ are gotten by reversing (extending) the algorithm. An easy way to organize this calculation is to rewrite the algorithm table except for the last line with the equations solved for the remainders. So we would have

$$\begin{aligned} 4 &= 14 - 10 \cdot 1, \\ 2 &= 10 - 2 \cdot 4. \end{aligned}$$

Put the expression in the first line, $4 = 14 - 10$, into the last line for 4 to get $2 = 10 - 2 \cdot (14 - 10)$ and simplify to $2 = 3 \cdot 10 - 2 \cdot 14$. From this expression, we find $10 \cdot 3 \equiv 2 (\text{mod } 14)$, so $x = 3$ and $c = 2$. Note that there are other solutions, such as $x = 10$ and $x = 17$. After we justify Méziriac's theorem, we will show how to find the rest.

Example 10.11

Let $S = \{rm + sn > 0 : r, s \in \mathbb{Z}\}$. This is the set of all integers which are positive linear combinations of m and n, and it is not void, since either m ($r = 1$ and $s = 0$) or $-m$ ($r = -1$ and $s = 0$) is in it. By well-ordering, this set must have a least element. Let us call the least element d. Since d is a linear combination of m and n, we can write $d = am + bn$. If we can show that $d = \gcd(m, n)$, we will be done. Try dividing m by d. By the division algorithm, we get $m = dq + r$, where – and this is critical to the proof – $0 \leq r < d$. Do you see that if we could somehow express $r > 0$ as a linear combination of m and n, then d would not be the least such element, and the resulting contradiction would force us to conclude that $r = 0$. Of course, that would mean $m = dq + 0$, or, in other words, d divides m. So suppose $r > 0$ and write $r = m - dq$, then substitute $d = am + bn$ to get $r = m - (am + bn)q = (1 - aq)m + (-bq)n$. This appears to be a linear combination of m

and n which is, because of the properties of r, positive and yet smaller than d. We have a contradiction, and the only way out is to conclude $r = 0$, hence d divides m.

Notice that m and n appear in a completely symmetrical way in this argument, so we could repeat the preceding line of reasoning and divide n by d, and conclude eventually that d divides n. So this would confirm that d is a common divisor of m and n and the only loose end is to show that it is the largest such divisor.

So suppose D is another common divisor and $D \geq d$. Since D divides both m and n, it must apparently also divide $am + bn = d$. But then $d \geq D$, and it must be the case that d is the greatest common divisor, or that $d = \gcd(m, n)$.

Example 10.12

From Example 10.11, there is always in principle a solution to $ax + my = d = \gcd(a, m)$. We can use the Euclidean algorithm (Example 10.10) to find the actual solution, and let us denote it as x and y. At this point, we are obviously not saying that x solves $ax \equiv b(\bmod m)$. But we are going to use x and y to bootstrap a solution to $ax \equiv b(\bmod m)$. First, multiply $ax + my = d$ across the board by $\frac{b}{d}$ to get $ax(\frac{b}{d}) + my(\frac{b}{d}) = d(\frac{b}{d})$ or $a(\frac{bx}{d}) + m(\frac{by}{d}) = b$. Applying the modulus m, we have $a(\frac{bx}{d}) \equiv b(\bmod m)$. So $\frac{bx}{d}$ certainly is a solution to $ax \equiv b(\bmod m)$. Let us denote this base solution $\frac{bx}{d}$ by x_0. Are there more solutions? If there are, we would want them to be incongruent to the base solution x_0 to consider them distinct.

Suppose x' is also a solution of the congruence. Then $ax' \equiv b \equiv ax_0(\bmod m)$ which by transitivity implies $ax' - ax_0 = km$ for some integer k. So m divides $a(x' - x_0)$. Since a and m have d as their largest common factor, consider dividing m and $a(x' - x_0)$ both by d. We have $\frac{m}{d}$ and $\frac{a}{d}(x_2 - x_1)$. As we noted, m divides $a(x' - x_0)$, so $\frac{m}{d}$ divides $\frac{a}{d}(x' - x_0)$. But $\frac{m}{d}$ has no nontrivial factors in common with $\frac{a}{d}$ (we made sure of this by dividing them out). Hence $\frac{m}{d}$ must divide $(x' - x_0)$, or $k \cdot \frac{m}{d} = x' - x_0$. From this, we have finally $x' = x_0 + \frac{km}{d}$ for $k \in \{0, 1, 2, \dots\}$. Note that as soon as $k = d$ the process of generating distinct solutions modulo m starts over. We conclude that if we find a base solution x_0 then all the incongruent solutions are of the form $x_k = x_0 + \frac{km}{d}$ for $k \in \{0, 1, 2, \dots, d-1\}$. In Example 10.10, $d = 2$, so we would expect two noncongruent solutions, 3 and $3 + \frac{1 \cdot 14}{2} = 10$. Note that 10 is not congruent to 3 modulo 14, but $17 \equiv 3 \pmod{14}$. So 17 does not count as a distinct (incongruent) solution modulo 14.

Example 10.13

The numbers are small enough, so we see $\gcd(42, 56) = 7$. Note that 7 divides 21, so there will be a solution. From the extended Euclidean algorithm,

$$
\begin{aligned}
56 &= 1 \cdot 35 + 21, \\
35 &= 1 \cdot 21 + 14, \\
21 &= 1 \cdot 14 + 7, \\
14 &= 2 \cdot 7 + 0.
\end{aligned}
$$

So the front end of the algorithm tells us something we already knew about the gcd. The back end is what we really want:

$$\begin{aligned} 21 &= 56 - 35, \\ 14 &= 35 - 21, \\ 7 &= 21 - 14, \end{aligned}$$

which gives us $7 = 21 - (35 - 21) = 2 \cdot (56 - 35) - 35 = 2 \cdot 56 - 3 \cdot 35$. More compactly, $7 = 2 \cdot 56 - 3 \cdot 35$. Multiplying by 3, we get $6 \cdot 56 - 9 \cdot 35 = 21$, so modulo 56 this reduces to $-9 \cdot 35 = 21$. Of course, $-9 \equiv 47 (\mathrm{mod}\, 56)$, so $x_0 = 47$. Checking: $47 \cdot 35 - 21 = 29 \cdot 56 \surd$. We are expecting $7 - 1 = 6$ additional incongruent solutions. From Example 10.12, $x_k = x_0 + \frac{km}{d}$, so in this case $x_k = x_0 + 8k$. The rest of the incongruent solutions are then 55, 7, 15, 23, 31, and 39.

Example 10.14

We run the Euclidean algorithm to find gcd(3184, 14083):

$$\begin{aligned} 14083 &= 4 \cdot 3184 + 1347, \\ 3184 &= 2 \cdot 1347 + 490, \\ 1347 &= 2 \cdot 490 + 367, \\ 490 &= 1 \cdot 367 + 123, \\ 367 &= 2 \cdot 123 + 121, \\ 123 &= 1 \cdot 121 + 2, \\ 121 &= 60 \cdot 2 + 1, \\ 60 &= 60 \cdot 1 + 0. \end{aligned}$$

Evidently, they are relatively prime and $\gcd(3184, 14083) = 1$. When this happens, we know immediately that the congruence will have a solution. Working the algorithm backwards, we solve for the remainders:

$$\begin{aligned} 1347 &= 14083 - 4 \cdot 3184, \\ 490 &= 3184 - 2 \cdot 1347, \\ 367 &= 1347 - 2 \cdot 490, \\ 123 &= 490 - 1 \cdot 367, \\ 121 &= 367 - 2 \cdot 123, \\ 2 &= 123 - 1 \cdot 121, \\ 1 &= 121 - 60 \cdot 2 \end{aligned}$$

and organize the backsubstitution in a separate bottom up list because of its length.

So

$$\begin{aligned}
1 &= 1579 \cdot 14083 - 6984 \cdot 3184,\\
1 &= 1579 \cdot 1347 - 668 \cdot 3184,\\
1 &= 243 \cdot 1347 - 668 \cdot 490,\\
1 &= 243 \cdot 367 - 182 \cdot 490,\\
1 &= 61 \cdot 367 - 182 \cdot 123,\\
1 &= 61 \cdot 121 - 60 \cdot 123.
\end{aligned}$$

Evidently, $(-6984) \cdot 3184 \equiv 1 \pmod{14083}$. We can substitute a least residue for -6984 and write the congruence as $3184 \cdot 7099 \equiv 1 \pmod{14083}$. Now the $\frac{b}{d}$ factor that we used in Example 10.12 to scale the coefficients that we found in the initial gcd calculation to match the b in the given congruence is $\frac{5573}{1}$. So we have $3184 \cdot (7099 \cdot 5573) \equiv 5573 \pmod{14083}$. Reducing $7099 \cdot 5573 \bmod 14083$ yields 3580. We can now say that the solution to $3184x \equiv 5573 \pmod{14083}$ is $x = 3580$. Since $d = 1$, there are no other incongruent solutions.

Example 10.15

$$m = 3,\ \begin{array}{cc} 1 & 1\\ 2 & 2 \end{array} \qquad m = 5,\ \begin{array}{cc} 1 & 1\\ 2 & 3\\ 3 & 2\\ 4 & 4 \end{array} \qquad m = 7,\ \begin{array}{cc} 1 & 1\\ 2 & 4\\ 3 & 5\\ 4 & 2\\ 5 & 3\\ 6 & 6 \end{array} \qquad m = 11,\ \begin{array}{cc} 1 & 1\\ 2 & 6\\ 3 & 4\\ 4 & 3\\ 5 & 9\\ 6 & 2\\ 7 & 8\\ 8 & 7\\ 9 & 5\\ 10 & 10 \end{array} \qquad m = 13,\ \begin{array}{cc} 1 & 1\\ 2 & 7\\ 3 & 9\\ 4 & 10\\ 5 & 8\\ 6 & 11\\ 7 & 2\\ 8 & 5\\ 9 & 3\\ 10 & 4\\ 11 & 6\\ 12 & 12 \end{array}$$

Note that you only have to actually calculate half the entries because of mutual inverses. Now for the congruences:

(i) $2 \cdot 2x \equiv 2 \cdot 1 \pmod 3$, so $x \equiv 2 \pmod 3$. Not too challenging.
(ii) $4 \cdot 4x \equiv 4 \cdot 3 \pmod 5$, so $x \equiv 12 \equiv 2 \pmod 5$.
(iii) $3 \cdot 5x \equiv 3 \cdot 2 \pmod 7$, so $x \equiv 6 \pmod 7$.
(iv) $7 \cdot 8x \equiv 7 \cdot 6 \pmod{11}$, so $x \equiv 42 \equiv 9 \pmod{11}$.
(v) $4 \cdot 10x \equiv 4 \cdot 8 \pmod{13}$, so $x \equiv 32 \equiv 6 \pmod{13}$.

Example 10.16

We check that $\gcd(3,5) = 1$, so we should have a solution. The congruence $x \equiv 2(\text{mod } 3)$ requires that $x - 2 = 3k_1$ for some k_1. So $x = 2 + 3k_1$. Plugging that into the congruence $x \equiv 3(\text{mod } 5)$ implies that $2 + 3k_1 \equiv 3(\text{mod } 5)$. Simplifying this, we have $3k_1 \equiv 1(\text{mod } 5)$. It looks like we are stuck trying to solve for k_1 because we cannot divide by 3, but the fact that $2 \cdot 3 \equiv 1\,(\text{mod } 5)$ comes to our rescue. So $2 \cdot 3k_1 \equiv 2 \cdot 1(\text{mod } 5)$ gives $k_1 \equiv 2(\text{mod } 5)$ and then $k_1 - 2 = 5k_2$. Now we have $k_1 = 2 + 5k_2$. Going back to $x = 2 + 3k_1$, we substitute for k_1 to get $x = 2 + 3(2 + 5k_2) = 8 + 15k_2$. So $x = 8 + 15k_2$ satisfies both congruences for any k_2. Setting $k_2 = 0$, we see that $x = 8$ works for $8 \equiv 2(\text{mod } 3)$ and $8 \equiv 3(\text{mod } 5)$. Moreover, 8 is the unique solution modulo $3 \cdot 5 = 15$.

Example 10.17

Let x be the number of men in the club. Then $x \equiv 1(\text{mod } 3)$, $x \equiv 2(\text{mod } 5)$, and $x \equiv 3(\text{mod } 7)$. The moduli are pairwise coprime, so we will have a solution. First, $x \equiv 1(\text{mod } 3)$ means, as usual, $x - 1 = 3k_1$ or $x = 1 + 3k_1$. Putting that into $x \equiv 2(\text{mod } 5)$ yields $1+3k_1 \equiv 2(\text{mod } 5)$, or $3k_1 \equiv 1(\text{mod } 5)$ which implies $k_1 \equiv 2(\text{mod } 5)$. Then $k_1 - 2 = 5k_2$, or $k_1 = 2 + 5k_2$. Then $x = 1 + 3(2 + 5k_2) = 7 + 15k_2$. Dropping that into $x \equiv 3\,(\text{mod } 7)$ gives $7 + 15k_2 \equiv 3\,(\text{mod } 7)$ or $k_2 \equiv 3\,(\text{mod } 7)$. So $k_2 - 3 = 7k_3$, which implies $k_2 = 3 + 7k_3$. Substituting this value of k_2 into $x = 7 + 15k_2$ implies $x = 7 + 15(3 + 7k_3) = 52 + 105k_3$. Finally, modulo $3 \cdot 5 \cdot 7 = 105$ we see that $x = 52$ is the unique solution. So we conclude the Ancient Order had at least 52 members and possibly $52 + 105k$ for $k \in \mathbb{N}$.

Example 10.18

Let $5x$ be the number. Then $5x \equiv 1(\text{mod } 2)$ or $x \equiv 1(\text{mod } 2)$, $5x \equiv 1(\text{mod } 3)$ or (since $2 \cdot 2 \equiv 1(\text{mod } 3)$) $x \equiv 2(\text{mod } 3)$, and $5x \equiv 1(\text{mod } 7)$ or (since $3 \cdot 5 \equiv 1(\text{mod } 7)$) $x \equiv 3(\text{mod } 7)$. From $x \equiv 1(\text{mod } 2)$, we have $x - 1 = 2k_1$, or $x = 1 + 2k_1$. Putting this in the second congruence gives $1 + 2k_1 \equiv 2(\text{mod } 3)$. Simplifying this yields $2k_1 \equiv 1(\text{mod } 3)$. So $k_1 \equiv 2(\text{mod } 3)$ and $k_1 - 2 = 3k_2$, or $k_1 = 2 + 3k_2$. Evidently, $x = 1 + 2(2 + 3k_2) = 5 + 6k_2$. Putting this into the third congruence gives $5+6k_2 \equiv 3(\text{mod } 7)$, which simplifies to $6k_2 \equiv 5(\text{mod } 7)$ and further (multiply both sides by 6) to $k_2 \equiv 2(\text{mod } 7)$ Now we have $k_2 - 2 = 7k_3$ or $k_2 = 2 + 7k_3$. Finally, $x = 5 + 6(2 + 7k_3) = 17 + 42k_3$. So 17 is the unique solution $\text{mod}(2 \cdot 3 \cdot 7)$ and $5 \cdot 17 = 85$ is the number we are looking for. Checking, $85 = 2 \cdot 42 + 1 = 3 \cdot 28 + 1 = 7 \cdot 12 + 1\checkmark$.

Example 10.19

Given the system of congruences $x \equiv a_i \bmod m_i$ for $i = 1, \ldots, n$, with the condition that the moduli are relatively prime in pairs, we seek an x which satisfies them all. The existence of a solution is usually proved by induction. The base case ($n = 1$) is immediate. Now suppose $n = 2$. From the first congruence, we have $x \equiv a_1 \bmod m_1$, so $x = a_1 + k_1 m_1$ for some $k_1 \in \mathbb{Z}$. Plugging this into the second congruence, we get $a_1 + k_1 m_1 \equiv a_2 \bmod m_2$. Rearranging, we find $k_1 m_1 \equiv (a_2 - a_1) \bmod m_2$, and we recognize this as a congruence in the variable k_1 where the coefficient m_1 is relatively prime to the modulus m_2. Hence there is a unique least residue solution, say t. So $k_1 = t + k_2m_2$, and then $x = a_1 + (t + k_2m_2)m_1 \equiv (a_1 + tm_1) \bmod m_1m_2$. Evidently, this x satisfies both $x \equiv a_1 \bmod m_1$ and $x \equiv a_2 \bmod m_2$.

Now suppose we have an r that solves $x \equiv a_i \bmod m_i$ for $i = 1, \ldots, n-1$. Specifically,

$$x \equiv r \bmod\left(\prod_{i=1}^{n-1} m_i\right).$$

Consider the system

$$x \equiv r \bmod\left(\prod_{i=1}^{n-1} m_i\right)$$

together with $x \equiv a_n \bmod m_n$. Now $\prod_{i=1}^{n-1} m_i$ and m_n are relatively prime, so the argument we used for the two congruences above can be employed again here. We just substitute $\prod_{i=1}^{n-1} m_i$ for m_1 and m_n for m_2. The conclusion is that the original system of n congruences has a unique least residue solution modulo $\prod_{i=1}^{n} m_i$.

Example 10.20

Let (i) be $x + 4y \equiv 2 \pmod 7$ and (ii) be $3x + 2y \equiv 3 \pmod 7$. Multiply (i) by 3 to get $3x + 5y \equiv 6 \pmod 7$ and subtract (ii). This leaves (iii) $3y \equiv 3 \pmod 7$. Use the table of inverses above to find the inverse of 3 is 5 modulo 7. So multiply (iii) by 5 to obtain $5 \cdot 3y \equiv 5 \cdot 3 \pmod 7$ which reduces to $y \equiv 0 \pmod 7$. Take this value of y back to (i) and immediately conclude $x \equiv 2 \pmod 7$. We have $x = 2 + 7k_1$ and $y = 7k_2$. The parameters k_1 and k_2 need not be the same since multiples of 7 are invisible mod 7.

Example 10.21

Let (i) be $x + y - z \equiv 3 \pmod{11}$, (ii) be $8x + 5y + 2z \equiv 3 \pmod{11}$, and (iii) be $4x + y - 6z \equiv 10 \pmod{11}$. Multiply (i) by 4 and subtract (iii) to give (iv) $3y + 2z \equiv 2 \pmod{11}$. Multiply (iii) by 2 and subtract from (ii) to yield (v) $3y + 3z \equiv 5 \pmod{11}$. Now multiply (v) by 4 to simplify it to (vi) $y + z \equiv 9 \pmod{11}$. Multiply (vi) by 2 and subtract from (iv) to isolate $2y \equiv 2 - 18 \equiv 6 \pmod{11}$, so $y \equiv 3$. Taking this back to (vi), we see that $3 + z \equiv 9 \pmod{11}$, or $z \equiv 6 \pmod{11}$. Finally, plug the values of y and z into (i) to get $x + 3 - 6 \equiv 3 \pmod{11}$, or $x \equiv 6 \pmod{11}$. So $x = 6 + 11k_1, y = 3 + 11k_2$, and $z = 6 + 11k_3$.

Example 10.22

We construct a table of integers x relatively prime to 30 and their corresponding squares $x^2 \bmod 30$. If x were not relatively prime to 30, then it would pass a factor of 30 to a and violate $\gcd(a, m) = 1$:

x	1	7	11	13	17	19	23	29
$x^2 \bmod 30$	1	19	1	19	19	1	19	1

Not too exciting, but we do see a symmetry in the bottom row. If $x_1 = 30 - x_2$, then $x_1^2 = (30 - x_2)^2 = 900 - 60x_2 + x_2^2$, so $x_1^2 \equiv x_2^2 (\bmod 30)$. This is clearly true in general.

Example 10.23

Certainly, $x^2 \equiv 0 \,(\bmod 2)$ has $x = 0$ as a solution and $x^2 \equiv 1 \,(\bmod 2)$ has $x = 1$ as a unique solution. Note $-1 \equiv 1 (\bmod 2)$. So by our condition that quadratic residues are to be nonzero, we would say that $x^2 \equiv a \,(\bmod 2)$ has one solution. Let us now assume the modulus is an odd prime p unless noted otherwise.

Example 10.24

Noting the symmetry in Example 10.22, we have $b^2 \equiv a \,(\bmod p)$ and $(p - b)^2 = p^2 - 2pb + b^2 \equiv b^2 (\bmod p)$. So if b is a solution, then modulo p, $p - b$ is also a solution.

Example 10.25

First, observe that there are $p - 1$ numbers from 1 to $p - 1$. Consider the squares of half of these numbers from 1 to $\frac{(p-1)}{2}$, namely $1^2, 2^2, \ldots, (\frac{p-1}{2})^2$. In Example 10.24 we showed that if $\frac{(p-1)}{2}$ was a QR, so was $p - \frac{(p-1)}{2} = \frac{(p+1)}{2}$. The numbers $\frac{(p+1)}{2}, \ldots, (p-2), (p-1)$ account for the other half of the overall numbers from 1 to $p - 1$ and their squares duplicate one-to-one the squares of the lower half. So we are led to conclude that there are no more than $\frac{(p-1)}{2}$ distinct squares in the sequence $1^2, 2^2, \ldots, (p-1)^2$. We cannot immediately say that there are exactly $\frac{(p-1)}{2}$ QR's because some of the squares from 1^2 to $(\frac{p-1}{2})^2$ might be duplicates. So for the sake of contradiction, suppose b and c are two numbers from 1 to $\frac{(p-1)}{2}$ with the property that $b^2 \equiv c^2 (\bmod p)$. This means that $b^2 - c^2 = (b-c)(b+c) = kp$ for some k. Note that p divides the difference of squares, so it must divide one of the factors. It cannot divide $(b + c)$ because $2 \le (b + c) \le p - 1 < p$. So it must divide $(b - c)$, hence $|b - c|$. But both b and c are between 1 and $\frac{(p-1)}{2}$, so that cannot happen either, unless $|b - c| = 0$, which forces $b = c$. Therefore the squaring map applied to the numbers 1 through $\frac{(p-1)}{2}$ is injective and there are $\frac{(p-1)}{2}$ distinct QR's from 1 to $(p - 1)$, leaving room for $\frac{(p-1)}{2}$ distinct QNR's.

Example 10.26

$p = 5$,

x	x^2
0	0
1	1
2	4
3	4
4	1

$p = 7$,

x	x^2
0	0
1	1
2	4
3	2
4	2
5	4
6	1

$p = 11$,

x	x^2
0	0
1	1
2	4
3	9
4	5
5	3
6	3
7	5
8	9
9	4
10	1

$p = 13$,

x	x^2
0	0
1	1
2	4
3	9
4	3
5	12
6	10
7	10
8	12
9	3
10	9
11	4
12	1

$p = 17$,

x	x^2
0	0
1	1
2	4
3	9
4	16
5	8
6	2
7	15
8	15
9	15
10	15
11	2
12	8
13	16
14	9
15	4
16	1

Example 10.27

(i) $3 \cdot 2x^2 \equiv 3 \cdot 4 (\text{mod } 5)$ or $x^2 \equiv 2 (\text{mod } 5)$. Since 2 is a QNR, there is no solution to (i).

(ii) $7 \cdot 5x^2 \equiv 7 \cdot 6 (\text{mod } 17)$ or $x^2 \equiv 8 (\text{mod } 17)$. Since 8 is a QR, both 5 and 12 are solutions.

Example 10.28

(i) $x^2 - 5x + 1 \equiv 0 (\text{mod } 13)$ implies $(x-4)(x-1) \equiv 0 (\text{mod } 13)$. So $(x-1) \equiv 0 (\text{mod } 13)$ gives $x \equiv 1 (\text{mod } 13)$, and $(x-4) \equiv 0 (\text{mod } 13)$ gives $x \equiv 4 (\text{mod } 13)$.

(ii) Make $3x^2 - 5x + 2$ monic by multiplying by $3^{-1} \equiv 4\ (\text{mod } 11)$. Then $4 \cdot 3x^2 - 4 \cdot 5x + 4 \cdot 2 \equiv 0 (\text{mod } 11)$, or $x^2 - 9x + 8 \equiv 0 (\text{mod } 11)$. Then we have $(x-8)(x-1) \equiv 0 (\text{mod } 11)$, giving solutions $x \equiv 8 (\text{mod } 11)$ and $x \equiv 1 (\text{mod } 11)$ as in (i).

Example 10.29

This is a new type of problem. The modulus is composite. We are looking for x such that $x^2 \equiv 71 (\text{mod } 7)$ and $x^2 \equiv 71 (\text{mod } 11)$. If x^2 satisfies both of these conditions, it will also satisfy $x^2 \equiv 71 (\text{mod } 77)$. Reducing these, we get $x^2 \equiv 1 (\text{mod } 7)$ and $x^2 \equiv 5 (\text{mod } 11)$. The good news is that 1 is always a QR for any modulus, and 5 is a QR mod 11. So $x^2 \equiv 1 (\text{mod } 7)$ has solutions $x = 1$ and 6, and $x^2 \equiv 5 (\text{mod } 11)$ has solutions $x = 4$ and 9. Apparently, we have four cases to consider. Let a be the residue mod 7 and b be the residue mod 11. Then we need to run the Chinese remainder theorem algorithm on the (a, b)

pairs $(1,4)$, $(6,4)$, $(1,9)$, and $(6,9)$. Eventually, we find two compatible pairs that give solutions $x \equiv 15 \pmod{77}$ and $x \equiv 48 \pmod{77}$.

Example 10.30

We want $(\frac{26}{73}) = (\frac{2}{73})(\frac{13}{73})$. Now $73 \equiv 1 \pmod 8$ so by Table 10.3(2) $(\frac{2}{73}) = 1$. So $(\frac{26}{73}) = (\frac{13}{73})$. Here is where reciprocity allows us to flip the Legendre symbol: $(\frac{13}{73}) = (\frac{73}{13})$ because $13 \equiv 1 \pmod 4$ by Table 10.4(3). Now $(\frac{73}{13}) = (\frac{8}{13})$ after reducing mod 13. Then $(\frac{8}{13}) = (\frac{4}{13})(\frac{2}{13}) = (\frac{2}{13})$ because 4 is a square by Table 10.4(4). Finally, $(\frac{2}{13}) = -1$ because $13 \equiv 5 \pmod 8$. So there is no point in going further because the congruence has no solution.

Example 10.31

Start with

$$\begin{aligned}\left(\frac{482}{1357}\right) &= \left(\frac{2}{1357}\right)\left(\frac{241}{1357}\right) = (-1)\left(\frac{241}{1357}\right) = (-1)\left(\frac{1357}{241}\right)\\ &= (-1)\left(\frac{152}{241}\right) = (-1)\left(\frac{4}{241}\right)\left(\frac{38}{241}\right) = (-1)\left(\frac{38}{241}\right)\\ &= (-1)\left(\frac{2}{241}\right)\left(\frac{19}{241}\right) = (-1)\left(\frac{19}{241}\right) = (-1)\left(\frac{241}{19}\right)\\ &= (-1)\left(\frac{13}{19}\right) = (-1)\left(\frac{19}{13}\right) = (-1)\left(\frac{6}{13}\right)\\ &= (-1)\left(\frac{2}{13}\right)\left(\frac{3}{13}\right) = (-1)^2\left(\frac{3}{13}\right) = (1)(1)\end{aligned}$$

from our table of QR's mod 13. So this congruence has a solution. Interestingly, Maple® comes to the same conclusion (a bit faster), but declines to actually solve the congruence, and so do we. Equality with the machine has been achieved.

Example 10.32

Consider $\gcd(a,b)$. This divides the left-hand side, so it must divide the right-hand side, namely c. So if $\gcd(a,b) \nmid c$, the equation has no solution. On the other hand, if $\gcd(a,b)|c$, then there are infinitely many solutions of the form

$$x_k = x_0 + \frac{b}{\gcd(a,b)}k \quad \text{and} \quad y_k = y_0 - \frac{a}{\gcd(a,b)}k,$$

where (x_0, y_0) is a solution gotten by any means (by inspection, algorithm, or divine revelation). Note what happens for arbitrary k. The equation becomes

$$a\left(x_0 + \frac{bk}{\gcd(a,b)}\right) + b\left(y_0 - \frac{ak}{\gcd(a,b)}\right) = ax_0 + by_0 + \left(\frac{abk}{\gcd(a,b)} - \frac{abk}{\gcd(a,b)}\right)$$

$$= ax_0 + by_0 = c.$$

So each k provides a distinct solution. We can put a bound on k to ensure only positive (or negative) solutions. Sometimes we may need to find a particular k that allows the solutions to satisfy some additional constraint (see Example 10.37 below).

Example 10.33

We first check $\gcd(5,7) = 1$ and 1 divides 19, so by Example 10.32 there will be a solution. We will run the Euclidean algorithm to find r and s such that $5r + 7s = 1$, then multiply across the board by 19 to get $x_0 = 19r$ and $y_0 = 19s$. So

$$\begin{aligned} 7 &= 5 \cdot 1 + 2, \\ 5 &= 2 \cdot 2 + 1, \\ 2 &= 2 \cdot 1 + 0. \end{aligned}$$

The remainders are

$$\begin{aligned} 2 &= 7 - 5 \cdot 1, \\ 1 &= 5 - 2 \cdot 2, \end{aligned}$$

so $1 = 5 - 2 \cdot (7 - 5) = 5 \cdot 3 - 7 \cdot 2$. We have $r = 3$ and $s = -2$, so $x_0 = 19 \cdot 3 = 57$ and $y_0 = 19 \cdot (-2) = -38$. It follows that $5 \cdot 57 - 7 \cdot 38 = 285 - 266 = 19$. Following Example 10.32, the general solution is $(x_k = 57 + 7k, y_k = -38 - 5k)$. The smallest positive solution is $(1, 2)$ when $k = -8$.

Example 10.34

Adding the two equations, we get $3x + 4z = 15$. The $\gcd(3,4) = 1$, so there will be a solution. We see right away that a base solution is $x_0 = 1$ and $z_0 = 3$. By Example 10.32, the general solution is $x_k = 1 + 4k$ and $z_k = 3 - 3k$. From the first equation, $y_k = 6 - x - z = 6 - (1 + 4k) - (3 - 3k) = 2 - k$. Since we want a solution with $x_k \geq 10$, we pick $k = 3$. Then $x_3 = 13, y_3 = -1$, and $z_3 = -6$. Checking the system, $13 - 1 - 6 = 6$ and $26 + 1 - 18 = 9\checkmark$.

Example 10.35

If $\gcd(a,b) = 1$, we can break the equation into two congruences: $ax \equiv c \pmod{b}$ and $by \equiv c \pmod{a}$. Since $\gcd(a,b) = 1$, we can use Méziriac's theorem to get integers r and s such that $ra + sb = 1$. Evidently, $ra \equiv 1 \pmod{b}$, hence, after multiplying through by c, $rac \equiv c \pmod{b}$. You can see that $a(rc) \equiv c \pmod{b}$, so $x = rc \pmod{b}$ solves $ax \equiv c \pmod{b}$. Hence $x - rc = kb$ for arbitrary $k \in \mathbb{Z}$, and we may take $x = rc + kb$ and insert it into the original equation, $a(rc + kb) + by = c$. Solving for y, we get the intermediate result $by = (c - arc) - akb$, and then we note that, since $a(rc) \equiv c \pmod{b}$, $\frac{(c-arc)}{b}$ is an integer, hence so is $y = \frac{(c-arc)}{b} - ak$. Then the complete solution to $ax + by = c$ is $x = rc + kb$

and $y = \frac{(c-arc)}{b} - ak$, where $k \in \mathbb{Z}$. Of course, r came from the Méziriac's theorem step, and we could also have used the congruence $by \equiv c \pmod{a}$ as the starting point, too. If $\gcd(a, b) = d > 1$ and d divides c, then go ahead and divide it out to give $a'x + b'y = c'$ where now $\gcd(a', b') = 1$ and we revert to the preceding case.

Example 10.36

Looking at $3x \equiv 7 \pmod{5}$, we reduce the congruence to $3x \equiv 2 \pmod{5}$, and it is obvious that $x \equiv 4 \pmod{5}$, or that $x = 4 + 5k$. Then $3(4 + 5k) + 5y = 7$ gives $5y = 7 - 12 - 15k$, or $y = -1 - 3k$. Checking our solution with $k = 0$, we see that $3(4) + 5(-1) = 7\checkmark$.

Example 10.37

Let p be the price of a peach in cents, then $p+20$ is the price of a pear. Let x be the number of peaches she bought, then $9 - x$ is the number of pears. Then the amount she spent on peaches is px, the amount she spent on pears is $(p + 20)(9 - x)$, and the total amount she spent is $px + 9p - px + 180 - 20x = 1000$. So $9p - 20x = 820$. Note that the $\gcd(9, 20) = 1$, so the Diophantine equation has a solution. A quick application of the Euclidean algorithm gives us $1 = 9(9) + 20(-4)$. Multiplying by 820, we get $9(7380) - 20(3280) = 820$. We identify $p_0 = 7380$ and $x_0 = 3280$. Appealing to Example 10.32, the general solution is $p_k = 7380 - 20k$ and $x_k = 3280 - 9k$. We need to pick k such that $0 \le x_k \le 9$. Letting $k = 364$ gives $x_{364} = 3280 - 9 \cdot 364 = 4$. The corresponding $p_{364} = 7380 - 20 \cdot 364 = 100$. We conclude that she bought 4 peaches at \$1.00 each and 5 pears at \$1.20 each. Checking: $4 \cdot \$1.00 + 5 \cdot \$1.20 = \$10.00\checkmark$.

Example 10.38

The continued fraction expansion of $\sqrt{7} = 2.64575\ldots$ is $[2; 1, 1, 1, 4, 1, \ldots]$, which is repeating, by the way, with period 4. The table of successive convergents $\frac{h_n}{k_n}$ is as follows:

n	−2	−1	0	1	2	3	4	5	6
a_n	x	x	2	1	1	1	4	1	1
h_n	0	1	2	3	5	8			
k_n	1	0	1	1	2	3			

It is a clue that the numerator and denominator of the convergent that appears just before a large partial denominator a_i might be a Pell's solution, since that convergent is a particularly good rational approximation. In this case $a_4 = 4$ is bigger than its predecessors and the previous convergent is $\frac{h_3}{k_3} = \frac{8}{3}$. So we try $8^2 - 7 \cdot 3^2$ and discover it equals 1, so the Pell's solution pair is $(8, 3)$. All other solutions are of the form $(8 + 3\sqrt{7})^k$.

Example 10.39

We have $\sqrt{92} = 9.591663047\cdots = [9; 1, 1, 2, 4, 2, 1, 1, 18, \dots]$. The 18 is a gift, as we shall see.

The table of convergents looks like this:

n	−2	−1	0	1	2	3	4	5	6	7	8
a_n	x	x	9	1	1	2	4	2	1	1	18
h_n	0	1	9	10	19	48	211	470	681	1151	
k_n	1	0	1	1	2	5	22	49	71	120	

The seventh convergent $\frac{1151}{120} = 9.591666667\dots$ is pretty close to $\sqrt{92}$, so we try the Pell's solution pair $(1151, 120)$. Indeed, $1151^2 - 92 \cdot 120^2 = 1324801 - 1324800 = 1$. Bingo!

Index

https://doi.org/10.1515/9783111506654-013

www.ingramcontent.com/pod-product-compliance
Lightning Source LLC
Chambersburg PA
CBHW080333090826
49567CB00029B/342
* 9 7 8 3 1 1 1 5 0 6 6 1 6 *